T0312667

# IP Telephony

## Deploying VoIP Protocols and IMS Infrastructure, Second Edition

# IP Telephony

## Deploying VoIP Protocols and IMS Infrastructure, Second Edition

**Olivier Hersent**
*CEO of Actility*

A John Wiley and Sons, Ltd., Publication

This edition first published 2011


First edition published 2005

*Registered office*
John Wiley & Sons Ltd, The Atrium, Southern Gate, Chichester, West Sussex, PO19 8SQ, United Kingdom

For details of our global editorial offices, for customer services and for information about how to apply for permission to reuse the copyright material in this book please see our website at www.wiley.com.



*Library of Congress Cataloguing-in-Publication Data*

Hersent, Olivier.
IP telephony : deploying VoIP protocols and IMS infrastructure / Olivier Hersent. — 2nd ed.
p. cm.
Includes index.
ISBN 978-0-470-66584-8 (cloth)
1. Internet telephony. 2. Convergence (Telecommunication) I. Title.
TK5105.8865.H47 20010
004.69′5–dc22

2010024553

A catalogue record for this book is available from the British Library.

Print ISBN 9780470665848 (H/B)
ePDF ISBN: 9780470973264
oBook ISBN: 9780470973080

Set in 10/12 Times-Roman by Laserwords Private Limited, Chennai, India
Printed and bound in Singapore by Markono Print Media Pte Ltd.

# Contents

# Abbreviations

| | |
|---|---|
| 3GPP | Third Generation Partnership Project |
| A-BGF | Access Border Gateway function |
| A-RACF | Access RACF |
| A/V | Audio-visual |
| AAD | Average Acknowledgement Delay |
| AAL2 | ATM Adaptation Layer 2 |
| ACD | Automatic Call Distribution |
| ACELP | Algebraic-Code-Excited Linear-Prediction |
| ACF | Admission Confirm |
| ACL | Access Control List |
| ACM | Address Complete Message |
| ADEV | Average Delay Deviation |
| ADPCM | Adaptive Differential Pulse Mode Modulation |
| ADSL | Asymmetric Digital Subscriber Line |
| AES | Advanced Encryption Standard |
| AF | Application Function |
| AGCF | Access Gateway Control Function |
| AMF | Access Management Function |
| AMR | Adaptive Multi-Rate |
| AMR-WB | Adaptive Multi-Rate (Wide Band) |
| AN-GW | Access Network Gateway |
| ANDSF | Access Network Discovery and Selection Function |
| ANM | Answer Message |
| ANSI | American National Standard Institute |
| AOC | Advice of Charge |
| AoR | SIP Address of Record |
| APDU | Application Protocol Data Unit |

| | |
|---|---|
| API | Application Programming Interface |
| ARF | Access Relay Function |
| ARJ | Admission Reject |
| ARQ | Admission Request |
| AS | Application Server |
| ASCII | American Standard Code for Information Interchange |
| ASF | Application Server Function |
| ASN-1 | Abstract Syntax Notation One |
| ASP | Application Service Provider |
| ASR | Automatic Speech Recognition or Answer Seizure Ratio |
| ATM | Asynchronous Transfer Mode |
| AUCX | Audit Connection |
| AUEP | Audit Endpoint |
| AVC | Advanced Video Coding |
| AVT | Audio/Video Transport |
| B2BUA | Back-to-back User Agent |
| BASIC | Beginners' All-purpose Symbolic Instruction Code |
| BBERF | Bearer Binding and Event Reporting Function |
| BCF | Bandwidth Confirm |
| BER | Basic Encoding Rule |
| BGCF | Breakout Control Gateway Function |
| BGF | Border Gateway Function |
| BICC | Bearer Independent Call Control |
| BNF | Backus-Naur Form |
| BRJ | Bandwidth Reject |
| BRQ | Bandwidth Request |
| BTF | Basic Transport Function |
| C-BGF | Core Border Gateway function |
| C-RACF | Core RACF |
| CA | Call Agent |
| CALEA | Communication Assistance for Law Enforcement Act |
| CallID | Call Identifier |
| CBC | Cipher Block Chaining |
| CC | CSRC Count |
| CCF | Charging Collector Function |
| CCIR | Consultative Committee for International Radio (ITU) |
| CDMA | Code Division Multiplex Access |
| CDR | Call Detail Record |
| CED | CallED |
| CELP | Code-excited Linear Prediction |
| CFB | Cipher Feedback |
| CFU | Call Forwarding Unconditional |
| CIC | Circuit Identification Code |
| CID | Conference Identifier |

| | |
|---|---|
| CIF | Common Intermediary Format |
| CLEC | Competitive Local Exchange Carrier |
| CLF | Connectivity Session Location and Repository Function |
| CLIP | Calling Line Identity Presentation |
| CLIR | Calling Line Identity Restriction |
| CMA | Call Management Agent |
| CMTS | Cable Modem Termination System |
| CND | Customer Network Device |
| CNG | CalliNG; Comfort Noise Generator |
| CNG | Customer Network Gateway |
| CNGCF | CNG Configuration Function |
| CO | Central Office |
| Codec | COder DECoder |
| CoIx | Connectivity-oriented Interconnection |
| COMEDIA | Connection-oriented Media Transport in SDP |
| COPS | Common Open Policy Service |
| CPE | Customer Premises Equipment |
| CPG | Call Progress (Message) |
| CPIM | Common Profile for Instant Messaging |
| CPL | Call Processing Language |
| CPN | Customer Premises Network |
| CPU | Central Processing Unit |
| CR | Carriage Return |
| CRC | Cyclic Redundancy Check |
| CRCX | Create Connection |
| CRLF | Carriage Return and Line Feed |
| CRV | Call Reference Value |
| CS-ACELP | Conjugate Structure, Algebraic Code-Excited Linear Prediction |
| CSRC | Contributing Source |
| CTI | Computer Telephony Integration |
| DCF | Disengage Confirm |
| DCME | Digital Circuit Multiplication Equipment |
| DCN | Disconnect |
| DCS | Distributed Call Signaling |
| DCT | Discrete Cosine Transform |
| DDNS | Dynamic DNS |
| DES/CBC | Data Encryption Standard, Cipher Block Chaining |
| DES | Data Encryption Standard |
| DHCP | Dynamic Host Configuration Protocol |
| DiffServ | Differentiated Services |
| DIS | Digital Identification Signal |
| DL | Downlink |
| DLCX | Delete Connection |
| DLSR | Delay Since Last Sender Report |

| | |
|---|---|
| DNS | Domain Name System |
| DNSSEC | Domain Name System Security Protocol |
| DOCSIS | Data over Cable Service Interface Specification |
| DoS | Denial of Service |
| DRJ | Disengage Reject |
| DRQ | Disengage Request |
| DSL | Digital Subscriber Line |
| DSMIP | Dual Stack Mobile IP |
| DSP | Digital Signal Processor |
| DSS1 | Digital Subscriber Signaling 1 |
| DTMF | Dual-Tone Multi-Frequency |
| DTX | Discontinuous Transmission |
| DVMRP | Distance Vector Multicast Routing Protocol |
| E-CSCF | Emergency-CSCF |
| E-UTRAN | Evolved-UTRAN |
| ECB | Electronic Code Book |
| ECF | Elementary Control Function |
| EFF | Elementary Forwarding Function |
| EFR | Enhanced Full Rate |
| ENUM | "Electronic Numbers" Protocol |
| EOL | End of Line |
| EOP | End of Procedure |
| EPCF | Endpoint Configuration Command |
| ePDG | evolved Packet Data Gateway |
| ETSI | European Telecommunications Standardisation Institute |
| ETSI TIPHON | ETSI Telephony and Internet Protocol Harmonization Over Networks |
| ETTB | Ethernet to the Building |
| ETTX | Ethernet to the <anything> (Curb, Home, Building) |
| FCF | Fax Control Field |
| FCS | Frame Check Sequence |
| FEC | Forward Error Correction |
| FIF | Fax Information Field |
| FIFO | First in First Out |
| FIPS PUB | Federal Information Processing Standards Publication |
| FR | Full-rate |
| FS | FastStart |
| GCF | Gatekeeper Confirm |
| GEF | Generic Extensibility Framework |
| GGSN | Gateway GPRS Support Node |
| GK | Gatekeeper |
| GOBs | Group of Blocks |
| GRJ | Gatekeeper Reject |
| GRQ | Gatekeeper Request |

| | |
|---|---|
| GSM | Global System for Mobile Communications |
| GTD | Global Transparency Descriptor |
| GTP | Generic Tunneling Protocol |
| HD | Hang Down (off-hook) |
| HDLC | High-level Data Link Control |
| HLR | Home Location Register |
| HLR/AuC | HLR Authentication Center |
| HR | Half Rate |
| HSPA | High Speed Packet Access |
| HSS | Home Subscriber Server |
| HTML | Hypertext Markup Language |
| HTTP | Hypertext Transfer Protocol |
| HU | Hang Up (on-hook) |
| I-BGF | Interdomain Border Gateway function |
| I-CSCF | Interrogating Call/Session Control Function |
| IAD | Integrated Access Device |
| IAM | Initial Address Message |
| IANA | Internet Assigned Numbers Authority |
| IARI | IMS Application Reference Identifier (IARI) |
| IBCF | Interconnection Border Control Function |
| ICID | IMS Charging Identifier |
| ICMP | Internet Control Message Protocol |
| ICSI | IMS Communication Service Identifier |
| IEC | ISO International Electrotechnical Commission |
| IETF | Internet Engineering Task Force |
| IF | Interface |
| IFP | Internet Fax Protocol |
| IFT | Internet Fax Transmission protocol |
| ILS | Internet Locator Service (Microsoft) |
| IM | Instant Messaging |
| IMCN | IP Multimedia Core Network |
| IMPI | IP Multimedia Private Identity |
| IMPP | Instant Messaging and Presence Protocol |
| IMPU | IP Multimedia Public Identity |
| IMS | IP Multimedia subsystem |
| IMTC | International Multimedia Teleconferencing Consortium |
| IN | Intelligent Network |
| INAP | Intelligent Network Application Protocol |
| IntServ | Integrated Services |
| IOI | Inter Operator Identifier |
| IP | Intelligent Peripheral |
| IP CAN | Internet Protocol Connectivity Access Network |
| IP-PBX | Internet Protocol–Private Branch Exchange |
| IPDC | Internet Protocol Device Control |

| | |
|---|---|
| IPR | Intellectual Property Rights |
| IPSec | Internet Protocol Security |
| IRC | Internet Relay Chat |
| IRQ | Information Request |
| IRR | Information Request Response |
| ISDN | Integrated Service Digital Network |
| ISP | Internet Service Provider |
| ISUP | ISDN USER PART protocol |
| ITSP | Internet Telephony Service Provider |
| ITU | International Telecommunications Union |
| IVR | Interactive Voice Response |
| IWF | Interworking Function |
| JFIF | JPEG File Interchange Format |
| JPEG | Joint Photographic Experts Group |
| LAN | Local Area Network |
| LCD | Liquid Crystal Display |
| LCF | Location Confirm |
| LD-CELP | Low-delay, Code-excited Linear Prediction |
| LDAP | Lightweight Directory Access Protocol |
| LF | Line Feed |
| LNP | Local Number Portability |
| LRJ | Location Reject |
| LRQ | Location Request |
| LSP | Line Spectral Pair |
| LSR | Last Sender Report |
| LTE | Long Term Evolution |
| M | Marker Bit (RTP) |
| mBone | Multicast Backbone of the Internet |
| MC | Multipoint Controller |
| MCF | Message Confirmation |
| MCU | Multipoint Control Unit |
| MD5 | Message Digest 5 |
| MDCX | Modify Connection |
| MEGACO | Media Gateway Controller |
| MGCF | Media Gateway Control Function |
| MGCP | Media Gateway Control Protocol |
| MGCP/L | MGCP Line |
| MGCP/T | MGCP Trunk |
| MGF | Media Gateway Function |
| MH | Modified Huffmann |
| MIME | Multipurpose Internet Mail Extension |
| MIP | Mobile IP |
| MIPS | Millions of Instructions Per Second |

| | |
|---|---|
| MME | Mobility Management Entity |
| MMS | Multimedia Message Service |
| MMUSIC | Multiparty Multimedia Session Control |
| MOS | Mean Opinion Score |
| MP | Multipoint Processor |
| MP-MLQ | Multipulse Maximum Likelihood Quantization |
| MPEG | Moving Picture Experts Group |
| MPLS | Multiprotocol Label Switching |
| MRFC | Media (or Multimedia) Resource Function Controller |
| MRFP | Media (or Multimedia) Resource Function Processor |
| MTP | Message Transfer Part |
| MTT | Minimum Transmission Time |
| MTU | Maximum Transmission Unit |
| MWI | Message Waiting Indication |
| MX | Mail Exchange |
| NACF | Network Access Configuration Function |
| NAPT | Network Address and Port Translation |
| NAPTR | Naming Authority Pointer Record |
| NAS | Network Access Server |
| NASS | Network Attachment Subsystem |
| NAT | Network Address Translation |
| NCS | Network Based Call Signaling Protocol |
| NFE | Network Facility Extension |
| NTFY | Notify |
| NTP | Network Time Protocol |
| NTSC | National Television System Committee |
| OFB | Output Feedback |
| OGW | Originating Gateway |
| OID | Object Identifier |
| OLC | Open Logical Channel |
| OO | On–off |
| OS | Operating System |
| OSP | Open Settlement Protocol |
| P-CSCF | Proxy Call/Session Control Function |
| P-frame | Prediction Frame |
| PAL | Phase-alternation-line |
| PBDF | Profile Data Base Function |
| PBX | Private Branch Exchange |
| PCC | Policy and Charging Control |
| PCEF | Policy and Charging Enforcement Function |
| PCM | Pulse Code Modulation |
| PCMA | Pulse Code Modulation A Law |
| PCMU | Pulse Code Modulation μ Law |

| | |
|---|---|
| PCRF | Policy and Charging Rule Function |
| PDF | Policy Decision Function |
| PDN-GW | Packet Data Network Gateway |
| PDU | Protocol Data Unit |
| PEP | Policy Enforcement Point |
| PER | Packed Encoding Rules |
| PES | PSTN/ISDN Emulation Subsystem |
| PGR | Pages Received (Fax) |
| PGS | Pages Sent (Fax) |
| PI | Progress Indicator |
| PIDF | Presence Information Data Format |
| PIM | Protocol-independent Multicast |
| PMIP | Proxy Mobile IP |
| POSIX | Portable Open System Interconnect |
| POTS | Plain Old Telephone Service |
| PSTN | Public Switched Telephone Network |
| PT | Payload Type |
| QCIF | Quarter CIFV (144*176) |
| QoP | Quality of Protection |
| QoS | Quality of Service |
| RACF | Resource and Admission Control Function |
| RACS | Resource and Admission Control Subsystem |
| RAI | Resource Availability Indicator |
| RAN | Radio Access Network |
| RAS | Registration, Admission, Status Protocol |
| RC | Reception Report Count |
| RCEF | Resource Control Enforcement Function |
| RCF | Registration Confirm |
| RD | Restart Delay |
| RED | Random Early Detection |
| RFC | Request for Comments |
| RGB | Red–green–blue |
| RGW | Residential Gateway |
| RLE | Run Length Encoding |
| RM | Restart Method |
| RQNT | Notification Request |
| RR | Resource Record |
| RRJ | Registration Reject |
| RRQ | Registration Request |
| RRs | Resource Records |
| RSA | Rivest, Shamir, Adleman (public key algorithm) |
| RSIP | Restart in Progress |
| RST | Reset |
| RSVP | Resource ReserVation Protocol |

| | |
|---|---|
| RTC | Return to Command |
| RTCP | Real-time Control Protocol |
| RTO | Retransmission Timeout |
| RTP/AVT | Real Time Protocol under the Audio/Video Profile |
| RTP | Real-time Protocol |
| RTP | Real-time Transport Protocol |
| RTSP | Real-time Streaming Protocol |
| S-CSCF | Serving Call/Session Control Function |
| S-GW | Serving Gateway |
| S/MIME Secure | Multipurpose Internet Mail Extension |
| SAP | Session Announcement Protocol |
| SBC | Session Border Controller |
| SCN | Switched Circuit Network |
| SCP | Service Control Point |
| SCTP | Stream Control Transport Protocol |
| SDES | Source Description RTP Packet |
| SDL | Specification and Description Language |
| SDP | Session Description Protocol |
| SECAM | Séquentiel Couleur à Mémoire |
| SGCF | Signaling Gateway Control Function |
| SGCP | Simple Gateway Control Protocol |
| SGF | Signaling Gateway Function |
| simcap | Simple Capability (SDP Declaration) |
| SIMPLE | SIP for Instant Messaging and Presence Leveraging Extensions |
| SIP | Session Initiation Protocol |
| SIPS | Session Initiation Protocol Secure |
| SLF | Subscription Locator Function |
| SMG | Special Mobile Group (of ETSI) |
| SMS | Short Message Service |
| SMTP | Simple Mail Transfer Protocol |
| SoIx | Service-oriented Interconnection |
| SP | Single Space |
| SPDF | Service Policy Decision Function |
| SQCIF | Sub-QCIF (128 × 96) |
| SR | Sender Report |
| SRV | Server DNS Record |
| SS | Supplementary Service |
| SS-CD | Supplementary Service: Call Deflection |
| SS-CFB | Supplementary Service: Call Forwarding on Busy |
| SS-CFNR | Supplementary Service: Call Forwarding on No Reply |
| SS-CFU | Supplementary Service: Call Forwarding Unconditional |
| SS-DIV | All Diversion Supplementary Services |
| SS7 | Signaling System 7 |
| SSF | Service Switching Function |

| | |
|---|---|
| SSL | Secure Sockets Layer |
| SSW | Softswitch |
| STP | Signaling Transfer Point |
| STUN | Simple Traversal of UDP through Network Address Translators |
| SUD | Single Use Device |
| TAPI | Microsoft Telephony API |
| TCAP | SS-7 Transaction Capabilities |
| TCF | Training Check Function |
| TCP | Transport Control Protocol |
| TCS | Terminal Capability Set |
| TCS=0 | NullCapabilitySet Call Flow in H.323 |
| TDM | Time Division Multiplexing |
| TE | Terminal Equipment Unit |
| TFTP | Trivial File Transfer Protocol |
| TGCF | Trunking Gateway Control Function |
| TGW | Terminating Gateway |
| TIA | Telecommunications Industry Association (USA) |
| TIPHON | Telephony and Internet Protocol Harmonization over Networks (ETSI) |
| TLS | Transport Layer Security |
| TLV | Type, Length, Value Format |
| TO | Timeout |
| TPKT | Transport Packet (RFC 1006) |
| TTL | Time to Live |
| TTS | Text to Speech |
| TURN | Traversal Using Relay NAT |
| UA | SIP User Agent |
| UAAF | User Access Authorization Function |
| UCF | Unregistration Confirm |
| UCS | Universal Character Set |
| UDP | User Datagram Protocol |
| UDPTL | UDP Transport Layer |
| UE | User Equipment |
| UICC | Universal Integrated Circuit Card |
| UII | User Input Indication |
| UL | Uplink |
| UMTS | Universal Mobile Telecommunication System |
| UPSF | User Profile Server Function |
| UPT | Universal Personal Telephony |
| URI | Uniform Resource Identifier |
| URJ | Unregistration Reject |
| URL | Uniform Resource Locator |

| | |
|---|---|
| URN | Uniform Resource Name |
| URQ | Unregistration Request |
| USH | Université de Sherbrooke |
| UTF | UCS Transformation Format |
| UTRAN | UMTS Terrestrial Radio Access Network |
| VAD | Voice Activity Detector |
| VAS | Value Added Services |
| VASA | Value Added Services Alliance |
| VLAN | Virtual Local Area Network |
| VoIP | Voice over Internet Protocol |
| VPIM | Voice Profile for Internet Messaging |
| VPN | Virtual Private Network |
| VSELP | Vector Sum-excited Linear Prediction |
| WAP | Wireless Application Protocol |
| WWW | World Wide Web |
| XML | eXtensible Markup Language |
| XMPP | eXtensible Messaging and Presence Protocol |

# Glossary

| | |
|---|---|
| Abstract Syntax Notation-1 (ASN-1) | Defined in ITU standard X.691. |
| Access Control List (ACL) | A packet filter on a router. |
| Admission Confirm (ACF) | A RAS message defined in H.225.0. |
| Admission Reject (ARJ) | A RAS message defined in H.225.0. |
| Admission Request (ARQ) | A RAS message defined in H.225.0. |
| Application Protocol Data Units (APDUs) | See H.450.1. |
| Associate Session | A related session. Two related sessions must be synchronized (e.g., an audio session can specify a video session as being related). The receiving terminal must perform lip synchronization for those sessions. |
| Backus–Naur Form (BNF) | See RFC 2234. |
| Bandwidth Confirm (BCF) | A RAS message defined in H.225.0. |
| Bandwidth Reject (BRJ) | A RAS message defined in H.225.0. |
| Bandwidth Request (BRQ) | A RAS message defined in H.225.0. |
| Basic Encoding Rule (BER) | See ASN.1. |
| Call Identifier (Call-ID) | A globally unique call identifier. |
| Call Reference Value (CRV) | A 2-octet locally unique identifier copied in all Q.931 messages concerning a particular call (see also conference identifier). |
| Conference Identifier (CID) | This is not the same as the Q.931 Call Reference Value (CRV) or the call identifier (CID). The CID refers to a conference which is the actual communication existing between the participants. In the case of a multiparty conference, if a participant joins the conference, leaves, and enters again, the CRV will change, while the CID will remain the same. |

| | |
|---|---|
| The Common Intermediary Format (CIF) | A video format which has been chosen because it can be sampled relatively easily from both the 525-line and 625-line video formats: 352 × 288 pixels |
| Contributing Source (CSRC) | When an RTP stream is the result of a combination put together by an RTP mixer of several contributing streams, the list of the SSRCs of each contributing stream is added in the RTP header of the resulting stream as CSRCs. The resulting stream has its own SSRC. |
| Disengage Confirm (DCF) | A RAS message defined in H.225.0. |
| Disengage Reject (DRJ) | A RAS message defined in H.225.0. |
| Disengage Request (DRQ) | A RAS message defined in H.225.0. |
| Dual-Tone Multi-Frequency (DTMF) | Tones composed of two well-defined frequencies that represent digits 0–9, *, #. The combination of frequencies has been selected to be almost impossible to reproduce with the human voice. DTMF tones are used to dial from analog phones and to control IVR servers. |
| Dynamic Host Configuration Protocol (DHCP) | Used by end points to acquire a temporary IP address and important TCP/IP parameters (router IP address, DNS IP address, etc.) from a server in the network. |
| End of Line (EOL) | The end of line sequence for group 3 fax (001H). |
| Energy | For an image on a particular color, the sum of the squared color values of the pixels is called the energy. |
| ENUM | An E.164 number resolution protocol defined in RFC 2916. |
| Fast-Connect | A procedure to eliminate media delays after the connection of the call introduced in H.323v2. Another name used for the same procedure is Fast-Start. |
| Fast-Start | See Fast-Connect. |
| Gatekeeper Confirm (GCF) | A RAS message defined in H.225.0. |
| Gatekeeper Request (GRQ) | A RAS message defined in H.225.0. |
| Gatekeeper Reject (GRJ) | A RAS message defined in H.225.0. |
| Information Request (IRQ) | A RAS message defined in H.225.0. |
| Information Request Response (IRR) | A RAS message defined in H.225.0. |
| Initial Address Message (IAM) | SS7 ISUP message initiating a call set-up. |
| Inter-mode | Refers to a video-coding mode where compression is achieved by reference to the previous, or sometimes the next, frame. |

| | |
|---|---|
| Interactive Voice Response server (IVR) | A machine accepting DTMF or voice commands, and executing some logic which interacts with the user using pre-recorded prompts or synthetic voice. |
| Internet Fax Transmission (IFT) | A protocol, see ITU recommendation T.38. |
| Internet Relay Chat (IRC) | The famous 'chat' service of the Internet, based on a set of servers mirroring text-based conversations in real-time. |
| Intra-mode | Refers to a video-coding mode where compression is achieved locally (i.e., not relatively to the previous frame). |
| IP-PBX | Private phone switch with a VoIP wide area network interface. Most IP-PBXs have an H.323 WAN interface. See also IPBX. |
| IPBX | Same as IP-PBX. Some use the term IPBX for private phone switches which use only VoIP (i.e., the phones are also IP phones), whereas an IP-PBX can be a traditional PBX with analog phones and only uses a WAN VoIP interface. See IP-PBX |
| Jitter | Statistical variance of packet interarrival time. It is the smoothed absolute value of the mean deviation in packet-spacing change between the sender and the receiver. The smoothing is usually done by averaging on a sliding window of 16 instantaneous measures. |
| jitter | Varying delay. |
| Location Confirm (LCF) | A RAS message defined in H.225.0. |
| Location Reject (LRJ) | A RAS message defined in H.225.0. |
| Location Request (LRQ) | A RAS message defined in H.225.0. |
| macroblock | For the H.261 algorithm, a group of four 8*8 blocks. |
| Maximum Transmission Unit (MTU) | The largest datagram that can be sent over the network without segmentation. |
| Multicast Backbone of the Internet (mBone) | Capable of sending one packet to multiple recipients. |
| Multipoint Control Unit (MCU) | An H.323 callable end-point which consists of an MC and optional MPs. |
| Multipoint Controller (MC) | The H.323 which provides the control function for multiparty conferences. |
| Multipoint Processor (MP) | The H.323 entity which processes the media streams of the conference and does all the necessary switching, mixing, etc. |

| | |
|---|---|
| Naming Authority Pointer Resource of the DNS (NAPTR) | Defined in RFC 2915 and used notably by ENUM, see ENUM. |
| Network Facility Extension (NFE) | Defined in H.450.1. |
| Network Time Protocol (NTP) | This defines a standard way to format a timestamp, by writing the number of seconds since 1/1/1900 with 32 bits for the integer part and 32 bits for the decimal part expressed as number of $1/2^{32}$ seconds (e.g., 0x800000000 is 0.5 s). A compact format also exists with only 16 bits for the integer part and 16 bits for the decimal part. The first 16 digits of the integer part can usually be derived from the current day, the fractional part is simply truncated to the most significant 16 digits. |
| P-frame | Prediction frame obtained by motion estimation or otherwise, and representing only the difference between this image and the previous one. |
| Packed Encoding Rules (PER) | See ASN.1. |
| Payload Type (PT) | As defined by RTP. |
| port | An abstraction that allows the various destinations of the packets to be distinguished on the same machine (e.g., Transport Selectors, or TSELs, in the OSI model, or IP ports). On the Internet, many applications have been assigned 'well-known ports' (e.g., a machine receiving an IP packet on port 80 using TCP will route it to the web server). |
| Prediction frame (P-frame) | Obtained by motion estimation or otherwise, and representing only the difference between this image and the previous one. |
| Private Branch Exchange (PBX) | A private phone switch. |
| Proxy server | An intermediary program that acts as both a server and a client for the purpose of making requests on behalf of other clients. Requests are serviced internally or by passing them on, possibly after translation, to other servers. A proxy interprets, and, if necessary, rewrites a request message before forwarding it. |
| Q-interface Signaling (QSIG) | Protocol used at the Q-interface between two switches in a private network. ECMA/ISO have defined a set of QSIG standards. |

| | |
|---|---|
| Real-time Control Protocol (RTCP) | See RFC 1889. |
| Real-time Transport Protocol (RTP) | As specified by RFC 1889. |
| Registration Confirm (RCF) | A RAS message defined in H.225.0. |
| Registration Reject (RRJ) | A RAS message defined in H.225.0. |
| Registration Request (RRQ) | A RAS message defined in H.225.0. |
| Registration, Admission, and Status (RAS) | The name of the protocol used between the gatekeeper and a terminal, and between gatekeepers for registration, admission, and status purposes. Defined in H.225.0. |
| Return To Command (RTC) | Six consecutive EOLs instructing a G3 Fax to return to command mode. |
| Sender Report (SR) | Used in RTCP and RTP. |
| Session ID | A unique RTP session identifier assigned by the master. The convention is that the value of the session ID is 1 for a primary audio session, 2 for a primary video session, and 3 for a primary data session. See Associate session. |
| Single Use Device (SUD) | See H.323 annex F. |
| SIP dialog | This was defined in RFC 3261 as a peer-to-peer SIP relationship between two UAs which persists for some time. A dialog is established by SIP messages, such as a 2xx response to an INVITE request. A dialog is identified by a call identifier, a local tag, and a remote tag. A dialog was formerly known as a call leg in RFC 2543. |
| SIP final response | A SIP response that terminates a SIP transaction (e.g., 2xx, 3xx, 4xx, 5xx, 6xx responses). See SIP provisional response. |
| SIP provisional response | A SIP response that does not terminate a SIP transaction, as opposed to a SIP final response (1xx responses are provisional). |
| SIP redirect server | A redirect server is a server that accepts a SIP request, maps the address into zero or more new addresses, and returns these addresses to the client. Unlike a proxy server, it does not initiate its own SIP request. Unlike a user agent server, it does not accept calls. |
| SIP registrar | A registrar is a server that accepts REGISTER requests. A registrar is typically co-located with a proxy or redirect server and may offer location services. |

| | |
|---|---|
| SIP server | A server is an application program that accepts requests in order to service requests and sends back responses to those requests. Servers are either proxy, redirect, or user agent servers or registrars. |
| SIP transaction | A SIP transaction occurs between a client and a server, and comprises all messages from the first request sent from the client to the server up to a final (non-1xx) response sent from the server to the client. A transaction is identified by the CSeq sequence number within a single-call leg. The ACK request has the same CSeq number as the corresponding INVITE request, but comprises a transaction of its own. |
| Stream Control Transport Protocol (SCTP) | Defined in RFC 2960. |
| Supplementary Services (SS-DIV) | Includes all diversion supplementary services, such as SS-CFU, SS-CFB, SS-CFNR, SS-CD. |
| Switched Circuit Network (SCN) | A generic term for the 'classic' phone network, including PSTN, ISDN, and GSM. |
| Synchronization Source (SSRC) | Source of an RTP stream, identified by 32 bits in the RTP header. All the RTP packets with a common SSRC have a common time and sequencing reference. |
| Talkspurt | A period during which a participant usually speaks, as opposed to silence periods. |
| Time Division Multiplexing (TDM) | The traditional voice transmission and switching technique based on assigning each communication a fixed "time slot" on a communication line between central offices. |
| TPKT | A TCP connection establishes a reliable data stream between two hosts, but there is no delimitation of individual messages within this stream. RFC 1006 defines a simple TPKT packet format to delimit such messages. It consists of a version octet ('3'), two reserved octets ('00'), and the total length of the message including the previous headers (2 octets). |
| Transport address | Combination of a network address (e.g., IP address 10.0.1.2) and port (e.g., IP port 1720) which identifies a transport termination point. |
| Transport Control Protocol (TCP) | The most widely used, reliable transport protocol for IP networks. |

| | |
|---|---|
| Transport Layer Security (TLS) | A secure protocol using TCP, defined in RFC 2246. |
| Trivial File Transfer Protocol (TFTP) | A very simple file transfer protocol over UDP, frequently used by IP appliances to download their initial configuration parameters. |
| Uniform Resource Identifier (URI) | Defines a uniform syntax and semantic convention for any resource. The URI is defined in RFC 2396. See also URL, URN. |
| Uniform Resource Locator (URL) | A specific type of URI identifying a resource by its primary network address. URLs are used by SIP to indicate the originator, current destination, and final recipient of a SIP request, and to specify redirection addresses. See also URI. |
| Uniform Resource Name (URN) | A specific type of URI required to be universally unique and persistent even if the resource ceases to exist. See also URI. |
| Unregistration Confirm (UCF) | A RAS message defined in H.225.0. |
| Unregistration Reject (URJ) | A RAS message defined in H.225.0. |
| Unregistration Request (URQ) | A RAS message defined in H.225.0. |
| User Agent Client (UAC) | Also known as a calling user agent. A user agent client is a client application that initiates the SIP request. |
| User Agent Server (UAS) | Also known as a called user agent. A user agent server is a server application which contacts the user when a SIP request is received and returns a response on behalf of the user. The response accepts, rejects, or redirects the request. |
| User agent | A SIP end system participating in a SIP transaction. See UAC, UAS. |
| User Datagram Protocol (UDP) | The most widely used unreliable transport protocol for IP networks. UDP only guarantees data integrity by using a checksum, but an application using UDP has to take care of any data recovery task. |
| Zone | An H.323 zone is the set of all H.323 end points, MCs, MCUs, and gateways managed by a single gatekeeper. |

# Preface

## VoIP 1998–2004, 6 YEARS FROM R&D LABS TO LARGE SCALE DEPLOYMENTS

Since 1998 Voice over IP, in short VoIP, has been the favorite buzzword of the telecom industry. In 1998, IP was not yet as established and dominant as it is today, and most telecom engineers still believed that only ATM technology would be able to support multimedia applications. Indeed at this time most of us experienced the Internet only through dial-up modems and most ISPs, unable to keep-up with the exploding demand for Internet connections, were providing a level of service that could hardly qualify even for 'best effort'.

But even in this context, the R&D teams that started to work on VoIP were not simply taking a leap of faith. Their bet on VoIP was backed by the last developments of packet networking theory, which proved that properly designed IP networks could provide an appropriate support for applications requiring quality of service. Knowing this, most of these teams felt confident that VoIP could be deployed on a wide scale in the future, and in the mean time tried to evaluate what could be the impact of VoIP, compared to previous technologies.

It took a relatively long time to understand the reasons that would lead a service provider to deploy VoIP instead of traditional switched voice networks. Initially VoIP was presented as a technology that could enable a service provider to transport voice 'for free' over the Internet, because IP transport was 'free', and calls could be routed to local breakout trunks on the far end. The first commercial applications of VoIP focused on prepaid telephony, which was a reasonable target given that potential buyers of prepaid card systems do care about costs, and they are much more tolerant to quality of service issues than any other market segment. VoIP prepaid telephony systems did have a great success—today the majority of international calling card services use VoIP—but not because of cheaper call termination costs (which are regulated independently of the technology in most countries), or cheaper transport costs (traditional voice compression systems are much more efficient than VoIP systems). The reason for the success was mainly because VoIP facilitates the trading of minutes between multiple networks without the constraint of establishing leased lines: on the Internet, virtually all VoIP service providers 'see' each other and can decide to exchange traffic immediately, or to stop as soon as better arbitrage opportunities exist. In addition the central switching system of a VoIP service provider does not process voice streams, but only signaling messages: a call initiated from a gateway in Paris can be routed to a gateway in London by a VoIP call controller located in NewYork with very few overhead costs. Only signaling messages make the round trip through the Atlantic, voice packets only cross the Channel.

It is now clear that the key reasons for the success of VoIP are:

- location independence: because of the unique characteristics of VoIP call controllers, or 'Softswitches', many functions that previously required multiple distributed points of presence can now be centralized, reducing administrative overheads and accelerating deployments
- simplification of transport networks: in the example above, service providers no longer need to establish leased lines dedicated to voice prior to exchanging traffic. But the use of standard IP data networks—configured appropriately—is a major breakthrough in many other circumstances: core transport networks no longer need to maintain the dedicated network that was required by SS7 signaling, enterprises moving to new offices can save the significant expenses required by dedicated telephony wiring and use virtual LANs instead
- the ability to establish and control multimedia communications, e.g. interactive audio and video calls, data sharing sessions, etc.

Because of these unique characteristics, VoIP technology is a very good choice every time a relatively complex call control function would require multiple points of presence close to the end-users in traditional switched technology, and can be centralized with an application softswitch:

- In *residential telephony*, new service providers can deploy centralized VoIP call control servers and use any IP networking technology. For instance FastWeb, in Italy, serves the Italian market from just two PoPs located in Milan and Rome. This is not possible with traditional technology using traditional (TDM) switches (even with V5.2/GR303 ATM gateways used at the edge of the network), because the voice streams need to be physically switched by the backplane of the TDM switch. In addition of course, VoIP technology makes it possible to introduce additional media, like video communications, which differentiate the service and help maintain the ARPU[1]
- *Informal, Distributed contact centers* also become much easier and cheaper to operate with VoIP: the centralized call distribution point no longer needs to switch the voice

[1] Average Revenue Per User

streams, and therefore tromboning[2] through the VoIP call distribution server is completely eliminated, which reduces communications costs and minimizes the required bandwidth for the connection of the call distribution platform
- In general, all applications which previously required a complex *intelligent network* architecture in order to minimize tromboning (call switching occurs at specific nodes in the network, and the applications can be located elsewhere), can be significantly simplified using a centralized call control server which controls voice signaling but optimizes the voice path through the IP network.

Today more and more service providers and enterprises, as they have become confident in the VoIP technology and quality of service of IP networks, deploy VoIP applications in order to enjoy the location independence and greater flexibility of the technology. With more successful deployments, VoIP is gaining in maturity, and the cost of VoIP gateways and IP phones is quickly dropping with the increased volumes.

## SCOPE OF THIS BOOK

In *IP Telephony*, we will also assume, like the pioneers of VoIP, that it is possible to carry multimedia data flows over an IP network with an appropriate quality (i.e. low latency and low packet loss), and we will focus only on the functional aspects of VoIP. Voice coding technology is presented as a 'black box', with enough information for an engineer who wants to use an existing coder in an application, but without describing the technology in detail. *IP Telephony* will be useful mainly in the lab (development platforms, validation platforms), when designing and troubleshooting new interactive multimedia applications.

The companion book *Beyond VoIP Protocols* becomes necessary when you deploy these applications in the field, over a real network with limited capacity. *Beyond VoIP protocols* contains an overview of the techniques that can be used to provide custom levels of quality of service for IP data flows, and guidelines to properly dimension an IP network for voice. It also delves into the details of voice coding technology, and the influence of the selected voice coder and the transmission channel parameters on perceived voice quality.

In theory, it is sufficient to read the VoIP standards in order to become an efficient VoIP engineer. Although reading the standards is always necessary at some point, these documents were never written to be read from A to Z. Not only the mere volume is a problem, hundreds of pages for each standard, but also the structure is inappropriate: all VoIP standards are written as umbrella documents, which point explicitly or implicitly to dozens of other more detailed documents. Sometimes, these documents are also misleading, because some of the recommended methods were discussed in a specific context

[2] 'Tromboning' refers to a non-optimal media path through the network, compared to the shortest path. It happens when the media streams have to 'zigzag' across multiple nodes, reminding of the shape of a bent trombone.

in the standard bodies, but this context was lost or not clearly expressed in the written recommendation (see for instance the issues presented in Chapter 7 for call redirection). Last but not least most standards are the result of 'diplomatic' agreements between firms, which often result in multiple alternate ways of doing the same thing, very long and cumbersome documents with many 'options' and unclear sentences designed to preserve the agreed compromise, while in practice after a few years, the market forces lead to a 'de-facto' standard choice, in general adopted from the practice of the dominant players.

We wrote *IP Telephony* because we believe it is much more efficient to gain first a general overview on VoIP, and only then go into the details of the standard documents, but only when needed and if clarification is required on a specific item. Initially this book was designed as an internal training tool within France Telecom, and over the years it developed by capturing the accumulated experience of the authors and their colleagues, in over 50 voice over IP deployments, among which some of the largest VoIP deployments worldwide: Orange and its multi-million livebox® deployment (well over 50% of the French telephony traffic now uses VoIP), FastWeb in Italy, Equant (the world's largest VoIP virtual private network), etc.

*IP telephony* begins by giving an overview of the techniques that can be used to encode media streams and transmit them over an IP network (Chapter 1). It focuses on the functional requirement of transmitting an isochronous data stream over an asynchronous network which introduces delay variations ("jitter"). The media encoding methods themselves are presented very briefly, with just enough details for an engineer who wants to use them and understand the main parameters required for the transmission of the resulting data.

The most popular VoIP standards are presented in Chapter 2 (H.323), Chapter 3 (SIP), and Chapter 5 (MGCP). In Chapter 4 we describe the IMS (IP multimedia subsystem), which has become the standard architecture for large scale residential VoIP networks (using the TISPAN profile, also described in Chapter 3) and will be the cornerstone of future LTE deployments. These chapters do not intend to fully replace the standards, but provide a detailed overview that should be sufficient for most engineers and pointers to relevant normative documents if further reference is required. The value of these chapters comes also from the many discussions on aspects of the standards that are still immature, and descriptions of calls flows or protocol extensions commonly used by vendors but not described in standard documents.

The 'advanced topics' chapters (Chapters 6 and 7), discuss two issues faced by all service providers when deploying public VoIP services (as opposed to custom services designed for a single enterprise). The first issue comes from the incompatibility of current VoIP protocols with Network Address Translation routers and firewalls, which change the addresses of IP packets on the fly but without properly translating the IP addresses contained in the VoIP messages carried by these packets. The second issue comes from the widespread confusion between private telephony techniques and public telephony techniques for call transfers. In both cases the chapter presents techniques that were deployed successfully, and explains the pros and cons of each possible method.

## CONCLUSION

When the first edition of this book was published, VoIP standards were beginning to mature, at the protocol level. VoIP products, which were using totally proprietary protocols before the year 2000, began to interwork first using H.323 and then MGCP and SIP also. Simultaneously, some operators began to deploy huge VoIP residential networks, reaching millions of users. In 2005, most deployments used standard protocols; however, the architectural details of the VoIP networks were still proprietary and specific to each VoIP network: network interconnection was possible, but roaming across networks was impossible. The need for a standard architecture became stronger as the size of deployments reached massive dimensions: the work of 3GPP on the IP Multimedia Subsystem architecture aimed at defining such a standard architecture. This was quite an ambitious and difficult challenge, but with the help of ETSI TISPAN which complemented the standard with specific functions required in fixed networks, the IMS architecture, in its release 8 ("Common IMS"), finally reached a level of maturity which makes real deployments possible.

In this new edition, we dedicate a full chapter to the IMS architecture, the underlying transport network architecture (Enhanced Packet System), and the TISPAN specific additions for fixed networks. We continue, however, to present in detail, protocols such as H.323 or MGCP which are not used inside the IMS system, but as peer networks or at the edges of the IMS network. These protocols are still used intensively in existing VoIP networks, and are still the best candidates in some situations, e.g., videoconferencing and ISDN PBX trunking for H.323, or business IP phone control for MGCP. It is likely that future evolutions of SIP and IMS will progressively alleviate the need for other protocols in VoIP; however, most VoIP operators will still need to support multiple VoIP protocols in the next 5 to 10 years.

We hope that this book will help network engineers to deploy, maintain, or upgrade their VoIP networks, using each protocol where it fits best, and with full awareness of the potential pitfalls and difficulties.

# 1

# Multimedia Over Packet

## 1.1 TRANSPORTING VOICE, FAX, AND VIDEO OVER A PACKET NETWORK

### 1.1.1 A Darwinian view of voice transport

#### *1.1.1.1 The circuit switched network*

The most common telephone system on the planet today is still analog, especially at the edge of the network. Analog telephony (Figure 1.1) uses the modulation of electric signals along a wire to transport voice.

Although it is a very old technology, analog transmission has many advantages: it is simple and keeps the end-to-end delay of voice transmission very low because the signal propagates along the wire almost at the speed of light.

It is also inexpensive when there are relatively few users talking at the same time and when they are not too far apart. But the most basic analogue technology requires one pair of wires per active conversation, which becomes rapidly unpractical and expensive. The first improvement to the basic 'baseband' analog technology involved multiplexing several conversations on the same wire, using a separate transport frequency for each signal. But even with this hack, analog telephony has many drawbacks:

- Unless you use manual switchboards, analog switches require a lot of electromechanical gear, which is expensive to buy and maintain.
- Parasitic noise adds up at all stages of the transmission because there is no way to differentiate the signal from the noise and the signal cannot be cleaned.

For all these reasons, most countries today use digital technology for their core telephone network and sometimes even at the edge (e.g., ISDN). In most cases the subscriber line remains analogue, but the analogue signal is converted to a digital data stream in the

*IP Telephony: Deploying VoIP Protocols and IMS Infrastructure, Second Edition* O. Hersent

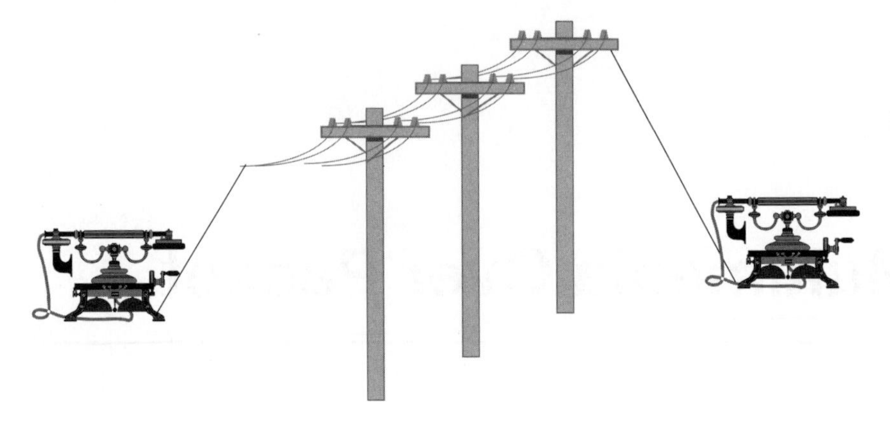

**Figure 1.1** Analog telephony, as old as the invention of the telephone, and still in use today at the edge of the network.

first local exchange. Usually, this signal has a bitrate of 64 kbit/s or 56 kbit/s (one sample every 125 µs).

With this digital technology, many voice channels can easily be multiplexed along the same transmission line using a technology called **time division multiplexing (TDM)**. In this technology, the digital data stream which represents a single conversation is divided into blocks (usually an octet), and blocks from several conversations are interleaved in a round robin fashion in the time slots of the transmission line, as shown in Figure 1.2.

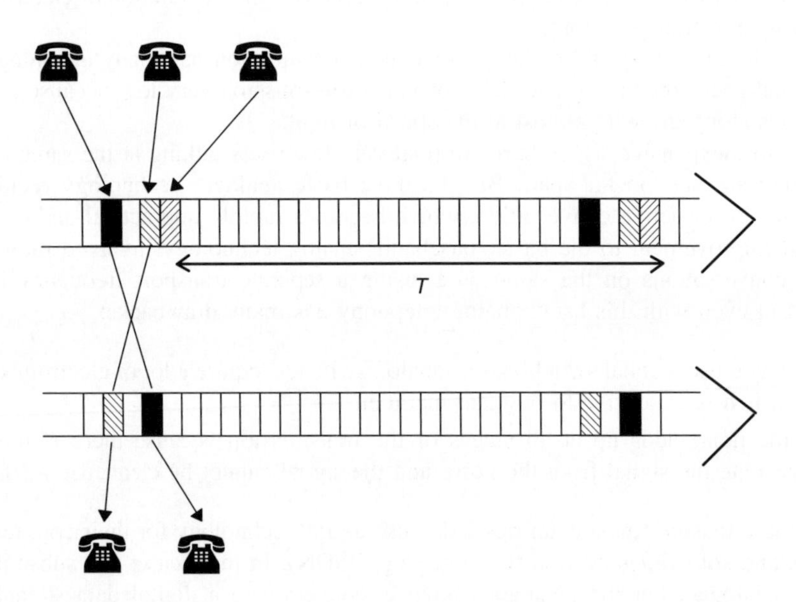


**Figure 1.2** TDM switching.

Because of digital technology, the noise that is added in the backbone does not influence the quality of the communication because digital 'bits' can be recognized exactly, even in the presence of significant noise. Moreover, digital TDM makes digital switching possible. The switch just needs to copy the contents of one time slot of the incoming transmission line into another time slot in the outgoing transmission line. Therefore, this switching function can be performed by computers.

However, a small delay is now introduced by each switch, because for each conversation a time slot is only available every $T$ µs, and in some cases it may be necessary to wait up to $T$ µs to copy the contents of one time slot into another. Since $T$ equals 125 µs in all digital telephony networks, this is usually negligible and the main delay factor is simply the propagation time.

#### 1.1.1.2 Asynchronous transmission and statistical multiplexing

Unless you really have a point to make, or you're a politician, you will usually speak only half of the time during a conversation. Since we all need to think a little before we reply, each party usually talks only 35% of the time during an average conversation.

If you could press a button each time you talk, then you would send data over the phone line only when you actually say something, not when you are silent. In fact, most of the techniques used to transform your voice into data (known as codecs) now have the ability to detect silence. With this technique, known as voice activity detection (VAD), instead of transmitting a chunk of data, voice, or silence every 125 µs, as done today on TDM networks, you only transmit data when you need to, asynchronously, as illustrated in Figure 1.3.

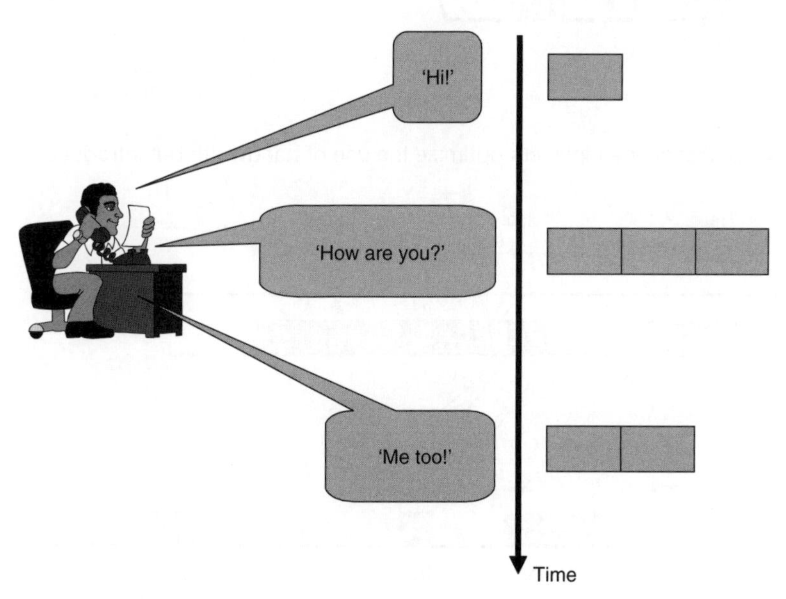


**Figure 1.3** Transmitting voice asynchronously.

And when it comes to multiplexing several conversations on a single transmission line, instead of occupying a fraction of bandwidth all the time, 'your' bandwidth can be used by someone else while you are silent. This is known as 'statistical multiplexing'.

The main advantage of **statistical multiplexing** is that it allows the bandwidth to be used more efficiently, especially when there are many conversations multiplexed on the same line (see companion book, *Beyond VoIP protocols* Chapter 5 for more details). But statistical multiplexing, as the name suggests, introduces uncertainty in the network. As just mentioned, in the case of TDM a delay of up to 125 μs could be introduced at each switch; this delay is constant throughout the conversation. The situation is totally different with statistical multiplexing (Figure 1.4): if the transmission line is empty when you need to send a chunk of data, it will go through immediately. If on the other hand the line is full, you may have to wait until there is some spare capacity for you.

This varying delay is caller **jitter**, and needs to be corrected by the receiving side. Otherwise, if the data chunks are played as soon as they are received, the original speech can become unintelligible (see Figure 1.5).

The next generation telephone networks will use statistical multiplexing, and mix voice and data along the same transmission lines. Several technologies are good candidates (e.g., voice over frame relay, voice over ATM, and, of course, voice over IP).

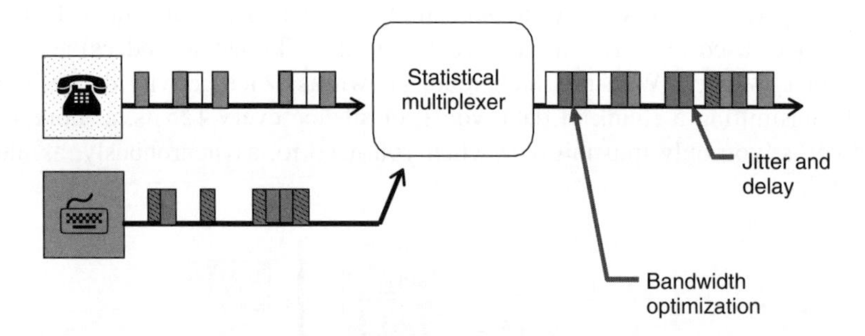


**Figure 1.4** Statistical multiplexers optimize the use of bandwidth but introduce jitter.

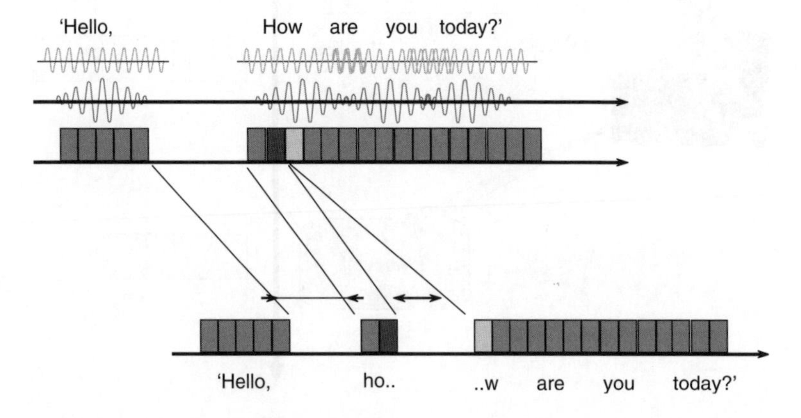


**Figure 1.5** Effects of uncompensated jitter.

We believe voice over IP is the most flexible solution, because it does not require setting up virtual channels between the sites that will communicate. VoIP networks scale much better than ATM or frame relay networks, and VoIP also allows communications to be established directly with VoIP endpoints: there is now a variety of **IP-PBXs** (private switches with a VoIP wide-area network interface), or IP phones on the market today that have no ATM or frame relay equivalent.

## 1.1.2 Voice and video over IP with RTP and RTCP

The Real-time Transport Protocol and Real-time Control Protocol, described in RFC 3550, are the protocols that have been used for the transport of media streams since the first conferencing tools were made available on the Internet. The **visual audio tool (VAT)** used RTP version 0. A description of version 1 is available at ftp://gaia.cs.umass.edu/pub/hgschulz/rtp/draft-ietf-avt-rtp-04.txt

Since then, RTP has evolved into version 2. RTPv2 is not backward compatible with version 1, and therefore all applications should be built to support RTPv2.

### *1.1.2.1 Why RTP/RTCP?*

When a network using statistical multiplexing is used to transmit real-time data such as voice, jitter has to be taken into account by the receiver. Routers are good examples of such statistical multiplexing devices, and therefore voice and video over IP have to face the issue of jitter.

RTP was designed to allow receivers to compensate for jitter and desequencing introduced by IP networks. RTP can be used for any real-time (or more rigorously isochronous) stream of data (e.g., voice and video). RTP defines a means of formatting the payload of IP packets carrying real-time data. It includes:

- Information on the type of data transported (the 'payload').
- Timestamps.
- Sequence numbers.

Another protocol, RTCP, is very often used with RTP. RTCP carries some feedback on the quality of the transmission (the amount of jitter, the average packet loss, etc.) and some information on the identity of the participants as well.

RTP and RTCP do not have any influence on the behavior of the IP network and do not control quality of service in any way. The network can drop, delay, or desequence an RTP packet like any other IP packet. RTP must not be mixed up with protocols like **RSVP** (Resource Reservation Protocol). RTP and RTCP simply allow receivers to recover from network jitter and other problems by appropriate buffering and sequencing, and to have more information on the network so that appropriate corrective measures can be adopted (redundancy, lower rate codecs, etc.). However, some routers are actually able to parse

IP packets, discover whether these packets have RTP headers, and give these packets a greater priority, resulting in better QoS even without any external QoS mechanism, such as RSVP for instance. Most Cisco® routers support the `IP RTP PRIORITY` command.

RTP and RTCP are designed to be used on top of any transport protocol that provides framing (i.e., delineates the beginning and end of the information transported), over any network. However, RTP and RTCP are mostly used on top of **UDP (User Datagram Protocol)**.[1] In this case, RTP is traditionally assigned an even UDP port and RTCP the next odd UDP port.[2]

#### 1.1.2.2 RTP

RTP allows the transport of isochronous data across a packet network, which introduces jitter and can desequence the packets. Isochronous data are data that need to be rendered with exactly the same relative timing as when they were captured. Voice is the perfect example of isochronous data; any difference in the timing of the playback will either create holes or truncate some words. Video is also a good example, although tolerances for video are a lot higher; delays will only result in some parts of the screen being updated a little later, which is visible only if there has been a significant change.

RTP is typically used on top of UDP. UDP is the most widely used 'unreliable' transport protocol for IP networks. UDP can only guarantee data integrity by using a checksum, but an application using UDP has to take care of any data recovery task. UDP also provides the notion of a '**port**', which is a number between 0 and 65,535 (present in every packet as part of the destination address) which allows up to 65,536 UDP targets to be distinguished at the same destination IP address. A port is also attached to the source address and allows up to 65,536 sources to be distinguished from the same IP address. For instance, an RTP over UDP stream can be sent from 10.10.10.10:2100 to 10.10.10.20:3200:

| Source IP address: 10.10.10.10 | Source port: 2100 | Destination IP address: 10.10.10.20 | Destination port: 3200 | RTP Data |
|---|---|---|---|---|

When RTP is carried over UDP, it can be carried by multicast IP packets, that is, packets with a multicast destination address (e.g. 224.34.54.23): therefore an RTP stream generated by a single source can reach several destinations; it will be duplicated as necessary by the IP network. (See companion book, *Beyond VoIP Protocols*, Chapter 6. IP multicast routing).

##### 1.1.2.2.1 A few definitions

- **RTP session**: an RTP session is an association of participants who communicate over RTP. Each participant uses at least two transport addresses (e.g., two UDP ports on the

[1] For streaming (e.g., RTSP), since there are is no real-time constraint on transmission delay, it can be used over TCP.

[2] But this is not mandatory, especially when RTP/RTCP ports are conveyed by an out-of-band signaling mechanism.

local machine) for each session: one for the RTP stream, one for the RTCP reports. When a multicast transmission is used all the participants use the same pair of multicast transport addresses. Media streams in the same session should share a common RTCP channel. Note that H.323 or SIP require applications to define explicitly a port for each media channel. So, although most applications comply with the RTP requirements for RTP and RTCP port sharing as well as the use of adjacent ports for RTP and RTCP, an application should *never* make an assumption about the allocation of RTP/RTCP ports, but rather use the explicit information provided by H.323 or SIP, even if it does not follow the RTP RFC guidelines. This is one of the most common bugs still found today in some H.323 or SIP applications.

- **Synchronization source (SSRC)**: identifies the source of an RTP stream, identified by 32 bits in the RTP header. All RTP packets with a common SSRC have a common time and sequencing reference. Each sender needs to have an SSRC; each receiver also needs at least one SSRC as this information is used for **receiver reports** (RRs).
- **Contributing source (CSRC)**: when an RTP stream is the result of a combination put together by an RTP mixer from several contributing streams, the list of the SSRCs of each contributing stream is added in the RTP header of the resulting stream as CSRCs. The resulting stream has its own SSRC. This feature is not used in H.323 or SIP.
- **NTP format**: a standard way to format a timestamp, by writing the number of seconds since 1/1/1900 at 0 h with 32 bits for the integer part and 32 bits for the decimal part (expressed in $\frac{1}{2^{32}}$ s (e.g., $0 \times 80000000$ is 0.5 s). A compact format also exists with only 16 bits for the integer part and 16 bits for the decimal part. The first 16 digits of the integer part can usually be derived from the current day, the fractional part is simply truncated to the most significant 16 digits.

#### *1.1.2.2.2 The RTP packet*

All fields up to the CSRC list are always present in an RTP packet (see Figure 1.6). The CSRC list may only be present behind a mixer (a device that mixes RTP streams, as defined in the RTP RFC). In practice, most conferencing bridges that perform the function of a mixer (H.323 calls them ‘multipoint processors’, or MPs) do not populate the CSRC list.

Here is a short explanation of each RTP field:

- Two bits are reserved for the **RTP version**, which is now version 2 (10). Version 0 was used by VAT and version 1 was an earlier IETF draft.
- A **padding bit P** indicates whether the payload has been padded for alignment purposes. If it has been padded (P = 1), then the last octet of the payload field indicates more precisely how many padding octets have been appended to the original payload.
- An **extension bit X** indicates the presence of extensions after the eventual CSRCs of the fixed header. Extensions use the format shown in Figure 1.7.
- The 4-bit **CSRC count** (CC) states how many CSRC identifiers follow the fixed header. There is usually none.

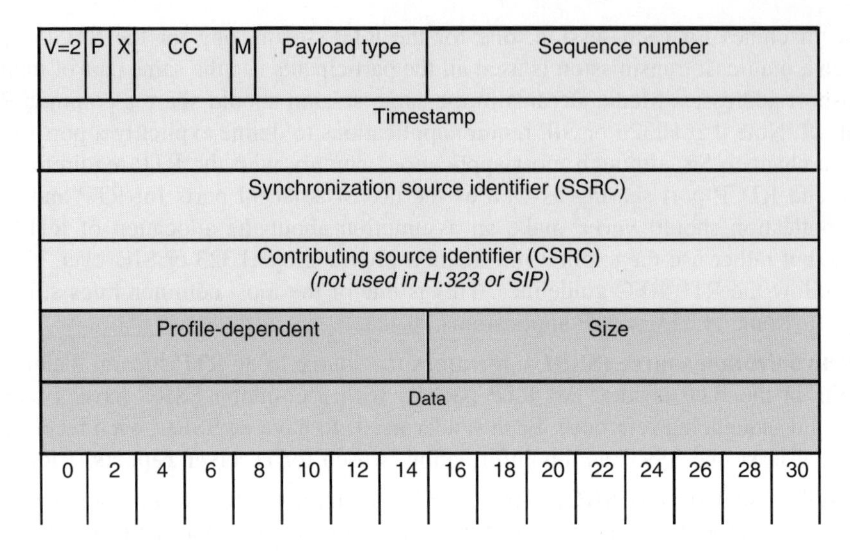


**Figure 1.6** RTP packet format.

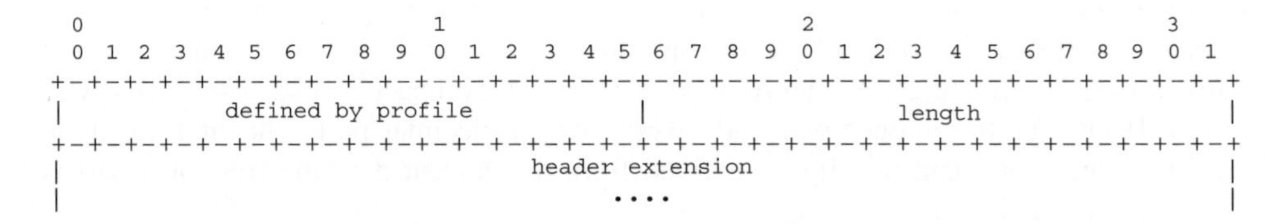


**Figure 1.7** Optional extension header.

- **Marker (M)**: 1 bit. Its use is defined by the RTP profile. H.225.0 says that for audio codings that support silence suppression, it must be set to 1 in the first packet of each talkspurt after a silence period. This may allow some implementations to dynamically reduce the jitter buffer size without running the risk of cutting important words (e.g., by trimming off some silence packets).
- **Payload type (PT)**: 7 bits. The payload of each RTP packet is the real-time information contained in the packet. Its format is completely free and must be defined by the application or the profile of RTP in use. It enables applications to distinguish a particular format from another without having to analyse the content of the payload. Some common identifiers are listed in Table 1.1; they are used by H.225 and SIP. These are called **static payload types** and are assigned by **IANA (Internet Assigned Numbers Authority)**; a list can be found at http://www.isi.edu/in-notes/iana/assignments/rtp-; parameters PT 96 to 127 are reserved for **dynamic payload types**. Dynamic payload types are defined in the RTP audio-visual (A/V) profile and are not assigned in the IANA list. The dynamic PT meaning is defined only for the duration of the session. The exact meaning of the dynamic payload type is defined through some out-of-band

**Table 1.1** Common static payload types

| Payload type | Codec | |
|---|---|---|
| **0** | PCM, $\mu$-law | Audio |
| **8** | PCM, A-law | |
| **9** | G.722 | |
| **4** | G.723 | |
| **15** | G.728 | |
| **18** | G.729 | |
| **34** | H.263 | Video |
| **31** | H.261 | |

mechanism (e.g., though Session Description Protocol parameters for protocols like SIP, H.245 OpenLogicalChannel parameters for H.323, or through some convention or other mechanism defined by the application). The codec associated with a dynamic PT is negotiated by the conference control protocol dynamically. Since RTP itself doesn't define the format of the payload section, each application must define or refer to a profile. In the case of H.323, this work is done in annex B of H.225.

- **A sequence number and timestamp**: the 16-bit sequence number and timestamp start on a random value and are incremented at each RTP packet. The 32-bit timestamp uses a clock frequency that is defined for each payload type (e.g., H.261 payload uses a 90-kHz clock for the RTP timestamp). For narrow-band audio codecs (G.711, G.723.1, G.729, etc.) the RTP clock frequency is set to 8,000 Hz. For video, the RTP timestamp is the tick count of the display time of the first frame encoded in the packet payload. For audio, the RTP timestamp is the tick count when the fist audio sample contained in the payload was sampled. Each RTP packet carries a sequence number and a timestamp. RTP timestamps do not have an absolute meaning (the initial timestamps of an RTP stream can be selected at random); even timestamps of related media (e.g., audio and video) in a single session will be unrelated. In order to map RTP packet timestamps to absolute time, one must use the information held in RTCP sender reports, where RTP timestamps are associated with the absolute NTP time. Depending on the application, timestamps can be used in a number of ways. A video application, for instance, will use it to synchronize audio and data. An audio application will use the sequence number and timestamp to manage a reception buffer. For instance, an application can decide to buffer 20 10-ms G.729 audio frames before commencing playback. Each time a new RTP packet arrives, it is placed in the buffer in the appropriate position depending on its sequence number. It is important to note that the protection against jitter allowed by RTP comes with a price: a greater end-to-end delay in the transmission path. If a packet doesn't arrive on time and is still missing at playback time, the application can decide to copy the last sample of the packet that has just been played and repeat it long enough to catch up with the timestamp of the next received packet, or use some interpolation scheme as defined by the particular audio codec in use. The sequence number is used to detect packet loss.

#### 1.1.2.3 RTCP

RTCP is used to transmit control packets to participants regarding a particular RTP session. These control packets include various statistics, information about the participants (their names, email addresses, etc.), and information on the mapping of participants to individual stream sources. The most useful information found in RTCP packets concerns the quality of transmission in the network. All participants in the sessions send RTCP packets: senders send '*sender* reports' and receivers send '*receiver* reports'.

##### 1.1.2.3.1 Bandwidth limitation

All participants must send RTCP packets. This causes a potential dimensioning problem for large multicast conferences: RTCP traffic should grow linearly with the number of participants. This problem does not exist with RTP streams in audio-only conferences using silence suppression, for instance, since people generally don't speak at the same time (Figure 1.8).

Since the number of participants is known to all participants who listen to RTCP reports, each of them can control the rate at which RTCP reports are sent. This is used to limit the bandwidth used by RTCP to a reasonable amount, usually not more than 5% of the overall session bandwidth (which is defined as the sum of all transmissions from all participants, including the IP/UDP overhead).

This budget has to be shared by all participants. Active senders get one-quarter of it because some of the information they send (e.g., CNAME information used for synchronization) is very important to all receivers and RTCP sender reports need to be very responsive. The remaining part is split between the receivers. The average sending rate is derived by the participant from the size of the RTCP packets that he wants to send and from the number of senders and receivers that appear in the RTCP packets it receives. This is clearly relevant for multicast sessions; in fact, many of the recommendations and features present in the RTP RFC are useless for most VoIP applications, which have a maximum of three participants in most cases. Even for small sessions, the fastest rate at

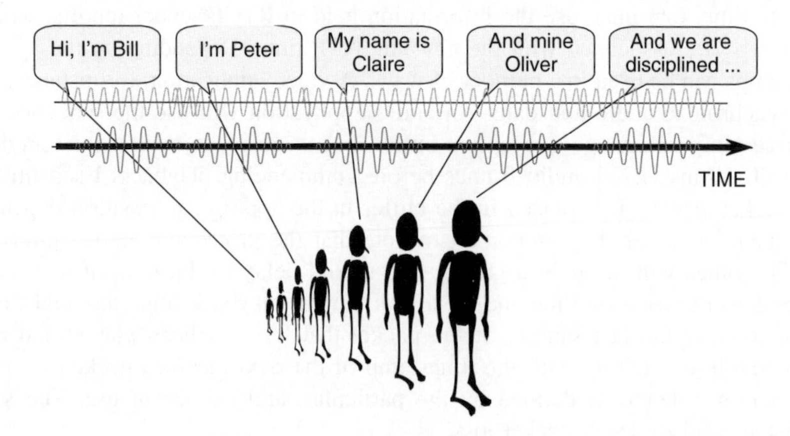


**Figure 1.8** Bitrate is self-limiting in audio conferences (at least among polite participants).

which a participant is allowed to send RTCP reports is one every 5 s. The sending rate is randomized by a factor of 0.5 to 1.5 to avoid unwanted synchronization between reports.

Most H.323 and SIP implementations actually use a simplified version of these guidelines, which is not a problem because there is no scaling issue. The RFC recommendations remain applicable for larger conferences, however, such as the conferences using the **H.332** protocol to broadcast information to multiple receivers.

#### *1.1.2.3.1.1 RTCP packet types*

There are various types of RTCP messages defined for each type of information:

- **SR**: sender reports contain transmission and reception information for active senders.
- **RR**: receiver reports contain reception information for listeners who are not also active senders.
- **SDES**: source description describes various parameters relating to the source, including the name of the sender (CNAME).
- **BYE**: sent by a participant when he leaves the conference.
- **APP**: functions specific to an application.

Several RTCP messages can be packed in a single transport protocol packet. Each RTCP message contains enough length information to be properly decoded if several of those RTCP messages are packed in a single UDP packet. This packing can be useful to save overhead bandwidth used by the transport protocol header.

#### *1.1.2.3.1.2 Sender reports*

Each SR contains three mandatory sections, as shown in Figure 1.9.

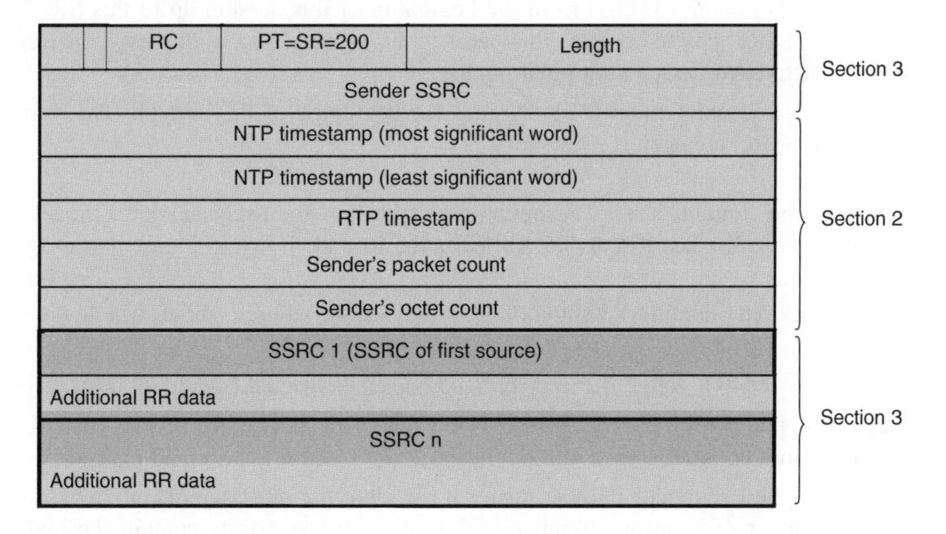


**Figure 1.9** RTCP packet format.

The first section contains:

- The 5-bit reception report count (RC), which is the number of report blocks included in this SR.
- The packet type (PT) is 200 for an SR. In order to avoid mixing a regular RTP packet with an SR, RTP packets should avoid payload types 72 and 73 which can be mistaken for SRs and RRs when the marker bit is set. However, normally a UDP port is dedicated to RTCP to eliminate this potential confusion.
- The 16 bit length of this SR including header and padding (the number of 32-bit words minus 1).
- The SSRC of the originator of this SR. This SSRC can also be found in the RTP packets that originate from this host.

The second section contains information on the RTP stream originated by this sender (this SSRC):

- The NTP timestamp of the sending time of this report. A sender can set the high-order bit to 0 if it can't track the absolute NTP time; this NTP measurement only relates to the beginning of this session (which is assumed to last less than 68 years!). If a sender can't track elapsed time at all it may set the timestamp to 0.
- The RTP timestamp, which represents the same time as above, but with the same units and random offset as in the timestamps of RTP packets. Note that this association of an absolute NTP timestamp and the RTP timestamps enables the receiver to compute the absolute timestamp of each received RTP packet and, therefore, to synchronize related media streams (e.g., audio and video) for playback.
- Sender's packet count (32 bits) from the beginning of this session up to this SR. It is reset if the SSRC has to change (this can happen in an H.323 multiparty conference when the active MC assigns terminal numbers).
- Sender's payload octet count (32 bits) since the beginning of this session. This is also reset if the SSRC changes.

The third section contains a set of reception report blocks, one for each source the sender knows about since the last RR or SR. Each has the format shown in Figure 1.10:

- SSRC_n (source identifier)(32 bits): the SSRC of the source about which we are reporting.
- Fraction lost (8 bits): equal to Floor(received packets/expected packets $*$ 256).
- Cumulative number of packets lost (24 bits) since the beginning of reception. Late packets are not counted as lost and duplicate packets count as received packets.
- Extended highest sequence number received (32 bits): the most significant 16 bits contain the number of sequence number cycles, and the last 16 bits contain the highest sequence number received in an RTP data packet from this source (same SSRC).

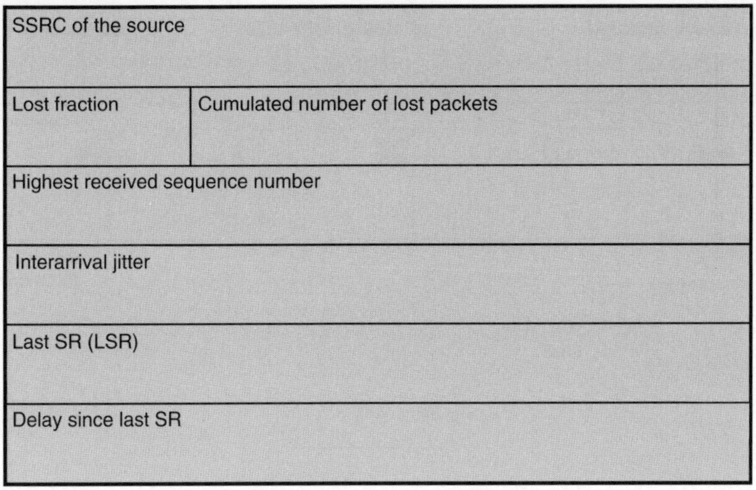

| SSRC of the source | |
|---|---|
| Lost fraction | Cumulated number of lost packets |
| Highest received sequence number | |
| Interarrival jitter | |
| Last SR (LSR) | |
| Delay since last SR | |

**Figure 1.10** Format of a reception report block.

- Interarrival jitter (32 bits): an estimation of the variance in interarrival time between RTP packets, measured in the same units as the RTP timestamp. The calculation is made by comparing the RTP timestamp of arriving packets with the local clock, and averaging the results (as shown in Figure 1.11).
- The last SR timestamp (LSR) (32 bits): the middle 32 bits of the NTP timestamp of the last SR received (this is the compact NTP form).
- The delay since the last SR arrived (DLSR) (32 bits): expressed in compact NTP form (or, more simply, in multiples of 1/65536 s). Together with the last SR timestamp, the sender of this last SR can use it to compute the round trip time.

#### *1.1.2.3.1.3 Receiver reports*

A receiver report looks like an SR, except that the PT field is now 201, and the second section (concerning the sender) is absent.

#### *1.1.2.3.1.4 SDES: source description RTCP packet*

An SDES packet (Figure 1.12) has a PT of 202 and contains SC (source count) chunks. Each chunk contains an SSRC or a CSRC and a list of information. Each element of this list is coded using the type/length/value format. The following types exist but only CNAME has to be present:

- CNAME (type 1): unique among all participants of the session; is of the form user@host, where host is the IP address or domain name of the host.
- NAME (type 2): common name of the source.
- EMAIL (type 3).

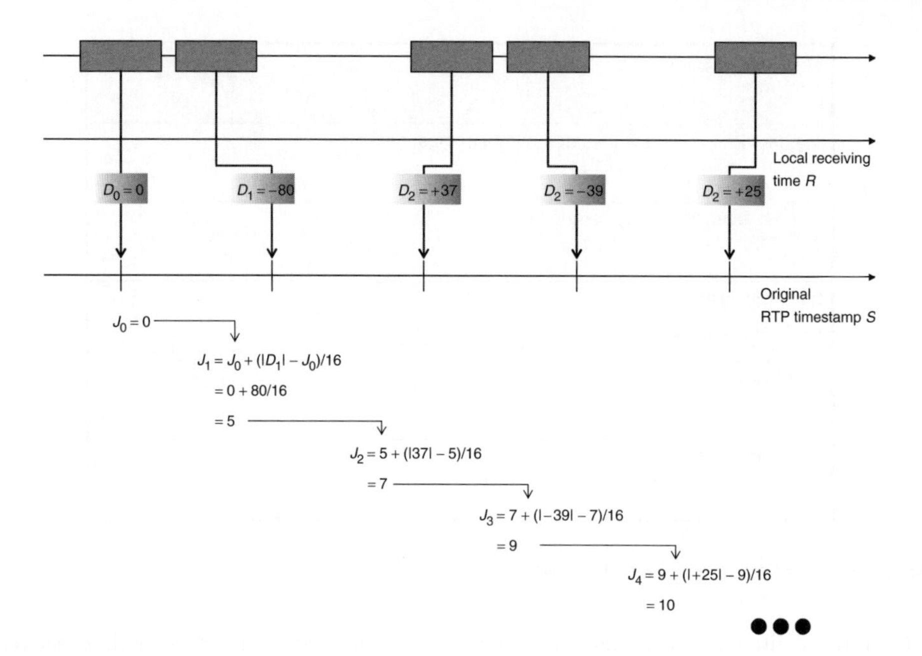


**Figure 1.11** Jitter evaluation.

```
 0                   1                   2                   3
 0 1 2 3 4 5 6 7 8 9 0 1 2 3 4 5 6 7 8 9 0 1 2 3 4 5 6 7 8 9 0 1
|V=2|P|    SC   |  PT=SDES=202  |             length            | header
|                          SSRC/CSRC_1                          | chunk
|                           SDES items                          |   1
|                              ...                              |
|                          SSRC/CSRC_2                          | chunk
|                           SDES items                          |   2
|                              ...                              |
```

**Figure 1.12** SDES message format.

- PHONE (type 4).
- LOC (type 5): location.

#### *1.1.2.3.1.5 BYE RTCP packet*

The BYE RTCP packet (Figure 1.13) indicates that one or more sources (as indicated by source count SC) are no longer active.

#### *1.1.2.3.1.6 APP: application-defined RTCP packet*

This can be used to convey additional proprietary information. The format is shown in Figure 1.14. The PT field is set to 204.

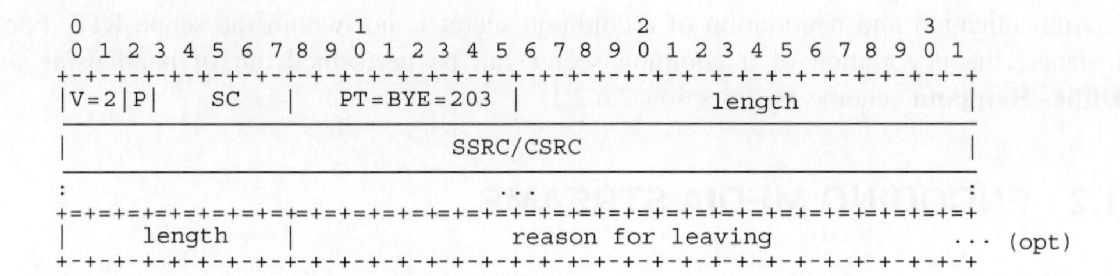


**Figure 1.13** BYE message format.

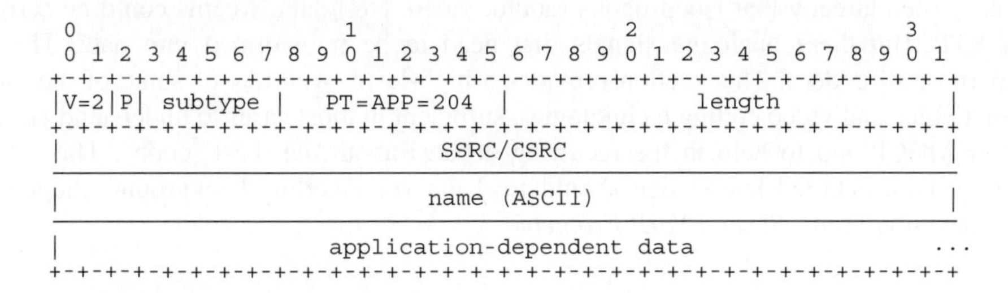


**Figure 1.14** APP message format.

#### *1.1.2.4 Security*

Security can be achieved at the transport level (e.g., using IPSec) or at the RTP level. The RTP RFC presents a way to ensure RTP-level privacy using DES/CBC (data encryption standard, cipher block chaining) encryption. Since **DES**, like many other encryption algorithms, is a block algorithm (for a more detailed description see Section 2.6.2 about H.235), there needs to be some adaptation when the unencrypted payload is not a multiple of 64 bits.

The most straightforward method, padding, is described in RTP (RFC 1889, Section 6.1). When this method is used the padding bit of the RTP header is set, and the last octet of the RTP payload contains the number of padding bits to remove (Figure 1.15). The last octet can be located because the underlying transport protocol must support framing. There are other encryption methods that do not require padding (e.g., ciphertext stealing); some of these alternative methods are described in Chapter 2 (on H.235).

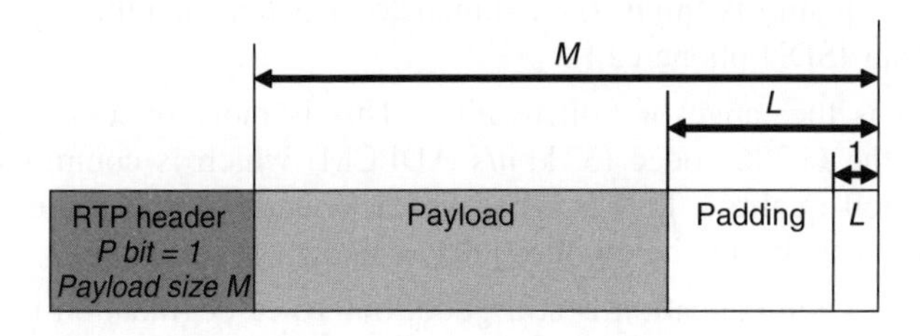


**Figure 1.15** RTP payload padding for encryption using block algorithms.

Authentication and negotiation of a common secret is not within the scope RTP. For instance, the negotiation of a common secret can be performed out of band using a **Diffie–Helmann** scheme (see Section 2.6.2.1).

## 1.2 ENCODING MEDIA STREAMS

### 1.2.1 Codecs

We have seen already that isochronous (audio, video, etc.) data streams could be carried over RTP. But these analogue signals first need to be transformed into data. This is the purpose of codecs. This section provides a high-level overview of some of the most popular voice and video coding technologies, sufficient in most cases to understand H.323, SIP, or MGCP and to help in the recurring debates about the 'best' codec. The reader wanting more detailed knowledge should read the voice-coding background chapter in the companion book, *Beyond VoIP Protocols*.

#### *1.2.1.1 What is a good codec?*

When the International Multimedia Telecommunications Consortium (www.IMTC.org) tried to choose a default low-bitrate codec, sufficient to promote interoperability, they faced a difficult issue because there was not common agreement about what constituted a good codec. The difficulty was so great that other bodies who are also trying to profile VoIP applications are reticent to enter into the debate at all.

Let's look at the criteria that must now be considered when evaluating a voice codec.

##### *1.2.1.1.1 Bandwidth usage*

The bitrate of available narrow-band codecs (approximately 300–3400 Hz) today ranges from 1.2 kbit/s to 64 kbit/s. Of course there is a consequence on the quality of restituted voice. This is usually measured by MOS (mean opinion score) marks. MOSs for a particular codec are the average mark given by a panel of auditors listening to several recorded samples (voice samples, music samples, voice with background noise, etc.). These scores range from 1 to 5:

- From 4 to 5 the quality is 'high' (i.e., similar to or better than the experience we have when making an ISDN phone call).
- From 3.5 to 4 is the range of 'toll quality'. This is more or less similar to what is obtained with the G.726 codec (32 kbit/s ADPCM) which is commonly taken as the reference for 'toll quality'. This is what we experience on most phone calls. Mobile phone calls are usually just below the 'toll' quality.
- From 3.0 to 3.5, communication is still good, but voice degradation is easily audible.

- From 2.5 to 3, communication is still possible, but requires much more attention. This is the range of 'military quality' voice. In extreme cases the expression 'synthetic', or 'robotic', voice is used (i.e., when it becomes impossible to recognize the speaker).

There is a trade-off between voice quality and bandwidth used. With current technology toll quality cannot be obtained below 5 kbit/s.

##### *1.2.1.1.2 Silence compression (VAD, DTX, CNG)*

During a conversation, we only talk on average 35% of the time. Therefore, silence compression or suppression is an important feature. In a point-to-point call it saves about 50% of the bandwidth, but in decentralized multicast conferences the activity rate of each speaker drops and the savings are even greater. It wouldn't make sense to undertake a multicast conference where there are more than half a dozen participants without silence suppression.

Silence compression includes three major components:

- **VAD (voice activity detector)**: this is responsible for determining when the user is talking and when he is silent. It should be very responsive (otherwise the first word may get lost and unwanted silence might occur at the end of sentences), without getting triggered by background noise. VAD evaluates the energy and spectrum of incoming samples and activates the media channel if this energy is above a minimum and the spectrum corresponds to voice. Similarly, when the energy falls below a threshold for some time, the media channel is muted. If the VAD module dropped all samples until the mean energy of the incoming samples reaches the threshold, the beginning of the active speech period would be clipped. Therefore, VAD implementations require some lookahead (i.e., they retain in memory a few milliseconds worth of samples to start media channel activation before the active speech period). This usually adds some delay to the overall coding latency, except on some coders where this evaluation is coupled with the coding algorithm itself and does not add to the algorithmic delay. The quality of the implementation is important: good VAD should require minimal lookahead, avoid voice clipping, and have a configurable hangover period (150 ms is usually fine, but some languages, such as Chinese, require different settings).
- **DTX (discontinuous transmission)**: this is the ability of a codec to stop transmitting frames when the VAD has detected a silence period. If the transmission is stopped completely, then it should set the marker bit of the first RTP packet after the silence period. Some advanced codecs will not stop transmission completely, but instead switch to a silence mode in which they use much less bandwidth and send just the bare minimum parameters (intensity, etc.) in order to allow the receiver to regenerate the background noise.
- **CNG (comfort noise generator)**: it seems logical to believe that when the caller isn't talking, there is just silence on the line and, when the VAD detects a silence period, it should be enough to switch off the loudspeaker completely. In fact, this

approach is completely wrong. Movie producers go to great lengths to recreate the proper background noise for 'silent' sequences. The same applies to phone calls. If the loudspeaker is turned off completely, street traffic and other background noise that could be overheard while the caller was talking would stop abruptly. The called party would get the impression that the line had been dropped and would ask the caller whether he is still on the line. The CNG is here to avoid this and recreate some sort of background noise. With the most primitive codecs that simply stop transmission it will use some random noise with a level deduced from the minimal levels recorded during active speech periods. More advanced codecs such as G.723.1 (annex A) or G.729 (annex B) have options to send enough information to allow the remote decoder to regenerate ambient noise close to the original background noise.

#### *1.2.1.1.3 Intellectual property*

End-users don't care about this, but manufacturers have to pay royalties to be allowed to use some codecs in their products. For some hardware products where margins are very low, this can be a major issue. Another common situation is that some manufacturers want to sell some back-end server, while distributing software to clients for free. If the client includes a codec, then again intellectual property becomes a major choice factor.

#### *1.2.1.1.4 Lookahead and frame size*

Most low-bitrate codecs compress voice in chunks called frames and need to know a little about the samples immediately following the samples they are currently encoding (this is called lookahead).

There has been a lot of discussion (especially at the IMTC when they tried to choose a low-bitrate codec) over the influence of frame size on the quality of the codec. This is because the minimal delay introduced by a coding/decoding sequence is the frame length plus the lookahead size. This is also called the algorithmic delay. Of course, in reality DSPs do not have infinite power and most of the time a fair estimate is to consider the real delay introduced as twice or three times the frame length plus the overhead (some authors improperly call this the algorithmic delay, although this is just an estimate of DSP power).

So, codecs with a small frame length are indeed better than codecs with a longer frame length regarding delay, when each frame is sent immediately on the network. This is where it becomes tricky, because each RTP packet has an IP header of 20 octets, a UDP header of 8 octets, and an RTP header of 12 octets! For instance, for a codec with a frame length of 30 ms, sending each frame separately on the network would introduce a 10.6-kbit/s overhead—much more than the actual bitrate of most narrow-band codecs!

Therefore, most implementations choose to send multiple frames per packet, and the real frame length is in fact the sum of all frames stacked in a single IP packet. This is limited by echo and interactivity issues (see companion book, Chapter 3 *Beyond VoIP Protocols*). A maximum of 120 ms of encoded voice should be sent in each IP packet.

So, for most implementations, the smaller the frame size, the more frames in an IP packet. This is all there is to it and there is no influence on delay. Overall, it is better to use codecs that have been designed for the longest frame length (limited by the acceptable delay), since this allows even more efficient coding techniques: the longer you observe a phenomenon, the better you can model it!

We can conclude that in most cases frame size is not so important for IP videoconferences when bandwidth is a concern. The exception is high-quality conferences where interactivity is maximized at the expense of the required bitrate.

##### *1.2.1.1.5 Resilience to loss*

Packet loss is a fact of life in IP networks and the short latency required by interactive voice and video applications does not allow us to request retransmissions. Since packets carry codec frames, this in turn causes codec frame loss. However, packet loss and frame loss are not directly correlated; many techniques such as FEC (forward error correction) can be used to lower the frame loss rate associated with a given packet loss rate. These techniques spread redundant information over several packets so that frame information can be recovered even if some packets are lost.

However, the use of redundancy to recover packet loss is a very tricky thing. It can lead to unexpected issues and, can even make the problem worse. To understand this, let's look at what some manufacturers could do (and have done!):

- You prepare a demonstration to compare your product and that of a competitor. You let it be known that you can resist a 50% packet loss without any consequence on voice quality.
- You simulate packet loss by losing one packet out of two.
- You put frames $N$ and $N - 1$ into your RTP packet.

Your product can recover all the frames despite one packet out of two being lost. Your competitor is restricted to emitting a few cracks. Bingo! The customer is convinced.

Well, the only problem is that packet loss on the Internet in not so neat. Packet loss occurs in a correlated way and you are much more likely to lose several packets in a row, than exactly one packet out of two. So, this simple RTP redundancy scheme will be close to useless under real conditions and still add a 50% overhead!

The effect of frame erasure on codecs should be considered on a case-by-case basis. If you lose $N$ samples from a G.711 codec (stateless coder) this will just result in a gap of $N * 125\,\mu s$ at the receiving end. If you lose just one frame from a very advanced codec it may be audible for much more than the duration of this frame, because the decoder will need some time to resynchronize with the coder. For a frame of 20 ms or so, this may result in a very audible crack of 150 ms. Codecs such as G.723.1 are designed to cope relatively well with an uncorrelated frame erasure of up to 3%, but beyond this quality drops off very rapidly. The effect of correlated loss is not yet fully evaluated. It is possible to reduce the occurrence of consecutive frame loss by interleaving codec frames across multiple RTP packets: unfortunately, this adds a lot of delay to the transmission and therefore can only be used in streaming media transmissions, not in the context of real-time communications.

Apart from the built-in features of the codec itself, it is possible to reduce the frame loss associated with packet loss by using a number of techniques.

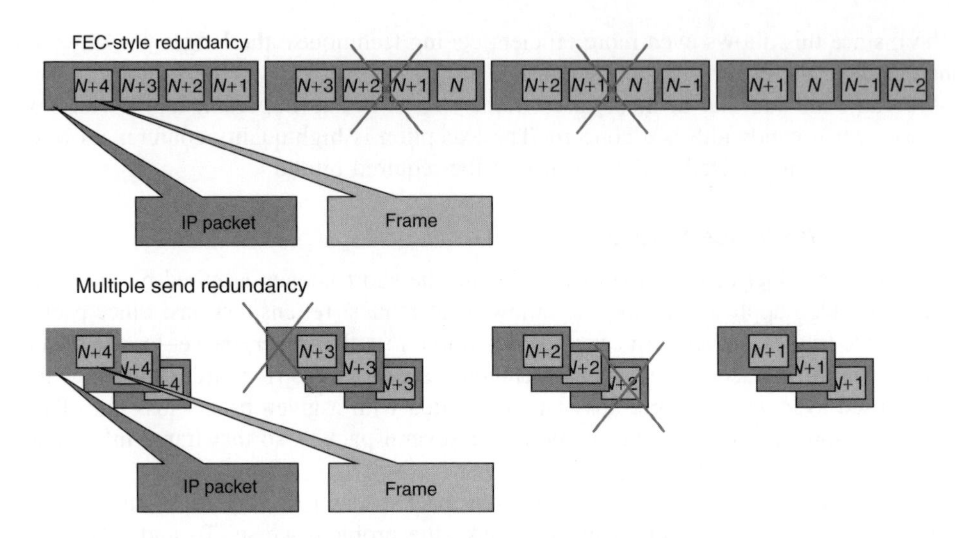


**Figure 1.16** Using redundancy to reduce codec frame loss.

FEC-style redundancy (Figure 1.16) can be used to recover from serious packet loss conditions, but it has a significant impact on delay. For instance, if you choose to repeat the same G.723.1 frame in four consecutive IP packets in order to recover from the loss of three consecutive packets, then the decoder needs to maintain a buffer of four IP packets, but this ruins the delay factor. More sophisticated FEC methods use XOR sums instead of simple repetitions, but have the same impact on delay.

It is also possible to send several copies of each frame immediately. But if one packet gets lost, probably all the copies will reach the same congested router at nearly the same time and might get lost as well.

An understanding of the different types of congestion is also important in deciding whether redundancy is useful and which type to use. The network can lose packets because a link is congested or because a router has to route too many small packets per second.

If a link is congested, then any type of redundancy will add to the congestion and increase the overall loss percentage of IP packets. But the frame loss rate of communicating devices that use FEC redundancy will still be reduced.

Some arithmetic proves this. Say we have congestion on a 2-Mbit/s line. It receives 2.2 Mbit/s and the average loss rate is $0.2/2.2 = 9\%$. Part of this is caused by someone using a codec producing a 100 kbit/s stream. The software detects a high loss and decides to use the FEC scheme described above. Now that same application produces a 400 kbit/s stream (the influence of headers is not taken into account for simplicity). The 2-Mbit/s line receives 2.5 Mbit/s and the packet loss rate is increased to 20% for all the users of the link. However, if we assume the congested link never causes the loss of four packets in a row (on average one packet in five is dropped), then the software will recover from *all* loss. However, this would be unacceptable behaviour, because it would be unfair to other users and could destabilize the network. Next-generation IP networks will probably include advanced techniques, such as RED, that will detect the greedy user and drop most of his packets.

If congestion is due to an overrun router (exceeding its packet/s limit), then FEC-style redundancy is not such a bad thing. It increases the size of packets but does not increase the average number of packets that the router has to forward per second. In this case increasing the size of the packets will not add to the congestion. The other type of redundancy (multiple simultaneous sending) will increase the number of packets through the router and would not work.

#### 1.2.1.1.6 Layered coding

There are several situations in which current codecs are not well suited. For instance, if you want to broadcast the same event to several listeners (H.332 type of conference), some will want high-quality reception (either because they have paid for it or because they have large IP pipes) and others will only be able to receive lower quality. You could send a customized data stream to each listener, but this is not practical for a large audience. The answer is to multicast the data stream to all listeners (for more information on multicast, refer to the multicast chapter of the companion book, *Beyond VoIP Protocols*). Current codecs include complete information in one data stream. If it is multicast, all participants will receive the same amount of data, so you usually have to limit the data rate to the reception capability of the least capable receiver.

Some codecs (most are still at experimental stage) can produce several data streams simultaneously, one with the core information needed for 'military quality' reception and the other data streams with more information as needed to rebuild higher fidelity sound or an image. A crude example for video would be to send black and white information on one channel and colour (chrominance) information on another.

Each part of the data stream can be multicast using different group addresses, so that listeners can choose to receive just the core level or the other layers as well. In a pay-for-quality scheme, you would encrypt the higher layers (this way you have the option of receiving a free low-quality preview and later of paying for the broadcast quality image).

Layered codecs are also very useful when it comes to redundancy: the sender can choose to use a redundancy scheme or a quality of service level for the core layer, so that the transmission remains understandable at all times for everyone, but leaves other layers without protection.

H.323v2 was approved with a specific annex on layered video coding (annex B: procedures for layered video codecs).

#### 1.2.1.1.7 Fixed point or floating point

We first need to say a few words about digital signal processors (DSPs). These are processors that have been optimized for operations frequently encountered in signal-processing algorithms. One such operation is $(a * b) + previous\ result$: one multiplication and addition. In a conventional processor, this operation would require multiple processor instructions and would be executed in several clock cycles. A DSP will do it in one instruction and a single clock cycle. Another example is the code book searches frequently used by vocoders. Some conventional processors also have extensions to accelerate signal-processing algorithms (e.g., MMX processors can execute a single instruction simultaneously on several operands as long as they can be contained in a 32-bit register and video algorithms can be accelerated by processing 4 pixels (8 bits each) simultaneously).

There are two types of DSPs: floating point DSPs, which are capable of operating on floating point numbers, and fixed point DSPs. Fixed point DSP operands are represented as a mantissa $n$ and a power $p$ of 2 (e.g., $12345678 * 2^5$), but the DSP can operate on two operands only if the power of 2 is the same for both operands. They are less powerful, but also less expensive, and chosen by many designers for products sold in large quantities. Some codecs have only been specified with fixed point C code. However, many implementations will run on processors or DSPs that are capable of floating point operation, and developers must develop their own version of floating point C code for the algorithm. This often results in interoperability problems between floating point versions.

Therefore, it makes sense for the codec to be specified in floating point C code as well, especially if the code has to run on PCs.

#### *1.2.1.2 Audio codecs*

##### *1.2.1.2.1 ITU audio codecs*

###### *1.2.1.2.1.1 Choosing a codec at ITU*

The choice of a codec at ITU WP3 is typically a very long process. This is not a bureaucracy problem, but rather a problem due to the stringent requirements of ITU experts.

Before a codec is chosen, the ITU evaluates MOS scores and usually requires quality that is equivalent to or better than G.726 ('toll quality'). The ITU also checks that this quality is constant for men and women, and in several languages. The ability to take into account background noise and recreate it correctly is also evaluated. The ITU pays special attention to the degradation of voice quality in tandem operation (several successive coding/decoding processes), since this a situation that is very likely to happen in international phone calls. Last but not least, if the codec has to be used over a non-reliable medium (a radio link, a frame relay virtual circuit, etc.), the ITU checks that the quality remains acceptable if there is some frame loss.

After checking all these parameters, it frequently occurs that no single proposal passes the test! Therefore, many ITU codecs are combinations of the most advanced technologies found in several different proposals. This leads to state-of-the-art choices, but, as we will see, this is a nightmare for anyone who needs to keep track of intellectual property.

###### *1.2.1.2.1.2 Audio codecs commonly used in VoIP*

The companion book, *Beyond VoIP Protocols* provides a detailed view on voice coder technology and discrete time signal processing in general. This section's purpose is to provide a quick reference to common VoIP coders found in VoIP systems for engineers uninterested in the details and theory of each coder.

(a) G.711 (approved in 1965)

G.711 is the grandfather of digital audio codecs. It is a very simple way of digitizing analogue data by using a semi-logarithmic scale (this is called 'companded PCM', and serves to increase the resolution of small signals, while treating large signals in the same way as the human ear does). Two different types of scales are in use, the A-law scale (Europe, international links) and the $\mu$-law scale (USA, Japan). They differ only in the choice of some constants of the logarithmic curve. G.711 is used in ISDN and on most digital telephone backbones in operation today.

A G.711-encoded audio stream is a 64-kbit/s bitstream in which each sample is encoded as an octet; therefore, the frame length is only 125 μs. Of course, all VoIP applications will put more than one sample in every IP packet (about 10 ms typically, or 80 samples).

Most sound cards are able to record directly in G.711 format. However, in some cases it is better to record using CD quality, which samples at 44.1 kHz (one 16-bit sample every 23 μs), especially if echo cancelation algorithms are used, since the full performance of some echo cancelation algorithms cannot be achieved with the quantification noise introduced by G.711.

The typical MOS score of G.711 is usually taken as 4.2; it is used as an anchor for other coder tests.

(b) G.722 (approved in 1988)

Although G.711 achieves very good quality, some of the voice spectrum (above 4 kHz) is still cut. G.722 provides a higher quality digital coding of 7 kHz of audio spectrum at only 48, 56, or 64 kbit/s, using about 10 DSP MIPS. This is an 'embedded' coder, which means that the rate can freely switch between 48, 56, or 64 kbit/s without notifying the decoder.

This coder is very good for all professional conversational voice applications (the algorithmic delay is only 1.5 ms). G.722 is supported by some videoconferencing equipment and some IP phones.

(c) G.722.1

This more recent wideband coder operates at 24 kbit/s or 32 kbit/s. It has been designed by Picturetel, which also sells a 16-kbit/s version (Siren™). The coder encodes frames of 20 ms, with a lookahead of 20 ms. The 16-kbit/s version is supported by Windows® Messenger.

(d) G.723.1 (approved in 1995)

In the early days of VoIP, the VoIP Forum chose the G.723.1 codec as the baseline codec for narrow-band H.323 communications. It is also used by the video cellphones of UMTS 99 (H.324M standard).

*(i) Technology*

G.723.1 uses a frame length of 30 ms and needs a lookahead of 7.5 ms. It has two modes of operation, one at 6.4 kbit/s and the other at 5.3 kbit/s. The mode of operation can change dynamically at each frame. Both modes of operation are mandatory in any implementation, although many VoIP systems have an incorrect implementation that works on only one of the two modes.

G.723 is not designed for music and does not transmit DTMF tones reliably (they must be transmitted out-of-band). Modem and fax signals cannot be carried by G.723.1.

**Table 1.2** Impact of frame-erasure and tandeming quality

| | G.723.1, 6.4 kbit/s | 32 kbit/s ADPCM |
|---|---|---|
| Clear channel, no errors or frame erasure | 3.901 | 3.781 |
| 3% frame erasure | 3.432 | — |
| Tandeming of two codecs | 3.409 | 3.491 |

G.723.1 achieves an MOS score of 3.7 in 5.3-kbit/s mode and 3.9 in 6.4-kbit/s mode. Table 1.2 compares the performance of G.723.1 (6.4 kbps) and the ADPCM released by Bell Labs in March 94.

The main effect of frame erasures is to desynchronize the coder and the decoder (they may need many more frames to resynchronize). In practice, networks should always have a frame error rate below 3% (and below 1% ideally).

G.723.1 is specified in both fixed point (where it runs at 16 MIPS on a fixed point DSP) and floating point C code (running on a Pentium 100, it takes 35% to 40% of the power of the processor). The fixed point implementation runs on VoIP gateways, while the floating point version runs on all Windows® PCs.

Beyond VoIP systems, G.723.1 is used in the H.324 recommendation (ITU recommendation for narrow-band videoconferencing on PSTN lines) and will also be used in the new 3G-324M (3GPP, 3GPP2 organizations) standard for 3G wireless multimedia devices.

*(ii) Silence compression*

G.723.1 supports voice activity detection (VAD), discontinuous transmission (DTX), and comfort noise generation (CNG) (defined in annex A of the recommendation).

Silence is coded in very small, 4-octet frames at a rate of 1.1 kbit/s. If silence information doesn't need to be updated, transmission stops completely.

*(iii) Intellectual property*

G.723.1 is one of the codecs that resulted from many contributions and, therefore, uses technology patented from several sources. About 18 patents currently apply to G.732.1 (the precise number is hard to keep track of), from eight different companies.

The main licensing consortium, which is made up of AudioCodes, DSP Group, FT/CNET, Université de Sherbrooke, and NTT, oversees the patents. The rights are managed by the DSP Group and SiproLabs (www.sipro.com) for all members of this consortium. Other patents are held by AT&T (1), Lucent (3), British Technology Group (1, formerly held by VoiceCraft), Nokia Mobile Phone (1, formerly held by VoiceCraft). Patent applications have also been made by Siemens, Robert Bosch, and CSELT. The source code is copyrighted by four companies.

There are typically several licensing agreements for this codec (the details hinge, of course, on the company involved), depending on whether the application is for a single user or multiple users, whether it is going to be a paying or free application, and on the volume licensed.

Of course, exact prices have to be negotiated with both patent owners and implementers, but some data can be gathered from conferences and newsgroups, *although they must be taken cautiously*. For instance, here are some price indications for acquisition of the intellectual rights of G.723.1:

- A license for a single-user client is said to be worth an initial payment of around $50,000 plus $0.8 per unit.
- A license for a server is said to be worth an initial payment of about $20,000 plus $5 per port.
- A license for unlimited distribution of a single-user application is said to be worth about $120,000.

Then, unless you do your own implementation (which is not recommended if you are not an expert!), be prepared to approximately double the previous fees to license a well-optimized implementation.

Here is a quote from a company trying to license these codecs, picked from a mailing list:

> *We have been trying to negotiate licensing arrangements with the patent holders for more than six months. As of today, we have received terms and conditions from six of the holders, and little to no response from the rest. The costs proposed by the first six strongly imply a substantial initial investment, and a per port cost in excess of $20.00.*
>
> *Our concern, however, extends far beyond the cost. The Internet's success is due to its readily available standards and lack of non-essential rules and constraints. The time requirements and logistics of establishing contact with 12 parties and negotiating licensing are significant barriers to growth in the industry. The legal risks associated with not doing so are an impediment to the rapid evolution of the industry.*

The reality is not quite so bad, as many IPR rights are now managed by Sipro Labs (www.sipro.com). The investments needed to produce the technology of standardized coders such as G.723.1 indeed justify a fee. But the question is how much is reasonable? When patented technology becomes a standard, the temptation is high to use this monopoly situation to maintain high licence prices. This underlies the so-called 'codec wars' that periodically break out in VoIP standard bodies.

(e) G.726 (approved in 1990)

G.726 uses an ADPCM technique to encode a G.711 bitstream in words of 2, 3, or 4 bits, resulting in available bitrates of 16, 24, 32, or 40 kbit/s.

G.726 at 32 kbit/s achieves a MOS score of 4.3 and is often taken as the benchmark for 'toll quality'. It requires about 10 DSP MIPs of processing power (full duplex) or 30% of the processing power of a Pentium 100. This is a low-delay coder: 'frames' are 125 μs long and there is no lookahead. There is also an embedded version known as G.727.

(f) G.728 (approved in 1992–1994)

G.728 uses an LD-CELP (low-delay, code-excited linear prediction) coding technique and achieves MOS scores similar to that obtained by G.726 at 32 kbit/s, but with a bitrate of only 16 kbit/s. Compared with PCM or ADPCM techniques, which are waveform coders (i.e., they ignore the nature of the signal), CELP is a coder optimized for voice (vocoder). These coders specifically model voice sounds and work by comparing the waveform to encode with a set of waveform models (linear predictive code book) and find the best match. Then, only the index of this best match and parameters like voice pitch are transmitted. As a result music does not transmit well on CELP coders, and it is only at 2.4 kbit/s that fax or modem transmission can succeed with G.728 compression. G.728 is used for H.320 videoconferencing and some H.323 videoconferencing systems.

G.728 needs almost all the power of a Pentium 100 and 2 Kb of RAM to implement. It is a low-delay coder (between 625 μs and 2.5 ms).

(g) G.729 (approved in 1995–1996)

*(i) Technology*

G.729 is very popular for voice over frame relay applications and V.70 voice and data modems. Together with G.723, it has become the most popular voice coder for VoIP, but is still not supported natively on the Windows® platform. It uses a CS-ACELP (conjugate structure, algebraic code-excited linear prediction) coding technique. G.729 is not designed for music and does not transmit DTMF tones reliably (they must be transmitted out-of-band). Modem and fax signals cannot be carried by G.729.

G.729 produces 80-bit frames encoding 10 ms of speech at a rate of 8 kbit/s. It needs a lookahead of 5 ms. It achieves MOS scores around 4.0. There are two versions:

- G.729 (approved in December 1996) requires about 20 MIPS for coding and 3 MIPS for decoding.
- G.729A (approved in November 1995): annex A is a reduced complexity version of the original G.729. It requires about 10.5 MIPS for coding and 2 MIPS for decoding (about 30% less than G.723.1).

*(ii) Silence compression*

Annexes A and B of G.729 define VAD, CNG, and DTX schemes for G.729. The frames sent to update background noise description are 15 bits long and are only sent if the description of the background noise changes.

*(iii) Licences*

Both G.729 and G.729A are the result of about 20 patents belonging to six companies: AT&T, France Telecom, Lucent, Université de Sherbrooke (USH, Canada), NTT, and VoiceCraft. NTT, France Telecom, and USH have formed a licensing consortium managed by SiproLabs, but not all patents (notably AT&T) are covered by this consortium. The source code is copyrighted by five companies.

As with G.723, there are several ways of getting a several licence for this codec, but the prices of the G.729 IPR pool managed by SiproLabs (www.sipro.com) are in the public domain.

#### *1.2.1.2.2 ETSI SMG audio codecs*

The ETSI SMG11 (European Telecommunications Standardization Institute Special Mobile Group) standardized the speech codecs given in Table 2.3. In addition the new AMR coder has been standardized for use in UMTS, but is not used yet in VoIP systems.

**Table 1.3** Performance of GSM coders in various environmental conditions

| Codec | MOS in clean conditions | Vehicle noise | Street noise |
|---|---|---|---|
| GSM FR | 3.71 | 3.83 | 3.92 |
| GSM HR | 3.85 | 3.45 | 3.56 |
| GSM EFR | 4.43 | 4.25 | 4.18 |
| Reference with no coding | 4.61 | 4.42 | 4.35 |

*Source*: TR 06.85 v2.0.0 (1998). Reproduced by Permission of the European Telecommunications Standards Institute – ETSI.

##### *1.2.1.2.2.1 GSM full rate (1987)*

GSM full rate, also called GSM 06.10, is perhaps the most famous codec in use today and runs daily in millions of GSM cellular phones. It provides good quality and operates well in the presence of background noise (see Table 1.3). It uses an RPE-LTP technique to encode voice in frames of 20 ms at a rate of 13 kbit/s. It needs no lookahead. GSM-FR achieves MOS scores slightly below toll quality.

GSM-FR is not extremely complex and requires only about 4.5 MIPS and less than 1 Kb of RAM.

The GSM full-rate patent is held by Philips and the licence is free for mobile phone applications.

##### *1.2.1.2.2.2 GSM half-rate (1994)*

Also called GSM 06.20, this coder aims at using less bandwidth while preserving the same or slightly lower speech quality as GSM-FR. This codec uses VSELP and encodes speech at a rate of 5.6 kbit/s. The frames are 20 ms long and there is a lookahead of 4.4 ms. The GSM-HR algorithm requires approximately 30 MIPS and 4 Kb of RAM. This coder has not been very successful, due to its high sensitivity to background noise. The patent is also held by Philips; ATT patents on CELP and NTT patents on LSP may also apply.

##### *1.2.1.2.2.3 GSM enhanced full rate (1995)*

This high-quality coder exceeds the G.726 'wireline reference' in clear channel conditions and in background noise. It is also called GSM 06.60. It was selected as the base coder for the PCS 1900 cellular phone service in the US and was standardized by TIA in 1996. This codec uses a CD-ACELP technique and encodes 20-ms frames at a rate of 12.2 kbit/s. Optional VAD/DTX functions with comfort noise generation have been defined and there is also an example implementation for error concealment.

AT&T patents for CELP and NTT patents for LSP may apply.

#### *1.2.1.2.3 Other proprietary codecs*

##### *1.2.1.2.3.1 Lucent/Elemedia SX7003P*

The SX7003P is another popular codec. Although used in Lucent hardware it is licensed to other manufacturers as well. This codec has a frame size of 15 ms, which contains 4 control octets and 14 data octets. Silence frames have 2 octets of data.

In many VoIP implementations, two frames are packed in each IP packet (overhead of 40 bytes), leading to an IP bitrate of 20.3 kbit/s during voice activity periods and only 13.6 kbit/s during silence periods.

##### *1.2.1.2.3.2 RT24 (Voxware)*

The RT24 is one of the ultra-low-bitrate coders. Unfortunately, it is spoiled by the IP overhead. It has a bitrate of 2,400 bit/s and achieves an MOS of 3.2. It has a frame size of 22.5 ms (54 bits) which results in a measured IP-level bitrate of 16.6/9.5/7.1/6 kbit/s with 1/2/3/4 frames per IP packet.

#### *1.2.1.2.4 Future coders*

Both the AMR and AMR-WB (G.722.2) coders (described in detail in the companion book, *Beyond VoIP Protocols*) will probably be implemented in VoIP systems. Their ability to dynamically reduce the bitrate to adapt to the conditions of the transmission channel is not as useful in VoIP as it is over radio links. Over radio links, there are bit errors, but AMR makes it possible to add redundancy information dynamically to the media stream without requiring more bandwidth when network conditions degrade. Because of this, the AMR coder can offer better voice quality than any of the current narrow-band coders, over a much wider range of transmission network quality.

Over IP transmission links, there are only frame erasure errors, because each IP packet (containing one or more coder frame) is protected by a CRC code.[3] Redundancy can only be added by repeating each frame in multiple packets (forward error correction and interleaving). This has a significant, often unacceptable impact on end-to-end delay. Therefore, AMR will mainly be used to avoid any transcoding (Tandeming) when communicating with UMTS and CDMA2000 systems, thereby improving end-to-end voice quality.

Usage of the AMR and AMR-WB coders will be closely associated with the deployment of UMTS and CDMA2000 3G systems, and in the future LTE (Long Term Evolution) systems. Both coders will not only require more DSP processing power, they will require more powerful DSPs to achieve the densities of current G.723.1/G.729 systems.

### *1.2.1.3 ITU video codecs*

#### *1.2.1.3.1 Representation of colours*

The representation of colours is derived from the fact that any colour can be generated from three primaries. From an artist's point of view, the three primaries are red, yellow, and blue. These colours are called subtractive primaries, because any colour can be generated from a white beam passed through a sequence of red, yellow, and blue filters. When an artist puts a layer of yellow paint of a sheet of paper, this layer acts as a filter

[3] Some VoIP networks have disabled the UDP checksum mechanisms in order to be more tolerant to bit ends. This could open the way to a more efficient use of the capabilities of AMR.

that allows most of the yellow component of the white light to be reflected, but filters out most other colours.

Video monitors use instead additive primaries: red, green, and blue. By mixing three beams of red, blue, and green light with various intensities, it is possible to generate any colour. Therefore, any colour can be represented by its barycentric coordinates (representing the intensity of each primary colour, not necessarily positive as illustrated in Figure 1.17!) in a triangle with a primary colour at each edge: this is the RGB (red–green–blue) format. The weight of each colour usually ranges from 0 to 255 in the RGB format: each pixel is described using 8 bits for each colour weight, which leads to 24 bits per pixel.

Another common representation is to use luminance (brightness represented by $Y$) and chrominance (hue represented by $U$ and $V$, or $Cr$ and $Cb$). Several conventions exist for this conversion (JFIF for JPEG, CCIR 601 for H.261, and MPEG). Figure 1.18 shows how JFIF converts from an RGB format to a $YUV$ format.

$Y$, $U$, and $V$ cover the range from 0 to 255 ($U$ and $V$ are often shifted to take values between $-128$ and $+127$). For CCIR this range is from 16 to 235.

Experiments have shown that the human eye is more sensitive to the luminance information. Because of this $U$ and $V$ values can be sampled at reduced frequency without inducing significant loss in the quality of the image. Typically, $U$ and $V$ are only sampled for a group of 4 pixels. Coding an image in this way leads to a 2:1 compression (i.e., instead of 24 bits per pixel, we now have 8 bits for $Y$ and $(8 + 8)/4$ pixels for $U$ and $V$).

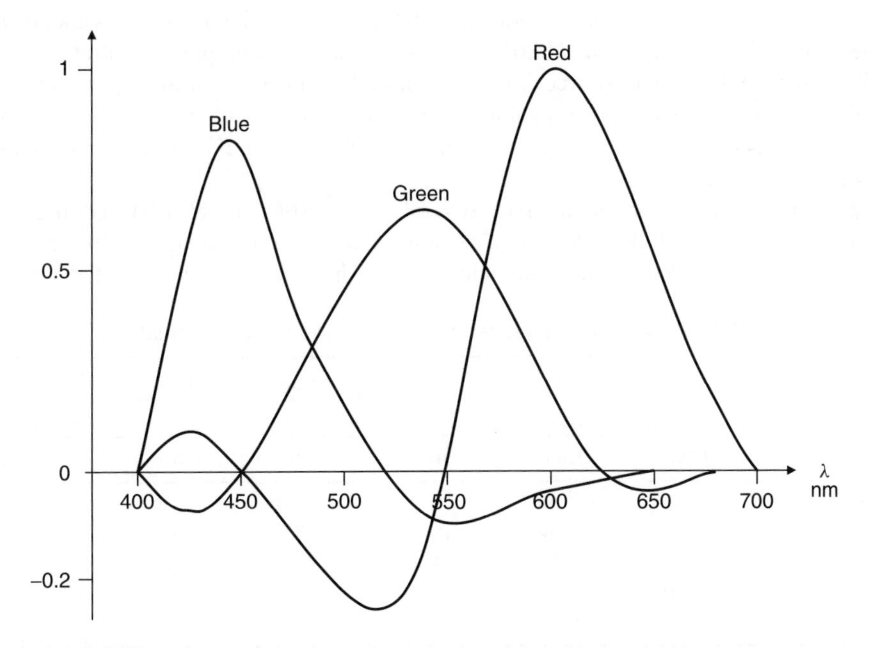


**Figure 1.17** Red–green–blue components of visible colours in the 400–700-nm wavelength range.

$$\begin{pmatrix} Y \\ U \\ V \end{pmatrix} = \begin{pmatrix} 0.299 & 0.587 & 0.114 \\ -0.1687 & -0.3313 & 0.5 \\ 0.5 & -0.4187 & -0.0813 \end{pmatrix} \begin{pmatrix} R \\ G \\ B \end{pmatrix}$$

**Figure 1.18** RGB to YUV conversion (JFIF).

#### *1.2.1.3.2 Image formats*

Several image formats are commonly used by video codecs. CIF (common intermediary format) defines a 352∗288 image. This size has been chosen because it can be sampled relatively easily from both the 525- and 625-line video formats and approaches the popular 4/3 length/width ratio.

In addition to the resolution of CIF being below that for TV quality, it is still relatively difficult to transmit over low-bandwidth lines, even with efficient coding schemes such as H.261 and H.263. For this reason two other formats with lower resolutions have been defined. At half the resolution in both dimensions, quarter CIF (QCIF) is for 176∗144 images, and SQCIF is only 128∗96.

For professional video application, CIF is clearly insufficient, and images need to be coded using 4CIF (704∗576) or 16CIF (1,408∗1,152) resolution (see Table 1.4).

#### *1.2.1.3.3 H.261*

H.261 is a video codec used in H.320 videoconferencing to encode the image over several 64-kbit/s ISDN connections, but in video over IP applications the bitstream is encoded in a single RTP logical channel. The H.261 codec is intended for compressed bitrates between 40 kbit/s and 2 Mbit/s. The source image is normally 30 (29.97) frames per second, but the bitrate can be reduced by transmitting only 1 frame out of 2, 3, or 4. The image formats shown in Table 1.5 can be encoded by H.261. The 4CIF and 16CIF formats are not supported by H.261.

The H.261 coding process involves several steps. After initial *YUV* coding of the original image using CCIR parameters, as described above, the image is divided in 8∗8 luminance pixels blocks for the luminance plane. The same surface is coded with only

**Table 1.4** Uncompressed bitrate for various video formats

| | Luminance | | Uncompressed bitrate (Mbit/s) | | | |
|---|---|---|---|---|---|---|
| | | | 10 frames/s | | 30 frames/s | |
| Picture format | Pixels | Lines | Grey | Colour | Grey | Colour |
| SQCIF | 128 | 96 | 1.0 | 1.5 | 3.0 | 4.4 |
| QCIF | 176 | 144 | 2.0 | 3.0 | 6.1 | 9.1 |
| CIF | 352 | 288 | 8.1 | 12.2 | 24.3 | 36.5 |
| 4CIF | 704 | 576 | 32.4 | 48.7 | 97.3 | 146.0 |
| 16CIF | 1,408 | 1,152 | 129.8 | 194.6 | 389.3 | 583.9 |

Grey images are obtained by transmitting only the *Y* luminance component. Colour images are obtained by also transmitting the *U*, *V* chrominance components sampled at half the resolution.

**Table 1.5** Video image sizes supported by H.261

| | | |
|---|---|---|
| SQCIF | 128∗96 | Optional |
| QCIF | 176∗144 | Required |
| CIF | 352∗288 | Optional |

4∗4 chrominance pixels for each chrominance plane. Four luminance blocks are grouped with two chrominance blocks (one for $U$, one for $V$) in a structure called a **macroblock**.

Each macroblock can be coded using the 'intra' method or the 'inter' method (Figure 1.19). The intra method codes by means of a local compression method (just using information relative to macroblocks that have already been encoded in the same image), while the inter method codes adjacent frames relatively in time. The coding method can be defined for a macroblock, for a 'group of blocks' (GOB's), or for a full frame. In general, video coders use the same method within a frame; hence the name intraframe (I-frame) or interframe (P-frame) frequently used when discussing video applications. The inter method is much more efficient, but leads to error accumulation; therefore, it is necessary to send intraframes intermittently.

Intraframes (I-frames) use a coding similar to the one used by JPEG, which involves DCT (discrete cosine transform), quantization, run length encoding, and entropy encoding (Figure 1.20).

For interframes (P-frames), the algorithm follows these steps:

- Motion detection: comparison of the image to be coded with the last coded image trying to find those parts of the image that have moved. This results in a representation of the difference between the motion-compensated image and the real one.

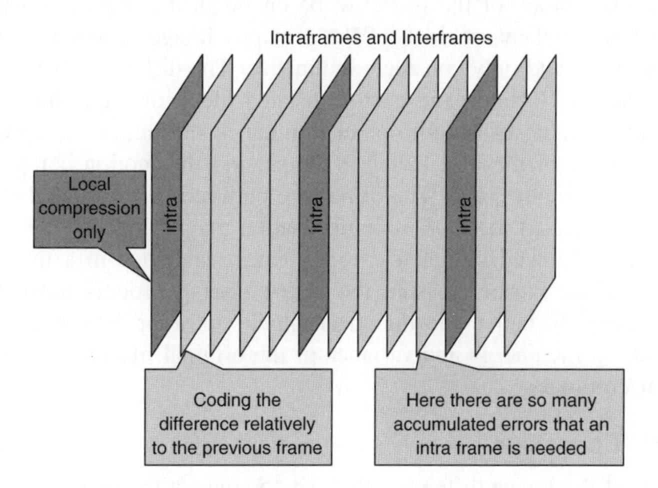


**Figure 1.19** Intra and inter coding methods.

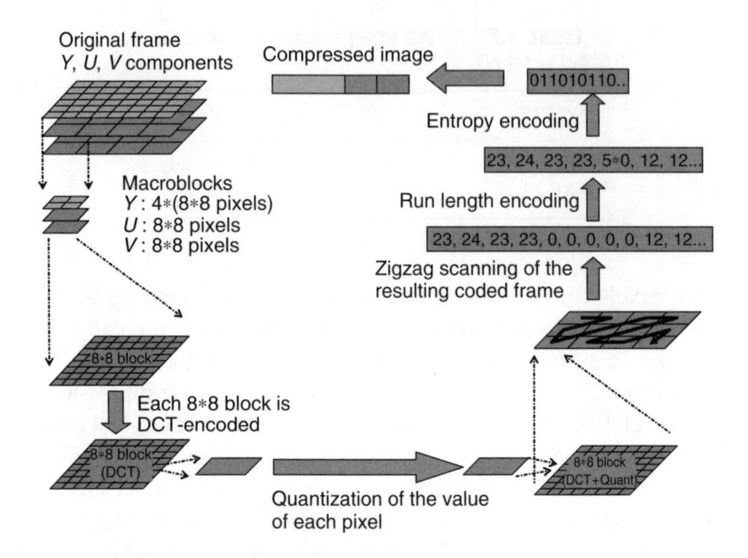


**Figure 1.20** JPEG-style encoding used for intra frames.

- Coding of the difference image using DCT transform and run length encoding.
- Entropy encoding to further reduce the image size.

#### *1.2.1.3.3.1 Motion detection*

The second stage of the H.261 P-frame coding process uses the fact that most images in a video sequence are strongly related. If the camera angle changes, many pixels will simply shift from one image to another. If an object moves in the scene, most of the pixels representing the object in a frame can be copied from the preceding frame with a shift. For each macroblock of the image to be encoded, the algorithm tries to discover whether it is a translated macroblock of the previous image. The search is done in the vicinity of $\pm 15$ pixels and only considers luminance. The difference between the original macroblock of the $n + 1$ frame and each translated block of the $n$ frame in the search area is the absolute value of pixel-to-pixel luminance difference throughout the block. The translation vector of the best match is considered the motion compensation vector for that macroblock (Figure 1.21). The difference between the translated macroblock and the original block is called the motion compensation macroblock.

If the image has changed completely (e.g., a new sequence in a movie), interframe coding is not optimal. Further reason, the H.261 coding process must decide at each frame which coding is better for the macroblock: intra or interframe. The decision function is based on the energy and variance of the original macroblock and the motion-compensated macroblock.

#### *1.2.1.3.3.2 DCT transform*

The pixel values of the image difference that we obtained at the previous step vary slowly within a macroblock. Let's take such a macroblock and repeat it in two dimensions so that

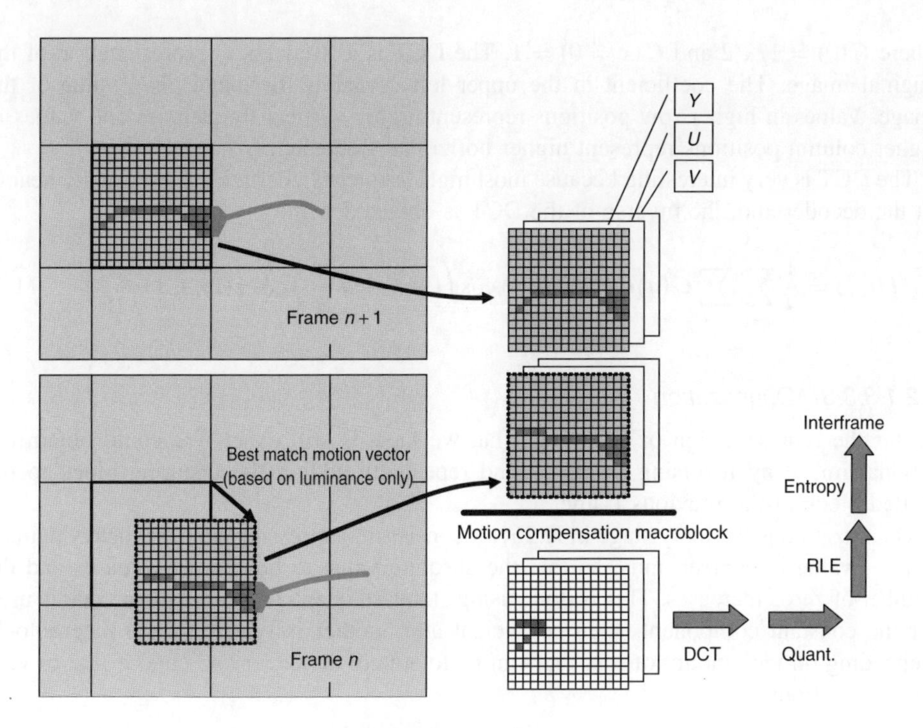


**Figure 1.21** Motion prediction and compensation of residual error.

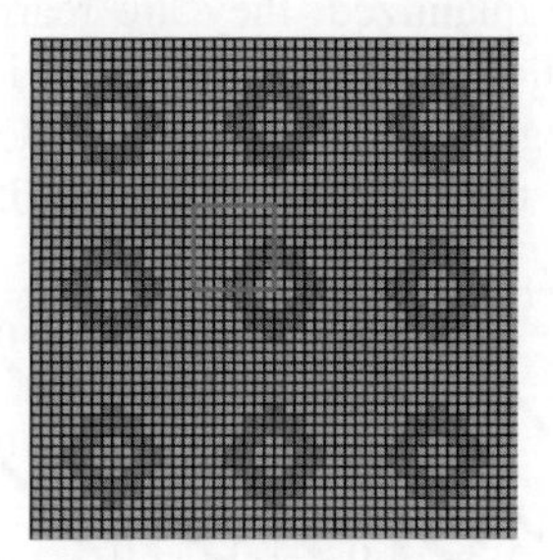

**Figure 1.22** Construction of the periodic function used for the DCT transform from the reference macroblock.

we obtain a periodic function (Figure 1.22). Such a function can be reproduced efficiently using just a few coefficients from its Fourier transform.

This transformation is called a bidimensional DCT. The formula used by H.261 to calculate the DCT of an 8∗8 block is

$$F(u,v) = \frac{1}{4}C(u)C(v)\sum_{i=0}^{7}\sum_{j=0}^{7} f(i,j)\cos\left((2i+1)u\frac{\pi}{16}\right)\cos\left((2j+1)v\frac{\pi}{16}\right) \quad (1.1)$$

where $C(0) = 1/\sqrt{2}$ and $C(x \neq 0) = 1$. The DCT is a 'frequency' representation of the original image. The coefficient in the upper left corner is the mean pixel value of the image. Values in higher row positions represent higher vertical frequencies and values in higher column positions represent higher horizontal frequencies.

The DCT is very interesting because most high-frequency coefficients are usually near 0. At the decoder and, the inverse of the DCT is obtained with:

$$f(i,j) = \frac{1}{4}\sum_{u=0}^{7}\sum_{v=0}^{7} C(u)C(v)F(u,v)\cos\left((2i+1)u\frac{\pi}{16}\right)\cos\left((2j+1)v\frac{\pi}{16}\right) \quad (1.2)$$

#### *1.2.1.3.3.3 Quantization*

So far the representation of the image that we have is still exact. We could obtain the original image by reversing the DCT and repeatedly adding the resulting block to the shifted block of the previous frame.

Quantization is the lossy stage in H.261; it consists in expressing each frequency domain $F(u,v)$ value in coarser units, so that the absolute value to be coded decreases and the number of zeros increases. This is done using standard quantization functions: one is used for the constant component (DC) coefficient and another is selected for a macroblock. Depending on the amount of loss that can be tolerated, the coder can choose fine or very coarse functions.

#### *1.2.1.3.3.4 Zigzag scanning and entropy coding*

Once the DCT coefficients are quantized, they are rearranged in a chain with the DC coefficient first and then they follow the sequence shown in Figure 1.23. This concentrates most nonzero values at the beginning of the chain. Because there are long series of consecutive zeros, the chain is then run length-encoded. This uses an escape code for

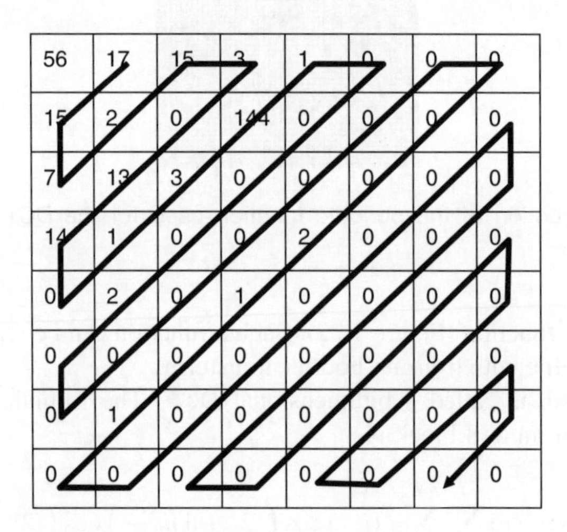


**Figure 1.23** Zigzag scanning.

the most frequently occurring sequences of zeros followed by a nonzero coefficient and variable escape codes for other less frequently occurring combinations.

This chain can be further compressed using entropy coding (similar to Huffmann coding), which creates smaller code words for frequently occurring symbols.

Huffmann coding first sorts the values to be encoded according to frequency of appearance, then constructs a tree by aggregating the two least frequent values in a branch, then repeating the process with the two values/branches that have the smallest occurrence values (counting the occurrence of a branch as the sum of the occurrences of its leaf nodes). Once the tree is complete, a '1' is assigned to each left side of any two branches and a '0' to each right side. Any value can be identified by its position in the tree as described by the sequence of digits encountered when progressing from the root of the tree to the value.

The output of the H.261 encoder consists of entropy-encoded DCT values. This bitstream can be easily decoded once the decoder has received the Huffmann tree. In the case of H.261, the tree calculation is not done in real time; the recommendation itself provides codes for the most frequently occurring combinations.

##### *1.2.1.3.3.5 Output format*

The H.261 bitstream is organized in GOBs (a group of blocks) of 33 macroblocks (each encoding 16∗16 luminance pixels and 8∗8 *U* and *V* pixels). A PAL CIF image has 12 GOBS, and a PAL QCIF image has 3 GOBS. A CIF picture cannot be larger than 256 kbits, and a QCIF picture cannot be larger than 64 kbits.

The output bitstream will consist of alternating inter-coded macroblocks and intra-coded macroblocks. The receiver can force the use of intra coding to recover from cumulative or transmission errors. Otherwise, a macroblock should be updated in intra mode at least once every 132 transmissions to compensate for error accumulation.

##### *1.2.1.3.3.6 Conclusion on H.261 video streams*

The description of H.261 found in the previous sections is not complete, but it is enough to allow a network expert to understand the nature of video traffic. The most important conclusion is that video traffic using H.261-style coding (this is also valid for H.263 and MPEG) is extremely bursty. A typical network load profile is represented in Figure 1.24. For instance, Microsoft Netmeeting sends an intraframe every 15 seconds. A videoconferencing MCU will send an intraframe for all macroblocks each time the speaker, and therefore the image, changes ('videoFastUpdate'). In other circumstances some implementations will not send all intra macroblocks simultaneously, in order to avoid the occurrence of large traffic peaks in the network.

It is also important to remember that H.261 only specifies a decoder. In fact, a very bad implementation could choose to use only intraframes if it was not capable of doing motion vector searches for interframes and still be H.261-compliant. This explains why not all video boards and not all videoconferencing software are equal, despite claiming they are using H.261 or H.263. A network engineer should always try to measure the actual bandwidth used by these devices.

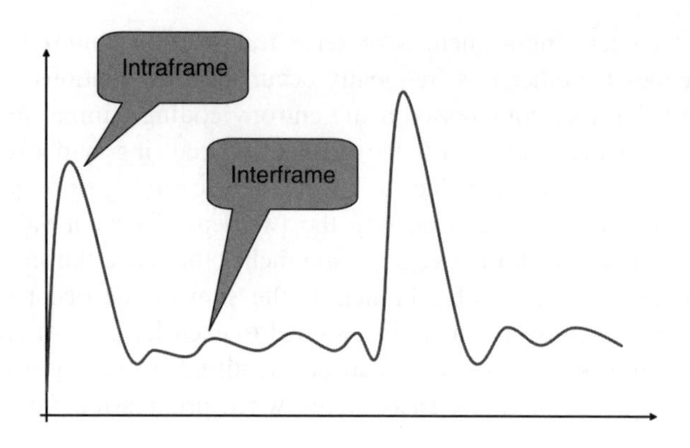


**Figure 1.24** Video traffic can be very bursty due to intraframes.

#### *1.2.1.3.4 H.263*

Table 1.6 lists the image formats that can be encoded with H.263. H.263 was designed for low-bitrate communication, as low as 20 kbits/s. The coding algorithm of H.263 is similar to that used by H.261, but involves some changes to improve performance and error recovery. H.263 is more recent, more flexible, and about 50% more bitrate-effective than H.261 for the same level of quality. It will replace H.261 in most applications. The main differences between H.261 and H.263 are:

- Half-pixel precision is used by H.263 for motion compensation, whereas H.261 used full-pixel precision and a loop filter. This accounts for much of the improved efficiency.
- Some parts of the hierarchical structure of the data stream are now optional, so the codec can be configured for a lower data rate or better error recovery.
- There are now four negotiable options included to improve performance: unrestricted motion vectors, syntax-based arithmetic coding, advance prediction, and forward and backward frame prediction (similar to MPEG) called P-B frames. Backward frames are added to allow motion vectors to refer not only to past frames, but also to future frames (e.g., when a partly hidden object becomes visible in a future frame).
- H.263 supports five resolutions. In addition to QCIF and CIF, which were supported by H.261, there is SQCIF, 4CIF, and 16CIF. SQCIF is approximately half the resolution of QCIF. 4CIF and 16CIF are 4 and 16 times the resolution of CIF, respectively.

**Table 1.6** Image sizes supported by H.263

| SQCIF | 128∗96 | Required |
|---|---|---|
| QCIF | 176∗144 | Required |
| CIF | 352∗288 | Optional |
| 4CIF | 704∗576 | Optional |
| 16CIF | 1,408∗1,152 | Optional |

Support of 4CIF and 16CIF means the codec can now compete with other higher bitrate video-coding standards, such as the MPEG standards.

With these improvements, H.263 is a good challenger to MPEG-1 and MPEG-2 for low to medium resolutions and bitrates. They have comparable features (such as B frames in MPEG and P-B frames in H.263) which are just as good for moderate movements. H.263 even has some options not found in MPEG, like motion vectors outside the picture and syntax-based arithmetic coding. MPEG has more flexibility, but flexibility means overhead. For videoconferencing applications, with little movement and a strong bandwidth constraint, H.263 is a very good choice.

#### *1.2.1.3.5 H.264*

H.264 is the latest ITU-standardized video coder and the latest video compression profile for MPEG-4 (part 10). Its production required more than 7 years of work. H.264, or **advanced video coding (AVC)**, requires only one-half to one-third of the video bandwidth necessary for an equivalent MPEG-2 channel when using all the possible optimizations of H.264 (Figure 1.25). Broadcast quality video becomes possible at a rate of 1.5 Mbit/s. This is likely to trigger an accelerated development of on-demand video over IP, in the same way that the 'MP3' format made musical applications popular on the Internet. With the traditional rate of 3.75 Mbit/s for MPEG2 movies, delivering video over ADSL is restricted only to the shortest copper lines and densely populated areas. Below 2 Mbps it is possible to add video streaming to many more ADSL lines. It also makes it easier to

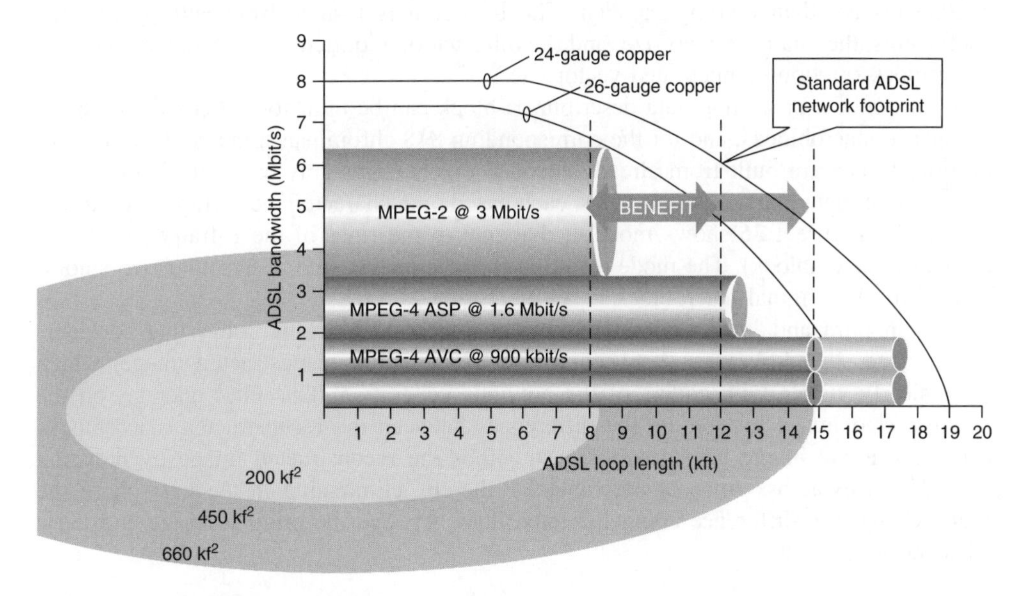


**Figure 1.25** H.264/AVC encoding extends the reach of video over ADSL. Reproduced with permission from Envivio, Inc.

provide video content over wireless links (H.264/AVC is one of the standard video coders of 3GPPv6).

With H.263, it is possible to have a business quality videoconference at about 386 kbit/s. With H.264, an equivalent conference can be achieved at about 192 kbit/s. The downside of H.264 is that it requires much more CPU power for compression than H.263. In 2005 the first implementations were able to run only on high end PCs, but now any computer, TV decoder and most video cameras support H.264. In addition, some of the optimizations introduced by H.264 (e.g., the ability to encode interframes referring to future frames) can only be used in non-real-time mode.

In line with the other MPEG standards, H.264 only describes the format of the encoded bitstream and gives no indication of the algorithms that should be used to generate the encoded data. Prediction, DCT, quantization, and entropy encoding are not fundamentally different from the previous standard, but they have been enhanced.

Each frame is processed in 16∗16-pixel macroblocks, each one being encoded in intra or inter mode. In intra mode, the encoded macroblock contains interpolation data using previously encoded macroblocks of the same frame. In inter mode, the encoded macroblock contains motion compensation information based on previous or future frames (up to two previous or subsequent frames).[4] One of the improvements of H.264 over its predecessors is that it allows intra or inter mode to be selected, not at each image, but in groups within the image called 'slices'.

In inter mode, the ability to refer to frames not immediately adjacent to the frame currently encoded is one of the major optimizations of H.264 compared with the previous generation of video coders. The difference between the predicted macroblock and the macroblock to encode is then computed, block-transformed, quantized, and the reordered coefficients are then entropy-encoded. The bitstream is formed from entropy-encoded coefficients, the quantizer step size, and the information required to recreate the predicted macroblock (motion-compensated vector, etc.).

In intra mode, prediction data describing a block can be built for either 4∗4 or 16∗16 luminance macroblocks, and for the corresponding 8∗8 chrominance macroblock. Prediction block data are built from already-encoded pixels (light gray bands in Figure 1.26), using one of eight extrapolation modes, each based on a characteristic extrapolation direction angle (Figure 1.26 shows mode 4, diagonal to the right of the P-frame, for a 4∗4 luminance macroblock). The mode resulting in the smallest sum of absolute errors compared with the original macroblock is selected.

Both in intra and inter modes, H.264 uses a new 'deblocking' filter that considerably reduces the differences between macroblocks in the reconstructed image, which were clearly visible with coders of the previous generation. This filter operates on the reconstructed image just before the differences between the reconstructed macroblocks and the original image are encoded: it smoothes the reconstructed image by reducing the differences across adjacent macroblocks, thereby eliminating in the next phase the brutal changes in difference compensation values between the original image and these adjacent macroblocks.

[4] In the baseline profile, which is more suitable for interactive videoconferencing, only P-frames and I-frames are supported (no backward prediction).

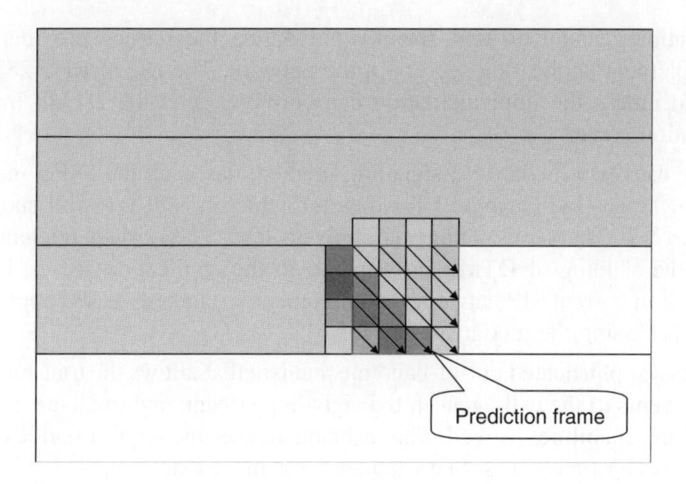


**Figure 1.26** Prediction in H.264's intra mode (mode 4).

### 1.2.2 DTMF

Strictly speaking, DTMF tones that are generated by a touchtone telephone when you press a key are part of the media stream. They are just another sound transmitted by the telephone. In the circuit-switched network this sound is digitized by the G.711 codec as part of the media stream and played back at the receiving end of the line. This does not cause any problem because G.711 does not assume that the signal is voice.

But, some narrow-band codecs that achieve much higher compression rates do use the fact that the signal is voice. Others do not assume the signal is voice, but distort it in such a way that the pure frequencies composing the DTMF tone cannot be correctly recognized when the signal is regenerated. DTMF will not get through these codecs.

Whenever a communication involves an IVR system, it is very important to be able to reliably transmit DTMF tones. In most cases the IVR system just asks a question and waits for a DTMF response. It just cares about which key has been pressed, the exact duration and timing of the tone is not so important. In other cases the IVR system will need more accuracy in the timing (e.g., when the system reads a list and asks you to press the star key when you hear something of interest).

In order to interwork properly with these IVR systems, it was necessary to develop special procedures to handle DTMF:

- H.323 generally uses the signaling channel (H.245 UserInputIndication) to convey DTMF tones (in fully decoded form). This method is sufficient in most cases and works with application servers that need to implement switching functions (e.g., contact centers), without accessing the media stream. Alternatively, since H.323v4 it is also possible to use RFC 2833, which transmits the DTMF tone in fully decoded form, but over the RTP channel. RFC 2833 mandates implementations to be able to handle this telephony event channel as a separate channel (i.e., it should not necessarily be sent to the destination address of other media streams). Unfortunately, most current implementations cannot do this, thereby preventing the service provider from being able to implement application servers in the network. The use of RFC 2833 should be discouraged unless the implementation can correctly send the DTMF information to the application server.

- SIP mainly uses two methods: a signaling method, based on the INFO message or the NOTIFY message (see Chapter 3 for details), which is still not well standardized, or RFC 2833/4733. The same comments apply to RFC 2833. Most implementations do not allow the sending of DTMF information to the application server. De facto it is very difficult in current SIP networks to implement a standards-based application server that is not accessing the media stream.
- MGCP uses a sophisticated out-of-band mechanism that allows the transmission of most telephony events to the call agent in the signaling stream, and implements filtering and accumulation capabilities as well. The mechanism uses the request notification (RQNT) and notify (NTFY) messages. See Chapter 5 for more details.

## 1.2.3 Fax

### *1.2.3.1 A short primer on Group 3 fax technology*

The purpose of facsimile transmission is to transmit one or several pages of a document across the telephone network. The first fax systems used Group 1 or Group 2 technologies, which scanned the document line by line and converted each line in black or white pixels. The data were then transmitted without compression over the phone line at the rate of 3 lines per second for Group 1 and 6 lines per second for Group 3.

Because this took over 3 minutes for an A4 document (1,145 lines of 1,728 bits) even in the best case, Group 3 technology was introduced. Group 3 faxes use a more efficient image-coding mechanism known as modified Huffmann coding (MH). MH coding uses the fact that each line is composed of large sequences of white pixels and large sequences of black pixels. Instead of sending data for each pixel MH coding just sends a short code for the sequence. Now the transmission time depends on the document, but is usually much shorter than 3 min: no wonder Group 3 faxes today rule the fax market.

With the advent of ISDN, Group 4 faxes have been introduced. The main difference from Group 3 is that ISDN can transmit raw data, so Group 4 technology need not care about the many hacks that are needed to carry data over an analog line. However, Group 4 has not succeeded in gaining a significant market, and the probability of having a Group 4 fax talking to another Group 4 fax is so low that this case has not so far been considered in the ITU's SG16 which is in charge of H.323.

#### *1.2.3.1.1 Transmitting a line (Group 3)*

Most faxes are physically linked to a printer. Because of compression, it is now possible to transmit a single line very quickly if the line is simple, so quickly that the receiving fax may not have enough time to print it. Of course the fax could buffer it in memory, but most faxes are very simple devices with very little memory. Therefore Group 3 supports

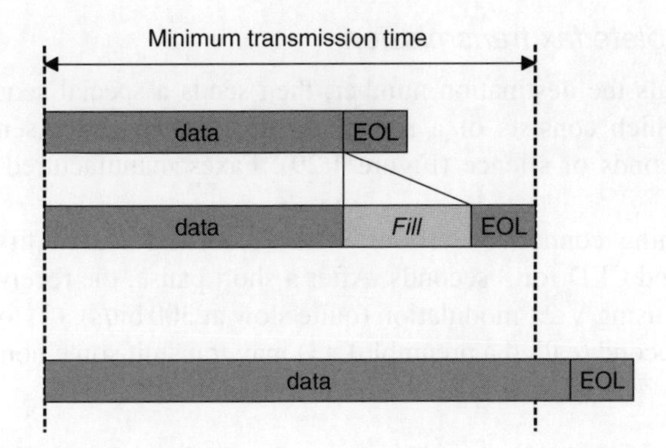


**Figure 1.27** Fax line transmission.

a minimum transmission time, as represented in Figure 1.27. If a line does not contain enough compressed data to take more than the MTT to be transmitted, a filling sequence of zeros will be added before the end of line sequence.

#### *1.2.3.1.2 Transmitting a page*

As Figure 1.28 shows, the transmission of a page is quite simple. Each line is transmitted in sequence, separated by an EOL, and the whole page is terminated by six consecutive EOLs, which means the fax has to return to command mode (RTC).

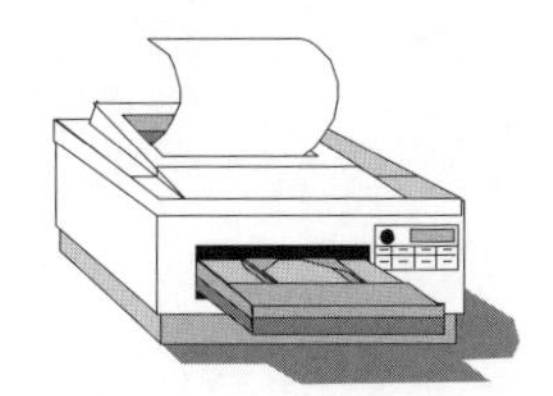

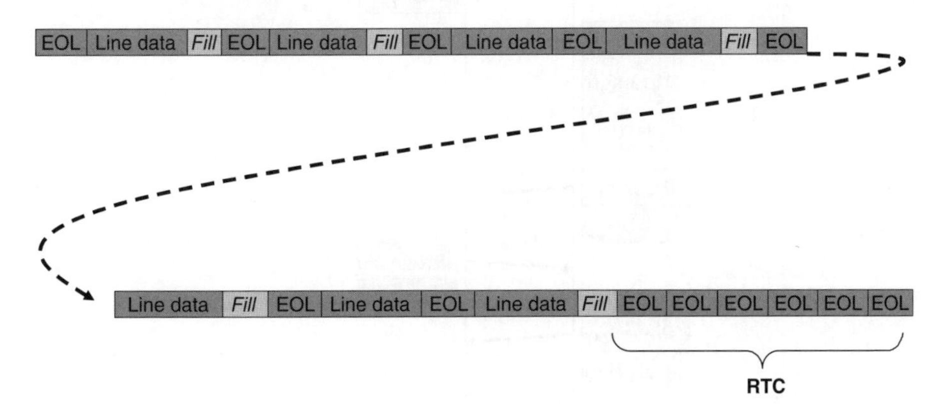


**Figure 1.28** Fax page transmission.

### 1.2.3.1.3 Complete fax transmission

The calling fax dials the destination number, then sends a special sequence called CNG (CalliNG tone), which consists of a repetition of 1,100-Hz tones sent for 0.5 seconds separated by 3 seconds of silence (Figure 1.29). Faxes manufactured before 1993 may not send this tone.

When an incoming connection arrives at the receiving fax, it first sends a special 2,100-Hz tone called CED for 3 seconds. After a short pause, the receiving fax (1) begins to send commands using V.21 modulation (quite slow at 300 bit/s), (2) to transmit synchronizing flags for 1 second (called a preamble), (3) may transmit some non-standardized data

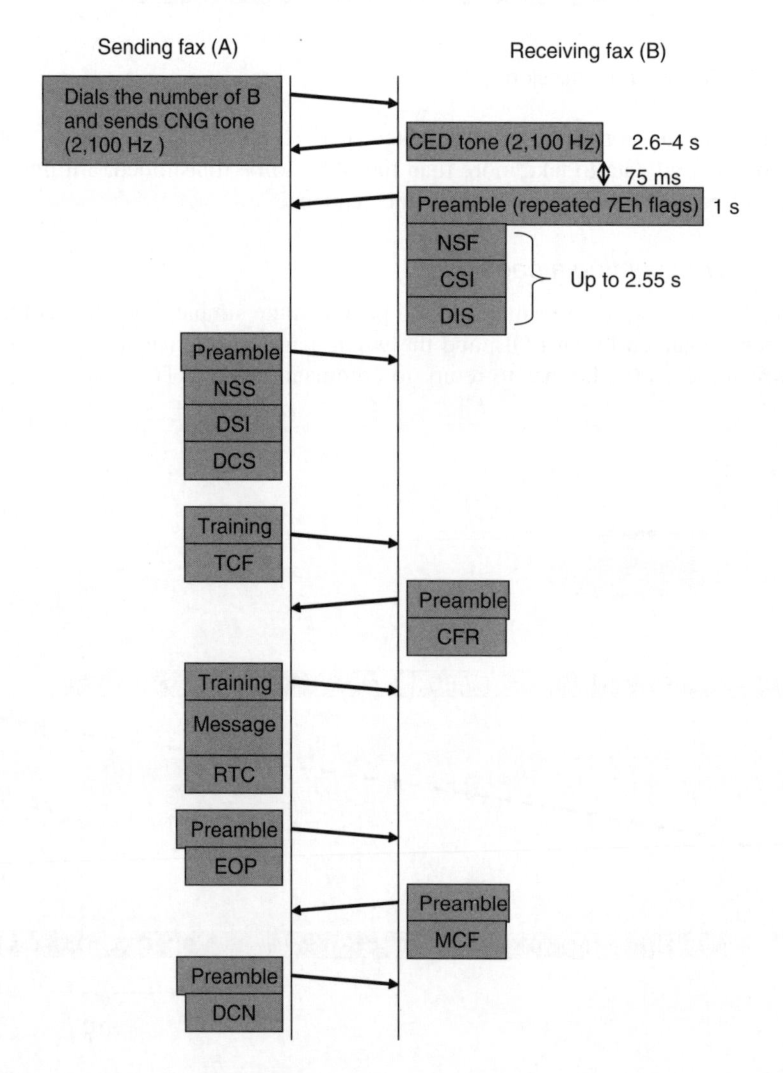


**Figure 1.29** Overview of a fax transmission.

(NSF) and its local identity (CSI), and (4) must transmit its capabilities (DIS, or digital identification signal). Each of these data elements is an HDLC frame that consists of:

- A starting flag (7Eh).
- An address field (always set to FFh).
- A command field which is set to C8h for a final frame and C0h otherwise.
- A fax control field (FCF): 02h for CSI, 01h for DIS, etc.
- A variable length fax information field (FIF).
- A checksum (FCS, or frame check sequence).

Transmitting NSF, CSI, and DIS may take up to 2.5 seconds.

The sending fax selects a mode of transmission (DCS) and replies by sending its own capabilities and its identity (TSI). As soon as the receiving fax is ready the sending fax begins the actual transmission phase and will use a faster modulation scheme, such as V.27 (4,800 bits/s) or V.29 (9,600 bits/s). This requires a training phase which is used by the receiving side to compensate for phase distortions and other issues. At the end of the training phase the sending fax sends zeros for 1.5 seconds (called a training check, or TCF). If the called fax receives this sequence correctly it considers the training phase successful and sends a CFR (ConFirmation to Receive) command to let the transmitting fax know it has succeeded. After another training sequence, the sending fax transmits the actual page data as formatted above. This takes approximately 30 seconds in V.29 mode and 1 minute in V.27 mode.

When this is finished, the modem can send an MPS (multi-page signaling) message to send another page or an EOP (end of procedure message) when it has transmitted the last page. The receiving fax acknowledges it with an MCF (Message ConFirmation) which means that the image data have been correctly received, and the sending fax sends a disconnection message DCN (DisCoNnect).

#### *1.2.3.1.4 Detection of fax for VoIP gateways*

It is important to reliably detect faxes on VoIP gateways, since fax modulation is not reliably transmitted across low-bitrate voice codecs. On the originating gateway, the T.30 calling tone can be detected, but it is an optional signal. Therefore, detecting CNG is not a reliable way to detect a fax signal. This can be resolved at the terminating gateway by detecting the V.21 preamble flag sequence which follows the called station identification tone (CED), when the CED is present. The CED itself cannot be used because it is also used by modems (V.25 ANS modem tone).

As soon as the signal is detected as a fax, the gateway should stop using regular audio encoding and switch to T.38 encoding.

#### *1.2.3.1.5 Error conditions*

If the training is not successful, the receiving fax can send an FTT command to ask for another try at a lower speed.

If an error is present in a line, the receiving fax will find it by counting how many pixels are present in the decoded line. If there are not exactly 1,728 (A4 format), the line is ignored or copied from the previous line, depending on manufacturer preference.

A fax can request the retransmission of a command at any time by sending a CRP command.

### *1.2.3.2 Fax transmission over IP (T.38 and T.37)*

#### *1.2.3.2.1 Store-and-forward fax and the challenge of real-time fax*

Sending faxes over the Internet is not something new. Many companies have been offering this service, called store-and-forward fax, for some time. The idea behind store-and-forward fax is quite simple. When computer *A* receives a fax, the fax data are represented as a set of bitmaps. This set of bitmaps is a file that can be transmitted to another computer (*B*) closer to the destination. Once this computer has received the file, it just needs to dial the receiving fax machine and emulate a fax machine to send the bitmap.

This technique is also used for bulk faxing, in which the original document is faxed once to a computer, and the computer is then provided with a list of fax numbers and sends a copy of this fax to each of them. Store-and-forward fax transmission is now standardized at ITU in recommendation T.37. However, since this book focuses on real-time applications we choose to put the emphasis on the real-time standard, T.38.

The problem with store-and-forward fax technology is that many faxes report back on the transmission of the document. Usually, they keep the result code in memory and then print it on demand. Many people tend to rely on these transmission reports: for fax-to-fax transmission this is confirmation that the fax has been correctly received, with a timestamp and the identity of the receiving fax machine.

When using store-and-forward, this report is only a confirmation that the fax has been sent, because the receiving machine is in fact computer *A*.

When it receives the file containing the document, computer *B* will dial the number indicated. But, the fax could be busy or, even worse, it could be a wrong number. So any company providing a store-and-forward service needs to report back to the sender, via email or fax. When receiving a negative acknowledgement the sender needs to know whether it is a problem with the receiving fax or the provider. This leads to potential conflicts and increases the cost of providing the service.

It is much easier for a service provider to be completely transparent in the transmission. In other words, the success report that is received by the originating fax machine should appear as a success report from the distant fax machine: such a service is real-time fax.

Real-time fax is much more complex than store-and-forward fax. There are many timers in the T.30 protocol. Once computer *A* has picked up the line, computer *B* has only a limited time budget to dial the other fax machine and get an answer. During the call, when *A*'s fax machine has sent a command, it expects a reply within 3 seconds. So, during this limited time budget *A* must send the command to *B* over the Internet, *B* must send it to the receiving fax machine, receive the reply, and forward it to *A*.

Fortunately, the ITU had a human operator in mind when setting the value of these timers; so, all are expressed in seconds. Moreover, as we saw in the preceding

Section 1.2.3.1.5 there are many ways of recovering from error conditions, which can be used to spoof the sending fax and get it to wait a little more if needed. These techniques are quite difficult to implement reliably with all brands of faxes. However, some manufacturers have built up a lot of experience and have announced they could transmit real-time faxes over IP networks with a round trip latency of up to 2 seconds!

Lately, many carriers have been tempted to do IP trunking without telling their customers. VoIP gateways make this quite simple. But, without support for real-time fax, whenever a subscriber tries to send a fax that is routed through this IP trunk, it will fail miserably. Of course, it is possible to tell subscribers not to use their faxes or even to dial a special prefix for faxes; but, this significantly complicates their lives. Real-time fax is the only appropriate answer to these issues: all VoIP gateways should be able to dynamically recognize a fax call and switch to T.38 transport mode.

##### *1.2.3.2.2 T.38*

###### *1.2.3.2.2.1 IFP*

T.38 is the approach of the ITU's SG16 to the problem of real-time fax. Its title is 'Procedures for real-time Group 3 facsimile communication between terminals using IP networks'. This recommendation is limited to Group 3 only and describes real-time fax transmission using VoIP gateways over an IP network, between faxes and computers connected on the Internet, or even between computers (the latter may not seem useful, but in some cases the receiving computer will be identified by an H.323 alias or even a phone number, and you may not know this is a computer). Usage of the T.38 protocol within the framework of H.323 is defined in H.323 annex D and was included in H.323v3.

T.38 uses a special transport protocol called IFP. IFP packets can be carried over TCP or UDP. Most gateways support UDP, but TCP transport has also been made mandatory in H.323v4. UDP transport includes a forward error correction mechanism.

IFP packets contain a type field and a data field (Figure 1.30) both encoded using ASN.1 syntax. The type field can have three values:

- T30_INDICATOR: the value of this indicator gives information on received CED and CNG tones, V.21 preambles, and V.27, V.29, and V.17 modulation training.
- T30_DATA: the value of this indicator tells us over which transport (V.21, V.17, or V.29) the data part of the message has been received.
- DISCONNECT: used to disconnect the session normally or after a failure, the value describes the error code.

The DATA part of the IFP message contains T.30 control messages as well as the image data. This DATA element is organized in fields that contain a field type and field data. Examples of these fields are:

- Type HDLC data: the data part of the field contains one, or part of an HDLC data frame, not including the checksum (FCS). This is coded as an ASN-1 octet.
- Type FCS OK: indicates that an HDLC frame is finished and the FCS has been checked. There are still other HDLC frames after an FCS OK.

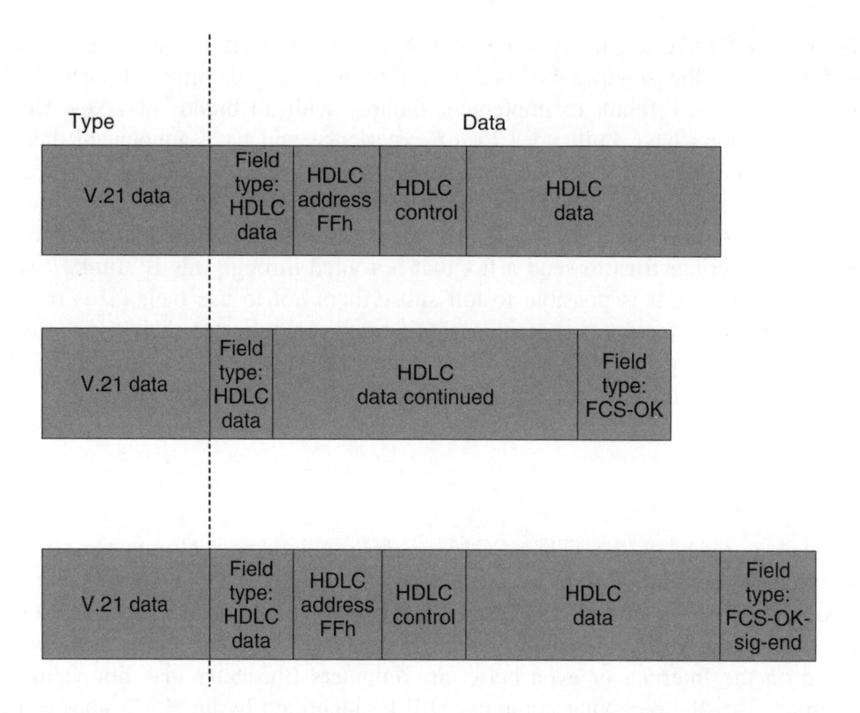


**Figure 1.30** IFP packet formats for the first, middle, and last HDLC frames.

- Type FCS OK-sig-end: same as FCS OK, except that this is the last HDLC frame.
- T4-non-ECM: the data part contains the actual image data including filling and RTC.

*1.2.3.2.2.2 IFP over TCP or UDP*

IFP messages can be carried as TCP payload or can be encapsulated in UDP, as shown in Figure 1.31.

An additional redundancy mechanism has been defined on top of UDP in order to make the delivery of IFP packets more reliable. As shown in Figure 1.31, the payload part contains one or more IFP messages, and the sequence number that appears in the header is the sequence number of the first IFP message in the payload, which is also called the primary message. The first message sent by a gateway should have a sequence number of 0. After this primary message other messages are inserted for error/loss recovery purposes; two modes can be used for this: redundancy mode and FEC (forward error correction) mode.

The control header indicates whether the secondary messages are redundancy messages (bit 3 set to 0) or FEC messages (bit 3 set to 1).

(a) Redundancy mode

In redundancy mode, copies of previous IFP messages are simply inserted after the primary message (Figure 1.32). The number of copies is the number of frames minus one. By

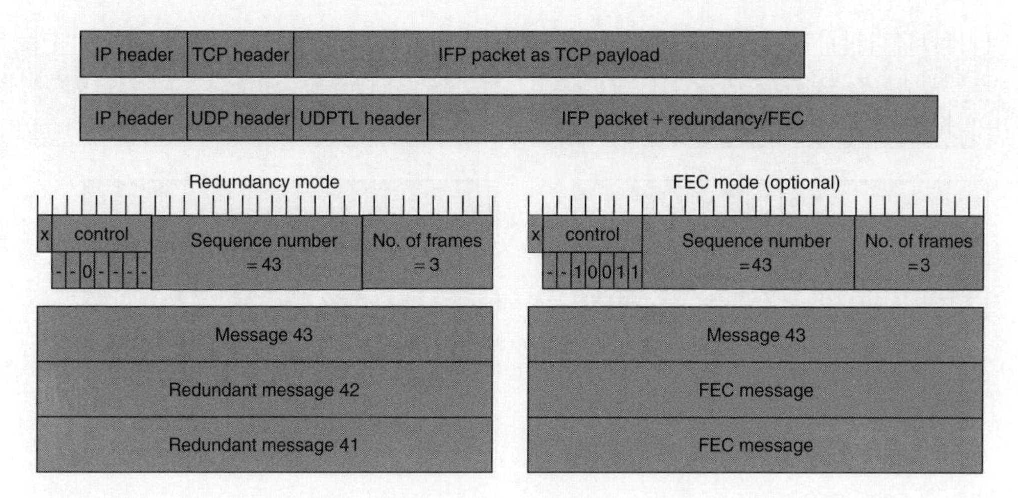


**Figure 1.31** IFP transport methods and error correction modes.

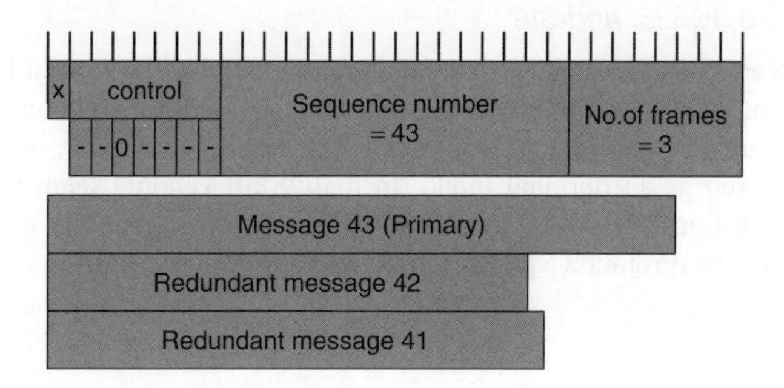


**Figure 1.32** Redundancy mode.

adding $n$ copies to the message, the transmission is protected against loss of up to $n$ consecutive packets.

A gateway is not required to transmit redundancy packets, and receiving gateways that do not support them may simply ignore the presence of redundancy packets.

(b) FEC mode

FEC mode is more complex. Each FEC message is the result of a bit-per-bit exclusive-OR performed on $n$ primary IFP messages. Before performing the OR, shorter messages are right-padded with zeros, so the resulting FEC message is as long as the longest $n$ primary message. The value of $n$ is indicated in the four last digits of the control field, '3' in Figure 1.33.

When several FEC messages are added, as in the left part of our example, the primary messages used for each FEC message are interleaved. When $n$ FEC messages are added, transmission is protected against the loss of $n$ consecutive UDP packets.

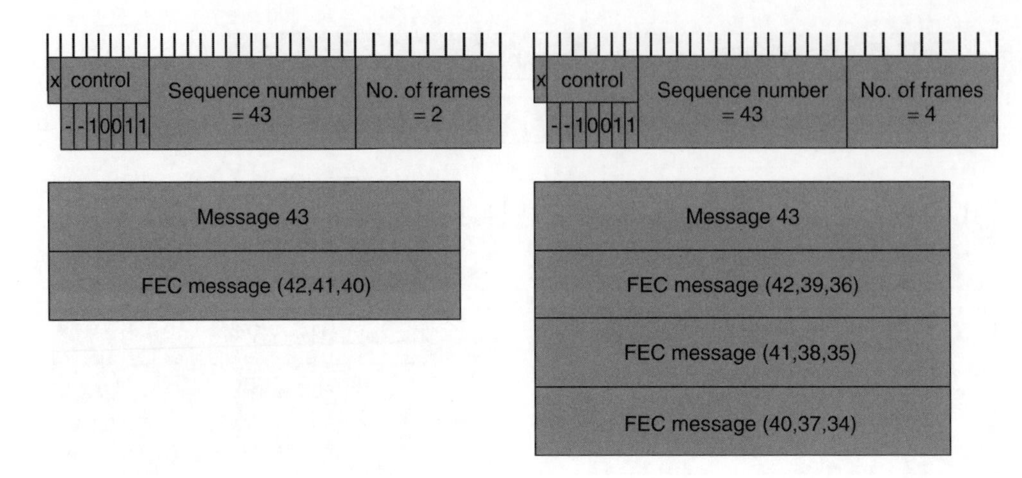


**Figure 1.33** FEC mode.

#### *1.2.3.2.2.3 T.38, H.323, and SIP*

The use of T.38 by SIP is explained in T.38 annex D 'SIP/SDP call establishment procedures'. The use of T.38 with the H.323 protocol is described in H.323 annex D and T.38 annex B. H.323 annex D mandates the use of IFP transport over TCP, but transport over UDP is still allowed as an optional mode. In reality, all vendors seem to use IFP over UDP. These capabilities (T38-TCP and T38-UDP) have been added in the DataApplicationCapability of DataProtocolCapability of H.245. IFP is transmitted over two logical channels (sender to receiver and vice versa).

# 2

# H.323: Packet-based Multimedia Communications Systems

## 2.1 INTRODUCTION

H.323 was the first VoIP standard to reach a sufficient level of maturity to be used in massive deployments. In 2011, H.323 is still used by over 10 million endpoints in fixed residential networks, by thousands of enterprises for PBX trunking, VPN or videoconferencing applications, and by 3G phones for videoconferencing (3G-324M is derived from H.323). Clearly, as the IMS architecture is now mature for residential VoIP networks, this is no longer an area where H.323 is, in general, the best choice. However there is more to H.323 than just legacy system support : there are still some domains where H.323 needs to be seriously considered and is more mature than SIP, for instance high-end videoconferencing or PBX trunking.

**H.323v1** had little ambition. Noting the growing success of IP, IPX, and AppleTalk-based local area networks in all kinds of companies, **Study Group 16** of **ITU-T** decided to create **H.323**, 'Visual telephone systems and equipment for local area networks which provide a non-guaranteed quality of service', a LAN-only standard for audiovisual conferences. SG16 leveraged the know-how of SG15, which had already acquired a lot of experience during the development of **H.320**, 'Multimedia conferencing for ISDN-based networks'. This background led to many benefits for H.323, such as seamless interworking with H.320 and H.324 systems (videoconferencing over POTS lines) and, in general, comprehensive support for interactive video, but it also led to some drawbacks in certain areas.

H.323 did not attract major interest from the market until VocalTec and Cisco founded the Voice over IP Forum to set the standards for VoIP products. At that time the focus in the VoIP Forum was given to the specification of endpoints using non-H.323, UDP-based

*IP Telephony: Deploying VoIP Protocols and IMS Infrastructure, Second Edition* O. Hersent

signaling protocols. When major software and hardware firms realized the potential of Internet telephony they pushed the VoIP Forum to become part of the IMTC (International Multimedia Teleconferencing Consortium) and simultaneously changed the focus of the VoIP activity group to profiling H.323 for use over the Internet, as opposed to creating a new protocol. Indeed, with a few minor adaptations, H.323 appeared to be just as usable over the Internet as on LANs.

Soon the ITU's SG16 acknowledged that the success of H.323v1 called for a much broader scope, and the title of H.323v2 was changed to 'Packet-based multimedia communications systems'.

### 2.1.1 Understanding H.323

***H.323*** is an umbrella specification that refers to many other ITU documents. It describes the complete architecture and operation of a videoconferencing system over a packet network. H.323 is not specific to IP. In fact, there are sections on the use of H.323 over IPX/SPX or ATM. The framework of H.323 is complete and includes the specification of:

- Videoconferencing terminals.
- Gateways between a H.323 network and other voice and video networks (H.320 videoconferencing, POTS, etc.).
- Gatekeepers, the control servers of the H.323 network, performing registration of terminals, call admission, and much, much more.
- **MCUs (multipoint control units)**, which are used for multiparty conferencing and include a control unit called an **MC (multipoint controller)**, and one or more media-mixing units called **MPs (multipoint processors)**.

#### *2.1.1.1 Core specifications*

In addition to the H.323 ITU recommendation itself, the H.323 standard references several other ITU recommendations and IETF RFCs. The most important normative documents are:

- **IETF RTP/RTCP** (Real Time Transport Protocol, Real Time Control Protocol) is described in RFC 1889. RFC 1889 describes a general framework enabling the transport of real-time (or, more precisely, isochronous) data over IP. RTP allows a level of tolerance for packet jitter and detection of packet loss by using sequence numbers and timestamps. See Chapter 1 for more details on RTP. Some profiling work is needed on top of RFC 1889 in order to build a specific application, as RFC 1889 does not describe the transport of specific media types within the RTP stream.
- **ITU recommendation H.225.0** does this profiling work in the context of H.323 videoconferencing applications (in fact, the entire specification of RTP/RTCP is annexed to it, as the ITU wanted to guarantee some stability in its references to the IETF standard). In particular, H.225.0 defines which identifiers are to be used for each type of codec

recognized by the ITU and discusses some conflicts and redundancies between RTCP and the H.245 media-control protocol that is used in H.323 to negotiate, open, and close media channels. H.225.0 also describes the **RAS (registration, admission, status)** protocol, which is used between a terminal and a gatekeeper (see Section 2.2.2). The RAS protocol is used mainly for the management of endpoint IP addresses and their mapping to aliases (e.g., telephone numbers—'registration'), but can also be involved in the call process, mainly for authorization and admission purposes. It can also be used to query the endpoint about some local statistics. Last but not least, H.225.0 defines the **call-signaling channel** protocol used in H.323. The call-signaling channel is used during the establishment and break-off phases of the call (see Section 2.2.1) and will look familiar to anyone used to **ISDN** networks. Like ISDN, it also uses ITU **Q.931** call control messages, but these messages are extended in order to support multimedia communications. The extended information is packed into the Q.931 user-to-user information element. It also describes how this information is to be transported over **TCP**, and more recently UDP and even SCTP (H.225.0v5).

- **ITU recommendation H.245** is mainly a library of **ASN-1** messages (ASN.1 has a formal syntax defined by the ITU for data structures and their serialization in messages) used by the H.245 control channel, which is opened at the beginning of the call to negotiate a common set of codecs and remains in use throughout the call to perform all media-related control functions. H.245 also defines the protocol-state machines that are used in H.323 and many other ITU standards for the management of media streams (in particular, video). H.245 is also used by the H.320 ISDN videoconferencing standard, the H.324 POTS videoconferencing standard, and the new videoconferencing standard for 3G mobile phones H.324M.

#### *2.1.1.2 Abstract Syntax Notation 1*

Use of the ASN.1 syntax by H.323 is the reason for its reputation of being a 'complex' protocol. ASN stands for 'abstract syntax notation'. ASN.1 is defined in ITU X.680 ('Abstract Syntax Notation-1'), and its serialization—the actual bit-level representation of the structured data for transport over a network used to code H.323 protocol data units (PDUs)—is defined in ITU X.691 ('ASN-1 encoding rules, specifications of packet encoding rules'). A small summary can be found at the end of H.245 specification.

ASN.1 has a very similar syntax to **XML** and can be used to describe almost any data structure. Although it is less popular than XML these days, ASN.1 is much more powerful than XML: it can describe a greater variety of types, has better support for constraints, and is much less ambiguous when used to specify data types. It is also a bit harder to learn than XML. Everything comes at a price!

ASN.1 also defines two ways of serializing data for transport over a network: **BER (basic encoding rules)** and **PER (packed encoding rules)**. BER is simple but not optimized, PER is complex but very efficient (typically data can be stored using ten times less space than data encoded in XML). The ASN.1 data description can be used by compilers that produce highly optimized BER/PER encoders automatically from the ASN.1 definition of the message set. ASN.1 is used extensively in telecommunication applications, as the use of ASN.1 automatic encoders greatly improves the robustness of applications, minimizing the sensitivity of telecom applications to malformed packets compared with manually designed parsers. ASN.1 data structure specifications are also free from any ambiguity, which facilitates interoperability across applications.

## 2.1.2 Development of the standard

### *2.1.2.1 H.323v1*

Work on H.323v1 began in May 1995, and this version was approved in June 96. This first version of H.323 still had a lot of issues; notably, the connection of audio streams was very slow and for the first few seconds of each H.323v1 phone call it was almost impossible to hear anything. Moreover, H.323v1, with its focus on LAN environments, lacked any mechanism for security. Despite these issues, H.323v1 still enjoyed great success due to the early introduction of NetMeeting® by Microsoft, an H.323v1-capable communication software. Unfortunately, too much flexibility was allowed for terminals implementing H.323v1 leading to interoperability issues, notably when endpoints also implemented the T.120 data-sharing protocol.

All these issues are still remembered today, and in many trade shows you can still hear complaints about H.323 delay or interoperability issues. This is very misleading, as all these issues have been solved by the more recent versions of H.323.

### *2.1.2.2 H.323v2*

**H.323v2** was approved in February 1998 and fixed the major issues of H.323v1. Post-connect audio delay was completely eliminated using a new procedure known as 'fast connect'. H.323v2 was extended to enable the use of enhanced security procedures. These procedures were defined in the new **H.235** standard, addressing the need for **authentication** (making sure people in a conference really are who they pretend to be), **integrity** (making sure the modification of the content of H.323 messages by a third party cannot go unnoticed), **non-repudiation** (the ability to prove that someone participating in a conference was there), and **privacy** (making sure that information exchanged between individuals remains unaccessible to third parties). The early deployment of H.323v1 had revealed other weaknesses that appeared in very specific call flows:

- In Germany and some other countries the size of phone numbers is not known in advance and there is a need to send the dialing digits one by one to the network until the network decides the number is complete. This requires a capability called **overlapped sending**.
- When calling some announcement servers in a network you may have to stop playing the ring-back tone to play specific announcements, without connecting the call first. The announcement server uses the **progress message** to inform the network of coming in-band prompts. The same interaction with interactive voice response servers required a better means of transporting **DTMF** tones.

- Even more fundamentally, people realized that the true benefit of VoIP was not so much 'voice compression', but the ability to control media channels from a server without requiring the server to be on the path of media channels. For instance, in a prepaid application, after the initial prompts to authenticate the user and ask for a destination, once the calling party *A* half-call has been connected to the called party *B*, the prepaid application server can cause media streams to be exchanged directly between endpoints *A* and *B*, while remaining in the call control path so that it can cut off the communication once credit has expired. In traditional telephony, the same server needs to relay these media streams for the entire duration of the communication. This key feature required a new procedure called the 'empty-capability-set' (or third-party rerouting, see Section 2.7.1.2.2) and makes VoIP much more scalable and easier to deploy than traditional TDM voice for services such as prepaid or hosted contact centres.

All these enhancements, and many more, were introduced in H.323v2, making it suitable for widespread deployment. At that time the new **H.450** standard series, providing supplementary services for H.323, was also introduced. H.450 is based on the QSIG extensions of ISDN for use by PBXs. **H.450.1** defined the general framework for exchanging supplementary service commands and responses for use by supplementary services, **H.450.2** defined the **call transfer** procedure (**blind call transfer** and **call transfer with consultation**), and **H.450.3** defined the **call diversion** procedures (*call forwarding unconditional, call forwarding on no answer, call forwarding on busy*, and **call deflection**).

Even though new versions of H.323 have been approved, H.323v2 is still the protocol powering the vast majority of voice over IP networks worldwide and serves this function very well.

#### 2.1.2.3 H.323v3

**H.323v3** was approved in September 1999. Of all the enhancements introduced in H.323v3, only one was really needed: the ability to support the CLIR (calling line identity restriction) in the same way as the traditional PSTN network. This was a legal requirement in most countries. Most of the other enhancements introduced in this version, such as the ability to reuse signaling connections, the ability to use UDP for the transport layer (annex E), an interdomain-routing protocol (annex G), have not really made it for commercial products, following the well-known pragmatic approach of 'if it works, don't fix it'.

The H.450 standard series was also expanded to include call hold, call park, call pickup, call waiting, and message-waiting indication (MWI). Out of these only MWI, H.450.7, is widely supported today by IP phones and residential gateways.

#### 2.1.2.4 H.323v4

**H.323v4** was approved in November 2000. It includes some useful modifications, such as the ability to start H.245 procedures in parallel with fast connect. Prior to this the fast-connect procedure (described in Section 2.3.3) was used to accelerate the establishment of media streams, but DTMF tones could not be transmitted before the call fully connected, which created interoperability issues in some call flows involving the PSTN intelligent network. Another very useful addition is the description of how protocols that cannot be fully mapped to H.323 can be transported in H.323. Annex M1 describes the encapsulation of QSIG (a supplementary services protocol for ISDN PBXs), and annex M2 describes the encapsulation of ISUP (the standard protocol for establishing calls in core telephone networks). H.323v4 also introduces an official format for an H.323 URL, which took almost 2 years to stabilize and be approved by all parties![1]

Throughout the life of the H.323 standard, each addition of a new feature required an edit of the ASN.1 description of control messages, as the new feature was explicitly described in ASN.1. This was powerful, as interoperability was guaranteed at least for the parsing of new features, but the growing ASN.1 syntax posed problems to developers wanting to support only a few of the features available. Since version 4, a generic extensibility framework allows the indication of features that are supported, desired, or required, without the need for further edits of ASN.1 syntax.

There were also a few more additions to the H.450 series (**H.450.8**: 'Name identification service'; **H.450.9**: 'Call completion'; **H.450.10**: 'Call offer'; **H.450.11**: 'Call intrusion'). As with the other members of the H.450 series, these additions have not received much support from the industry. They are only being employed in some private networks of IP-PBXs that use H.323 as a PBX-to-PBX protocol, where the H.450 series plays the role of QSIG (which is used in private networks of PBXs connected through ISDN).

In fact, instead of attempting to precisely define a standard for each feature of a business phone, the industry has now taken another approach: phones offer a stimulus-based control protocol for all of their user interface components (screen, buttons, lamps) and media streams. The network optimises use of these resources to provide value-added services. H.323 proposes two approaches for this:

- Using an HTTP control channel, which provides an arbitrary user interface, and making the network responsible for the execution of services.
- Annex L (stimulus control).

The industry has not adopted these methods and seems unlikely to do so in the medium term. Instead MGCP has become the de facto standard for stimulus control of IP phones. This will be covered in detail when we discuss the MGCP protocol in Chapter 5.

#### *2.1.2.5 H.323v5*

H.323v5 was approved in July 2003. This version does not introduce any major new feature but does correct some remaining problems with the former H.323 specifications, such as the missing 'hop count' parameter that prevents call loops, the much awaited H.460.6 (extended fast connect), which makes it possible to redirect and renegotiate media streams while in fast-connect mode. Another nice addition is the concept of digit maps (H.460.7), borrowed from MGCP, which makes it possible to reduce the post-dial

[1] For instance, see h323:someone@domaine.com.

delay of a call without using the overlapped sending procedure, when the digit pattern of the expected number is known in advance (e.g., in virtual private networks).

Most of the new additions use the generic extensibility framework that was introduced in H.323v4; this helps stabilize the ASN.1 files.

Although conversion of existing systems is likely to take some time, H.323v5 adds the possibility of using SCTP as a transport protocol, in addition of TCP and UDP, providing a very robust option with combined latency control and reliability.

## 2.1.3 Relation between H.323 and H.245 versions, H.323 annexes, and related specifications

Where are we today? Essentially, the core H.323 protocol is complete and only minor additions are required from time to time to cover specific details required in the context of a specific service. Since H.323v3, there has been a growing divide between the status of the standard and the reality of its deployment. No one is in any hurry to implement every detail of the newer standards. Rather, these documents are viewed by vendors as a pool of standard approaches to certain issues and implemented as and when customers request them. This book will cover in detail many of the H.323v2 and v3 features, and select only the features of H.323v4 that really have something to offer for real-life deployment.

Each version of H.323 corresponds to a version of the H.225.0 call control protocol and must be used with specific versions of the H.245 media control protocol (see Table 2.1). The protocol version is indicated in the protocolIdentifier information element of the messages (e.g., {itu-t (0) recommendation (0) h (8) 2250 version (0) 2}). The H.245 version can change dynamically during a call if third-party rerouting is used.

### *2.1.3.1 H.323 annexes*

Multiple annexes to H.323 have been defined, each specifying additional details for specific needs (Table 2.2).

### *2.1.3.2 H.323-related specifications*

Beyond the core set of specifications—H.225.0, H.245, and the annexes—many other specifications exist which relate to specific applications or aspects of H.323:

**Table 2.1** Relationships between versions H.323, H.225, and H.245

| H.323 | H.225 | H.245 |
|---|---|---|
| v1 | v1 | v2 |
| v2 | v2 | v3 or higher |
| v3 | v3 | v5 or higher |
| v4 | v4 | v7 or higher |
| v5 | v5 | v9 or higher |

Table 2.2 List of annexes to H.323

| | |
|---|---|
| Annex A* | H.245 messages used by H.323 endpoints |
| Annex B* | Procedures for layered video codecs |
| Annex C* | H.323 on ATM |
| Annex D* | Real-time facsimile over H.323 systems |
| Annex E* | Framework and wire protocol for multiplexed call-signaling support |
| Annex F* | Simple endpoint types |
| Annex G* | Text conversation and text set |
| Annex J* | Security for H.323 annex F |
| Annex K* | HTTP-based service control transport channel |
| Annex L* | Stimulus control protocol |
| Annex M1* | Tunelling of QSIG in H.323 |
| Annex M2* | Tunelling of ISUP in H.323 |
| Annex M3 | Tunelling of DSS1 through H.323 |
| Annex N | Quality of service |
| Annex O | Use of DNS |
| Annex P | Transfer of modem signals over H.323 |
| Annex Q | Far-end camera control |
| Annex R | Robustness methods for H.323 entities |

*These annexes are now in the main H.323 document.

- ***H.235*** specifies a secure mode of operation for H.323 terminals and refers to the SSL (secure sockets layer) specification.
- ***H.246*** describes in more detail the operation of H.323 gateways and specifies how to map SS7 ISUP call-control messages onto H.323 messages to maximize transparency of call flows initiated and terminated on a traditional SS7 network but traversing an H.323 network.
- **H.332** (loosely coupled conferencing) profiles H.323 and extends it for use in the context of a large conference with few speakers but a large audience. H.332 is a bridge between the world of conferencing and the world of broadcasting (see the companion book, *Beyond VoIP Protocols*, Chapter 6 on multicast technology).
- ***H.450*** is a series of standards defining messages and call flows for supplementary services, such as call transfers or how to set up the message-waiting indication on an IP phone (Table 2.3). These supplementary services mimic the services of QSIG and most are targeted for private telephony networks.

H.460 is a series of more recent recommendations, all of which use the generic extensibility framework (GEF) introduced in H.323v4:

- H.460.1: 'Overview of the generic extensibility framework and "author's guide"'.
- H.460.2: 'Number portability (GEF)'.

**Table 2.3** List of H.450 specifications and services

| | | |
|---|---|---|
| H.450.1 | 02/1998 | Generic functional protocol for the support of supplementary services |
| H.450.2 | 02/1998 | Call transfer supplementary service |
| H.450.3 | 02/1998 | Call diversion supplementary service |
| H.450.4 | 05/1999 | Call hold supplementary service |
| H.450.5 | 05/1999 | Call park and call pickup supplementary services |
| H.450.6 | 05/1999 | Call-waiting supplementary service |
| H.450.7 | 05/1999 | Message-waiting indication supplementary service |
| H.450.8 | 02/2000 | Name identification service supplementary service |
| H.450.9 | 03/2001 | Call completion supplementary service |
| H.450.10 | 03/2001 | Call offer supplementary service |
| H.450.11 | 07/2001 | Call intrusion supplementary service |

- H.460.3: 'Circuit status map (GEF)'.
- H.460.4: 'Call priority designation (GEF)'.
- H.460.5: 'Transport of duplicate Q.931 IEs (GEF)'.
- H.460.6: 'Extended fast connect (GEF)'.
- H.460.7: 'Digit maps (GEF)'.
- H.460.8: 'Querying for alternate routes (GEF)'.
- H.460.9: 'QoS monitoring and reporting (GEF)'.

## 2.1.4 Where to find the documentation

All **ITU** documents can be purchased on the ITU website (www.itu.int). However, H.323 is a living standard and the latest specifications only become available some time after they have been approved. For those needing detailed and up-to-date technical information, the best option is to read the working documents of SG16, together with the interesting discussions of standard details and implementation guides at http://www.packetizer.com. This excellent site is well maintained and really presents all the useful information a developer needs to start an H.323 project. Another related site has a similar focus on H.323: http://www.h323forum.org. It offers interesting discussions and forums on the ongoing development of the H.323 standard.

Many more standard bodies are involved in VoIP; for a more comprehensive view of the most active organizations on VoIP and voice quality of service over packet networks, see Figure 2.1.

H.323 is a complex standard. Although it is well understood and very well defined, there is still room for new interpretations. Much of the material presented in this chapter was gathered from discussions during standard body meetings and real-deployment experience.

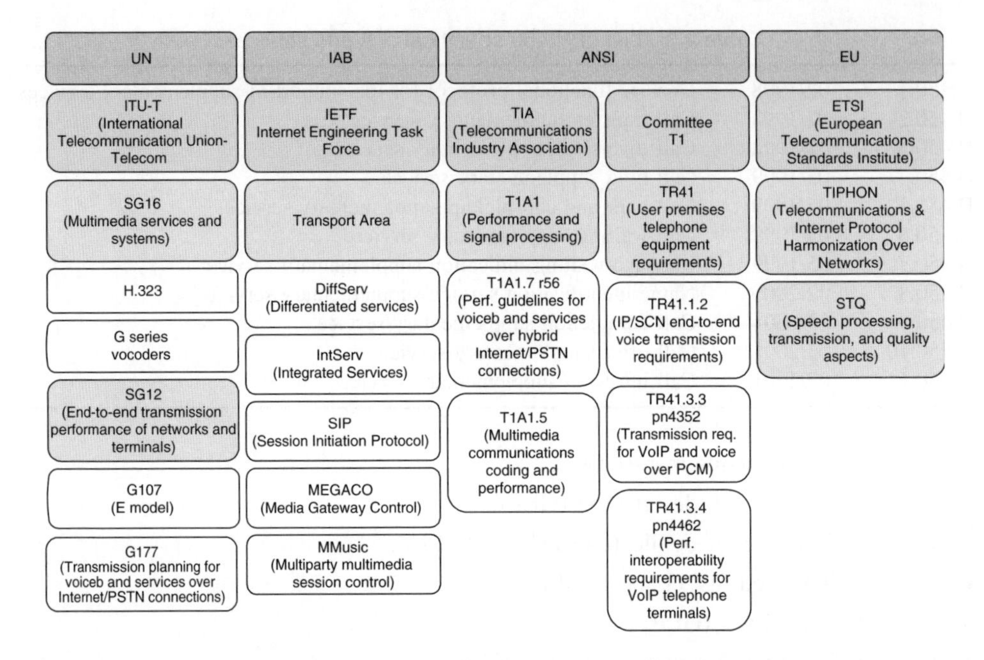


**Figure 2.1** Some of the many organizations involved in VoIP standardization.

## 2.2 H.323 STEP BY STEP

H.323 did not invent videoconferencing over IP. Researchers and students did this for years on the mBone network using RTP/RTCP (refer to the companion book, *Beyond VoIP Protocols*) multicast chapter. However, RTP/RTCP has very basic signaling capabilities, as we saw in **Chapter** 1, and cannot be used for common telephony.

H.323 mainly defines the signaling needed to set up calls and conferences, choose common codecs, etc. RTP/RTCP is still used to transport isochronous streams and get feedback on the quality of the network, but fancy RTCP features like email alias distribution are not normally used by H.323.

As H.323 tackles a very complex problem, it is consequently complex itself, as we have already stated. The set of documents that an H.323 engineer needs as a reference (Q.931, H.323, H.225, H.245, H.235, H.332, ETSI TIPHON and other profiles, etc.) is extensive and takes a while to read. Therefore, we have chosen here not to paraphrase H.323, but rather to illustrate the behaviour of H.323 entities in various configurations.

### 2.2.1 The 'hello world case': simple voice call from terminal A to terminal B

For our first example we assume that two users would like to establish a voice call, both using IP endpoints with fixed and well-known IP addresses. This is an important assumption, because most of the time IP addresses are dynamic and cannot be used directly to reach a user. Calls can also be established with regular phones not directly connected on IP: this more general situation will be studied in Section 2.2.2. We will be using the basic H.323v1 connection sequence, without security and without any of the optimizations of H.323v2, v3, or v4.

Establishment of a point-to-point H.323 call requires two TCP connections between the two IP terminals: one for call set-up and the other for media control and capability exchange:

- Call set-up messages are sent during the initial TCP connection established between the caller and a well-known port (defined by the standard, usually port 1720) at the callee endpoint. This connection carries the call set-up messages defined in H.225.0 and is commonly called the Q.931 channel, or call-signaling channel.
- Media-control messages are carried within a second TCP connection. On receipt of the incoming call, the callee allocates a dynamic TCP port for the media control connection, communicates this port to the caller in the call set-up response, and waits for the incoming media control connection request. The caller then establishes the second TCP connection, dedicated to media control messages, to the indicated port.

The second connection carries the control messages defined in H.245 and is therefore called the H.245 channel. It is used by the terminals to exchange audio and video capabilities and to perform a 'master–slave' determination; this is useful in very specific call flows (i.e., the simultaneous opening of a bidirectional data-sharing channel) which require a notion of priority of one endpoint over the other to resolve the race condition. The H245 channel is then used to signal the opening of 'logical channels' for audio and video streams (each corresponding to an RTP session), fax data (the media is then exchanged using the IFP protocol described by T.38), or even a data-sharing T.120 channel. The H.245 channel remains open for the duration of the conference.

Once the H.245 channel is established, the first connection is no longer necessary and may in theory be closed by either endpoint, and re-opened only for sending additional call control messages (e.g., to bring the call to an end). In practice, though, since TCP connections take significant resources and time to get established, we do not know of any endpoint in the market that closes call control connections.

#### 2.2.1.1 First phase: initializing the call

H.323 uses a subset of the Integrated Service Digital Network (ISDN) Q.931 user-to-network interface that signals messages for call control. The following messages belong to the core H.323 and must be supported by all terminals:

- SETUP.
- ALERTING.
- CONNECT.

- RELEASE COMPLETE.
- STATUS FACILITY.

Other messages, such as CALL PROCEEDING, STATUS, STATUS ENQUIRY, are optional. Support for the Q.931 PROGRESS message has been added in H.323v2 to support the interworking of call flows with the PSTN, notably when the PSTN signals the presence or absence of in-band media before making the connection. Regarding supplementary services, only the FACILITY message is supported; all others, such as HOLD, RETRIEVE, SUSPEND, are forbidden (they have been replaced by H.450 equivalents). Moreover, the ISDN RELEASE and DISCONNECT messages are not supported in H.323.

As we will see in Section 2.2.1.6, each time an ISDN message has been removed to make H.323 simpler, it was subsequently found to be a mistake and the message was either added later on (PROGRESS) or other messages were extended to support an equivalent feature (e.g., DISCONNECT is in some cases replaced by a PROGRESS message).

In our example John, logged on terminal *A*, wants to make a call to Mark, knowing Mark's IP address (10.2.3.4). Terminal *A* sends to terminal *B* a SETUP message on the well-known CallSignalingChannel port (port 1720 as defined by H.225.0 appendix D), using a TCP connection (see Figure 2.2). This message is defined in H.225.0 and contains the following fields, which have been borrowed from Q.931:

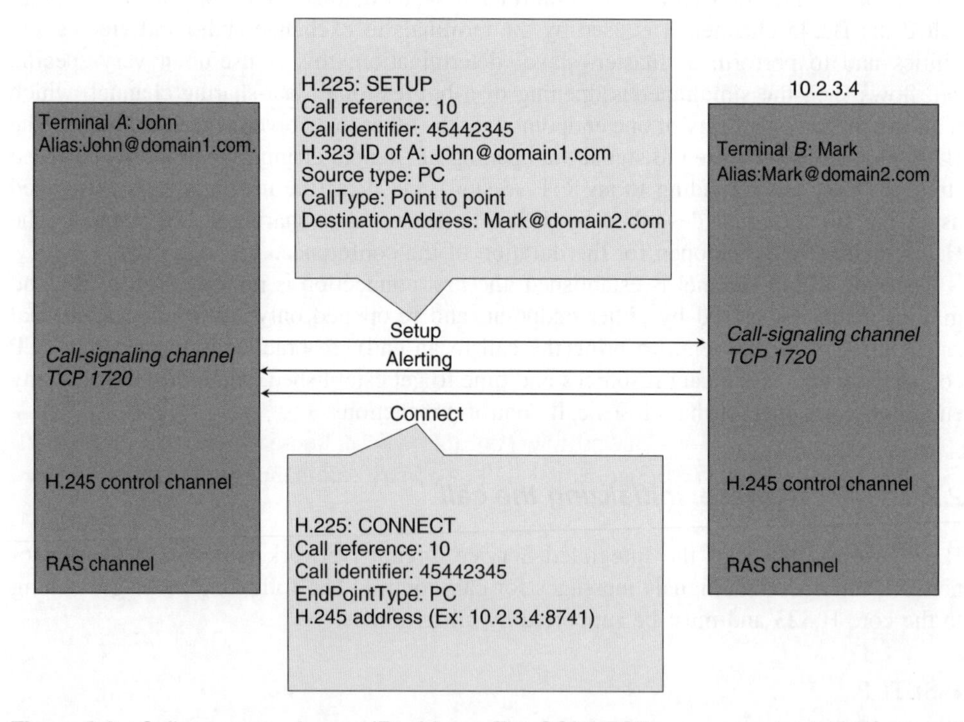


**Figure 2.2** Call set-up to a known IP address. The CONNECT message returns the transport address for H.245 signaling.

- A protocol discriminator field set to 08h (Q.931 defines this as a user network call-control message).
- A 2-octet, locally unique call reference value (CRV) chosen by the originating side which will be copied in each further message concerning this call. Here John's terminal has picked CRV = 10.
- A message type (05h for SETUP as specified in Q.931 Table 4.2).
- A bearer capability, a complex field that can indicate, among other things, whether the call is going to be audio-only or audio and video. ISDN gateways can place in this field some elements copied from the ISDN SETUP message.
- A called party number and sub-address, which must be used when the address is a telephone number. This field contains a numbering plan identification. When it is set to 1001 (private numbering plan) it means that the called address will be found in the user-to-user information element of the SETUP message (see below). If John knows Mark by his transport address only (10.2.3.4:1720), the numbering plan will be set to 1001.
- A calling party number and sub-address, which will be present if the caller has a telephone number.
- A user-to-user H.323 PDU (H323-UU-PDU) which encapsulates most of the extended information needed by H.323. In this case it is a SETUP information element that contains:
  - A protocol identifier (which indicates the version of H.225.0 in use).
  - An optional H.245 address if the sender agrees to receiving H.245 messages before connection. In the normal procedure, as used in the example, the callee allocates a TCP port for H.245 and waits for a H.245 connection from the caller.
  - A source address field listing the sender's aliases (e.g., John@myhouse.uk) (as indicated above; in case the sender only has an E.164 phone number then it should be in the Q.931 calling party information element).
  - A source information field can be used by the callee to determine the nature of the calling equipment (MCU, gateway, ...).
  - A destination address which is the called alias address(es). Several types are defined in H.323v2: E.164 which is a regular phone number using only characters in the set ≪0123456789#∗,''; H323-ID which is a unicode string; url-ID (a URL like those you can type on your browser, but this type is unused in practice); transport-ID (e.g., 10.2.3.4:1720), and Email-ID (e.g., Mark@domain.org). H.323v4 renamed type 'e164' into 'dialedDigits', as E.164 refers to a precise number format (country code, plus national number) which in general will not be used by end-users, who use their national numbering conventions or private numbers. H.323v4 also added a specific format for an H.323 URL, which must begin with "h323:" followed by a username and hostname (e.g., h323:mark@mydomain.org).
  - A unique Conference identifier (CID). This is not the same as the Q.931 CRV described above or the call identifier described below. The CID refers to a conference which is the actual communication existing between the participants. In the case of a multiparty conference, all participants use the same CID, and if a participant joins

the conference, leaves and enters again, the CRV and CallID will change, while the CID will remain the same. Refer to Section 2.4 for more details.

- A conferenceGoal which indicates if the purpose of this SETUP message is to create a conference, invite someone in an existing conference, or join an existing conference. In this simple scenario, we simply want to create a conference.
- A call identifier (CallID) which is set by A, and should be the globally unique identifier of the call, not only locally unique like the Q.931 CRV. It is also used to associate the call-signaling messages with the RAS messages (RAS is used in the next call scenario, see Section 2.2.2). In the gatekeeper scenario (also in the next example), the call leg to the gatekeeper and from the gatekeeper to the called endpoint should have the same CallID.

Note that TCP is a stream-oriented protocol and does not provide framing (delimitation of individual messages). For this reason the Q931 messages are not transported directly over TCP, but are first framed using a 'length data' type of structure known as **TPKT** and defined in RFC1006 (ISO transport service on top of the TCP). This structure can be seen in the network capture of Figure 2.3, and in Figure 2.4.

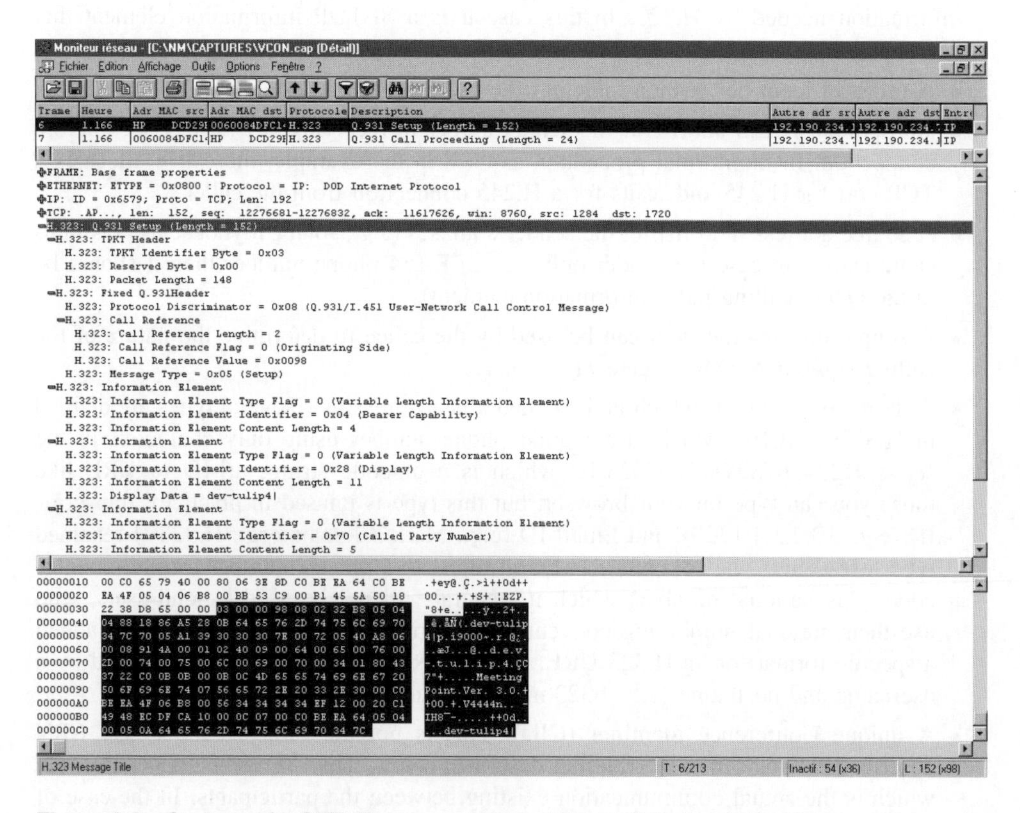


**Figure 2.3** Capture of a SETUP message (using Microsoft Network Monitor).

| Vrsn (8 bit) = 3 | Reserved (8 bit) | Packet length (16 bit) | Data |
|---|---|---|---|

**Figure 2.4** RFC 1006 framing using TPKT structure.

Either CALL PROCEEDING, ALERTING, CONNECT, OR RELEASE COMPLETE must be sent by Mark's terminal immediately on receipt of a SETUP message. One of these must be received by John's terminal before its set-up timer expires (in general, 4 s). After Alerting is sent, indicating that 'the remote phone is ringing', the user has up to 3 min to accept or refuse the call.

Finally, as Mark picks up the call, his terminal sends a CONNECT message with:

- The Q.931 protocol discriminator, the same call reference (10), and message type 07h.
- In the H323-UU-PDU there is now a CONNECT user-to-user information element with:
  - The protocol identifier.
  - The IP address and port that B wishes A to use to open the H.245 TCP connection.
  - Destination information, which allows A to know if it is connected to a gateway or not.
  - A conference ID copied from the SETUP message.
  - The call identifier copied from the SETUP message.

Note that, *the procedure we just described is called the 'en bloc' procedure. The destination address information is sent at once. This method is always used when the destination address is not a phone number (email alias, IP address, etc.). When the destination address is a phone number the 'en bloc method' is also used by cellular phones that have a 'send' button. For a normal phone without a 'send' button, however, it is not obvious to know when the number is complete and should be sent in the SETUP message. Most IP phones use a timer, which fires a few seconds after the last digit key is pressed. If this waiting time is inconvenient, or when the calling device is an existing PBX, a more sophisticated procedure exists in ISDN and H.323: 'overlapped sending'. With overlapped sending, the calling endpoint sends partial numbering information in the SETUP message (with a canOverlapSend flag), and if the number is incomplete the gatekeeper (see the next example for more information on routing the signaling messages through the gatekeeper) will respond with a SETUP ACKNOWLEDGE message instead of a CALL PROCEEDING or ALERTING message. The calling device then continues to send digits in 'INFO' messages, until it receives a CALL PROCEEDING message, meaning that enough digits have been accumulated.*

*Since H.323v5 (H.460.7), the 'DigitMap' function enables the gatekeeper to configure the endpoint with a set of patterns that can trigger an 'en bloc' call immediately the pattern is recognized, resolving the timer problem.*

### 2.2.1.2 Second phase: establishing the control channel

#### 2.2.1.2.1 Capability negotiation

Media control and capability exchange messages are sent on the second TCP connection, which the caller establishes to a dynamic port on the callee's terminal. The messages are defined in H.245.

The caller opens this H.245 control channel immediately after receiving the ALERTING, CALL PROCEEDING, or CONNECT message, whichever specifies the H.245 transport address to use first. It uses a TCP connection which must be maintained throughout the call. Alternatively, the callee could have set up this channel if the caller had indicated an H.245 transport address in the SETUP message. The H.245 control channel is unique for each call between two terminals, even if several media streams are involved for audio, video, or data. This channel is also known as logical channel 0.

The first message sent over the control channel is the TerminalCapabilitySet (Figure 2.5), which carries the following information elements:

- A sequence number.
- A capability table, which is an ordered list of codecs the terminal can support for the reception of media streams, each codec being identified by an integer, the CapabilityTableEntryNumber. Up to 256 codecs can be described. Not all combinations

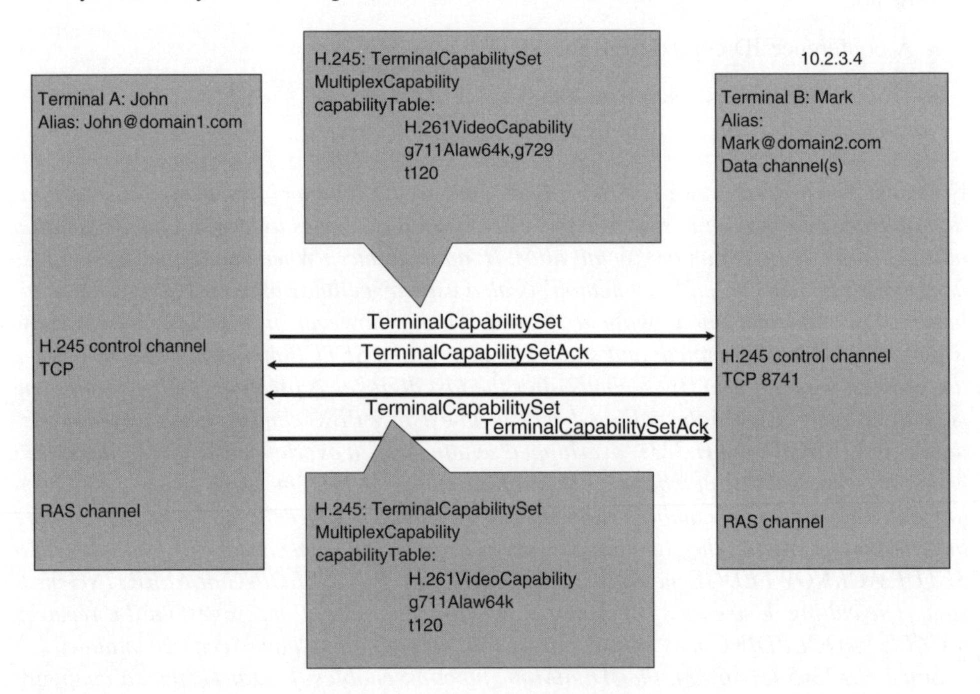


**Figure 2.5** Capability negotiation over the H.245 channel using TerminalCapabilitySet messages.

of codecs can be supported, and the CapabilityDescriptors structure describes which combinations of codecs can be supported.

- CapabilityDescriptor. This is a rather complex structure (Figure 2.6) which describes precisely the combinations of codecs a terminal can support. The **CapabilityDescriptor** structure is a list of supported codec configurations. Each supported codec configuration is of the form (Codec 1 *or* Codec 2 *or* Codec 3) *and* (Codec 4 *or* Codec 5) *and* ... where *or* is exclusive. The *and* structure is called a **SimultaneousCapabilities** block, and the *or* substructures are called **AlternativeCapabilitySets**. Each codec is represented by its number in the capability table.

For instance, a terminal could declare the following for its capability descriptors:

(1) (G.723 *or* g729) *and* T.120.

(2) G.711 *and* T.120 *and* (H.261 *or* H.263).

This would mean that the endpoint has a limited CPU and cannot support video compression (H.261 or H.263) simultaneously with audio compression (G.723 and G.729). If video is used, then only simple voice coders (G.711) can be used. In all cases, T.120 data sharing can be used.

This structure is also very useful for simultaneous presence video applications, where the capabilities structure can be used to indicate how many instances of the video decoder can be used simultaneously: the video codec is repeated in a SimultaneousCapabilities structure, (e.g., 'H263 and H263 and H263').

The terminals send this **terminalCapabilitySet** message to each other simultaneously (a common bug in early H.323 endpoint implementations was to wait for the other endpoint to send its capabilities before sending its own) and must acknowledge the reception of the other endpoint capabilities with a *terminalCapabilitySetAck* message.

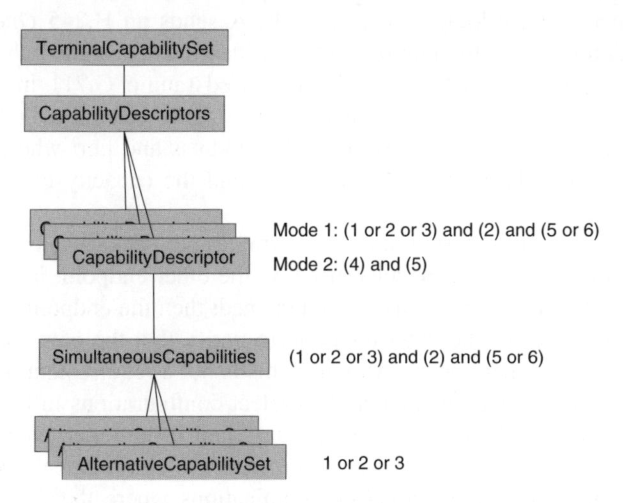


**Figure 2.6** TerminalCapabilitySet structure.

When troubleshooting audio problems on an H.323 network, the terminalCapabilitySet is one of the most useful messages to look at, in conjunction with the subsequent openLogicalChannel messages and the RTP streams. The problem is most likely a mismatch between the codec parameters (codec type, frame size) advertised by the terminalCapabilitySet, the parameters chosen by the OpenLogicalChannel, and the actual parameters streamed in the RTP flow, caused by a wrong parsing or use of the H.245 messages.

#### *2.2.1.2.2 Master/slave determination*

The notion of master and slave is useful when the same function or action can be performed by two terminals during a conversation and it is necessary to choose only one (e.g., when choosing the active MC on the opening of bidirectional channels). In H.235, the master is responsible for distributing the encryption keys for media channels to other terminals.

The determination of who will be the master is done by exchanging **masterSlaveDetermination** messages which contain a random number and a **terminalType** value reflecting the terminal category: multipoint control units, the H.323 name for a multimedia conferencing bridge (**MCU**); gatekeeper; gateway; simple endpoint. The terminalType values specified in H.323 prioritize MCUs over gatekeepers over gateways over terminals, and multipoint control (MC, multipoint conference-signaling control features)+multipoint processor (MP, media-mixing feature) capable units over MC-only units over units with no MC or MP.

### *2.2.1.3 Third phase: opening media channels*

Now terminal A and terminal B need to open media channels for voice, and possibly video and data. The digitized media data for these media channels will be carried in several '**logical channels**' which are unidirectional except in the case of T.120 data channels.

In order to open a voice-logical channel to B, A sends an H.245 *OpenLogicalChannel* message which contains the number that will identify that logical channel, and other parameters like the type of data that will be carried (audio G.711 in our example of Figure 2.7). In the case of sound or video, which will be carried over RTP, the *OpenLogicalChannel* message also mentions the UDP address and port where B should send RTCP receiver reports, the type of RTP payload, and the capacity to stop sending data during silences.

The codec type and configuration (number of frames per packet), *must* be selected from one of the supported configurations advertised by the other endpoint in its terminalCapabilitySet message. If prior channels have been opened, then the endpoint should check the SimultaneousCapabilities of the other endpoint to verify that the new coder is supported in conjunction with the other coders. Although this is not a requirement in the standard, it appears that most implementations attempt to select configurations in the order in which they appear in the CapabilitiesDescriptor structure, and if the other endpoint has already opened channels to this endpoint it also attempts to use symmetrical coders. This is in no way mandatory, and asymmeterical communications where the A to B and B to A streams use different coders are valid.

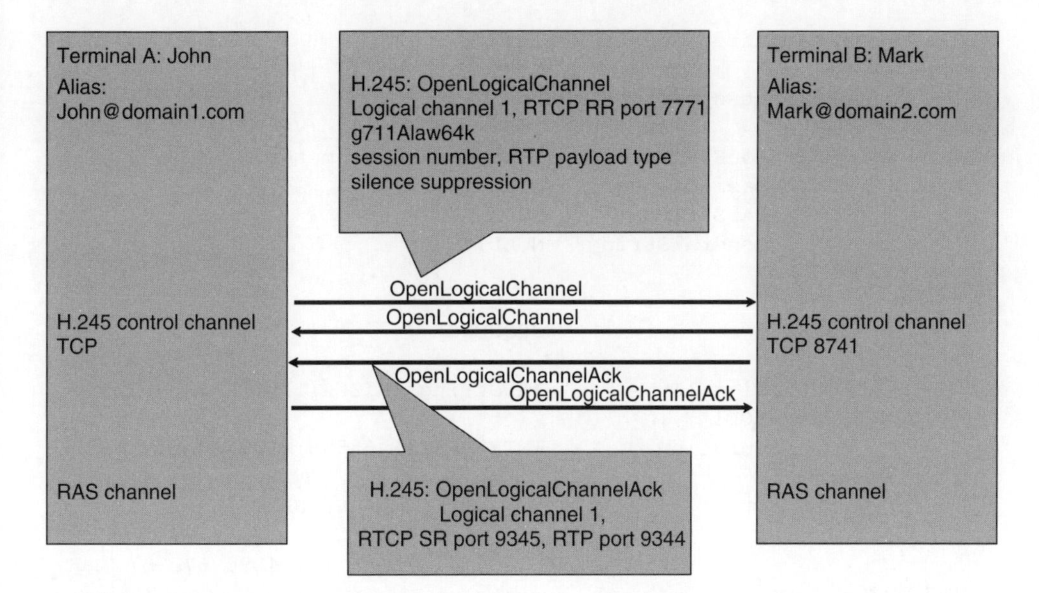


**Figure 2.7** Opening media channels using H.245 OpenMediaChannel messages.

B sends an *OpenLogicalChannelAck* for this logical channel as soon as it is ready to receive data from A. This message contains the IP address and UDP port number where A should send the RTP data and the UDP port where A should send RTCP sender reports.

Meanwhile, B also opens a logical channel to A following the same procedure.

### 2.2.1.4 Handling of DTMF tones

In H.323, there are several ways to transport DTMF tones:

- The special H.245 User Input Indication (UII) message, which must be supported by all H.323 systems. It has the advantage of using a reliable TCP connection, and therefore the message cannot be lost. But because TCP will try to retransmit the packet if it has been lost in the network, information might get delayed and get to the receiver too late. Two modes can be used: 'alphanumeric' and 'signal'. The most widely used mode is alphanumeric, this can be taken as the default in most gateways and H.323 phones. The UII message in this mode can carry all numeric characters, 'A', 'B', 'C', 'D', '*' and '#'. In H.323v2, the UserInputIndication message was updated to also include other information, such as the length and signal level of a tone, and synchronization information with the RTP stream: this is the signal mode. Here is an extract of the H.245 User Input Indication ASN.1 definition showing the added parameters:

```
UserInputIndication ::=CHOICE
{
        nonStandard NonStandardParameter,
```

```
        alphanumeric GeneralString,
        ...,
        userInputSupportIndication CHOICE
        {
                nonStandard        NonStandardParameter,
                basicString        NULL,
                iA5String          NULL,
                generalString      NULL,
        ...
        },
        signal                     SEQUENCE
        {
                signalType         IA5String (SIZE (1) ^ FROM
("0123456789#*ABCD!")),
                duration           INTEGER (1..65535) OPTIONAL,  --
milliseconds
                rtp                                 SEQUENCE
                {
                        timestamp        INTEGER (0..4294967295)
OPTIONAL,
                        expirationTime   INTEGER (0..4294967295)
OPTIONAL,
                        logicalChannelNumber   LogicalChannelNumber,
                        ...
                } OPTIONAL,
                ...
        },
        signalUpdate   SEQUENCE
{
                        duration         INTEGER (1..65535), --
milliseconds
                        rtp                                 SEQUENCE
                        {
                                logicalChannelNumber
LogicalChannelNumber,
                                ...
                        } OPTIONAL,
                        ...
        }
}
```

- The ISDN 'Keypad Facility' Information Element, which can be included in the SETUP or INFORMATION message. This is inherited from ISDN and used only in conjunction with the overlapped sending call flow (see note in Section 2.2.1.1).
- More recently (i.e., since H.323v4), H.323 can also use RFC 2833 for DTMF signaling (see Chapter 3 for details on RFC 2833); this requires H.245v7 and is an optional call flow. RFC 2833 can be used in conjunction with UserInputIndication (in this case the UserInputIndication message should have an **rtpPayloadIndication** flag). RFC 2833

also decodes the DTMF tone and includes it in a packet, but this time the packet is an RTP packet, not a signaling link packet. RFC 2833 can encode many telephony events (e.g., 'flash-hooks'), in addition to just DTMF tones. Note that RFC 2833 requires implementations to be able to send the telephony events to a destination that may not be the destination of the rest of the media streams. The inability to do so is a serious bug, as it prevents any DTMF-driven application from being built without accessing the media channel (e.g., a contact center). Unfortunately, this is a very frequent bug.

- A special RTP logical channel can be opened to carry the RTP DTMF payload. This payload is formatted as indicated in Figure 2.8. The unit used for the duration is the same as the unit used for the timestamp. If a separate logical channel is opened the sampling rate will be considered to be 8,000 Hz. For more details on RFC 2833, see Chapter 3. This method had been proposed for H.323v2 by the VoIP Forum. This was before H.245v7 was published. The original opening procedure was described as follows: It is possible to insert a DTMF RTP packet in the same logical channel as voice. In this case the payload type should be formed as follows to avoid confusion with dynamic or fixed RTP PT (these should be less than 128): 'chosen voice PT (e.g., 8)' 'DTMF PT' + 128. This PT should be used in the OpenLogicalChannel. If the remote terminal doesn't understand this meta-type, it means it doesn't support this method. This method should be scorned and replaced by the H.323v4 and H.245v7 procedure.

Overall, it seems that the User Input Indication method is preferred, since packet loss is typically very small for signaling links on well-engineered networks, and very few IVRs are sensitive to DTMF timing. In our various deployment experiences, this method always worked correctly. However, for international calls with large round trip times and time-sensitive IVR systems, it might prove necessary to use the second method. RFC 2833

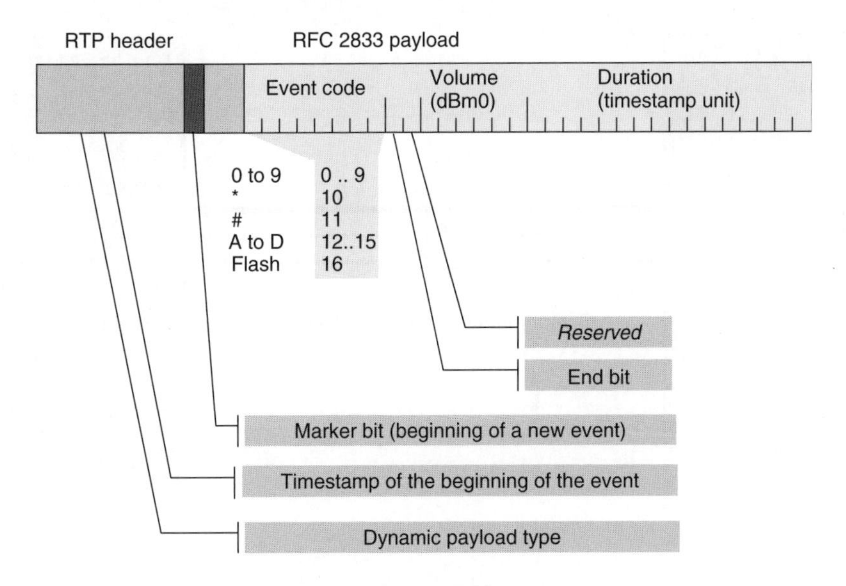


**Figure 2.8** RFC 2833 RTP packet format for DTMF transport.

implementations which are not capable of sending telephony events to an application server should be avoided. In addition, gateways should be extremely careful to mute in-band DTMF and convert it to an H.245 UserInputIndication or a special RTP payload type, since simultaneous transmission of in-Band DTMF and the special H.245 or RTP messages might cause the egress gateway to first render the RTP DTMF packet or H.245 UserInputIndication and then transmit the DTMF tone contained in the audio stream, duplicating the original tone.

### 2.2.1.5 Fourth phase: dialog

Now John and Mark can talk, and see each other if they have also opened video-logical channels. The media data are sent in RTP packets as shown in Figure 2.9.

RTCP receiver reports (RRs) enable each endpoint to measure the quality of service of the network: RTCP messages contain the fraction of packets that have been lost since the last RR, the cumulative packet loss, the inter-arrival jitter and the highest sequence number received. In theory, H.323 terminals should respond to increasing packet loss by reducing the sending rate, possibly by changing the audio coder dynamically . . . but, in practice, RTCP information is not used by most endpoints.

Note that H.323 mandates the use of only one RTP/RTCP port pair for each session. There can be three main sessions between H.323 terminals: the audio session (session id 1), the video session (session id 2) and the data session (session id 3), but nothing in the standard prevents a terminal from opening more sessions.

For each session there should be only one RTCP port used (i.e., if there are simultaneously RTP flows from A to B and from B to A, then the RTCP sender reports and receiver reports for both flows will use the same UDP port).

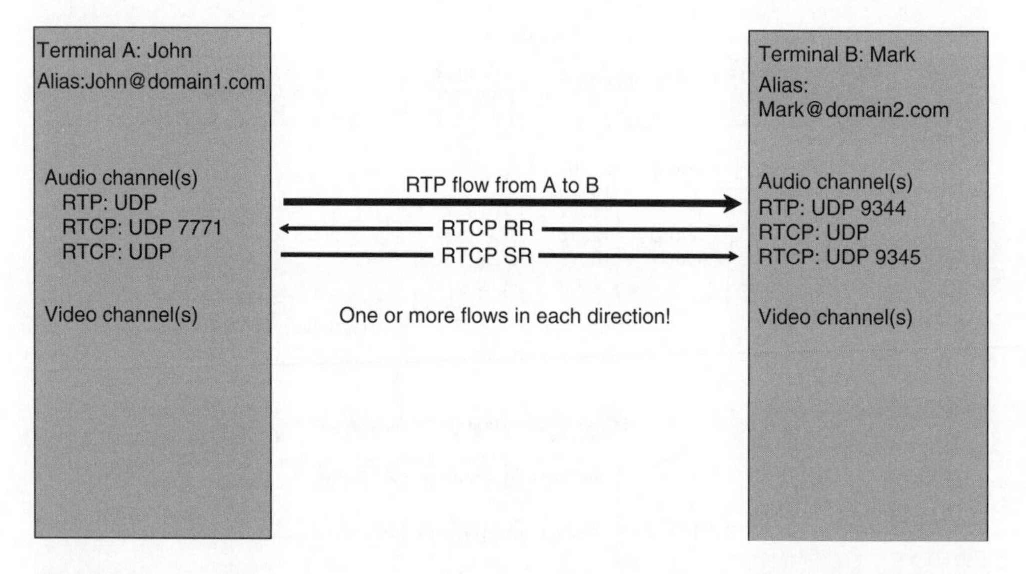


**Figure 2.9** RTP flows and associated RTCP receiver report and sender report messages.

#### 2.2.1.6 Clearing the call

How do we go about ending an H.323 call? Well, it is not that simple. If John hangs up, terminal A must send an H.245 **CloseLogicalChannel** message for each logical channel that A opened. B acknowledges those messages with a CloseLogicalChannelAck.

After all logical channels have been closed A sends an H.245 **endSessionCommand**, waits until it has received the same message from B (B will also close all its media channels before sending the endSessionCommand) and closes the H.245 control channel.

Finally, A and B must send an H.225 **ReleaseComplete** message over the call-signaling channel if it is still opened, then close this H.225 channel. The call is now cleared.

Needless to say, many software endpoints are not so polite, and terminate rather than close calls.

Note also that the call release sequence is different in ISDN. An ISDN endpoint releasing a call would first send a DISCONNECT message, which would be acknowledged at the other end by a RELEASE message, and the call would be over after the releasing endpoint sends a RELEASE COMPLETE. H.323 takes a short cut approach and sends only a RELEASE COMPLETE message.

In most cases this is fine, but this causes interoperability problems with the PSTN when the PSTN, instead of just releasing the call, wants to send an audio message to the caller first (this situation occurs frequently when calling mobile phones). An example of such an announcement is 'The party you are calling is currently not reachable on the network'. In this case the ISDN **DISCONNECT** message may contain a 'progress indicator' information element, with value 1 or 8 meaning that audio information is being provided to the caller. After a timer value of about 30 s or if the calling party hangs up before this timer, the calling party will send the **RELEASE** message to the network. In H.323, in order to respect the standard, a possible solution is to convert the DISCONNECT (progress indicator = 1 or 8) message into an H.323 PROGRESS (same **progress indicator** value) message in the PSTN gateway, with a special indication that this is really a release indication and the call should not be maintained longer than 30 s. This solution has been implemented in Cisco® gateways, for instance (see http://www.cisco.com/warp/public/788/voip/busytone.html for details). It is important to signal in the PROGRESS message that this is really a DISCONNECT, in order to provide the proper information to automated equipment that cannot interpret the in-band tone.

This progress indicator can also occur in other phases of the call, within the SETUP, ALERTING, CALL PROCEEDING, PROGRESS, CONNECT messages, but for these messages it does not cause problems because H.323 can transport it (these messages all exist in H.323). The only caveat is to make sure, when testing H.323 gateways, that the PI is properly supported. The PI can have the following values:

- Progress indicator = 1: the call is not end–end ISDN. Further call progress information may be available in-band.
- Progress indicator = 2: destination address is non-ISDN. This may be found in PROGRESS and CONNECT messages.
- Progress indicator = 3: origination address is non-ISDN. This is used in a SETUP ISDN message to signal that the calling party device is expecting in-band messages.

This is the case for most devices using analogue or CAS connections. A VoIP gateway receiving an ISDN SETUP with PI = 3 should provide in-band ring-back, as the calling device is unable to generate the ring-back locally.

- Progress indicator = 8: in-band information or an appropriate pattern is now available.
- No progress indicator in a message assumes that the originating device will provide the appropriate tone signaling to the calling party.

The commonly used progress indicators are '1' and '8'. A good test is to send a PROGRESS message in state 'alerting' with the progress indicator = 8. The originating gateway should play the in-band audio and stop ringing. If a new PROGRESS message is sent with no PI, since the state is 'alerting', local ring-back should be provided.

### 2.2.2 A more complex case: calling a public phone from the Internet using a gatekeeper

In the simple case described above, Mark called John directly on his current IP address 10.2.3.4. This situation is very convenient to show the basics of H.323, but very unlikely to happen in reality. If nothing else, a plain IP address is very hard to remember. In many cases it will even change—most ISPs allocate a dynamic IP address to their subscribers.

Our next example is more realistic: Mark now wants to call his grandmother, who only owns a regular phone and doesn't have the slightest idea of what an IP address is. This example will show the need for a new H.323 entity, called the **gatekeeper**.

The gatekeeper is the most complex component of the H.323 framework. It was first introduced in H.323v1, but at that time most people didn't really understand how useful it would be. At best, the gatekeeper was considered to be a sort of directory mapping friendly names to IP addresses. Some companies found alternative ways of doing this: some now-obsolete 'H.323-compliant' software and hardware used proprietary mechanisms ranging from IRC servers to LDAP servers to find the transport address of another VoIP phone or gateway.

H.323v2 has clarified the role of the gatekeeper, and now it is widely acknowledged that the gatekeeper is responsible for most network-based services (i.e., services which need to be performed independently of the terminal or when the terminal is turned off). These services include registration (the ability to know that someone has logged on and can be reached at a particular terminal, sometimes called 'presence'), admission (checking the right to access resources), and status (monitoring the availability of telephone-related network resources, such as gateways and terminals). Finding the transport address to use to reach a particular alias is naturally also part of the gatekeeper's role, since this transport address might depend on the status of the called party (e.g., if the person is not logged on, the call should be redirected to an answering machine or a regular phone through a gateway), the identity of the caller (not everybody might be allowed to call Mark, such as in the case of a do-not-disturb service), or the status of a particular resource (if all ports on a gateway are busy, then it might be necessary to use another gateway). Therefore, the gatekeeper is also in charge of routing all VoIP calls in the H.323 network, and the implementation of services like call forward on no answer.

The set of all H.323 endpoints, conference servers (MCUs) or gateways managed by a single gatekeeper is called a *zone*. In our example John's terminal and the gateway belong to the same zone.

In this Section 2.2.2 we consider that the caller has access to a gatekeeper, and show some of the gatekeeper features in action. The terminal and the gatekeeper use a specific protocol for registration, admission, and status purposes, which has logically been named RAS. This protocol is also defined in H.225.0.

### 2.2.2.1 Locating the gatekeeper

In simple configurations, the gatekeeper's IP address might simply be configured manually or automatically in the VoIP terminal. This is the most frequent case in real-life H.323 networks. This IP address is usually acquired when the VoIP endpoint boots: it first acquires an IP address and basic configuration parameters through the Dynamic Host Configuration Protocol (DHCP), one of the configuration parameters is the name of a TFTP server and a configuration file. The endpoint then downloads a configuration file using the TFTP protocol, which specifies, among other parameters, the address of a gatekeeper (and most of the time, a back-up gatekeeper). If such a configuration mechanism cannot be used or is not suitable, H.323 has developed a mechanism to dynamically find a gatekeeper on the network. This has a number of advantages (e.g., when someone has got a laptop and roams between several office locations). This mechanism also provides a way to introduce redundancy and load balancing between several gatekeepers in the network.

In order to find a gatekeeper, a H.323 terminal should send a multicast **Gatekeeper Request (GRQ)** to the group address 224.0.1.41 on UDP port 1718 (for more information on multicast, see companion book, *Beyond VoIP Protocols*). Within the GRQ message, it can specify whether it is willing to contact a particular gatekeeper. The terminal also mentions its aliases, allowing a gatekeeper to reply only to specific groups of terminals. Eventually, a GRQ can also be sent in unicast to port 1718, or preferably 1719, this is the default for unicast RAS messages, but obviously in this case the endpoint should know the possible gatekeeper IP addresses in advance.

The GRQ message should be sent with a very low TTL (time to live) initially in order to reach the gatekeepers on the local network first, and then use expanding ring search (Figure 2.10). This GRQ message tells the GK on what address and port the terminal expects to receive the answer, which type of terminal it is and what the terminal alias(es) is (are).

Each gatekeeper should be a member of group 224.0.1.41 and listen on port 1718. Therefore, one or more of these gatekeepers will reply on the address specified by the terminal with a **Gatekeeper Confirm (GCF)** message which indicates the name of the gatekeeper, and the unicast address and port that this gatekeeper uses for RAS messages. It can also include the names and transport information of other back-up gatekeepers.

The use of multicast for gatekeeper discovery has raised much controversy. In fact, not many IP networks support multicast today. Multicast routing is not activated by default on

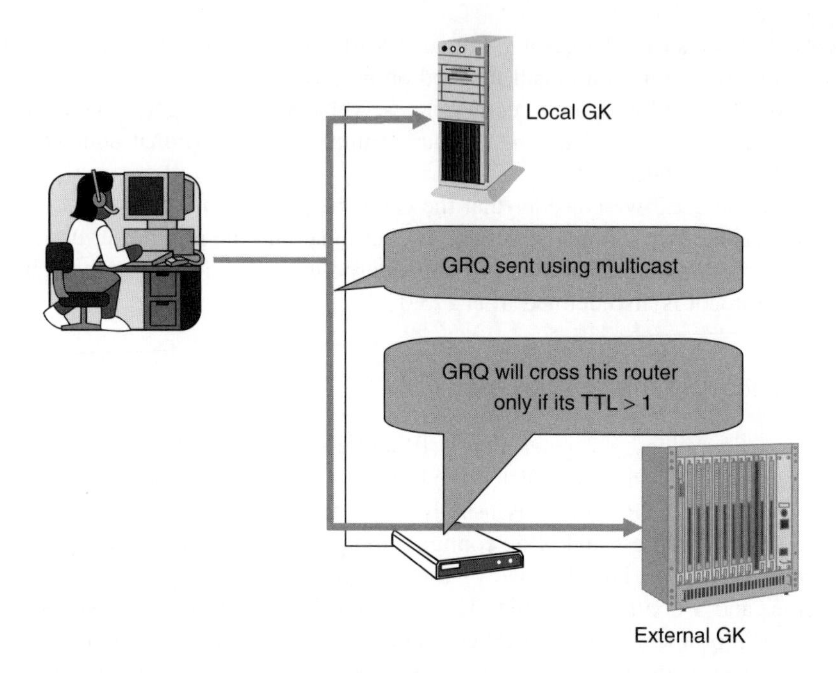


**Figure 2.10** Locating the gatekeeper using multicast GRQ messages.

routers, and many network administrators feel comfortable with static routes and are not really willing to experiment with a dynamic multicast routing protocol, such as DVMRP or PIM. Multicast discovery of the Gatekeeper is hardly ever used in practice, except perhaps in some private installations, e.g., videoconferencing inside a company.

#### *2.2.2.2 Registration*

If it received more than one answer from a gatekeeper in the discovery process, the terminal chooses one and registers this with the selected gatekeeper by sending a unicast **Registration Request (RRQ)** message (usually on UDP port 1719). This message carries an important piece of information: the transport address which is to be used for call signaling. The registration can be 'soft state' if the terminal so desires, in which case it also specifies a **time to live** for the registration and will refresh its registration periodically by sending more RRQs, also called **lightweight RRQs** or **keep-alive RRQs**. These RRQs have a special parameter 'keepAlive' set and do not include the full registration information.

The gatekeeper replies with a **Registration Confirm (RCF)** message in which the gatekeeper assigns a unique identifier to this terminal which must be copied in all subsequent RAS messages. The GK can also assign an alias to the requesting endpoint in this RCF.

Whether or not the terminal chose to use the 'keepAlive' registration, the gatekeeper can also request keepAlive RRQs by specifying a maximum time to live in its response.

Since the advent of H.323v4 it is also possible to use *additive registrations*, in order to register many aliases which would not fit in a single RRQ message (this can be used by IP-PBXs when registering many extensions to a core network). Such RRQs have a specific 'additiveRegistration' flag. They are also acknowledged by an RCF.

#### 2.2.2.3 Requesting permission to make a new call

Now that John's terminal has found a gatekeeper and is registered, John still needs to request a permission from the gatekeeper for each call he wants to make. In this case he wants to reach his grandma at +33 123456789.

His terminal will first send an **Admission Request (ARQ)** message to its gatekeeper. The ARQ message includes:

- A sequential number.
- The GK-assigned terminal identifier.
- The type of call (point to point).
- The call model that the terminal is willing to use (direct or gatekeeper-routed—see Section 2.2.2.2).
- The destination information (in this case the E.164 address +33 123456789 of grandma, but it could also have been Mark's email alias). Note that we used '+' to denote the country code, but this character is not transported in the ARQ message. In reality John would probably use the local dialing convention, and not a full number in international format.
- A Call Reference Value (CRV), which should be copied in the SETUP message.
- A globally unique CallID.
- An estimation of the bidirectional bandwidth that will be used for this call for media streams. This includes audio and video that will be sent from the called party and is measured excluding network overhead. This is a *very* rough estimation in most cases, since the codecs will be negotiated later. For instance, an audio-only terminal might indicate 128 kbit/s as a worst case if the two terminals negotiate a G.711 codec (64 kbit/s) for the incoming and outgoing audio logical channels. The endpoint may use **Bandwidth Request (BRQ)** messages later to ask for additional bandwidth (e.g., if it needs to open video channels).

The two possible call models refer to the way the call-signaling channel (carrying Q.931 messages) and the H.245 channel are set up between the endpoints. The calling endpoint can establish these channels directly with the called endpoint (the **direct mode**), or it can establish these channels with the gatekeeper which will relay the call-signaling and call control information to the called endpoint (there might be several gatekeepers routing the Q.931 and H.245 channels between the two endpoints). The later mode is the gatekeeper **routed mode**.

In this example we will use the direct model; we will discuss the GK-routed mode later. As we will see, the GK-routed mode is much more powerful and is the only model that works in carrier-class deployments.

If it decides to accept the call, the gatekeeper replies with an **AdmissionConfirm (ACF)** message which specifies:

- The call model to use (regardless of what was previously indicated by the calling endpoint).
- The transport address and port to use for Q.931 call signaling. This address can be the IP address of the called terminal directly (or the IP address of a gateway when calling a regular phone number) in the direct model, or it might be the gatekeeper itself if it decides to route the call. In our example the gatekeeper replies with the IP address of a gateway.
- The allowed bandwidth for the call.
- The GK can also request the terminal to send IRR (Information Request) messages from time to time to check whether the endpoint is still alive.

Note that this admission phase is really redundant if the gatekeeper wishes to use the routed mode, because the gatekeeper keeps full control of the call in routed mode. In H.323v2, the admission phase using ARQ/ACF messages can be skipped if the gatekeeper grants the **preGrantedARQ** right to the endpoint during the registration phase (see Section 2.3.6).

#### *2.2.2.4 Call signaling*

The Admission Confirm message has provided John's terminal with the information it needed to complete the call (Figure 2.11). Now the terminal can establish a call-signaling connection to the call-signaling address and port specified by the gatekeeper, in our case a gateway to the phone network, and send a Q.931 SETUP message. Before proceeding, the gateway may itself be required to ask the gatekeeper if it is authorized to place the call using an ARQ/ACF sequence. The ARQ will mention both the calling endpoint alias/call-signaling address and the called endpoint alias/call-signaling address, and a field indicating that this is an ARQ related to the termination of a call. In this receive side ARQ, the CRV (Call Reference Value) will be locally generated and, therefore, will differ from the CRV of the calling side ARQ. But the CallID should be copied from the SETUP message.

The gateway knows from the called party number information element of the H.323 SETUP message which phone number it must call. If it is connected to an ISDN phone line, it will simply send an ISDN Q.931 SETUP message on the D channel to initiate the connection on the ISDN. If it is connected to an analog line, it will go off-hook and dial the number using DTMF. If it is sending the call to an SS7 ISUP network, it will convert the SETUP message to an ISUP **Initial Address Message (IAM)**. Note that the

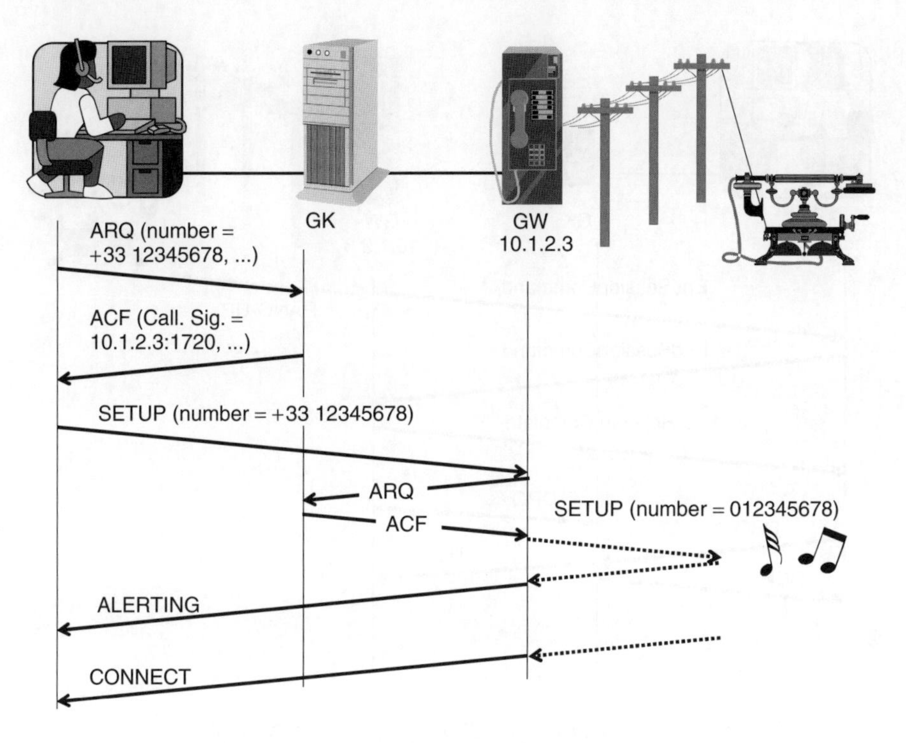


**Figure 2.11** Direct mode call authorized by the gatekeeper.

format of the phone number may need to be changed (e.g., the country code may need to be removed). In the direct mode, this needs to be done by the gateway, whereas in the routed mode this would typically be done by the gatekeeper, centralizing the numbering plan and routing management.

The gateway will send an H.225 ALERTING message to the caller as soon as it has received an indication from the phone network that Grandma's phone is ringing, and send the CONNECT message as soon as she has picked up the handset. If the gateway was connected through an ISDN line, these events will be signaled by the phone network using similar Q.931 ALERTING and CONNECT messages. If it is an analog line, the gateway needs to detect the appropriate ring/busy/connect conditions.

The ALERTING or CONNECT message contains a transport address to allow John's terminal to establish an H.245 control channel on which it can negotiate codecs and open media channels. This procedure is identical to the procedure used above when John was calling Mark. The media channels are then opened between the gateway and John's terminal as in the previous example.

### *2.2.2.5 Termination phase*

Whoever hangs up (e.g., the gateway in Figure 2.12) first needs to close its logical channels using the H.245 CloseLogicalChannel message. The gateway then sends an

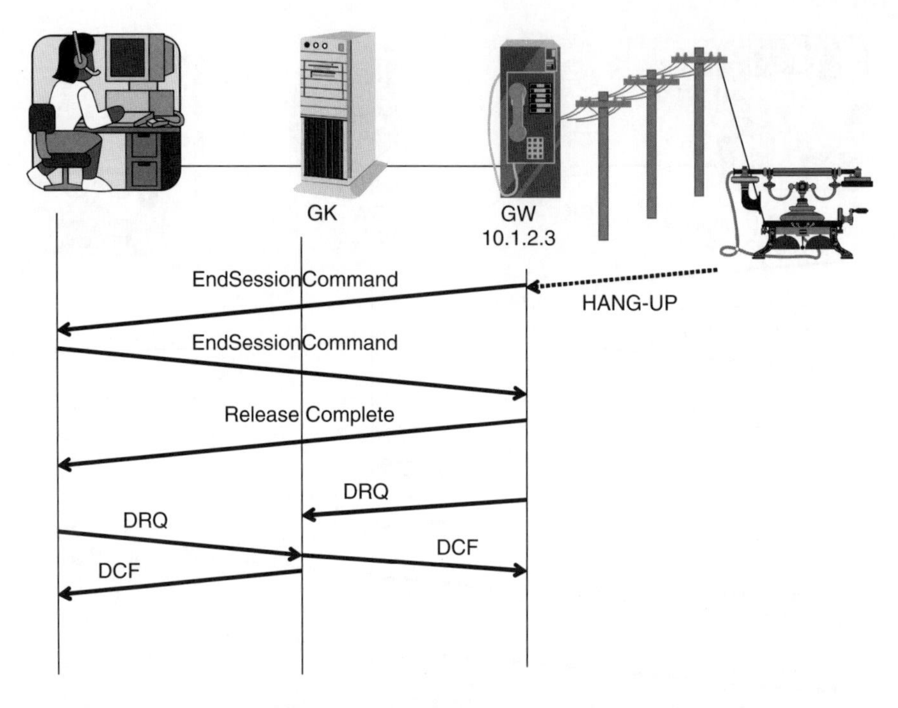


**Figure 2.12** Call released at H.245 and H.225 level (Q.931 and RAS).

H.245 **endSessionCommand** message to John's terminal and waits to receive the same message from John's terminal. The gateway then closes the H.245 channel. If the Q.931 channel is still open, each terminal must send a Q.931 ReleaseComplete message before closing it, then the terminal and the gateway must send a **Disengage Request** (**DRQ**) message to the gatekeeper, enabling the gatekeeper to know that the call and associated network resources have been released. The gatekeeper replies to each with a **Disengage Confirm** (DCF). At this stage, if the terminal or the gateway had been sending IRR messages to the gatekeeper, they must stop.

If there is more than one gatekeeper and the gateway and the terminal are registered to different gatekeepers, each one sends a DRQ to its own gatekeeper.

The terminal or the gateway have no reason to unregister (done by sending an **UnregistrationRequest** or **URQ** to the gatekeeper), unless, for instance, John decides to close his IP telephony software. A terminal should remain registered as long as it can make or receive calls.

If during the communication the gatekeeper wants to clear the call it can also send a DRQ to one or both endpoints. On receiving the DRQ the endpoint must send an H.245 endSessionCommand to the other endpoint, wait to receive an endSession command, close the Q.931 channel with a release complete, and send a DCF to the gatekeeper. Of course, this is dependent on the endpoint implementation and cannot be used reliably for applications like prepaid calls if at least one side of the call is not a trusted device. For such applications, if the endpoints are not fully trusted, the routed call model must be used.

In order to prevent a terminal from pretending it is closing a connection with a gateway without sending an endSessionCommand/release complete to the gateway, when a gatekeeper receives a DRQ from the terminal, it will wait until it has received a DRQ from the gateway before replying with a DCF. If the gatekeeper receives a DRQ from the gateway (as in our example), it will wait until it has received a DRQ from the terminal before sending a DCF to the gateway. In case the gatekeeper doesn't receive a DRQ from the terminal within a few seconds, it will ask the terminal to disconnect by sending a DRQ. The terminal is supposed to disconnect and send back a DCF, but if it doesn't within a few seconds (the PC might have crashed), the gatekeeper will send a DCF to the gateway anyhow. This procedure minimizes fraud and unwanted operation due to unstable or non-conformant terminals, but the routed mode is still more reliable.

In the direct mode, Call Detail Records must be generated by the gateways, the RAS information available at the gatekeeper level is not accurate enough for most purposes. For instance, it does not have access to the call release causes (a Q850 release code is provided by the network for each released call, specifying the reason for dropping the call: normal clearing, network congestion, user busy, switch failure, etc.). Also, the timing information is loosely coupled with the timing of the actual call release, and the start and stop information are available at different machines in the case of a network with multiple gatekeepers. This makes gatekeeper-level accounting records approximate at best. In real-life deployments with multi-gatekeeper networks, where unfortunately gateways or routers sometimes crash or show unexpected behaviour, the direct mode also makes it quite difficult to detect so-called 'zombie calls', calls that for some reason remain active in the network, out of control, for days or months.

All these issues are resolved by using the gatekeeper-routed mode.

### 2.2.3 The gatekeeper-routed model

Initially, virtually all gatekeeper implementations were using the direct call model. This model, where the gatekeeper is used really only as a sort of directory, seems very attractive at first glance:

- Very simple implementation, very few messages must be supported.
- The implementation can be made almost stateless if the accounting functions are external.
- The established calls are not affected if the gatekeeper fails.
- And, more importantly for marketing purposes, since the gatekeeper really does not do much, the manufacturer can claim the great performance figure of several hundred calls per second!

In fact the direct model has many shortcomings that do not allow VoIP networks to get to the same level of quality of service as traditional TDM networks. Direct mode is really acceptable only for simple enterprise networks.

### 2.2.3.1 Major issues of the direct mode

#### 2.2.3.1.1 Poor termination rates

In direct mode, the calling endpoint and the called endpoint communicate directly with one another, once the IP address of the called endpoint has been discovered. This is fine as long as the call succeeds. But if the first attempt to terminate the call fails (Figure 2.13), then the call is released.

A call attempt can fail for many reasons:

- Instability of gateways, resulting in their unavailability when the call arrives.
- Congestion of gateway resources.
- Congestion somewhere in the terminating PSTN network (as shown in Figure 2.13).

In the same situation, if a traditional TDM network had been used, then one of the class 4 central offices of the service provider in the path of the call would have detected the failure by analysing the **Q.850 release cause** included in the ISDN or SS7 release message. It would not have released the call on the calling side, but would have rerouted the terminating leg to other trunks. It is only in the unlikely situation where no trunk in the network can terminate the call that the call would have been released; and, even then, instead of just dropping the call, the call would have been routed to an announcement server explaining to the calling party that a temporary failure is occurring. Such a situation

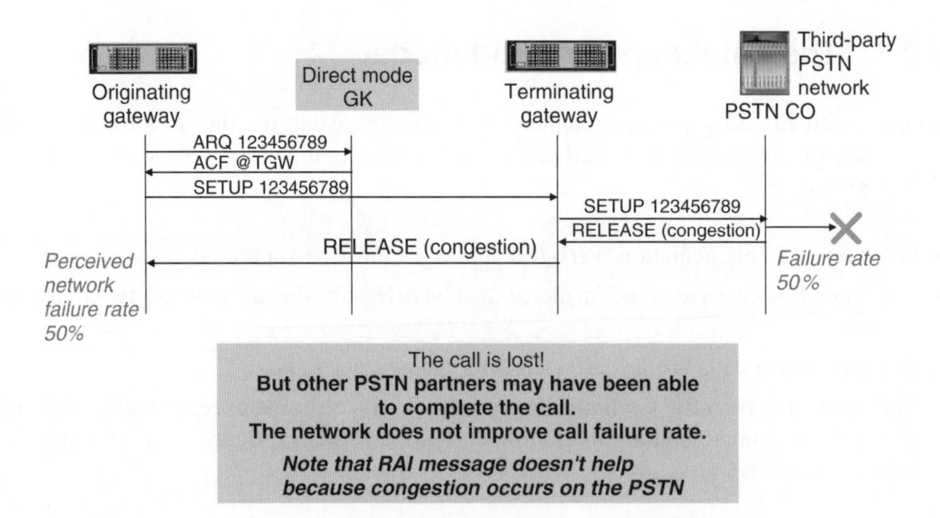


**Figure 2.13** Direct mode gatekeeper cannot improve call termination rates.

would also generate alarms at the service provider supervision center, and someone would verify the network dimensioning.

By comparison, the direct model in VoIP is not only very poor, it is in fact completely unacceptable as soon as some real traffic is carried. Many VoIP networks started by just providing low-quality prepaid termination, a market segment not particularly noted for its quality of service. But, as soon as the traffic started to diversify, many service providers were faced with complaints from users that the termination rate was poor. In fact, this poor termination rate quickly become a showstopper because it provided traffic termination for professional users, one of the most profitable segments of the market.

##### *2.2.3.1.2 Attempts to improve the direct model: Resource Availability Indicators (RAIs)*

Since the routed model is significantly more complex than the direct model, the initial response of the H.323 developer community to the poor performance of the direct mode was to attempt to avoid some of the causes for failed calls. For this purpose, the new RAI (Resource Availability Indicator) was introduced. The goal of this message was to let the gatekeeper know when a gateway was becoming congested. Above a certain threshold, the gateway will indicate to the gatekeeper that it is 'almost out of resources', and the direct mode gatekeeper is expected to divert traffic to other termination gateways.

This seems a good fix at first glance, but does it really solve the problem? Unfortunately, it doesn't:

- As we have just seen, most of the congestion situations occur in the PSTN, not locally at the gateway. For some destinations where the telephone network is not well developed the congestion rate can be as high as 50%! Also, some niche service providers specialized in low-cost termination have a poor quality of service. In order to save a termination fees, it is nice to be able to route traffic to them, but only if failures can be recovered by routing calls to alternative service providers in the event of a failure. Obviously, the RAI message only monitors resources at the gateway level and does not help for PSTN congestion.
- The RAI doesn't really help either for gateway congestion. Let's take two extreme situations: if the gateway average usage level is very low, say 50%, the RAI threshold level can be put very low (60%), despite obviously not needing the RAI to avoid gateway congestion. On the other hand, if the gateway usage rate is very high (a desirable situation given the cost of gateways), say 95%, then RAI on–off thresholds will be very high (e.g., 95% RAI 'OK' and 98% for RAI 'out of resources'). Unfortunately, a race situation occurs between RAI messages and the incoming calls from the PSTN. As each gateway has few T1/E1 ports, there will be an average of about two new call events and two call release events per second, when the difference between the two RAI thresholds represents only about four calls. This means that the RAI will continually change status, and the RAI status may be obsolete as soon as it is sent to the gatekeeper, if new calls arrive. Therefore, the RAI improves the situation only in networks where gateway usage is not above 80%, which is not very good from a capital utilization

perspective. If you have a low-cost service provider where you make a margin of a fraction of a cent a minute, and an alternative service provider where your margin may be negative, you really want the gateway to the low-cost service provider to be used at 100% capacity at all times!

- The RAI is an RAS message that is not routed if there are multiple gatekeepers; therefore, it is only useful at the last hop (last gatekeeper). But in many situations you would like rerouting to occur before the last hop.
- As a consequence of the previous limitation, the RAI doesn't work across administrative boundaries. If you are exchanging traffic with another VoIP service provider, it is almost certain that the other service provider will have its own gatekeeper, and you will not receive any RAI indication.

Less importantly, RAI is an H.323-only message with no SIP equivalent. If you deploy a mixed H.323/SIP network you will end up with a management of resources that is different between H.323 and SIP devices, which can quickly lead to some serious headaches; and, if you plan to migrate from H.323 to SIP, you will have to completely redesign network routing and congestion management.

If you have no other choice, you can use RAI when you can, but you should not expect major improvements of your network quality. RAI only works in marginal cases. As we will see in the coming paragraphs, the real solution to the issue is nothing new; it is the same solution as used on current TDM networks: full routing of the signaling messages by the switches (not the media streams in the case of VoIP), analysis of the Q.850 release codes which are also present in H.323 (and have SIP equivalents), and dynamic rerouting of calls.

#### *2.2.3.1.3 Centralized routing*

Although most gateways have some internal call-routing logic, using these capabilities quickly becomes very hard to manage as the number of gateways increases. A network of five gateways will need at least five routes to be configured on five gateways, a network of 100 gateways will need 100 routes on 100 gateways. Entering these 10,000 routes is a daunting task for a network manager.

Using a direct mode gatekeeper to control the routing of calls significantly simplifies the management process, but is still not ideal:

- Most gateway internal routing engines can fall back from one destination gateway to another in the case of congestion on other cause of call failure. This feature disappears when using the direct mode gatekeeper, possibly resulting in a reduced perceived quality of service by network users.
- Centralized routing really covers two tasks: selecting the proper destination, and changing the format of call aliases. A call initiated in San Jose, California to +1 212 xxx xxxx must be rewritten as a call to xxx xxxx if the destination gateway is in New York. Similar changes must be made to the calling party number. The direct mode gatekeeper

can manipulate the destination alias with the **CanMapAlias** feature of H.323, but very few gateway vendors support it. In addition, the source alias cannot be changed. As a consequence, it is fair to say that as soon as the service becomes complex, with multiple vendors, or requires manipulation of the calling party number (if the number presentation service is required), with the direct model the alias format management must remain distributed at gateway level (all gateways must convert local alias formats to/from an agreed network-wide 'pivot' format).

#### 2.2.3.1.4 Centralized accounting

Another frequent issue faced by service providers is the management of accounting information. In first-generation VoIP networks, the accounting information was generated by the edge gateways. It was either collected by batch processes by a central accounting function, or sent in real time by gateways using protocols, such as Radius.

While this works well for closed VoIP networks built from a single vendor, it becomes problematic if:

- The network is open to partners (clearing houses, termination partners, etc.) who do not provide access to their gateways.
- The network is open to customers (IP-PBXs, ASPs, etc.), who obviously cannot be trusted for billing information.
- The network uses multiple vendors, each having its own format for CDRs (Radius is only the transport protocol, the actual accounting information is always proprietary to each vendor).

A direct mode gatekeeper has only limited access to call information: it knows approximately the timing of the call start by using the ARQ messages and the timing of the call stop through the DRQ message. It does not know the call release causes (Q.850). Obviously, if the network involves multiple direct mode gatekeepers, this model also becomes complex because part of the RAS information is provided to different gatekeepers. It also does not work if the edge devices cannot be trusted (they could potentially send DRQ messages while continuing a conversation). These limitations do not allow the direct mode gatekeeper to be a reliable device to generate accounting records centrally in a network.

#### 2.2.3.1.5 Security issues

The last issue of the direct mode in an open network relates to **security**. Since the direct mode gatekeeper lets endpoints exchange signaling directly, any endpoint on the network can learn the IP addresses of other devices (this in itself is not a security problem), but more importantly can send signaling at any moment to any endpoint. This makes denial-of-service attacks trivial. Because of this, VoIP networks using direct mode gatekeepers cannot be opened up to third parties. They cannot be used to connect IP-PBXs and cannot send traffic directly to other VoIP networks.

### 2.2.3.2 *The gatekeeper-routed model*

A gatekeeper using the routed model handles all call-signaling information and does not let endpoints establish calls directly. Some gatekeepers can be configured to use the routed model or the direct model on a per-route basis.

The routed model is exactly identical to the way traditional TDM switches handle phone calls, with one exception: when using the routed model, the media streams are still exchanged directly by endpoints. The routed model provides all the advantages of full class 4 routing (ability to analyse release causes, reroute calls, better security), while still not requiring dedicated telecom hardware since no TDM switching matrix is required. Because of this the density- and hardware-related cost of softswitches is far better than their TDM counterparts.

All the issues described above for the direct mode are solved:

- Congestion, whether at the gateway level or anywhere in the PSTN network, is detected by analysing the Q.850 release cause. The call can be dynamically rerouted to other termination routes (Figure 2.14). This works regardless of the number of softswitches and across administrative boundaries (clearing houses or terminating VoIP partners can be used). Since the calls are rerouted dynamically in the event of congestion, the least costly routes can be used at 100% capacity without affecting the perceived quality of service of the network. With a routed mode gatekeeper, the failure rate perceived by call sources is equal to the product of the failure rates of all termination routes for a given destination. If the network has two partners each experiencing a 50% failure rate to a country, the perceived failure rate seen by service provider customers is only 25% (compared with 50% in the direct model). This drops to a 13% perceived call loss with three

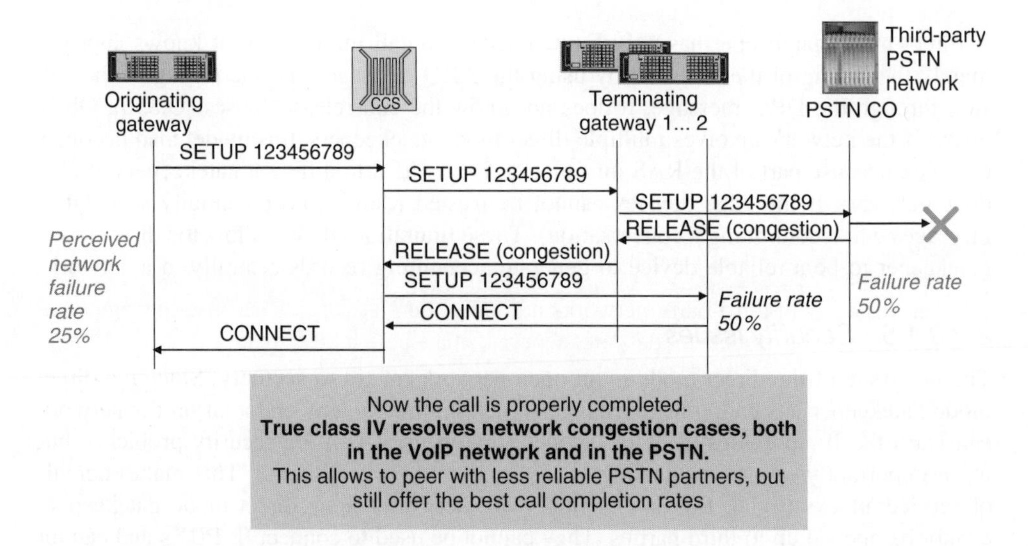


**Figure 2.14** The gatekeeper can interpret Q.850 release causes and redirect the call as appropriate on the fly.

partners each losing half of the calls. If the routed mode gatekeeper has the least costly routing features, a low-cost partner route losing 20% of the calls can be used at 100% capacity, while a high-cost partner losing only one call in a thousand can be used only in the event a call is dropped by the low-cost partner. This optimizes costs, while still providing a perceived call failure rate of less than one in a thousand to service provider customers. Note that with this model gateways do not need to support the RAI feature. In fact, the RAI message becomes completely useless with a routed mode gatekeeper.

- If calls cannot be completed due to congestion or any other reasons, they can be routed to a network announcement server (simply defined as the last-resort route for all destinations), terminating calls gracefully rather than just dropping them.
- Centralized routing now handles properly not only the selection of the proper destination, but also the conversion of alias formats. The gateways only need to support the basic H.323 call flow, with no local logic for routing or the manipulation of call aliases. Everything is provisioned centrally in the routed mode gatekeeper-routing engine. Since both the source and the destination alias can be manipulated, the calling line ID features can be provided. The routed mode gatekeeper has complete access to the alias information, which also contains the caller ID blocking status (**Q.931 octet 3A**): it can provide caller ID blocking for certain routes (e.g., international routes to ensure privacy), and caller ID forced delivery for emergency calls. The routed mode also enables more sophisticated features (e.g., virtual private networks) if the gatekeeper can translate between private and public numbering plans. This does not require any capability at the endpoints besides support for an H.323 basic call and can be provided to any endpoint, including IP phones or IP-PBXs.
- Centralized accounting information can be provided by the routed mode gatekeeper. The gatekeeper now has access to all signaling information including call release causes. Gateway-level accounting features can be disabled. The endpoints do not need to be trusted, as the gatekeeper can provide reliable accounting for IP-PBXs or simple IP phones. This enables service providers to provide VoIP business trunking services, replacing traditional E1/T1 lines connected to PBXs with VoIP-enabled broadband connections. With such a service, IP-PBXs do not need a local PSTN gateway in the customer premises: the service provider routed mode gatekeeper is defined as the default route and appears as a regular gateway to the IP-PBX. The only requirement is that the IP-PBX should support H.323 connections toward the public network, but this is the case of most IP-PBXs on the market today.
- Connectivity with third-party networks and customers is secured because the signaling is relayed by the routed mode gatekeeper. It may be useful to use a dedicated gatekeeper for connections with third parties. If it is attacked, the worst that can happen is that connectivity with those partners may be lost, but the rest of the network is not compromised. Note that media streams (RTP) can still flow directly between partners. With proper access lists on edge routers (RTP filters, UDP ports above 1024 only, anti-spoofing filters), this is secure. Some firewall vendors recommend relaying media streams on dedicated devices in core networks; this is very costly, degrades quality of service (added delays), and affects IP network design (tromboning is introduced). These techniques should be reserved for very specific situations (e.g., clearing houses wanting

to hide the identity of their partners, or when there are incompatible IP-addressing plans that need to be converted).

Besides resolving all the issues that cannot be addressed with direct mode gatekeepers, routed mode gatekeepers offer many more possibilities. For instance, they can act as multiprotocol softswitches acting both as an H.323 routed mode gatekeeper and as a SIP proxy with access to enough information to convert between signaling protocols (e.g., H.323 and SIP). Note that this requires SIP to support true out-of-band DTMF signaling through **INFO** or **NOTIFY** messages (major SIP gateway vendors already support these messages).

## 2.2.4 H.323 calls across multiple zones or administrative domains

As the initial title of H.323 implied, the first version of H.323 did not consider issues that would occur in a wide area environment. It was more or less assumed that the gatekeeper would get a complete view of the network and would be controlling all endpoints and gateways. In this context there was not much effort spent on defining the call flows to be used if the network was controlled by multiple gatekeepers or if multiple VoIP networks were connected. The VoIP industry had to solve this issue very quickly because real VoIP networks required multiple gatekeepers to scale . . . and soon a de facto inter-gatekeeper call flow emerged. Without much prior debates in standard bodies.

As we discussed in Section 2.2.3.2, most VoIP networks today still use direct mode gatekeepers. In order to be compatible with direct mode gatekeepers, the de facto inter-gatekeeper call flow uses the **Location Request (LRQ)** RAS message. It is probably not the optimal choice: using the ARQ/DRQ message would have facilitated the correlation between the RAS messages exchanged between direct mode gatekeepers and the Q.931 messages exchanged between endpoints (SETUP, CONNECT, etc.). But this call flow is so widely deployed today that the usage of the LRQ message is not likely to change. What is happening instead is that most vendors are adding (in a proprietary way, within the H.323 extension tokens) the information that is missing in LRQ messages, notably the call identifier: all messages used in H.323 can easily be extended.

Between routed mode gatekeepers, the most efficient call flow is to simply forward Q.931 messages between gatekeepers. This is identical to the call flow used between class 4 central offices in the TDM networks. RAS messages are unnecessary, but can be used if desired: some routed mode gatekeepers will send a Q.931 SETUP message directly to the next hop gatekeepers (this assumes the prior knowledge that the next hop gatekeeper is also a routed mode gatekeeper), and some will begin by sending an LRQ message, in case the next hop gatekeeper is using direct mode only and cannot handle Q.931.

### *2.2.4.1 Direct call model*

#### *2.2.4.1.1 Call set-up*

In the direct call model, only RAS messages are routed by the gatekeepers. Now, John wants to call his grandma using a gateway managed by a service provider, Cybercall. The service provider has its own gatekeepers. Therefore, John's terminal and the gateway will be located in different zones. John's terminal will register to his own gatekeeper and the gateway will be registered to the service provider's gatekeeper.

When John became a customer of Cybercall, his gatekeeper IP address was configured in the routing tables of Cybercall's gatekeeper, and vice versa. Therefore, these gatekeepers know about each other. Security is usually based on identifying the IP addresses of both gatekeepers, but can be enhanced by adding security tokens in LRQ messages.

The admission request is sent by John's terminal to the gatekeeper to which it has registered (Figure 2.15). This gatekeeper knows that all calls to the PSTN are handled by Cybercall. Therefore, it sends a Location Request (LRQ) to the gatekeeper of Cybercall, the LRQ message queries the Cybercall gatekeeper for a next hop IP address where the Q.931 signaling can be sent for a specific destination. Because the LRQ comes from a gatekeeper that is known, and assuming John is authorized to make the call, Cybercall's gatekeeper will accept it and returns a **Location Confirm (LCF)** to John's gatekeeper. The LCF message contains the IP address of the gateway where John's terminal should send the SETUP message. John's gatekeeper has still not replied to the initial ARQ, because it did not have enough information to do so. Now, with the IP address contained in the LCF, the gatekeeper knows where the call should be routed and sends this information to John's terminal in an ACF. If this is taking too long, the gatekeeper can send **Request In Progress (RIP)** messages to John's terminal to prevent any timeout that could cause John's terminal to reject the call or resend an ARQ.

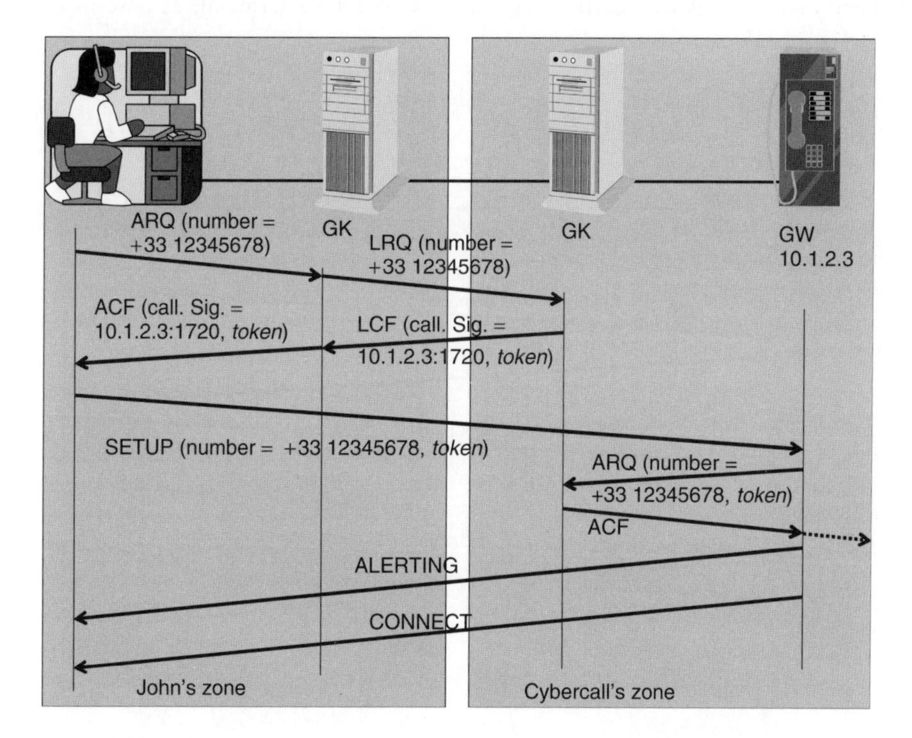


**Figure 2.15** Direct call model across two domains, using LRQ/LCF messages.

Cybercall can also include a **token** in the LCF. A token is an optional parameter that consists of a 'bag of bits'. Unless it knows of this specific token, an H.323 entity should simply pass it along transparently. Here, the token serves as a secret which will be copied by John's terminal in the SETUP message. Cybercall's gatekeeper has put in this token a digital signature of some important aspects of the call, such as the destination and the current time. When it receives a SETUP message including this token, the gateway can now verify that the call has been previously authorized by the gatekeeper. However, Cybercall, in order to centralize security management, has not given the gateway enough information to decode and verify the token locally. The gateway will simply pass this token to the gatekeeper in the receive side ARQ, Cybercall's gatekeeper will check it, and return an ACF if the token is correct. Otherwise, the call would be rejected with an ARJ (Admission Reject) message, and the gateway would release the call with a Q.931 RELEASE COMPLETE message.

When it receives the ACF, the gateway will set up the call on the PSTN side and send a CONNECT message to John as soon as Grandma picks up the phone.

John then establishes the H.245 control channel to the gateway using the address and port specified by the gateway in the CONNECT message. Then, logical channels are established using OpenLogicalChannel messages, and John can talk.

#### *2.2.4.1.2 Call tear-down*

This time (Figure 2.16) if Grandma hangs up first, the gateway will send an EndSession-Command and RELEASE COMPLETE message to John's terminal, as described in the

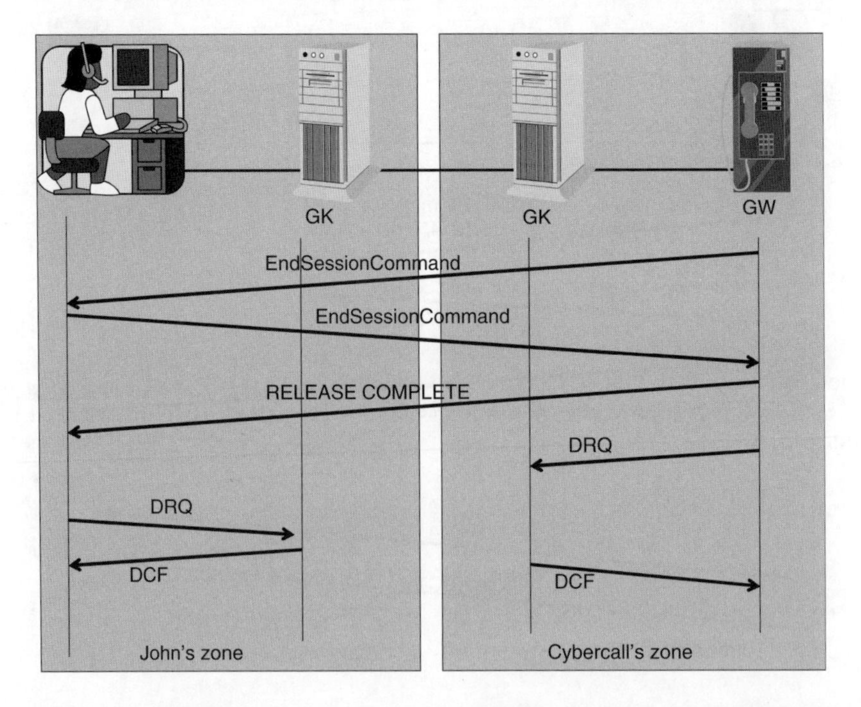


**Figure 2.16** Call released end to end at H.245 and Q.931 level, locally at RAS level.

first H.323 examples. Then, the gateway sends a DRQ message to Cybercall's gatekeeper, and John's terminal sends a DRQ message to its own gatekeeper. Note that because the LRQ message exchanged between the two direct mode gatekeepers is a stateless query message, no message is exchanged between the direct mode gatekeepers when the call is released. This illustrates the problem arising from the use of the LRQ message, as opposed to the ARQ and DRQ, for inter-gatekeeper communications.

#### 2.2.4.2 Gatekeeper-routed model

There are many reasons that Cybercall would like to have finer control over John's communication. With the direct model, Cybercall doesn't know what occurs during the call (e.g., if grandma's phone is busy Cybercall's gatekeeper will see it simply as a very short call). This forces Cybercall to do all accounting at the gateway level, which may be a pain if the Cybercall domain has several dozens of gateways. Cybercall may also want to protect its domain and prevent John from potentially initiating denial-of-service attacks on the gateways; signaling ports. This is impossible to do using the direct model.

These are just a few of the reasons the gatekeeper-routed model—or a mixture of direct and gatekeeper-routed model—will be preferred in most situations where the network involves several administrative domains.

##### *2.2.4.2.1 Call set-up*

In the example shown in Figure 2.17, Cybercall's gatekeeper decides to route the call by putting its own IP address (10.1.2.2) in the LCF call-signaling address (as we saw in Section 2.2.2.3, this call flow can be optimized using the preGrantedARQ procedure). John's gatekeeper also decides to route the call by putting its own IP address in the ACF call-signaling address. But John's gatekeeper could also have used the direct model by copying the call-signaling address provided in Cybercall's LCF in its own ACF: in this case John's terminal would have sent the set-up message directly to Cybercall's gatekeeper, this would be a call using a mixed model. If John's gatekeeper knows in advance that Cybercall is always using the routed model, then the LRQ is unnecessary and a direct SETUP can be sent to Cybercall's gatekeeper IP address.

You probably remember that one of the most important information elements of the ALERTING or CONNECT message is the H.245 call control channel address that John's terminal must use to establish the call control channel. Here, the H.245 channel will also be routed because both Cybercall's and John's gatekeepers have put their own IP addresses in the call control transport address field of the ALERTING message. It is also possible to route the Q.931 messages but let the H.245 control channel be established directly between the endpoints.

What about the media channels? They could be routed too, but there would be very little to gain from doing so, since all the significant events of the call are signaled using H.245 or Q.931 messages.[2] But unless there is a very specific need to do so, *media channels*

[2] An exception could be fax, because the entire T.30 protocol is encapsulated in a media channel; therefore, the gatekeeper needs to have access to the media channel to know how many pages have been transferred.

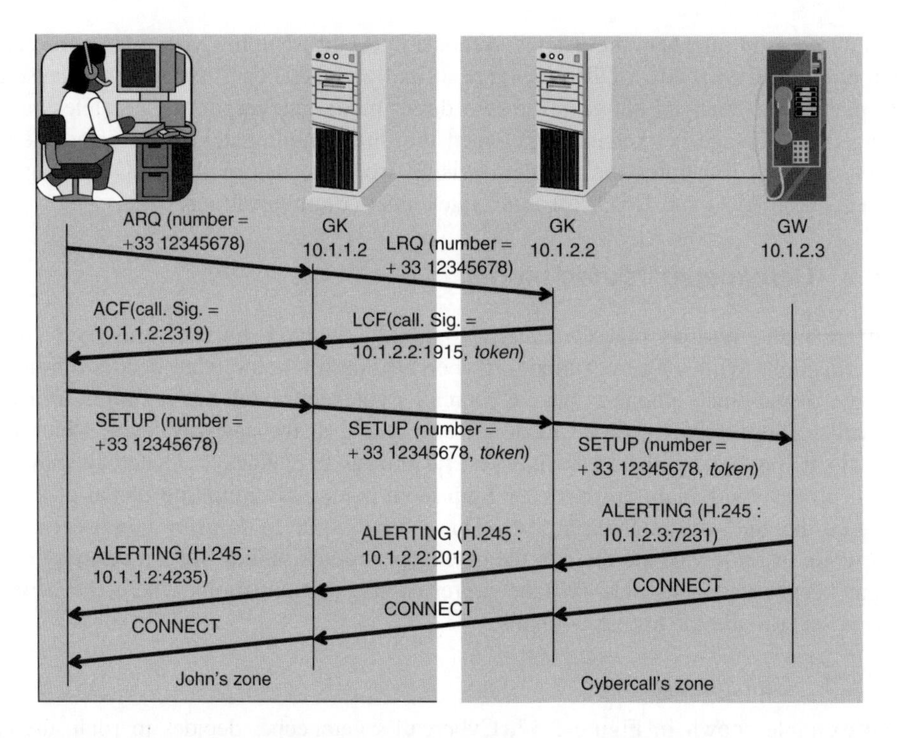


**Figure 2.17** Routed call model across two domains.

*flow directly between the endpoints*, even if the gatekeeper-routed model is used. Doing otherwise and routing the media channels would remove most of the scalability benefits of VoIP over TDM.

By letting media streams flow directly between endpoints, the media latency is optimized, even if call-signaling has to go through many gatekeepers, because IP shortest path routing protocols will be used to route RTP packets. This gives VoIP a very interesting 'location-independent' property, which allows customers to be served from remote gatekeepers, thereby reducing the number of points of presence required to offer the service. The 'Voice for IP VPN' service from service provider Equant, which serves over a hundred multinational companies over a VoIP network, operates from only two VoIP gatekeepers, one located in the US, one in Europe.

Some service providers are concerned about security issues that could occur using the RTP stream. Although most VoIP networks worldwide let RTP flow through transparently, we have never heard of such problems. In order to secure such a VoIP network the following protection should be configured:

- **Access Control Lists (ACLs)** on edge routers should allow VoIP signaling information only toward the routed mode gatekeeper.

- Other ACLs should allow only UDP RTP traffic to ports higher than 1024[3] (checking the proper RTP patterns in UDP packets is possible on most routers) and only to VoIP endpoints (the easiest is to allocate well-defined subnets to the gateways).
- The only possible attack is denial of service, because RTP doesn't have much logic in it! Gateways are expecting a lot of traffic on RTP ports, so bringing them down with RTP traffic requires significant bandwidth, making the attack detectable. Furthermore, gateways will accept the RTP traffic only if the logical channel has been opened properly by the routed mode gatekeeper... in this case the identity of the attacker is known, which acts as a determent to such attacks. The last remaining possibility is a DoS attack on closed UDP ports, causing the gateway to reply with ICMP IP-level error messages. Filtering these can give an early warning of the attack; once again, the attack would require significant bandwidth as sending back an ICMP message is not CPU-intensive on the gateway.
- As in any IP network, anti-spoofing (preventing anyone from injecting in the network packets with the source IP address belonging to someone else) should be taken very seriously, as it is the only real protection against DoS attacks.
- Finally, because we are only expecting RTP traffic and know what bandwidth to expect, if per-flow traffic policing is available on the edge routers, it should be used. DoS attacks will exceed the allowed bandwidth and be rejected by the edge routers.

If you still want to relay media streams, devices exist that do just this at the edge of a network ('border session controllers'). But, by forcing RTP packets to go through these devices without care, you may significantly reduce the QoS of the VoIP network (e.g., if the device is in New York, a San Francisco to San Jose call may have its streams relayed through New York, instead of flowing directly between the two cities using IP shortest path routes).

#### 2.2.4.2.2 Call tear-down

The call tear-down is very similar to the direct model case, except of course that Q.931 messages and optionally H.245 messages are routed through the gatekeepers.

#### 2.2.4.2.3 More LRQ usage scenarios

When a gatekeeper is used at the interface between two administrative domains, LRQ call flows can be more complex. Gatekeepers at the edge of a domain need to manage:

- Multiple simultaneous LRQ targets.
- The sequencing of LRQ and Q.931 messages.

---

[3] Blocking ports below 1024 makes unreachable most applicative ports that could potentially be opened, and subject to attack. VoIP applications use ports higher than 1024.

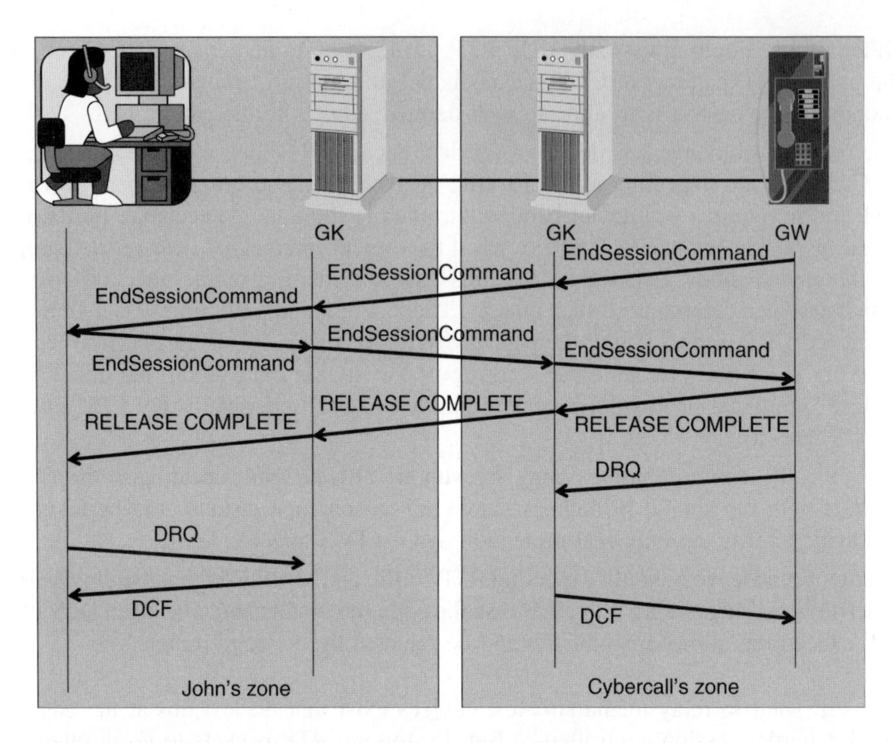


**Figure 2.18** Call release scenario, under the routed model.

#### *2.2.4.2.3.1 LRQ blast*

If a call to a certain destination can be sent to multiple termination networks, each with its own gatekeeper, it may be interesting to check the availability or willingness to accept the call of all partners. In order to do this, multiple LRQ messages can be sent (simultaneously, or in sequence), to all these potential termination networks. This is sometimes called an **LRQ blast**. Among the LCFs received, one will be selected by the source gatekeeper.

Note that it is tempting to do the same with SETUP messages (some SIP vendors do this with the INVITE message[4]), but only the sending of multiple SETUPs in *sequence* is allowed. A call establishment message should never be duplicated. This is because the PSTN network can send announcements before CONNECT (200 OK in SIP). If multiple calls receive network announcements, the softswitch would be unable to properly relay them to the caller.

[4] When using SIP, duplicating INVITE messages should be allowed *only* if the expected answer is a redirect message or the expected media is not voice. Some vendors use it for a 'simultaneous ringing' feature . . . although this is a cool demo, it simply does not work on real telephony networks. For this reason the 3GPP Group defining the UTMS 3G standard has decided not to use SIP forking for now.

#### 2.2.4.2.3.2 *Proper LRQ sequencing*

When an edge routed-mode gatekeeper at the interface between several administrative domains receives an LRQ, it can choose between the following call flows:

(1) Reply immediately with an LCF, then receive the SETUP message from the calling device, then send an LRQ message to the potential termination zones, then forward the SETUP to the selected termination device.
(2) Forward the LRQ to the potential termination zones, wait for the LCF for the termination softswitches that will accept routing the call, and only then reply with an LCF with its own IP address. When the edge gatekeeper receives the SETUP, it routes it to the selected softswitch.

Both call flows seem equivalent, but they are not when *all* potential termination gatekeepers reject the call.

In call flow (1), the SETUP has already been routed to the edge gatekeeper, so the edge softswitch is responsible for finding a fallback route for the call.

In call flow (2), the edge gatekeeper can reject the call by sending back an LRJ (Location Reject) to the calling party gatekeeper. This gives the possibility of rerouting the call to the initiator.

Call flow (1) reduces call latency, but may not be appropriate if a service provider connected to the edge gatekeeper wants to keep the possibility of rerouting calls. This is the case with most clearing houses.

Call flow (2) solves this issue, but introduces more latency in the call.[5]

In any case, both call flows are useful, and a gatekeeper used as an edge device should offer the possibility of choosing between the two modes for each route.

### 2.2.4.2.4 *Some issues with the LRQ message*

The calls flows for interdomain calls really should have used the ARQ message, not the LRQ between gatekeepers. Unfortunately, the first implementation of the call flow by Cisco Systems used the LRQ, and then the whole industry followed. There are mainly two information elements which are missing in the LRQ message that would really be useful: a call reference identifier and a hop counter.

#### 2.2.4.2.4.1 *The missing call reference identifier*

The LRQ misses a unique call reference identifier, typically the CallID. This is the main difference between an LRQ and an ARQ. The absence of this unique call reference number makes it impossible to correlate the LRQ and the subsequent SETUP message. There are many cases where this correlation would be useful. For instance, when a routed

[5] In addition, because the LRQ/LCF is stateless (no resource is reserved when replying with an LCF), it should also reinitiate an LRQ when the SETUP arrives. This doubles the number of LRQs. LCFs could be cached for a short time, but this is a violation of the standard.

mode gatekeeper is used as an edge element between multiple domains (clearing house function), then the owner of the clearing house would like to be able to easily identify each connected domain in the CDRs generated by the gatekeeper. The CDRs generated from the SETUP messages will include the source IP address of the device that initiated the SETUP to the edge gatekeeper and destination IP addresses of the SETUP message sent from the edge gatekeeper. If each connected domain also uses a routed mode gatekeeper, then these IP addresses will be the IP addresses of each gatekeeper, and they allow easy identification of the administrative domains. But many VoIP service providers still use the direct model. In this case the IP addresses will be those of the PSTN gateways. It is very time-consuming to keep track of all these IP addresses and correlate them to a service provider. Since the direct mode gatekeeper of each service provider will send an LRQ before each call, it would be nice to include the LRQ source IP address in the CDRs. Unfortunately, because the LRQ cannot be correlated to the SETUP message, this is impossible. Another case where the presence of a call identifier in an LRQ would be useful was described in Section 2.2.4.2.3.2. If the edge gatekeeper is required to completely proxy the LRQ message before accepting the SETUP message, then two LRQ messages will be generated for each call, because the LRQ is stateless. Correlating the LRQ to a specific call would make it easier to keep the LCF response of edge domains in cache, knowing that these edge domains can now reserve resources for the coming call.

###### *2.2.4.2.4.2 The missing hop counter. Discussion of call loops*

The second element that would be useful in an LRQ message is a hop counter, to prevent loops in the VoIP domain. Note, however, that this is nothing more than a useful tool, because it would still be possible to loop calls using SETUP messages without RAS and also because call loops can include a PSTN hop that would reset the counter. The only way to completely prevent loops in VoIP networks is to not only include a counter in LRQ but also SETUP messages, and to take into account the SS7 ISUP hop counter if there is a PSTN hop. This is possible if the edge gateways support the H.246 encapsulation of ISUP information, or H.323 annex M2, or H.323v5, which adds such a hop counter to standard SETUP messages. Cisco Systems also proposed a mechanism called **Global Transparency Descriptor (GTD)**, where ISUP national information elements are passed and stored in a uniform way within a data structure in H.323 Q.931 messages. GTD is much more powerful than H.246 (or its SIP equivalent SIP-T) because it proposes a uniform coding of the ISUP information, as opposed to transporting national ISUP 'flavors' as is. If the proposal becomes a standard it will certainly be the best way to address the loop problem, among many other interworking issues. Even with these improvements, call loops remain possible if edge devices connected through user interfaces (analog, ISDN) are allowed to loop calls back to the network, because in this case the hop counter is reset. This is one of the reasons the call forward of external calls back to the PSTN is usually forbidden as part of the certification program of edge devices.

## 2.3 OPTIMIZING AND ENHANCING H.323

### 2.3.1 Issues in H.323v1

#### *2.3.1.1 Call set-up time*

One of the major weaknesses of H.323 in its version 1 was the time required to actually establish the media channels for a new call. Even in the simple cases, H.323v1 procedures involve:

- One message round trip for the ARQ/ACF sequence.
- One message round trip for the SETUP-CONNECT sequence.
- One message round trip for the H.245 capabilities exchange.
- One message round trip for the H.245 master slave procedure.
- One message round trip for the set-up of each logical channel.

This looks bad enough, but the real situation is even worse since the Q.931 and H.245 channels use TCP connections which must also be set up.

Each TCP connection needs an extra round trip to synchronize TCP window sequence numbers (Figure 2.19). In a WAN environment where each round trip may take several hundred milliseconds, this can lead to unacceptably long set-up delays, especially when using the gatekeeper routed-call model where a TCP connection needs to be established between each gatekeeper.

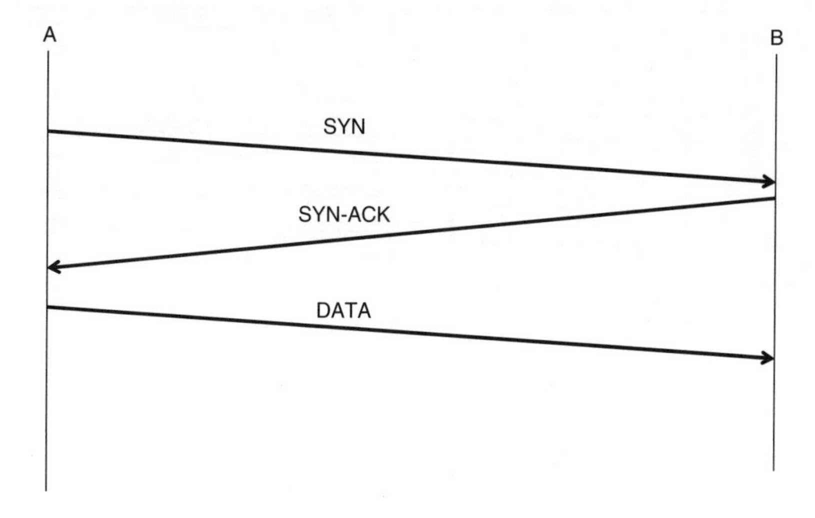


**Figure 2.19** TCP connection three-way handshake.

### 2.3.1.2 The TCP slow-start issue

The use of TCP causes at least one unnecessary round trip due to the SYN/ACK handshake. In fact, the situation can be worse if the SETUP message is larger than the maximum transmission unit (MTU) or if the first segment is lost.

If the SETUP message is larger than the MTU, the sender must send the SETUP in two or more TCP segments. The problem is that most TCP implementations are designed to be friendly to the network and follow RFC 2001, which mandates a slow-start procedure (Figure 2.20). In our case, after sending the first segment, the sender must wait until it has received an ACKNOWLEDGE before transmitting the next segment. Only then it can increase the window size and send two segments at once.

Because of this, large SETUP messages may cause one additional round trip. A good practice is to try to limit the size of the SETUP message below 576 octets, as this is the minimal IPv4 MTU, or at least below 1,500 octets (the Ethernet MTU), but this may not always be possible.

During an active TCP connection, the TCP stack dynamically estimates the round trip time (RFC 793) and uses this value to detect lost packets. But, for the first segment, TCP starts by using a worst case estimate. For the initial connection, the timeout value is normally set to the average round trip time *A* (initialized to 0 seconds), plus twice the deviation *D* (initialized to 3 s). If the first segment is lost, most TCP implementations

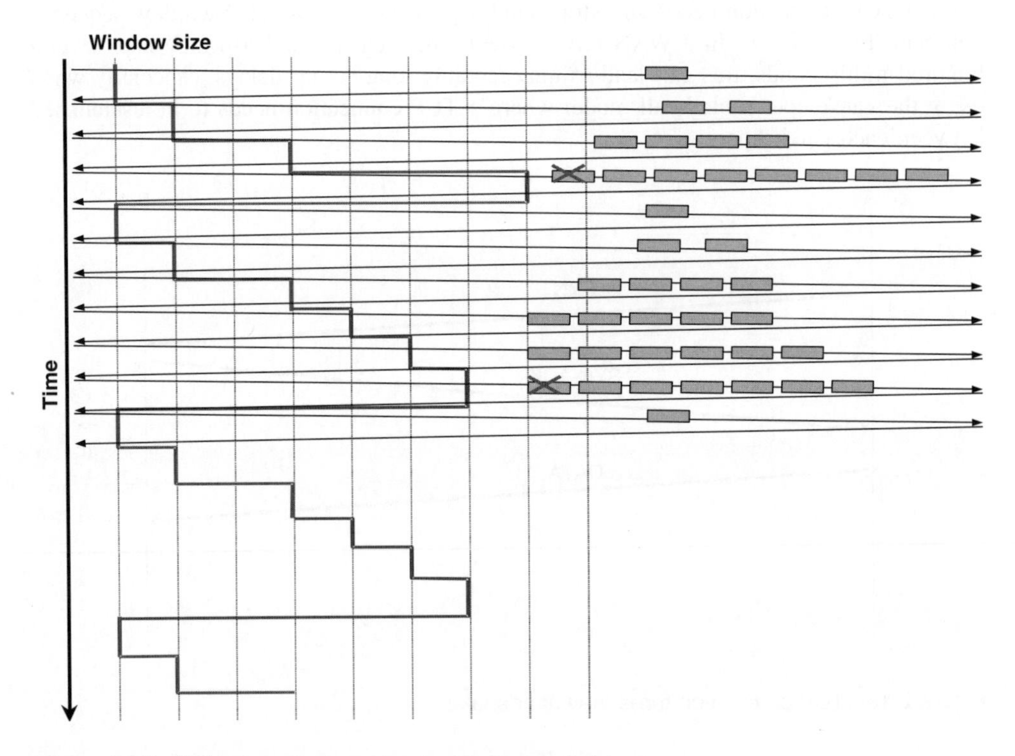


**Figure 2.20** TCP slow-start and back-off after packet loss.

will therefore wait for up to 6 s (retransmission timeout $= A + 2D$) before retransmitting the first segment. Some operating systems (e.g., Microsoft Windows®) only allow this value to increase (this is used for satellite connections). Once the first segment has been lost, things get even worse. The RTO (retransmission timeout) after the first segment is calculated as $A + 4D$ (12 s with the default values), and after each segment loss the RTO is doubled. Therefore, if the first segment is lost and retransmitted, the timeout value for this second segment will be 24 s! If the first two attempts fail, there will be 30 s between the first of the call set-up attempts, and the third retransmission.

Gatekeepers, gateways, and IP phones should tune the TCP stack settings:

- To send signaling TCP packets on a higher quality of service (e.g., using DiffServ, see the companion book, *Beyond VoIP Protocols*) in order to minimize packet loss.
- To aggressively retransmit TCP segments. This can be achieved by lowering the initial setting for $D$ for the connection establishment and lowering the 500-ms granularity of the RTO used on many operating systems. The maximum value of the RTO before the connection is lost—64 s—should also be lowered. On Linux, the number of retransmissions can be controlled with the/proc/sys/net/ipv4/tcp_syn_retries and/proc/sys/net/ipv4/tcp_synack_retries parameters.
- To increase the initial window size of the TCP slow-start mechanism up to two to four segments instead of one.
- To disable any buffering of information before packets are sent. This is Nagle's algorithm, which was designed to optimize transmissions for keyboard input; it delays transmission of a packet until a sufficiently large transmission buffer is accumulated, or 200 ms have elapsed. This can be done by setting the TCP_NODELAY socket option.
- To reduce the number of failures that are necessary to notify the application that 'something is wrong', or to clear the faulty socket. On Linux, this is configured by setting the/proc/sys/net/ipv4/tcp_retries1 and/proc/sys/net/ipv4/tcp_retries2 parameters, respectively.
- To reduce the time the gatekeeper waits for client confirmation when it closes a socket. This avoids accumulating sockets in the half-closed state. Most operating systems will wait up to 7 RTOs before closing the socket, which can exceed 15 min! On Linux, this is the/proc/sys/net/ipv4/tcp_orphan_retries and/proc/sys/net/ipv4/tcp_fin_timeout parameters. This setting is extremely important for VoIP networks with PC based softphones, due to the number of occurrences of abrunt disconnections of PCs (crashes, physical disconnections, loss of modem connections due to a call-waiting signal, etc.). Each 'orphan' connection also uses memory, which can lead to relatively simple denial-of-service attacks (the number of orphan connections in Linux can be controlled by the/proc/sys/net/ipv4/tcp_max_orphans parameter).
- To reduce the amount of time the gatekeeper waits for an acknowledgment of sent data if the socket has been closed (typical problem when sending the RELEASE COMPLETE message). This requires using the SO_Linger option (disabling linger or using a small linger timeout): after the timeout, if the ACK of the sent data has not been received (graceful close), the socket is forcibly closed with a TCP RST packet.

- Try to use the selective acknowledge feature of some TCP stacks. This option (SACK) is negotiated during the three-way handshake. On Linux, this is the/proc/sys/net/ipv4/tcp_sack option. This helps to speed up recovery of lost fragments and avoids retransmitting segments that have not been lost.[6]

This tuning makes the performance of TCP comparable with the performance of UDP-based retransmission schemes on most VoIP networks.

If TCP tuning is not enough in specific cases, H.323v3 introduced a new transport mode allowing the use of UDP signaling instead of (or simultaneously with) TCP signaling. This is in annex E of the standard. In H.323v4, a new possibility was introduced of keeping the TCP channel open and reusing it for the H.225 signaling of multiple calls. This option was designed to facilitate the design of routed mode gatekeepers and large-scale gateways on operating systems that have a low limit to the maximum number of TCP connections, or have scalability problems on the 'poll()' function required to detect incoming events on multiple sockets (this is a well-known issue with Linux, but it can be solved by using a modified version of the poll function). This option is not widely implemented and should be considered with great care because it causes head-of-line blocking (i.e., if one call is blocked for any reason, no further event related to other calls will be transmitted). Overall, the option of using multiple TCP sockets is much more robust and should be preferred.

The best solution is to use the new SCTP, which is optimized for telecom applications and offers the best of UDP and TCP simultaneously . . . this option has just been introduced in H.323v5!

#### 2.3.1.3 Network-generated prompts

Another problem was discovered by network experts after H.323v1 had been standardized. In the switched circuit network there are situations in which a message is played to the caller before it receives a connect:

- When the SCN is congested and the call cannot be established, you can get a prompt saying 'Due to congestion, your call cannot be connected, please call later'. This prompt is generated by the network itself at the local exchange, and, because it does not originate from the called endpoint, no CONNECT message is ever sent.
- In some applications the Intelligent Network can also generate network messages (e.g., for televoting applications: you dial a phone number and you get back a message saying 'Thank you for voting YES, the current status is 34% YES, 66% NO'). Similar pre-connect prompts are also used in many countries to implement prepaid calling card services in the network: the destination number and the PIN code are requested before connect, and the Q.931 CONNECT message is sent only when the call connects to the final destination.

With H.323v1, it is impossible to send a voice message to the calling party before sending a CONNECT message, because the media channels are not yet established.

[6] The default TCP ACKNOWLEDGE is cumulative, which means all the packets since the last acknowledged packet will be retransmitted even if only one segment has been lost.

### 2.3.2 The 'early H.245' procedure

The early H.245 procedure is used when the H.323 SETUP message contains an H.245 address, which is available to the called party if it wants to start connecting to the H.245 channel immediately. Alternatively, if the calling party has not proposed an H.245 connection address, the called party can make its own proposal by including an H.245 address in the call control messages sent before the CONNECT message (CALL PROCEEDING, ALERTING). The early H.245 helps the H.245 procedure to start as early as possible in the call, sometimes even before it actually connects. In most cases, unless the call connects right away, it makes the inherent delays due to the multiple message exchanges required by H.245 invisible to the call participants. It also solves the network prompt before connect issue explained in Section 2.3.1.3. All this while preserving full compatibility with all H.323 features, including out-of-band DTMF transmission, third-party call control using the TCS = 0 procedure (see more details see Section 2.7.1.2.2), and sophisticated video and conference control procedures.

The early H.245 procedure is so useful that it is should be one of the most important criteria for the selection of any H.323 equipment. This call flow should be used whenever possible.

### 2.3.3 The 'fast-connect' procedure

The fast-connect procedure was introduced in H.323v2 to enable unidirectional or bi-directional media channels to be established immediately after the Q.931 SETUP message and eliminate any post-connect delay in the audio path.

The usefulness of the fast-connect procedure is questionable, as the early H.245 procedure (described in Section 2.3.2) also solves these issues. In the early days of H.323 there was still some confusion on which was the best method to solve the delay issue, and all possible solutions were welcome. Fast connect has one little advantage over early H.245: it removes any post-connect audio delays even in the case of an immediate call connection. It also has major drawbacks compared with the early H.245 procedure: for instance, it does not allow the out-of-band transmission of DTMF information[7] and it does not provide a third-party call control feature before H.323v5 (this version adds this possibility in the H.460.6 extension). It is tempting to use both early H.245 and `fastStart` (see next paragraph) at the same time, which in fact many vendors are currently doing. Since H.323v4, this is officially possible, but the actual H.245 communication should not transmit anything but the endpoint capabilities and master/slave information before the completion of the initial fast-connect negotiation.

An endpoint which decides to use the fast-connect procedure will include a new parameter, called `fastStart`, in the SETUP user-to-user information element. This parameter includes a description of all the media channels that the endpoint is prepared to receive

[7] RFC 2833 may be used, but does not allow feature servers to act on DTMF commands without also relaying the media stream.

and all the media channels that the endpoint offers to send. This description includes the codecs used, the receiving ports, etc.

If the called endpoint cannot or doesn't want to use the fast-connect procedure, it will not return the `fastStart` element in subsequent Q.931 messages. In this case the normal procedure involving H.245 will take place.

If the called endpoint supports the fast-connect procedure, then it will return, in the CALL PROCEEDING, PROGRESS, ALERTING or CONNECT Q.931 message, a `fastStart` element that selects from among the media offered by the caller.

The `fastStart` element is always inserted in the H323-UU-PDU of the user-to-user element (its use with the SETUP message is shown in bold):

- Protocol discriminator field (08H).
- Call Reference Value (CRV).
- A Message Type (SETUP).
- ...
- Called party number and subaddress.
- Calling party number and subaddress.
- The H.323 user-to-user element which contains the SETUP user-to-user information element in which we find:
  - The protocol identifier.
  - ...
  - The sender's aliases.
  - The destination address.
  - The CID and CallID.
  - `fastStart`: used only in the fast-connect procedure, `fastStart` is a sequence of **OpenLogicalChannel** structures. Each **OpenLogicalChannel** structure (Figure 2.21) describes *One* media channel that the caller wants to send (**forwardLogicalChannelParameters** within the **OpenLogicalChannel** structure) or receive (**reverseLogicalChannelParameters**). All proposed **OpenLogicalChannels** can be selected simultaneously, unless they share a common **sessionID** value in the **H2250-LogicalChannelParameters** of the **OpenLogicalChannel** structure, in which case they are considered alternative options for the same channel.
  - The **`mediaWaitForConnect`** boolean.

The calling terminal can select one or more acceptable **OpenLogicalChannel** structures within the offered `fastStart` parameter and return them in a `fastStart` parameter within an H.225 CALL PROCEEDING, PROGRESS, ALERTING, or CONNECT message. The selected logical channels are considered open after this.

Note that the network can send media to *any* of the receiving channels mentioned in the SETUP message of the caller, immediately after the calling terminal has sent this message, unless MediaWaitForConnect is true. Therefore, even if the calling terminal

```
OpenLogicalChannel ::=SEQUENCE
{
        forwardLogicalChannelNumber    LogicalChannelNumber,
        forwardLogicalChannelParameters SEQUENCE
        {
                portNumber  INTEGER (0..65535) OPTIONAL,
                dataType    DataType,
                multiplexParameters    CHOICE
                {
                        h2250LogicalChannelParameters H2250LogicalChannelParameters,
                        none            NULL

                },
                ...,
                forwardLogicalChannelDependency    LogicalChannelNumber OPTIONAL,
                replacementFor          LogicalChannelNumber OPTIONAL
        },

        reverseLogicalChannelParameters    SEQUENCE
        {
                dataType    DataType,
                multiplexParameters    CHOICE
                {
                        h2250LogicalChannelParameters H2250LogicalChannelParameters
                } OPTIONAL, -- Not present for H.222
                ...,
                reverseLogicalChannelDependency  LogicalChannelNumber OPTIONAL,
                replacementFor          LogicalChannelNumber OPTIONAL
        } OPTIONAL, --Not present foruni-directional channel request

        ...,
        separateStack          NetworkAccessParameters OPTIONAL,
        encryptionSync         EncryptionSync OPTIONAL -- used only by Master
}
```

**Figure 2.21** OpenLogicalChannel ASN.1 structure.

plans to use only one of these channels for regular conversation, as indicated in the `fastStart` response, it *must* be prepared to receive media on any one of these channels (before the response). Although most H.323 vendors have implemented the `fastStart` procedure, many of them actually do not support this requirement and are not able to receive audio before the remote endpoint has selected a media channel proposal in its own `fastStart` element. This is because most implementations do not have enough memory to load multiple voice coders at the same time, and the vendor selects the right coder to load when it receives the remote `fastStart`. In the best implementations that fully support `fastStart`, the first RTP packet that is received can be used to load the codec, without waiting for the remote `fastStart` element. An example of a case where supporting the full `fastStart` requirement is important is network-based redirection announcements: multiple announcement servers may have to send audio to the calling endpoint, before it connects to the party redirected to. In this case the `fastStart` element will be sent only by the last endpoint, but the announcement servers still need to stream audio toward the calling party before this happens.[8]

[8] Some vendors can receive multiple `fastStart` elements in sequence and always take into account the last one received. This greater flexibility makes it possible to activate media reception only after having received a `fastStart` response, without restricting the feasible services. Unfortunately, this way of interpreting the `fastStart` response, which creates a form of third-party media control, is not yet taken into account in the standard.

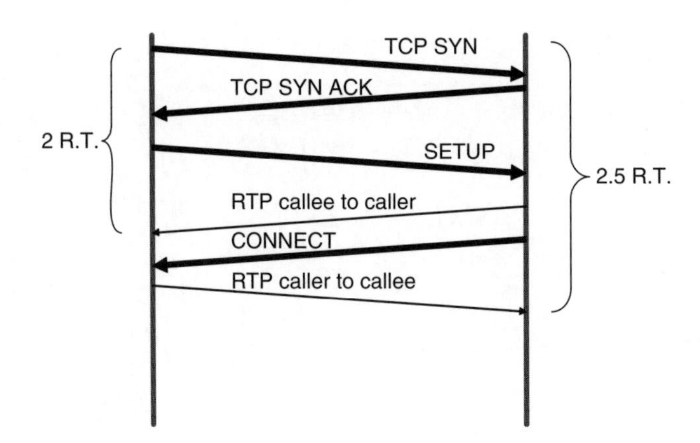


**Figure 2.22** Audio path delays with the fast-connect procedure.

Usually, in a normal ISDN call the called endpoint does not send media before the CONNECT message has been sent. It is possible to force this behaviour with the `fastStart` procedure by setting the `mediaWaitForConnect` element of the Q.931 SETUP to true.

As shown in Figure 2.22, the fast-connect procedure dramatically improves the number of round trips required to set up the conversation, and eliminates all post-connect audio delays.

Since the fast-connect procedure solves a major flaw of H.323v1 regarding interworking with the traditional TDM networks, it has been made a core feature in the H.323 profile of the ETSI TIPHON project.

Fast-connect procedure usage becomes a bit subtle in certain circumstances (e.g., when H.323 is used to provide class 5 services). The interactions between the call forward on no answer service and the use of `fastStart` are extremely complex, and not well studied by the standard (see Chapter 6). In addition the `mediaWaitForConnect` Boolean is a bit too simple to fully account for the variety of call flows found in the PSTN. The sending of audio information before connect is controlled by the in-band audio indicator in Q.931 messages, including the PROGRESS message. Some call flows can become extremely complex, as in the following example where ring-back tones alternate with redirection prompts.

Such call flows today are possible (the previous scenario is currently used in Milan, Italy), but need prior vendor agreement on the exact handling of **in-band audio indicators**. Also, some call flows really would require the sending of multiple `fastStart` elements to update the RTP logical channel information before CONNECT. H.460.6 has refined the `fastStart` procedure and allows the refreshing of `fastStart` elements before CONNECT.

Using just the fast-connect procedure, DTMF transmission is not possible as H.245 UserInputIndication is not available (no H.245 channel is opened). Because of this limitation, early H.245 (establishing the H.245 channel before CONNECT) should be used in conjunction with `fastStart` in most cases, as it is officially allowed since H.323v4. Many types of calls will require DTMF information before CONNECT. For instance, this is what would happen typically when calling an IN-based card telephony service. An IVR

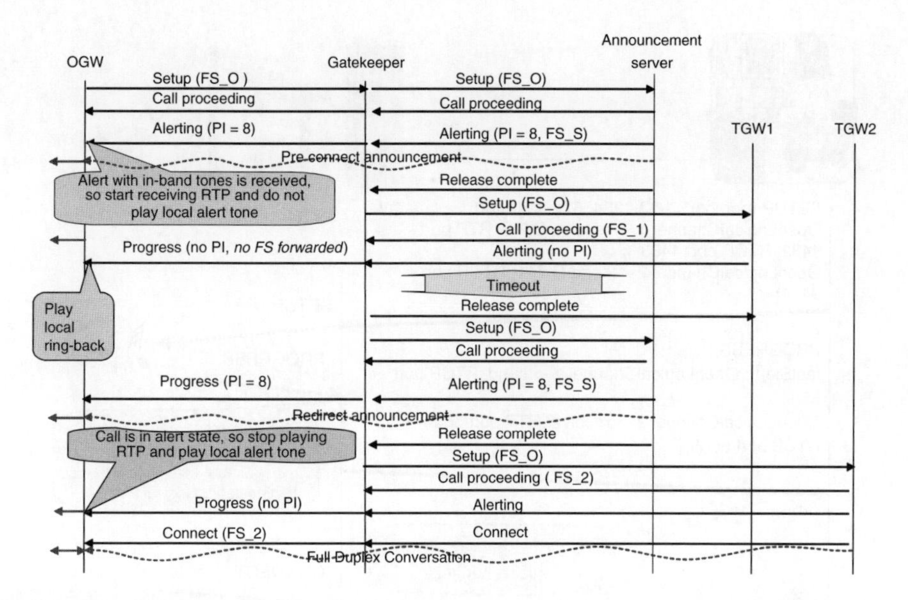


**Figure 2.23** Complex call flow with pre-connect audio. Note: there is no DTMF capability before connect.

server requests the card code before sending a CONNECT Q.931 message because the call is not yet charged. The CONNECT message is sent only after the code has been checked and the destination party has answered the call (as shown in Figure 2.23).

***Caveat***: The addition of fast-connect mode to H.323 has made it possible to manufacture a simpler, yet H.323-compliant terminal. In fact, this was one of the goals of the initial `fastStart` proposal, at a time when SIP began to claim simplicity compared with H.323. By supporting only H.323 in fast-connect mode, developers can avoid implementing H.245 in simple appliances like IP phones in this way. However, most of the potential of H.323 comes from the conferencing and third-party call control features which are enabled only by H.245. Simple H.323 terminals without H.245 will not be able to participate in such conferences. Moreover, DTMF is normally carried using H.245: simple terminals will have to use in-band DTMF coding which does not work as soon as complex services like prepaid servers or contact centres are implemented. Overall, such 'simplified' terminals would not meet even the basic requirements for telephony and would make the design of services on networks with such endpoints extremely challenging. Interestingly, facing the same issues, SIP has become significantly more complex over time, notably adding third-party call control and out-of-band DTMF capabilities to the basic call, and is now virtually identical to the H.323 protocol, with `fastStart` and H.245 tunneling enabled.

## 2.3.4 H.245 tunneling

Most H.323 devices today use two separate TCP connections for each call: one for the Q.931 messages (SETUP, ALERTING, CONNECT, etc.), and one for the H.245 messages

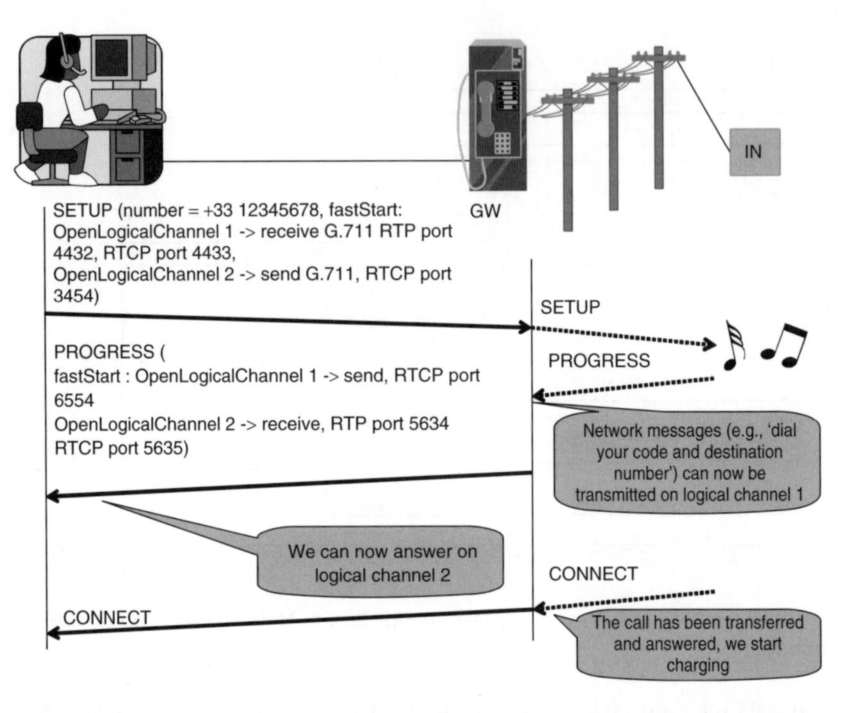


**Figure 2.24** Use of a PROGRESS message for pre-connect audio.

(OpenLogicalChannel, TerminalCapabilitySet, etc.). This may become a problem for some gatekeeper or gateway implementations that run on operating systems with low limits on TCP connections and that do not use distributed designs. It also unnecessarily doubles the issues associated with the use of TCP.

H.323v2 offers a way of using a single TCP connection by encapsulating H.245 messages in Q.931 messages: this is called H.245 tunneling.

An endpoint which wants to use H.245 tunneling must set the h245Tunneling element of the SETUP message and all subsequent Q.931 messages to TRUE. A called endpoint also indicates its willingness to accept H.245 tunneling by setting this same element to TRUE in all Q.931 messages.

The calling endpoint simply encapsulates one or more H.245 messages in the h245Control element of any Q.931 message. If the called endpoint is also capable of receiving it, all H.245 messages can be exchanged in this way and there is no need to open a separate TCP connection for the H.245 channel. Otherwise, if the called endpoint has not set the h245Tunneling to TRUE in the first Q.931 message it sends back (it could be CALL PROCEEDING, PROGRESS, ALERTING, or CONNECT), the calling endpoint knows this is not supported and the normal procedure for opening an H.245 channel is followed. Q.931 messages are modified as shown in bold:

- Protocol discriminator field (08H).
- Call Reference Value (CRV).

- A Message Type (SETUP, ALERTING, . . .).
- . . .
- The H.323 user-to-user element (H323-UU-PDU) now contains:
  - The **H245Tunneling** boolean.
  - **H245Control**: a sequence of ASN.1 octet strings representing ASN.1 PER-encoded H.245 PDUs. The H323-UU-PDU type can be the usual SETUP, CALL PROCEEDING, CONNECT, ALERTING, USER INFORMATION, RELEASE COMPLETE, FACILITY, or PROGRESS, or a new NULL value called *empty*, which is explained later.

When using H.245 tunneling the Q.931 channel needs to remain open during the entire duration of the call. If an H.245 message needs to be sent when no Q.931 message is pending to be sent, then the H.245 message will be encapsulated in a Q.931 FACILITY message. In this case the Q.931 FACILITY message is sent, but the H.323 user-to-user element only contains the H245Tunneling boolean and the H.245 PDUs encoded in H245Control: in the next paragraph the H323-UU-PDU type isn't the usual FACILITY type, but is set to the new 'empty' value. Such a need to use FACILITY messages may also occur in the gatekeeper-routed model (Figure 2.25).

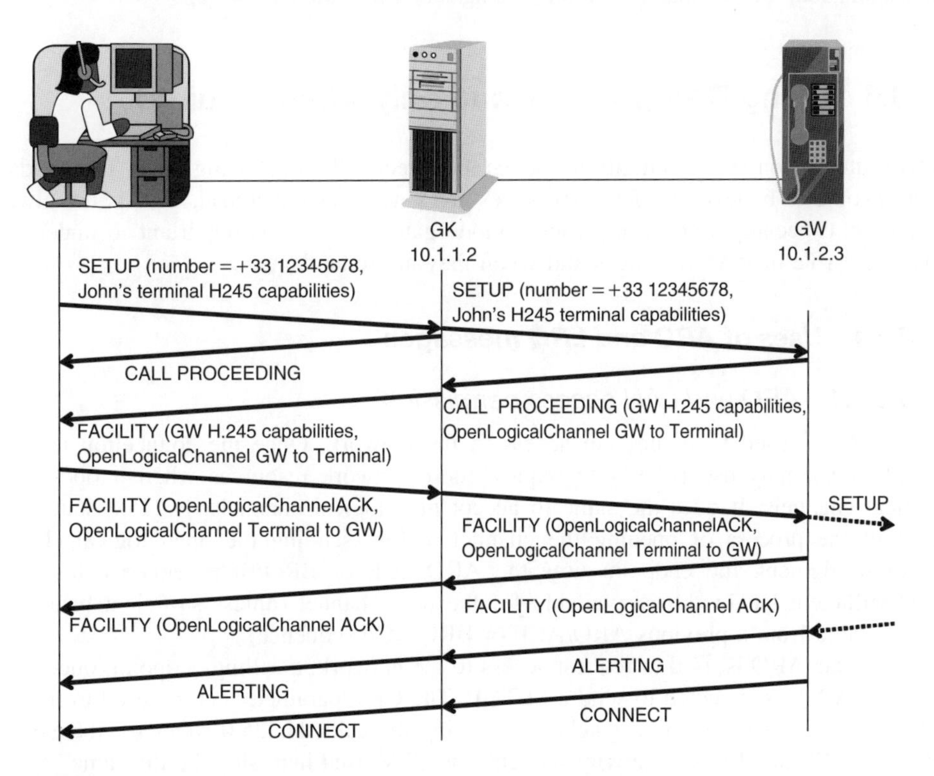


**Figure 2.25** Use of FACILITY messages for H.245 tunneling.

A terminal can signal that it wants to use H.245 tunneling in a Q.931 message which already contains a fastStart element with OpenLogicalChannels parameters by setting H245Tunneling to TRUE. In H.323 prior to version 4, it was forbidden to encapsulate H.245 messages in the same Q.931 messages as the fast-connect procedure (as they would supersede the indications found in the OpenLogicalChannels parameter). This was clarified in version 4, where capability messages and master/slave messages can be sent during the fast-connect negotiation. All other H.245 messages must wait until the fast-connect exchange has occurred.

### 2.3.5 Reverting to normal operation

In some cases a terminal using fast-connect and/or H.245 tunneling may need to use a separate H.245 control channel in the middle of an established call (e.g., when a terminal that has opened an audio connection in fast-connect mode needs to open a new media channel). In this case the terminal can send a FACILITY message to the other terminal indicating it wishes to establish a separate H.245 channel and proposing a transport address for it. The terminal which receives the facility message must establish a new TCP connection for the H.245 channel using this transport address. Once the new connection is established, the terminals must stop using the H.245 tunnel.

### 2.3.6 Using RAS properly and only when required

Most tutorials on H.323 initially introduce the direct call model using RAS. This tends to lead people to believe that RAS messages are a necessary overhead. This is not always the case, especially when using routed mode gatekeepers. It is important to understand the exact role of RAS messages and when they are redundant.

#### *2.3.6.1 Uses of ARQ and LRQ messages*

##### *2.3.6.1.1 ARQ and LAN access permission*

The ARQ is used to request an access to the network. Once the connection is open, the terminal may use a BRQ to request more network resources when it opens new logical channels. It does the same to accept new logical channels. H.225.0 states: ‘As part of the process of opening the channel, before sending the open logical channel acknowledgment, the endpoint uses the ARQ/ACF or BRQ/BCF sequence to ensure that sufficient bandwidth is available for the new channel (unless sufficient bandwidth is available from a previous ARQ/ACF or BRQ/BCF sequence).’

Since the ARQ is used to request access to the network, a calling endpoint, once it has sent an ARQ, is expected to send a SETUP. The CRV parameter can be used to link the two messages (e.g., within a gatekeeper routing the call). H.225.0 states for the SETUP message: ‘If an ARQ was previously sent, the CRV used here shall be the same.’

##### *2.3.6.1.2 Address resolution*

Both the ARQ and the LRQ messages can be used for the address resolution function (i.e., obtaining a destination IP address from a destination). The name 'location request' seems to imply that the LRQ message is the right message to use, but when comparing the semantics of the LRQ and the ARQ, it becomes obvious that the ARQ can also serve this purpose. In fact, an ARQ is a superset of an LRQ, requesting not only an address translation but also LAN access. This is good because, otherwise, terminals might need to first do an LRQ, then an ARQ, and finally a SETUP! Note that this address resolution can be:

- A pure alias to transport address resolution (e.g., from an email alias the gatekeeper indicates the IP address to send the SETUP).
- An alias translation, if the terminal specifies it, can map aliases (canMapAlias parameter of the ARQ); in this case the gatekeeper can also force the terminal to change the destination alias in the SETUP.

##### *2.3.6.1.3 Conclusion*

Usually, a terminal needs to do an address resolution just before it sends a SETUP. In this case the ARQ can be used to request LAN access as well as an address resolution, and thus the LRQ is redundant. This is reflected by the fact that H.225 mandates the support of ARQs by H.323 endpoints, whereas the support of LRQs is optional. Overall, the LRQ message is not very useful, except that it is now the de facto standard for inter-gatekeeper call flows (the ARQ would have been more appropriate here too). The most legitimate use of the LRQ message is when a call control gatekeeper needs to query an access gatekeeper for the current location of an endpoint. Such a 'split gatekeeper' architecture is used in networks with hundreds of thousands of endpoints, in order to distribute the registration function (see Section 2.3.6.1.1 for more details).

#### *2.3.6.2 Disabling the ARQ/ACF sequence*

If the gatekeeper is used in routed mode, it has the possibility of authorizing or blocking the call when it receives the SETUP message, and, since it has access to all aspects of call signaling, keeps complete control over the call for the entire duration of the conversation. In this case, the ARQ message is really unnecessary and only adds an extra round trip to the call SETUP delay. The gatekeeper can instruct an endpoint to send a SETUP directly, without a prior ARQ/ACF, by using the **preGrantedARQ** parameter that is contained in the RCF (registration confirm) message.

If preGrantedARQ is not configured, the terminal is required to send an ARQ to the gatekeeper before each call SETUP.

If preGrantedARQ is configured in the RCF, the gatekeeper can give one or more of the following privileges to the terminal:

- Initiate a call without first sending an ARQ.
- Initiate a call without first sending an ARQ, but *only* if it sends the SETUP to the gatekeeper.
- Answer a call without first requesting permission from the gatekeeper.
- Answer a call without first requesting permission from the gatekeeper *only* when the SETUP message comes from the gatekeeper.

Using preGrantedARQ is an excellent way to optimize call SETUP time. In most endpoints, the ARQ can also be disabled manually using the configuration tool—sometimes indirectly (e.g., on some endpoints the routed mode gatekeeper should be declared as a 'gateway').

## 2.4 CONFERENCING WITH H.323

### 2.4.1 The MCU conference bridge, MC and MP subsystems

There are two distinct functions that may be present in any conference. The first one is the control function, which decides who is allowed to participate or not, how new participants are introduced in existing conferences, how the participants synchronize on a common mode of operation, who is allowed to broadcast media, etc. This role is assumed in H.323 by a functional entity called the **multipoint controller (MC)**.

When many people talk in an audio conference, they might simply multicast their audio, and all terminals can do the mixing of individual media streams themselves. In most cases, however, individual terminals will have limited capabilities, or it might be impractical to multicast all media streams (especially in the case of video). If multicast cannot be used, an entity in the network needs to do the mixing or switching of incoming media streams, and send only the resulting processed outgoing stream to each terminal. In the case of video it can be the image from the last active speaker, in the case of audio each terminal will receive a stream resulting from the addition of all streams from other speakers in the conference (plus some of its own, but attenuated). In H.323, this functional entity is called the **multipoint processor (MP)**.

A dedicated callable endpoint, which contains an MC and optionally one or more MPs, is called a **multipoint control unit (MCU)**. It is not just MCUs that have the MC functionality, however, a terminal or gatekeeper with sufficient resources can also have the capability to act as an MC and may be able to do some media mixing locally. However, the MC functional entity in a terminal or a gatekeeper cannot be called directly, but will be included in the call when it becomes multiparty.

A conference is called a **centralized conference** when a central MP is used to mix or switch all media streams for participating endpoints. When each terminal sends its media streams to all other participating terminals (in multicast or multi-unicast), it is called a decentralized conference.

### 2.4.2 Creating or joining a conference

#### 2.4.2.1 Using an MCU directly

Most of the time, people will decide to create a conference and name it, for instance, myconference@conferencerooms.com. So, the participants know from the beginning that it is going to be a conference call. The easiest way to create such a conference is simply to call an MCU and send a SETUP message with the `conferenceGoal` parameter set to `Create` and a globally unique CID. It may also include the alias of the conference (myconference@conferencerooms.com). So far, nothing differs from a regular call.

If the MCU decides to accept the call (this can be based on previous reservation done through a website), it replies with a CONNECT.

The endpoint and the MCU exchange their TerminalCapabilitySets. Then the master–slave procedure begins, the MCU always wins and becomes *the* active MC of the conference. The MCU indicates it is the active MC of the conference by sending a MCLocationIndication message to the calling endpoint. It can also assign an 8-bit number to the terminal with a terminalNumberAssign message (the terminal must copy those 8 bits in the low 8 bits of the SSRC field of all its RTP datagrams).

##### 2.4.2.1.1 Inviting new participants

Once a terminal is in a conference, it may invite others (e.g., terminal C) to participate by sending a SETUP message to the active MC with a new CRV, the CID of the conference and the `conferenceGoal` parameter set to `invite` The destination address and, optionally, the destination call-signaling address of the SETUP message must be those of terminal C.

When it receives this message (Figure 2.26), the MCU will send a SETUP message to terminal C with the CID of the conference and `conferenceGoal=invite` Terminal C accepts by sending a CONNECT message. At this point the MCU sends a RELEASE COMPLETE to the inviting terminal. The active MC establishes an H.245 control channel with terminal C using the transport address provided in the connect. They exchange their TerminalCapabilitySets. The MC signals during the master–slave procedure that it is already the active MC and may send an MCLocationIndication message. When this is done, the MC sends a `multipointConference` message to the inviting and invited terminals. If there were already other terminals in the call, the MCU will send them a `terminalJoinedConference` H.245 message to make them aware of the new entrant.

Because the incoming terminal might have capabilities which are incompatible with the existing media channels in place in the conference, the MCU must send a `communicationModeCommand` to all terminals specifying the new set of allowed transmitting modes for each stream. All media channels that happen not to conform must be closed.

At this point the MC can begin to send OpenLogicalChannel messages to the endpoints. The endpoints should also wait, when they have received a `multipointConference` message, until they receive a `communicationModeCommand` message to open logical channels. All endpoints must send the openLogicalChannel messages to the MC.

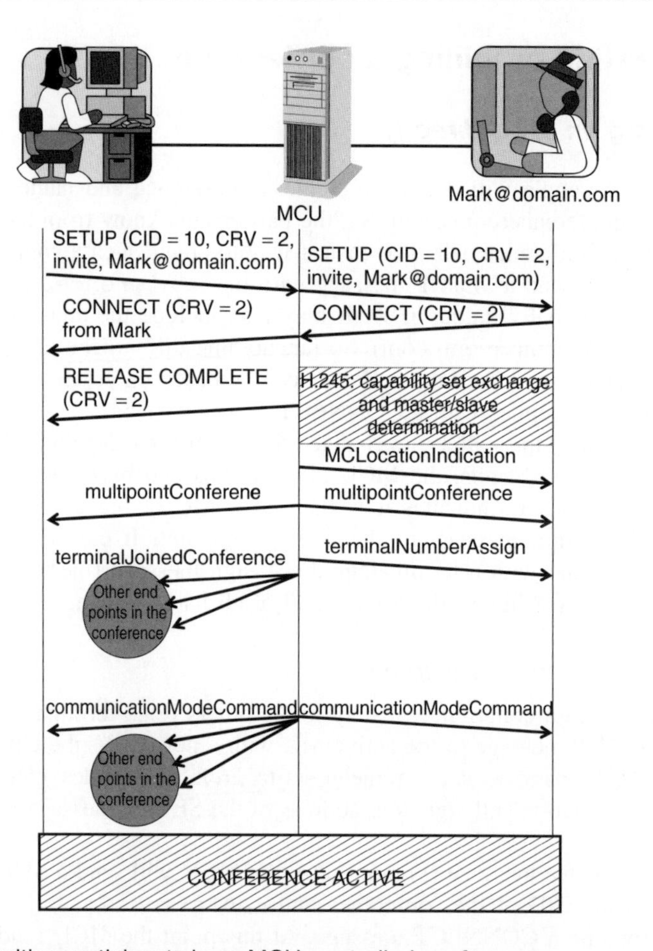


**Figure 2.26** Inviting participants in an MCU-controlled conference.

The MCU can also initiate the invitation (e.g., if the invitation is not done from an H.323 terminal, but from a Web interface of the MCU). In current H.323 commercial products, this is the most widely used model, because it does not require support for any specific H.323 call flow on the participating endpoints.

### *2.4.2.1.2 Joining an existing conference*

A terminal can easily join an existing conference by sending a SETUP message to the MCU with the CID at the conference and `conferenceGoal=join`. If the terminal only knows the Alias of the conference, it must provide it and leave the CID at 0. Most commercial MCUs use this simpler model, where all participants calling the same number are automatically bridged into the same conference. This model does not suppose any support for the H.323 conferencing features at the endpoint. In order to secure the conference and prevent any random user from calling the bridge directly, the call can be routed through a routed mode gatekeeper, which decides on the fly if the participant is allowed and may translate the initial called party number of the SETUP message into a conference number (this is the solution used by the France Telecom eVisio IP videoconferencing service). Other vendors have an embedded interactive voice response server in the MCU, which authenticates users by prompting for a PIN code.

##### *2.4.2.1.3 Browsing existing conferences*

The MCU can in theory provide a list of existing conferences that a terminal could join by sending a conferenceListChoice H.245 message to a terminal. This can be used, for instance, when an alias that has been used in the SETUP message is in fact the name of a group of conferences (e.g., H323support@conferences.com might be a group name for Q931support, H245support, and RASsupport@conferences.com).

Again most commercial MCUs use a simpler Web-based administrator interface to browse for ongoing conferences.

#### *2.4.2.2 Ad hoc conferences*

When two endpoints (John and Mark) have started a call as a point-to-point call, they still might want to include someone else in the conversation. Someone might call one of the two parties, or suddenly they might need to talk to someone else to solve a problem. In these cases the call has not been set up directly using a MCU because John or Mark had no idea, when first placing the call, that it would become a conference call. This type of conference is called *ad hoc*.

##### *2.4.2.2.1 John invites Mary*

During the discussion, John and Mark decide to go to the cinema, but they need to talk to John's wife, Mary, so that she can choose the movie.

###### *2.4.2.2.1.1 If John and Mark are using the direct call model*

In this case, either John or Mark's terminal needs to have an MC functionality and the MC will basically behave as the MCU in Section 2.4.2.2.1. For instance, if Mark's terminal has a MC, John will send to Mark's terminal a SETUP message with a new CRV (not the one used for the point-to-point call between John and Mark), `conferenceGoal=invite`, and the alias of Mary. The rest is exactly as in Section 2.4.2.2.1.

If neither John nor Mark have an MC-capable terminal, they must clear the call and call Mary again via an MCU... not very user-friendly! Fortunately, many H.323 phones now include three-way conferencing capabilities.

###### *2.4.2.2.1.2 If John and Mark are using the gatekeeper-routed model*

Now, with the gatekeeper-routed model, if no terminal has an MC, John and Mark can still invite Mary if the gatekeeper has an MC capability. In this case the gatekeeper behaves as a MCU in the INVITE example above. In some cases the gatekeeper will not have an MC in the same box, but it can easily redirect all conference-related messages to an external MC entity since it routes all Q.931 and H.245 messages.

#### 2.4.2.2.2 *Mary calls John*

When Mary calls John, she has no reason to know that he is already in a call, so she sends a regular SETUP message.

##### 2.4.2.2.2.1 *If John and Mark are using the direct call mode*

When John's terminal receives this SETUP message, it will probably propose a menu to John asking whether he wants to:

(1) Reject the call.

(2) Put the call with Mark on hold and talk with Mary.

(3) Include Mary in the call with Mark.

If either John's or Mark's terminals have an MC capability, John can choose (3). For our example we consider that only Mark's terminal has an MC capability.

Mary's terminal will receive a FACILITY message indicating that the call should be routed to John's terminal (a parameter `routeCallToMC` will be present in the facility message) which is the MC-capable terminal. This message also indicates the existing CID of the call with Mark. Mary's terminal releases the call with John (RELEASE COMPLETE) and sets up a new call with Mark.

Now, the SETUP message sent by Mary's terminal contains the same CID as the ongoing call between Mark and John, and the parameter `conferenceGoal=Join`. Then, the call flow continues as if Mark was the MCU in the JOIN example of Section 2.4.2.2.2.

This example, which follows the theory of H.323 conferences, is very unlikely to occur in practice. First, most H.323 endpoints designed for use with a direct mode gatekeeper have an internal MC function. Another reason is that source-based redirections (initiated by a message sent to the terminal that should redirect his call) only work on private networks: on public networks, the dialing convention of the redirecting terminal and the redirected terminal may be different, and therefore, the number given by John to Mary in the FACILITY message may be unusable by Mary. There are many other reasons that make such a redirection scenario impossible to use over public networks (see Chapter 6).

##### 2.4.2.2.2.2 *If John and Mark are using the gatekeeper routed model*

This is very similar to the direct call case, except that now the FACILITY message sent by John's terminal will contain the address of the gatekeeper, which will be responsible for the MC function, directly or by invoking an external server. Since the call from Mary arrived through the gatekeeper, the FACILITY message can be intercepted by the gatekeeper and the gatekeeper can use the third-party rerouting procedure (see Section 2.7.1.2.2), instead of propagating the FACILITY message all the way to Mary's endpoint. This eliminates the problem of incompatible dialing conventions and all other issues associated with public networks.

#### 2.4.2.3 Conferences and RASs

In most cases, terminals will only know the alias of the conference they want to join. Therefore, the initial ARQ will contain only the conference alias in the `destinationInfo` parameter of the ARQ. The CID parameter will be set to 0, which means it is unknown. The callIdentifier must be set by the caller as usual. The gatekeeper will return the Q.931 transport address of the endpoint containing the MC (the MCU, or an endpoint with MC capability) in the ACF.

In theory, as soon as the caller knows what is the exact value of the CID (after receiving a connect from the MC), it must inform the gatekeeper using an IRR RAS message; but, in practice, we don't know of any endpoint doing this.

### 2.4.3 H.332

The conference model described in Section 2.4.2.3 is **tightly coupled**: all the participants maintain a full H.245 control connection with the MCU. This is very resource-intensive and this model breaks when the number of participants increases beyond a few dozen.

Conferences with a large number of participants tend to be organized with a panel of several speakers (less than 10, typically) and a large audience that listens most of the time and speaks only when requested by the moderator. H.332 describes the electronic equivalent of a panel conference (Figure 2.27), called a **loosely coupled conference**, and is designed to scale to thousands of participants. H.332 is a mix of a usual tightly coupled conference (used by permanent speakers) and a multicast RTP/RTCP conference (as known on the mBone) for passive listeners. The RTP/RTCP-only listeners must know which codec is used and other details (UDP ports, ...). H.332 uses the syntax of the IETF **Session Description Protocol (SDP)** to encode the value of those parameters. A new SDP type (a = type:H332) is defined to let the RTP listener know that this is an H.332 conference. The information can be conveyed using the IETF **Session Announcement Protocol (SAP)** or a simple file on a webpage or sent by email.

Due to the large number of participants, the highly coupled conference among panel members is subjected to several constraints: the codecs used should remain stable. If a new member forces a new capability negotiation and triggers a change of codecs, a new SDP announcement must be created. Spreading this information using SAP or otherwise takes time, and most RTP listeners are left out until they have been notified of the new announcement.

The difficult part is to allow the panel to invite a listener to talk, to let listeners request and be granted the right to ask a question. In order to join or be invited by the panel, the RTP listener must also have some H.323 capabilities. Simple RTP/RTCP terminals can only listen. In order to join the panel, a listener must use the regular H.323 conference join and must know the address of the MC that is provided in the SDP.

Similarly, the panel needs to know the callable address of the terminals to be able to invite them to the conference. This is possible because conference listeners periodically transmit information elements, such as their name on email address, as RTCP SDES

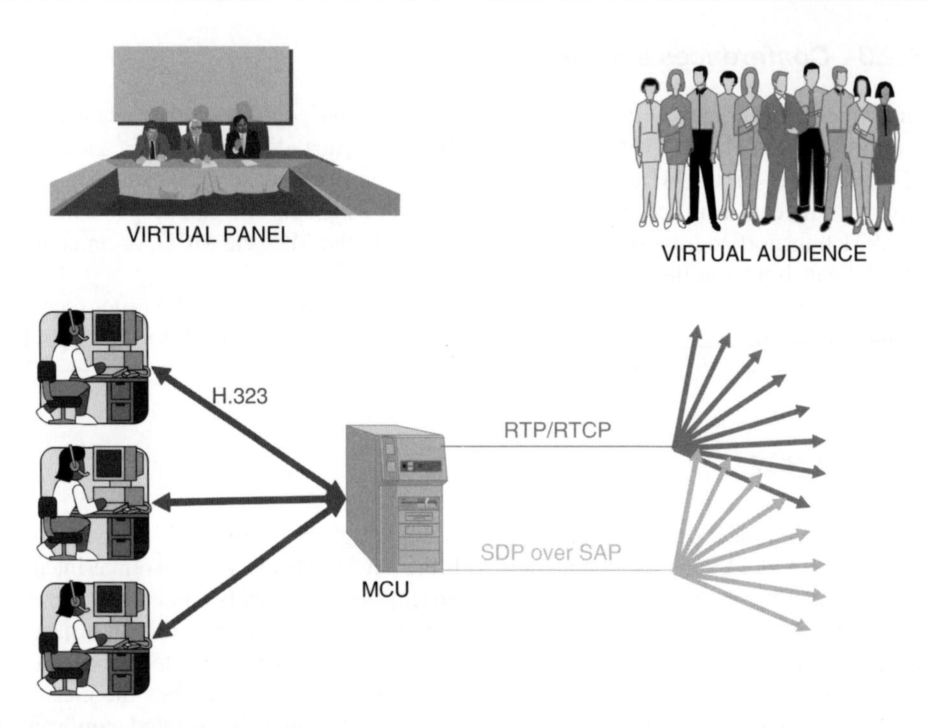


**Figure 2.27** H.332 conference.

packets (see Chapter 1). A new information element, RTCP SDES item 'H323-CADDR', conveys the H.323 callable address of the terminal. Since bandwidth reserved for RTCP traffic is limited, it takes some time to build a complete list of listeners, and therefore a listener may become callable only some time after he has joined the conference.

## 2.5 DIRECTORIES AND NUMBERING

### 2.5.1 Introduction

In the early days of IP telephony (this wasn't so long ago!), one of the major problems was to call someone using a dial-up connection, since the IP address of such users was allocated dynamically. Early solutions all used the same scheme: when the IP telephony software was started, it immediately contacted a central server on a preconfigured IP address and sent a message with the name of the phone user and the current IP address. There were many implementations, ranging from **Microsoft ILS/ULS** to solutions running on top of IRC servers...

H.323 makes these solutions completely obsolete. A terminal implementing RAS properly has to register to a gatekeeper, and the RAS message contains all the necessary information—in particular, the current IP address—needed to contact the terminal from a known alias.

## 2.5.2 Contacting an email alias with H.323 and the DNS

### 2.5.2.1 Resolution algorithm

The gatekeeper can be queried for the current network location (as a call-signaling transport address) of the aliases within its zone, using an LRQ or an ARQ if a call follows immediately. Therefore, when trying to reach an alias, the first step is to find the gatekeeper responsible for this alias. A possibility in small environments is to multicast the LRQ until someone answers, but this is obviously not scalable.

If the alias used is an email alias, like someone@domain.org, then a much better strategy is described in the informational annex of H.225.0. Much information about a domain name can be found by using the Internet **Domain Name System (DNS)**.

The DNS was originally invented to help resolve names into IP addresses. When your computer needs to communicate with another computer named othercomputer.domain.org, it uses the DNS to find its IP address. This involves several steps:

- First, the computer asks a well-known master server which second-level DNS server has the information about domains ending with. org. In fact, this information is likely to be cached locally, and this step is probably skipped.
- Then, the computer queries the appropriate second-level DNS about domain.org, and the reply tells it the IP address of the third-level DNS that stores the information about all names within the domain.org name space.
- Finally, your computer queries this third-level DNS server and obtains the IP address of othercomputer.domain.org. (There are more steps to resolve host names that have more than three hierarchical levels.)

In fact, the DNS holds much more information about each domain, such as the name of the administrator, the address of the mail server for this domain, etc. All this information is stored in 'DNS **resource records (RRs)**'. For instance, the DNS record:

Othercomputer IN A 10.0.1.1

means that the computer named 'othercomputer' can be reached at the IP address 10.0.1.1. The DNS record:

Domain.org IN MX 10 10.0.1.2

means that the mail server (Mail eXchange) for domain.org can be reached at 10.0.1.2.

How can we use this? A special DNS record of type TXT can hold any text. So, we can use it to store the location of the gatekeeper handling the alias resolution for the entire domain. The syntax used is the following:

Domain.org IN TXT

**ras** [**<gk id>**@]**<domain name>**[: **<portno>**][**<priority>**]

**<domain name>** can be the DNS name of the host running the GK software or its IP address. The other fields are optional, <portno> specifies a non-standard RAS port,

<gk id> can be used if multiple logical GKs are running on the same host and, therefore, the name of the host cannot be used as the GK ID. Priority can be used if there are multiple gatekeepers in this domain, smaller numbers have precedence. For instance, valid strings for gatekeeper TXT records could be:

ras10.0.1.3

ras gatekeeper.mydomain.com : 1234 10

Now, when trying to call someone@domain.org, a computer can first locate the appropriate DNS for domain 'domain.org' as explained above, and then retrieve all the TXT records for domain.org. If a TXT record begins with 'ras', then the IP address or server name that follows is the name of the gatekeeper for this domain. There can be several RAS records; therefore, a zone can be served by several gatekeepers and a domain can be used by several H.323 zones.

Once the gatekeeper has been found, the caller can learn the transport address to send the SETUP by sending an LRQ or an ARQ. This call flow is supported by most commercial gatekeepers, making it simple to organize the VoIP network routing for email aliases.

#### 2.5.2.2 H.323 URL

In H.323v4, the syntax for an H.323 URL was added. The URL should begin with 'h323:' and be followed by a user name and optionally a server name, separated by the '@' sign. The server name can be an IP address, but in general it will be a DNS name. The procedure described above will be used to locate the relevant gatekeeper.

The H.323 URL can be located within a web page to cause a browser to make a call to the indicated address, if a properly configured H.323 VoIP softphone has been installed on the PC. Note that NetMeeting® does not support H.323v4 and will not react to such an URL. Microsoft uses a proprietary URL scheme to trigger NetMeeting calls.

### 2.5.3 E164 numbers and IP telephony

#### 2.5.3.1 A country code for the Internet

If IP telephony becomes a successful technology, more and more people will have an IP phone or IP telephony software running on a computer. How can they be called from the PSTN (Public Switched Telephone Network)? It is not possible to use email aliases.

Obviously, it is possible to call a gateway with an interactive voice response system that will ask which person must be called on the IP network. For instance, it could ask for a subscriber identifier and then place a VoIP call. This is not very practical. Many VoIP service providers buy large blocks of numbers from traditional carriers and allocate these numbers to VoIP subscribers.

The problem with all these solutions is that there is no way to know in advance that these numbers are reachable over IP; therefore, incoming calls are routed over the PSTN all the way to the network owning these numbers, then to a VoIP gateway. This may not be the optimal route, in many cases it would be much better to route the call sooner over IP.

The problem is that no specific numbering resource has been allocated to IP telephony: in 1997 discussions began within ETSI TIPHON about the numbering resource that would be most appropriate for IP telephony. Many solutions were proposed:

- A special prefix to be chosen in each country (e.g., in Japan all numbers beginning with 050 followed by 8 digits are allocated for IP telephony endpoints).
- A global service code for IP telephony. An example of such a global service code is 800 for free phone calls. From most countries in the world, it is possible to dial the international access code, followed by 800 and the service number.
- A global network code. Providers offering services in several countries can request to have a 5-digit network code allocated to them (e.g., a satellite phone company could be allocated the network number 99999 and you would reach its subscribers by dialing the international access code + 99999 + the subscriber number).
- A country code for IP telephony.

The first solution can be implemented very easily in each country if the local carrier wishes to and if local regulations allow it. This solution, adopted in several countries (e.g., Japan and Norway), may allow some countries to implement VoIP on a large scale easily; but, it has several drawbacks:

- The chosen prefix will be different in each country, and make IP phone numbers less easily recognizable.
- Although these IP phone numbers are now grouped under a single prefix, as opposed to being allocated at random according to each specific VoIP carrier request, this is still a large number of prefixes to manage on a global scale. It makes it difficult for telecom carriers to detect whether these calls should be routed over IP as soon as possible: in most cases the calls will still be routed to the user's country via regular telephone lines. If the call is not recognized as a VoIP call, which is very sensitive to media transcoding (tandeming) and delays, it may also get routed through DCME equipment or satellite links, thereby reducing call quality significantly.

The three last solutions are technically identical, the difference lies in the ITU rules for allocating each type of global code. IP telephony falls between these rules:

- It is not a geographic country.
- It is not a private network.
- It is not a specific service.

There were even proposals to present the need for an 'Internet country code' as an interworking requirement.

Still, the need is there and a global code would present the advantage of enabling VoIP users to connect anywhere in the world and always receive calls over the shortest IP path, as opposed to receiving their calls through their home country. A global code for VoIP is coherent with the global nature of the Internet. For instance, when receiving a call that originated in France, a US citizen temporarily working in France would receive a call that was routed over the French IP network, as opposed to a call routed first to the US over PSTN, then back to France over IP. This would justify lower prices for communications to IP phones behind such a country code. But, there are also some regulatory issues linked with such a solution, such as number portability (is it possible to port a VoIP number to PSTN and vice versa?) and legal interception (should the US CALEA apply to a US citizen working in France... and therefore do we want all these calls routed through the US?).

Some countries have already started to allocate a special prefix for IP telephony: on 14 December 1998, Norway allotted prefix 850 to Telenor Nextel for its VoIP service 'Interfon PC' (http://www.totaltele.com/view.asp?ArticleID=20742), and Japan allocated one million telephone numbers behind the '050' prefix.

Meanwhile, several technical proposals have been made to support the address resolution of a telephone number to a call-signaling IP transport address.

All proposals use a database (flat or hierarchical) to find the home gatekeeper handling the resolution of the phone number into a call-signaling address that can be used to send the SETUP message.

The number is not resolved directly into the IP call-signaling address of the endpoint because this IP address may change very often: for a dial-up user, a new address will be dynamically allocated each time he connects to the Internet. In addition, supplementary services based on the gatekeeper may redirect the call to different terminals based on time-based or other rules.

#### 2.5.3.2 UPT

The ITU already has a framework for a service called Universal Personal Telephony (UPT). UPT is based on a special access code and presents many similarities to the solutions presented above. UPT calls are routed to a 'serving exchange' which resolves the original number into an E.164 number. UPT includes several models:

- Model 3a is a flat numbering scheme behind access code +878.
- Model 3b is a numbering scheme behind access code +878 that substructures the numbering space with country codes.

In order to extend the scope of UPT beyond classical voice, the UPT model would only need a modification enabling the UPT information to include not only the address of one or more localization servers, but also the technology to use in order to reach them (classical voice, VoIP, protocol information, etc.).

The reader will not be surprised to learn that UPT 3a faces similar problems to the IP country code requests. There is no agreement on how to administer the flat space, and many political implications are far more complex to solve than all the technical problems put together.

Some ongoing UPT VoIP trials use UPT numbers that are first routed to a national switch of the operator owning the UPT number, then handed off to a VoIP system: the calls are not routed to a VoIP network at the source.

#### 2.5.3.3 DNS-based number resolution

##### 2.5.3.3.1 History

This proposal was first presented to ETSI TIPHON. It uses the network part of the IP address as the first part of the phone number. For instance, a user managed by a gatekeeper having an IP address allocated from class C 192.190.132.xxx could have +999192190132678 as a telephone number, where 999 is the country code allocated to IP telephony (only one thing is certain for the moment: 999 cannot be allocated for IP telephony!).

Readers familiar with IP addressing will probably be shocked by the last three digits, which clearly are outside the 1–254 range. We chose these numbers on purpose, because they do not need to respect the IP addressing rules.

According to this proposal, when a gatekeeper routes a call to +999192190132678, it will decide by analysing the first digits whether the network part of the phone number is class C (see Companion book, *Beyond VoIP Protocols* multicast chapter for more details on IP addresses class). Therefore, the part of the number that is an IP network identifier is 192.190.132.

Then, the gatekeeper will locate the DNS that has information about this network by doing an operation called a REVERSE LOOK-UP. During this operation, the network address is mapped to the DNS name 132.190.192.in-addr.arpa.

Once the proper DNS is located using the regular hierarchical DNS procedure, the gatekeeper queries the DNS server for information on 678.132.190.192.in-addr.arpa. At this stage 678 is just a name for the local DNS server, so there is no need to follow the rules for IP addresses. There should be an SRV record or a TXT record with the IP address or DNS name of a 'home gatekeeper'.

Once it has obtained this information, the gatekeeper can route the call to the home gatekeeper, or query this gatekeeper using an LRQ.

At first glance, this proposal was attractive: it was simple and enabled the use of distributed databases, solving the 'who owns the database' issue. However, it also had a number of showstopper issues that made it impossible to adopt.

(1) It is a hierarchical dialing plan. It is impossible to assign blocks of numbers of arbitrary sizes, which leads to rapid exhaustion of the numbering space.

(2) It is unfair. Some US universities have a class A of their own (255∗999∗999∗999 potential phone numbers) while entire countries have only a few class C networks (999 numbers). This is the unfortunate result of careless IP address allocation in the

past, but it is unlikely that people would accept this historical artifact to be replicated for telephony.

Finally, although one of the key features of this proposal is the ability to distribute the management of the database, considering features like number portability, and taking into account the fact that the very concept of network classes is now obsolete (a single 'class' is distributed over many owners in most IP networks[9]), leads to the conclusion that the whole database should be centrally managed.

#### *2.5.3.3.2 ENUM*

The key idea of the previous technique, using the DNS to resolve a telephone number into a call control resource information, has been refined and expanded by the Telephone Number Mapping working group of the IETF, resulting in RFC 3761. 'The E.164 to URI Dynamic Delegation Discovery System Application', the use of ENUM in the context of H.323 is outlined in RFC 3762 ('ENUM service registration for H.323 URL') and in the context of SIP in RFC 3824 ('Using E164 numbers with the Session Initiation Protocol'). This procedure is more commonly called 'ENUM'.

ENUM decomposes any telephone number into a pseudo host name. The number is first written in international format including the country code (e.g. +46-8-9761234 for a number in Sweden), then all non-digit characters (– , .) are removed, and finally the number is written in reverse order (to respect the right to left hierarchical nature of the DNS system), to form a pseudo host name in the new domain name E164.arpa. For instance, the previous number becomes:

```
4.3.2.1.6.7.9.8.6.4.e164.arpa
```

When a system needs to locate the appropriate resource to reach +46-8-9761234, it must query the DNS for the **Naming Authority Pointer Record** (**NAPTR**, DNS-type code 35, defined in RFC 2168 and RFC 2915) corresponding to the pseudo host name `4.3.2.1.6.7.9.8.6.4.e164.arpa.`

The NAPTR record is used to attach a rewrite rule, based on a regular expression, to the DNS domain name. Once rewritten, the resulting string can be interpreted as a new domain name for further queries, or a URI (Uniform Resource Identifier) which can be used to delegate the name look-up. The syntax of the NAPTR RR is as follows:

```
Domain TTL Class Type Order Preference Flags Service Regexp
    Replacement
```

Domain, TTL, and Class are standard DNS fields; Type is set to 35 in the case of the NAPTR. The order and preference field specifies the order in which records must be processed when multiple NAPTR records are returned in response to a single query. The ordering is lexicographic, order is used first, then preference.

[9] IP addresses are now allocated in blocks of arbitrary size (as long as it a power of 2). The mechanism of reverse DNS look-up can be adapted for classless IP addresses, but most DNS servers do not support it and very few of those that do have been properly configured.

The 'S', 'A', 'P', and 'U' flags indicate how the next query should be processed. The next query should request SRV records for flag 'S', A records for flag 'A'. Flag 'P' indicates a protocol-specific algorithm, in this case the 'replacement' field will be used as the new name to fetch the corresponding resource record. If the flag is 'U', the regular expression (an expression composed of a series of symbols each defining a specific modification to a string, and defined in POSIX) specified in the RegExp field should be applied to the domain name to get an absolute URI. This last option is used by ENUM.

The service field defines the protocol that will be used for the next step of the resolution (H323, LDAP, SIP, TEL), and the type of service that will be provided ('E2U' in the case of ENUM).

The input of the regular expression is the E.164-encoded telephone number (international format), with a leading '+' sign, and only digits (e.g., +4689761234).

For instance, the following record transforms +4689761234 into sip:info@company.se and mailto:info@company.se, respectively. The preferred method is H.323, then SIP, then the Simple Mail Transfer Protocol (SMTP):

```
$ORIGIN 4.3.2.1.6.7.9.8.6.4.e164.arpa.
  IN NAPTR 100 10 "u" "h323+E2U"    "!^.*$!h323:info@company.se!".
  IN NAPTR 100 20 "u" "sip+E2U"     "!^.*$!sip:info@company.se!"
  IN NAPTR 102 10 "u" "mailto+E2U" "!^.*$!mailto:info@company.se!".
```

LDAP could also be used to continue the query:

```
$ORIGIN 6.4.e164.arpa.
 *IN NAPTR 100 10 &u" &ldap+E2U" &!^+46(.*)$!ldap://ldap.se/cn=01!".
```

At present, ENUM is still very much theoretical, as it raises complex administrative and technical issues. The delegation of the e164.arpa domain is a very sensitive issue, being discussed between the Internet Architecture Board and the ITU Study Group 2 in charge of numbering issues.

The technical issues are many, prominent among which is latency, as the DNS resolution can be very low. This is a characteristic of any database relying on hierarchy to scale. Another issue is the timing of record updates: the caching mechanism used by DNS times cache records out after the time-to-live period, but this is not sufficiently precise for telephony use, notably for number portability which would require resource record updates simultaneously by all service providers.

Another major concern is security, because many denial-of-service attacks or, even worse, phone number-hacking schemes are made possible in ENUM. DNSSEC (RFC 2535) solves some of the security issues, at the expense of a more complex administration and possibly increased resolution delays.

In order to be used in operational networks, ENUM will probably require further work. It seems likely that the question of which protocol to use to request a phone number resolution and which back-end database to use will need to be more clearly separated.

#### 2.5.3.4 Dialing plan distribution

The previous discussions apply mainly to the issue of reaching an IP terminal from its phone number allocated from a E.164 numbering space. This is the problem of the 'last hop'.

But there will probably be several transit IP networks offered to reach a particular VoIP alias, and similarly many IP telephony providers will offer outgoing gateways to call regular phones. Therefore, another interesting problem is the distribution of reachability information and any related data (prices, QoS levels, etc.) that may help a service provider to choose among several possibilities when routing a call to a regular PSTN number or an H.323/SIP alias.

In H.323v3, H.225 annex G ('Communications between administrative domains'), was introduced as a way to exchange reachability and pricing information between administrative domains. Essentially, the protocol is very similar to a simple routing protocol, like RIP in IP networks, but distributes reachable phone numbers instead of reachable IP networks. Annex G really takes a simplistic approach to exchanging reachability and cost information, and shares many limitations and drawbacks also found on simpler IP routing protocols. There is a big difference though, VoIP calls are charged and any error, even temporary, can cost millions or ruin the reputation of a carrier. The potential issues that can be introduced by annex G, from call loops to denial of service (by announcing low prices on fake routes) have caused all carriers so far to back off. Therefore, there hasn't been any operational deployment of annex G by any significant VoIP carrier.

Exchange of route and price information between carriers is still done in the old-fashioned fax-and-paper way, or using file transfers, in most cases, and routing information is entered in the central routing systems under human supervision. Essentially, nothing has changed from the days of traditional carriers.

#### 2.5.3.5 Conclusion

The dream of many engineers who worked on DNS usage for phone number resolution or automated phone-routing protocols was to make the phone network as informal as the Internet is today, where a new service provider can get connected in minutes just by listening to the routing protocol advertisements of neighbouring service providers. The introduction of a country code also aimed at facilitating easy identification of calls to IP phones and promoting lower prices to such destinations, justified by the fact that calls would be routed over IP directly from the source.

It is likely that no global code will be allocated to VoIP, service providers will instead negotiate bilateral agreements to exchange traffic over IP. This happens more and more because many traditional carriers now exchange minutes over VoIP. When routing a call to a VoIP service provider, a carrier with a VoIP-capable transit network will certainly send the call directly over IP, and not through the PSTN. Therefore, calls will be routed over IP straight from the source, but charged as regular telephone calls since the absence of a global code identifies these calls as calls to IP phones.

The use of DNS-like mechanisms for telephone networks is also proving more difficult than anticipated. The key difference between IP networks and telephone networks is number portability. In most countries, it is no longer possible to know the service provider owning a number from a prefix in the number. In fact, most telephone service providers already have to carry out phone number resolution today, not to find a home gatekeeper, but to find the carrier owning the destination phone number. Today, number resolution schemes in the PSTN roughly fall in three categories:

- Onward routing: every service provider is responsible for resolving a block of numbers, even if these numbers get ported out. Call routing in the network remains hierarchical, calls are always routed first to the owner of the block identified by a well-known prefix (e.g., 123xxxxxx is always routed to Acme Telecom). The technique used by each service provider for number resolution of incoming calls is arbitrary, as no external query protocol is used: the call is simply rerouted to the new owner based on a local resolution. For instance, Acme Telecom finds that 123555555 has been ported out to Competitor Telecom and reroutes the call to this carrier, usually using a special routing prefix (e.g., D000991234555555). This technique can be used by VoIP 'as is'; in fact, it already works in several networks today (Italy, Germany, France). The major advantage of VoIP is that media tromboning is avoided. This technique does not require ENUM or anything else, since each service provider is free to choose its local resolution mechanism (many use LDAP).
- All-call query: a central database is queried by all service providers for each call in order to obtain a new routing number, identifying the new owner of the number. This scheme can also be extended easily to include technology information (this call should be H.323, SIP, . . .). All-call query is used in the US and several other countries.
- Synchronized databases: each service provider is required to have a locally synchronized database of all numbers and their current owner. National authorities define the database synchronization primitives, which are rather complex in order to take into account number portability (numbers can have a portability request pending, be technically ready to be ported out, technically ready to be ported in, etc.). This technique is used in Denmark. The 'synchronized database' approach would require a protocol that has nothing to do with the current DNS query protocol.

The DNS query protocol of ENUM could be used to replace the current 'all-call query' mechanism (based on SS7 TCAP), with the same or even better functionality, but this will in many cases require a centralized database, as opposed to the distributed DNS database foreseen by ENUM. This is because the DNS database is hierarchical, and the DNS delegation model supposes that you delegate 'blocks' of numbers to a service provider, which is no longer true with number portability. In order to use the DNS delegation model, the original owner of a number would need to be required to manage the associated resource record indefinitely, even if the number is ported out (this is the equivalent of the onward routing method for DNS requests). If this is not acceptable, then the DNS delegation model can be used only at the country code level, but all national resolution servers would need to be owned and managed by a single entity. In addition, ENUM would also need to clarify the use of caching mechanisms in the context of number portability (DNS caching with a TTL value of $T$ creates an interval of time of length $T$

where servers can respond to queries with the new or the old record) and would need to take into account non-E164 numbers (e.g., national 800 numbers and short service numbers that do not have an international form).

Overall, VoIP will not change much the way service providers work today:

- Network interconnections and call-routing decisions, including the choice of transport technology and protocol, are negotiated on a bilateral basis, not via a routing protocol.
- The number resolution mechanism will depend on the country. Number resolution performed by onward routing may be enhanced by requiring all carriers to maintain an ENUM database for their blocks of numbers. Alternatively, a national database can be used, queried via SS7/TCAP (an IP-based query protocol), the ENUM DNS query protocol, or anything else.

## 2.6 H.323 SECURITY

### 2.6.1 Typical deployment cases

Building a telephony trial based on IP technology is easy; it is much more difficult to build a production-grade VoIP network, providing scalable and secure operations despite the vast heterogeneity of connected devices and networks.

A security and authentication issue exists whenever the VoIP core network is connected to untrusted IP networks or untrusted VoIP devices.

This section presents typical deployment cases, and the combination of softswitch and network features that can be used to solve security and authentication issues.

#### *2.6.1.1 Carrier-to-carrier connections*

Interconnection between **Internet Telephony Service Providers (ITSPs)** is required in order to:

- Send traffic to another ITSP for least cost routing application.
- Terminate traffic in the local VoIP network for another ITSP.
- Route traffic from ITSPs to other ITSPs according to a least cost routing policy (arbitrage, clearing house).

These call flows present multiple security challenges.

##### *2.6.1.1.1 Third-party dependence for security*

Some ITSPs run their internal call routing using direct mode softswitches (also called light class 4). These softswitches are in fact used as simple directories, and reply to VoIP gateway queries with the IP address of the target VoIP gateway used for each call. In the direct call model, call-signaling is set up directly between gateways, in a peer-to-peer model. This model presents a number of issues even in a closed VoIP network (e.g., no ability to reroute calls on PSTN congestion, no ability to connect untrusted devices such as IP-PBXs, gateway-based billing, etc.), but presents an unacceptable security risk when used across different VoIP domains. For each A to B call, the peer-to-peer model supposes that gateways from domain A will be given (by the direct mode gatekeeper of domain B) the IP address of the gateway in domain B where the call SETUP message should be sent.

This implies that any gateway in domain A should be able to reach the signaling ports of any gateway in domain B. In H.323 this means opening TCP port 1720 (Q.931), and even all TCP ports over 1024 if H.245 tunneling is not used, to all gateways in network A. Since the IP addresses of gateways are not known or frequently changing, this means opening access to all gateways in network B from the entire IP network A. If network A security is compromised, network B security becomes compromised as well. As the amount of network peering grows, the security of the overall network becomes weaker.

The VoIP gateways signaling ports should always be protected from the outside. Most VoIP gateways cannot sustain more than a few calls per second and, therefore, are very easy to break using relatively light denial-of-service attacks.

Whenever possible, the routed call model, which does not authorize call control messages to be exchanged directly between gateways, should be used. In router access control lists (ACLs), only the call control communications between the gateways and the routed mode softswitch (e.g., TCP port 1720 for H.323) should be opened. The softswitch sitting between the ITSPs acts as a 'fuse' between networks: any attack may bring down the edge softswitch, isolating the two networks, but will not compromise the protected network.

#### *2.6.1.1.2 Authentication*

Arbitrage requires the ITSP to generate bills for each connected ITSP and, therefore, to be able to determine which calls are coming from which ITSP. This authentication must be done in a way that is as independent as possible from the IP telephony vendor used in the connected networks, because ITSPs use a wide variety of gateways and softswitches, many implementing a number of proprietary extensions. Some methods using cryptography have been proposed (e.g., the **Open Settlement Protocol** on **OSP** which uses SSL HTTP requests), but they cannot in general be deployed in real networks due to the interoperability challenges caused by the cryptographic token formats used, the variety of VoIP protocol flavours, and the performance issue of TCP-encrypted links. The most efficient and flexible way of authenticating calls coming from a given ITSP is to validate the source IP address of the signaling against the IP network allocated to that ITSP. When the routed mode is used the IP address of the ITSP gatekeeper can be preconfigured. IP address forging is impossible, since VoIP requires messages to be sent to and received by each ITSP. Therefore, ITSPs that could forge their source IP addresses, even if this wasn't blocked by the router ACLs, would not be able to place calls. Having a routed mode softswitch which supports source IP address validation and reports the source IP address in its CDRs, combined with IP spoofing protection at the router level, ensures that inter-carrier billing is both reliable and scalable, and can be deployed in a multi-vendor, multi-protocol context.

##### *2.6.1.1.3 Confidentiality*

ITSPs running an arbitrage or a clearing house business may not want to expose the IP addresses of their partners' VoIP gateways, for fear that third parties may discover the names of their termination partners and establish direct business relationships bypassing the clearing house. This issue is highly theoretical since termination tariffs are generally negotiated by clearing houses based on the aggregate call volume of all their customers and would not be accessible to individual customers anyway. Clearing houses are really protected by their purchasing power more than anything else. Still, exposing termination partners' IP addresses is a concern for some clearing houses.

Most routed mode softswitches natively include IP address-hiding features in all the messages that are relayed by the softswitch. If these options are activated, all signaling-related IP addresses of the network hidden by the routed mode softswitch are replaced by the IP address of the softswitch. However, the softswitch does not see the media stream and, therefore, cannot by itself hide the IP addresses of media streams. In most cases, hiding the IP address of signaling messages is deemed sufficient, as it will make it impossible for the connected ITSP to see the hidden IP addresses in all of its common reporting tools: CDRs and traffic reports will only indicate the IP address of the softswitch. Learning media-level IP addresses would imply using network sniffers on a regular basis, which is unlikely in any serious deployment environment. If media-level IP address hiding is necessary, it will be necessary to use RTP relays, which have in some cases a significant impact on network engineering, quality of service and scalability.

##### *2.6.1.1.4 Scalability and quality of service*

Considering the potential problems exposed above, some ITSPs have decided to peer using only traditional TDM switches connected to gateways. Ignoring the cost, this poses two issues: audio delays will be at least doubled, probably exceeding acceptable limits for most users, because both gateways will have audio jitter delays and coding delays. In addition, the gateways on both sides of the TDM switch may use different coders for each portion of the call (e.g., GSM on one side and G723.1 on the other, converted to G.711 in the middle because the TDM switch can only process this codec: this creates codec-tandeming issues further degrading the audio quality). Another solution is to use an IP/IP VoIP gateway, which terminates all signaling and RTP media flows, thereby hiding the entire network behind its IP address. Such a solution significantly improves audio quality compared with the previous technique, because the RTP stream does not need to be decoded and recoded, but still it prevents the RTP stream from being routed along the shortest path between the calling party and the called party, and obviously introduces additional delay compared with solutions which route only the signaling data.

Note that it is generally a bad idea to use a general purpose firewall to relay RTP media streams, because RTP packets are significantly smaller than average data packets, leading to unusually high packet-per-second rates. Most firewalls will at least require hardware add-ons to properly relay RTP streams without introducing packet loss or jitter on carrier size links.

### 2.6.1.2 Class 5 residential networks

Residential telephony services over an IP network are becoming increasingly popular. Next-generation service providers want to provide bundled video data and telephony services that can only be provided over IP, and consumers may also be willing to use the software IP phones provided with the Microsoft Windows® OS. One of the key challenges posed by residential VoIP services is the heterogeneity of endpoints that can be used:

- Analogue telephone adaptors, with or without modems.
- Voice-enabled cable modems.
- Softphones.
- IP phones.

Most of the time these endpoints are registered to an access gatekeeper or SIP registrar (using RAS messages in H.323, Register in SIP), while call-related signaling is handled by a routed mode softswitch as shown in Figure 2.28.

Class 5 application adds the security requirements given in Sections 2.6.1.2.1–2.6.1.2.5.

#### 2.6.1.2.1 Authentication

Residential phones or gateways need to be authenticated, and the authentication method needs to be as independent as possible of the phone manufacturer.

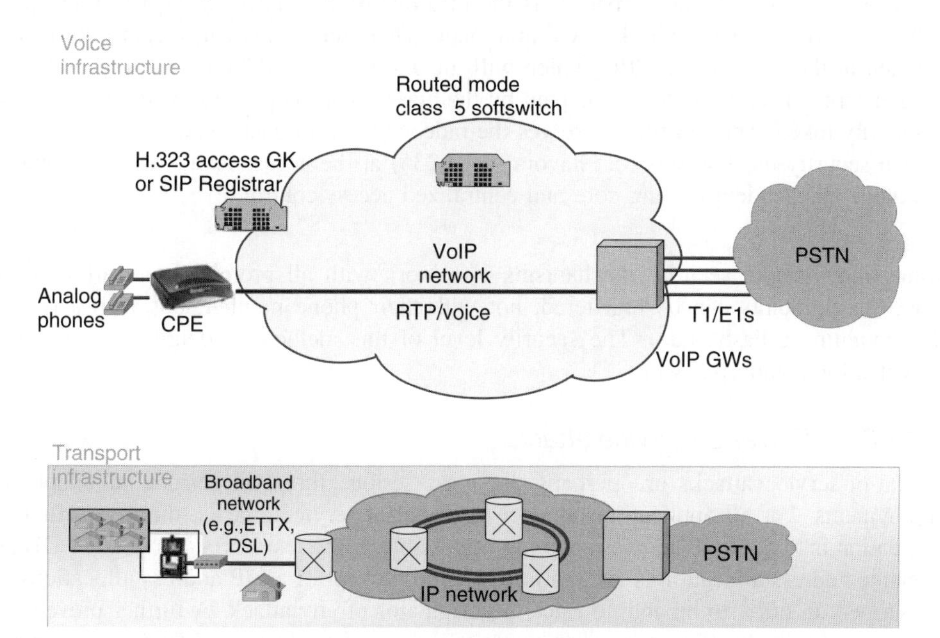


**Figure 2.28** Large-scale residential network combining a set of access gatekeepers and one or more routed mode call control gatekeepers.

Unfortunately, strong authentication methods implemented in VoIP edge devices are still mostly proprietary at this time. VoIP standards define frameworks for implementing security (e.g., H.235 for H.323), but fail to describe the exact implementation of security mechanism, or propose too many options. In order to implement strong security mechanisms in a class 5 context, the separation of access functions and call control functions in the class 5 architecture allows access gatekeepers or registrars from the same manufacturer as the CPEs to be used, while using a third-party softswitch for centralized call control. This enables the use of multiple proprietary authentication schemes at the edge, while maintaining a unique call control device.

In H.323 this can work as follows:

- If all the CPEs and trunk gateways are from the same manufacturer, then when the CPE is cleared to make the call by the access gatekeeper, it receives in the ACF message a cryptographic token which is copied in the SETUP message. The authentication and authorization policy can be used by the access gatekeeper. The token is then carried transparently by the class 5 call control switch, and when it arrives at the trunk gateway the trunk gateway will validate this token again using a receive side ARQ sent to an access gatekeeper. In the case the token is not present or valid, the call is rejected.
- If the CPEs and network gateways are from different manufacturers, then a hierarchical authentication mechanism can be used. Once the access gatekeeper has validated the call attempt from the CPE, it sends an LRQ message to the class 5 call control gatekeeper. The class 5 gatekeeper signs the information contained in the LRQ and puts this electronic signature in the LCF in a token. Then, the access gatekeeper returns this token in the ACF to the CPE, which will include it in the SETUP sent to the class 5 gatekeeper. The call will be accepted by the class 5 gatekeeper only if there is a valid security token. This scheme preserves the independence of each vendor to design their own security mechanisms (or 'flavors' of H.235) at the access level, while keeping a vendor-independent routing core and centralized access control.

Some other simple security mechanisms also work with all protocols and all vendors (e.g., the endpoints can be registered, not with their phone number, but with a secure ID containing a hash code). The security level of this method is identical to the level provided by a static password.

#### *2.6.1.2.2 Denial-of-service attacks*

Denial-of-service attacks are perhaps the most serious threat in residential telephony deployments. For all applications subject to denial-of-service attacks, the most efficient prevention is to make IP address spoofing impossible in the residential network. If DHCP dynamic address allocation is used, a trace should be kept of all IP address allocations to subscribers in order to be able to trace the originator of any attack. A further prevention made possible if the IP access layer supports it (e.g., if sophisticated IP-aware Ethernet switches are used in Ethernet-to-the-building or ETTB deployment) is to configure token bucket rate control at the edge on all signaling flows. For instance any data flow going to

TCP port 1720 of the gatekeeper from a residential H.323 subscriber should be allocated a very low average bitrate.

Any routed mode softswitch by definition routes all signaling, and therefore acts as a fuse in case a DoS attack should occur. Therefore, an independent routed mode softswitch should be placed at network boundaries potentially subject to DoS attacks, notably those where anti-spoofing is not in place. When such an attack occurs, the CDRs provide immediate information on the IP address of the calling party (except for the distributed DoS attacks, against which there is no known prevention today for any application). Some softswitches have built-in protection against DoS: preservation of a minimal level of service, immediate detection of the attack, and automatic recovery as soon as the attack stops.

#### *2.6.1.2.3 Billing*

Customer premises' devices can never be trusted, therefore billing records should never depend on them (e.g., Radius records generated from a CPE gateway cannot be trusted). Again, the solution is to use a routed mode softswitch, which processes all H.323/SIP call control signaling:

- No call can be made to the network unless the softswitch allows it.
- All call signaling events, including call establishment, call termination, abnormal abort, are known to the softswitch and taken into account in the billing records.

Prepaid communications are a special case, as they require a dynamic call cut-off capability. The routed mode softswitch may support this dynamic call cut-off feature by injecting call release messages in the call-signaling path: the two half-calls (from the caller and to the callee) are released by the softswitch, causing the network-side device (e.g., a VoIP gateway) to close all signaling and media ports associated with the call. Even if the residential endpoint ignores the call release message and continues to send audio packets, the device at the other end will not transmit these audio packets and will reply with ICMP error packets.

This architecture ensures that no call lasts longer than actually measured by CDRs; but, the opposite problem exists as well, some calls may be shorter than indicated in the CDRs.

In all VoIP networks, but even more importantly when software IP phones or residential devices are used, it is very difficult to get a reliable indication of call termination. Such devices may suddenly stop responding without properly releasing calls, due to a failure or a network connectivity problem. This can create very significant business issues if customers get billed for minutes they did not use. The softswitch should implement a dead endpoint detection algorithm ensuring that calls from/to any non-responding endpoint will be cleared within seconds. The transport layers used by traditional network protocols include a 'keep-alive' mechanism that has been forgotten in most VoIP protocols and must be recreated. There are several ways of doing this, either at the transport level (e.g., in H.323 sending malformed TCP segments from time to time and checking that the other end rejects them as expected, this is called '**TCP keep-alive**'), or at the protocol level by sending asynchronous queries from the softswitch to the endpoint (e.g., SIP OPTIONS message or MGCP AUEP). The use of the new **Stream Control Transmission Protocol** (**SCTP**, RFC 2960) for VoIP would help solve this issue in a more systematic way, because it includes a built-in keep-alive mechanism.

##### *2.6.1.2.4 Regulatory features*

National regulations require the availability of the following services for all IP to PSTN, PSTN to IP, and IP to IP calls:

- Legal call interception.
- Malicious call identification.
- **Calling Line Identity Presentation (CLIP)**.
- **Calling Line Identity Restriction (CLIR)**.

Many VoIP networks today support these features only through PSTN switches: they are supported only for IP to PSTN or PSTN to IP calls, not for IP to IP calls. This is not acceptable for deployments, as regulatory agencies require these features for all calls regardless of the technology used. The softswitch should implement all of these features in the network and for all call flows, including IP to IP calls. The legal intercept feature in particular is always a bit tricky to implement in VoIP networks, as it should not introduce any noticeable delay in the audio path. It is impossible to use traditional conference bridges (which add delays due to RTP decoding and jitter buffers): dedicated devices must be designed which duplicate RTP packets on the fly without decoding them.

##### *2.6.1.2.5 Is media encryption required?*

Sometimes, the availability of technologies creates the need. This is what happened with media encryption for telephony. It is often heard that IP telephony is not secure if media streams are not encrypted. Such a statement is exaggerated. Indeed, in traditional residential or business networks the call can be intercepted at the following places:

- On the phone wire, simply by connecting a loudspeaker (or a passive sniffer for ISDN and digital phones).
- On the link between the user site and the service provider, simply by connecting a loudspeaker for analogue lines and T1/E1 network sniffers for digital lines.
- In the service provider network, where it is necessary to be on the path on the call and to use the ISUP SS7 information to find the TDM time slots and circuits used to transport the call voice stream. This is close to impossible without the collaboration of the service provider.

So far this level of security has been sufficient for most uses. Standard VoIP offers a similar or superior level of security:

- If LAN switches (not hubs) are used in the customer premises, the call can be wire-tapped only by using a network sniffer with VoIP capabilities, *and placing it on the Ethernet wire directly connected to the IP phone*. This is because switches do not broadcast packets, which therefore are routed only on the shortest wire path between the source and the destination. The level of security is therefore comparable with that of standard home or business telephony networks. There is no way you can listen to your neighbour's conversation; you do not even have access to the IP packets.
- On the link between the service provider and the company/user the call can be wire-tapped by using a WAN network sniffer. If IPSec is used between the customer router and the network router, this becomes impossible without government-level encryption-cracking technology.
- In the carrier network the H.323/SIP signaling of the call must be correlated with the IP addresses of the end-to-end media stream, then with the IP routing information, to locate the path of call IP packets. This is a lot more complex than in traditional TDM networks.

In short, standard unencrypted VoIP is more secure than traditional technology. Even governments are having a hard time trying to break it down, even with softswitch vendors and service providers' active co-operation in implementing wiretapping. Media-level encryption at the phone level should be reserved only for niche markets (e.g., for military use—VoIP is very popular in the armed forces, as it significantly simplifies and accelerates wiring in battlefields).

Even VoIP over wireless LAN will not require any specific development, as the air link is already secured by layer 2 mechanisms.

### 2.6.2 H.235

H.235 aims at providing application-level privacy (no eavesdropping) and authentication (assuring that people are really who they pretend to be) to H.323 communications and more generally all protocols using H.245. Because H.323 can be used on the open Internet, it inherits the reputation of the Internet; some say that it is less secure than regular telephony. In fact, even without H.235, it is much more difficult to listen to an H.323 phone call than to wiretap a phone line, because you need to implement not only sophisticated network-sniffing tools, but you also need to be on the path of IP packets and to implement the proper voice codec algorithm. Legal call interception, for instance, is much more complex with voice over IP than traditional phone networks. With H.235, IP telephony becomes *much* more secure than regular telephony. Security based on H.235 can be implemented at several levels. With the strictest options, it becomes virtually impossible, even for someone having free access to the IP network, to listen to any conversation that has been secured by H.235 or even to know the number that is being called. In most cases though, H.235 will be used only to make sure that users do not forge their identities. For other aspects the security level provided by standard H.323, comparable with or better than the security level of the TDM networks, is deemed sufficient.

#### 2.6.2.1 A short introduction to cryptography

This chapter has been written purposefully to avoid any reference to the complex notions of algebra, but unfortunately we cannot avoid it altogether. It is possible to read H.235 by considering each encryption function as a black box, but then many parameters, random numbers here and there, remain obscure. We have chosen to describe the cryptographic algorithms used by H.235 in a simple way, but it doesn't mean that they are simple. The real complexity of cryptography is in the detail: How do you choose a random number? How do you calculate a large prime? And so on. So, this chapter will probably seem crude to cryptography experts, but we hope it will help those just wanting to have an overview.

##### 2.6.2.1.1 Common terms

Cryptography is a set of techniques and mathematical algorithms which address one or several of the following needs:

- Privacy: the need to keep the content of a piece of information unknown to anybody except a controlled set of individuals.
- Authentication: the need to check and verify identities.
- Non-repudiation: the ability to attribute with certainty a document, a call, or any piece of information to an author.
- Integrity: the need to preserve the original content of a document from any modification or falsification.

##### 2.6.2.1.2 Cryptographic techniques

Two main techniques are in use today:

- The first one (called symmetric cryptography, shared key cryptography or secret key cryptography) is probably as old as civilization. Caesar was already using it to send messages to Rome.
- The second one (called asymmetric cryptography or public key cryptography) is much more complex and is based on elaborate mathematical algorithms.

###### 2.6.2.1.2.1 Secret key cryptography

(a) Simple algorithms

Secret key cryptography relies on a shared secret between the sender of a message and the receiver. The shared secret can be the algorithm used to encode the message (e.g., a given permutation of letters), or a 'key' used as a parameter in a well-known algorithm.

Simple algorithms, such as letter permutation, are very weak unless the permutation changes frequently: a message can usually be cracked by examining only about 40 letters of the cryptogram when a fixed permutation is used.

A refinement of this algorithm, called **one time pad** was described by Vernam in 1926: if a message is encrypted by adding a completely random key of the same length (e.g., doing an XOR with a random bitstream), then the cryptogram contains absolutely no information for anyone not knowing the random key. Nothing can distinguish it from a random message. In other words, the security of this system is perfect and mathematically proven, provided of course the random string is *really* random and unknown to anyone except the two parties exchanging information. A pseudo-random string can be used instead, but then the security of the system then depends completely on the quality of the pseudo-random generator.

One time pad has, however, a serious drawback: it needs to send an extremely long random key in advance to the recipient of the messages in a secure way. The advent of the CD-ROM has made this relatively easy, and this system is still used today for military or diplomatic communications.

(b) DES and its successors

The most widely used secret key algorithm in use today is the **Data Encryption Standard (DES)**. DES is a consequence of a consultation by the US Commerce Department in 1971 asking for a secure, yet easily implementable encryption algorithm. They requested a publishable algorithm (i.e., security could not rely on the fact that the algorithm was unknown).

It was only in 1974 that an appropriate proposal was submitted: IBM's Lucifer algorithm. The proposed algorithm was modified and finally resulted in the algorithm that was standardized in 1976 as DES. The current standard is Federal Information Processing Standards Publication 46-2 (**FIPS PUB 46-2**) of 1993.

DES is a block algorithm that can code a message of 64 bits into a cryptogram of 64 bits using a 56-bit key (the actual key has 64 bits, but 8 are used just for error detection). Regarding patents, IBM grants under certain conditions free licenses for devices using DES.

Coding 64 bits is not extremely useful, and an additional standard (FIPS PUB 81) describes how to extend the use of DES to data of arbitrary size:

- **Electronic Code Book** (**ECB**) is the direct application of DES on a message split into 64-bit chunks using the same key repeatedly. It is not very secure because similar sequences in the initial message will also appear in the coded message, leading to several potential attacks.
- **Cipher Block Chaining** (**CBC**) avoids this weakness of ECB by using the result of the encryption of a block $n$ to perform an XOR (eXclusive OR) with block $n + 1$ before encrypting it. A transmission error on one block of a CBC-encoded file will prevent the proper decoding of both this block and the next one.
- **Cipher Feedback** (**CFB**) is more appropriate for coding sequences of less than 64 bits.
- **Output Feedback** (**OFB**) uses DES to generate a pseudo random bit sequence that is added (XOR) to the message to be encoded. OFB can in theory code small messages of less than 64 bits but is considered more secure when coding messages over 64 bits. OFB is not subject to error propagation and for this reason is quite appropriate for coding audio or video. (The coding used in the GSM cellular standard is derived from OFB.)

Because CBC, CFB, and OFB all use chaining, the sender and the receiver must be provided with a common initialization vector, in addition to the key. When using H.235, the key is carried in the **EncryptionSync** parameter. H.235 also describes how to construct an initialization vector for CBC, CFB, and OFB.

With the power of computers ever increasing, the safety provided by DES has been questioned. Triple DES simply chains three individual DES blocks using different keys, which raises the complexity of the algorithm from a $2^{57}$ equivalent codebook to $2^{112}$ (because of some theoretical reasons, the exponent is increased only by a factor 2, not 3). The justification for three stages instead of two is quite involved, but in short it has been proven that using just two stages does not increase the security of basic DES.

There are many other efficient algorithms using a shared secret key:

- RC2 codes blocks of 64 bits with a key of 40, 56, or 128 bits.
- RC4 is flow-oriented and uses a 40- or 128-bit key.
- The brand new Rijdael algorithm, which will be the successor of DES.

The RC2 and RC4 algorithms have been implemented by many US companies, as they once were easier to export than DES-based solutions.

###### *2.6.2.1.2.2 Asymmetric cryptography*

**Asymmetric cryptography** is based on a new pragmatic way to consider the security of information: a piece of information is not secure only if you don't know how to extract the information; it is also secure when you do know how to extract the information but you cannot practically do it because it would require too much time to run the extraction algorithm even for the fastest computer.

(a) One-way functions, 'hash' functions

Asymmetric cryptography uses many so-called 'one-way' functions. A function $F$ is a one-way function when it is extremely difficult to find $x$ knowing $F(x)$. The idea is that if we have such a function mapping a set of messages $M$ to another set of messages $C$, it is possible to code a message $m$ from $M$ by using $F(m) = c$ as a cryptogram. For instance, the following function is 'one way':

$$Z/pZ \rightarrow Z/pZ$$

$$x \rightarrow q^x \bmod p$$

where $p$ is a very large prime number (typically with over 100 digits). It is possible to demonstrate that in this case there is at least one number $q$ that is 'primitive' (i.e., for any element $E$ of $Z/pZ$ it is possible to find an element $x$ such as $q^x = E$. If $q$ is chosen to be primitive, there is a one-to-one mapping between the initial message $m$ and the cryptogram $F(m)$. There are classic methods to find a primitive element of $Z/pZ$ once $p$ has been fabricated to have some 'good' properties.

*In theory*, it is possible to find the initial message from the cryptogram by calculating $F(x)$ for each possible $x$ and compare the result with the cryptogram. But, there are about

$10^{100}$ possible values for $x$ and each calculation is very costly—just imagine how long it takes to calculate, say $12343323223^{654654654654}$! Even being smart and trying to optimize the calculation (e.g., trying to calculate $((((((q^2)^2)^2)^2)^2)^2 \ldots)$ there will still be about log(654654654654) multiplications. But, why not 'just use the inverse function'. The fact is we don't know how to invert this function efficiently. There is no other way than to try each possible solution until a match is found!

A straightforward application of one-way functions is password storage. Instead of storing the clear form of a password $p$, we store $F(p)$ on the authentication server. This way the password file cannot be used to find the original passwords. When a user logs on with a password $p'$, the system can simply check the validity of the password by verifying that $F(p')$ equals the stored value $F(p)$. In this case the function does not need to provide one-to-one mapping: we can tolerate having several passwords mapping to the same code, if of course the number of possible codes remain very high; this is called a hash function. Hash functions are also frequently used to 'summarize' and electronically sign information (e.g., in H.323 the information contained in H.235 ClearTokens (Section 2.6.2.2.1.1) can include a hash code parameter, which is the result of $F$(token information, secret)). This prevents anyone from easily modifying the token information, because they cannot recalculate the proper hash code without the secret.

(b) How to negotiate a shared secret with the Diffie and Hellman algorithm

This algorithm allows two persons, say Bob and Mary, to negotiate a common secret over a *public* link. First, Bob and Mary need to agree on a large prime $p$ and an integer $q$. It is not a problem if other people know this choice as well. Then, Bob and Mary execute the following steps:

(1) Bob secretly chooses an element of $Z/pZ$: $a$. Mary secretly chooses an element of $Z/pZ$: $b$.
(2) Bob sends $q^a \bmod p$ to Mary. Mary sends $q^b \bmod p$ to Bob.
(3) Bob and Mary choose $S = q^{ab} \bmod p$ as a common secret. They can calculate it easily because $S = (q^a \bmod p)^b = q^{ab} \bmod p = q^{ba} \bmod p = (q^b \bmod p)^a$!

There is no known way to calculate $S$ knowing only $q^a$ or $q^b$! Bob and Mary have managed to negotiate a common secret on a public link and can use any symmetric cryptography method using this secret to exchange messages. The Diffie–Hellman algorithm is in the public domain.

(c) Public key encryption with the El Gamal algorithm

The public key encryption system presented below was authored by El Gamal and derives directly from the Diffie–Hellman method. Again, we use the discrete logarithm function $F : x -> q^x \bmod p$, in which $q$ and $p$ are known to both the sender $B$ and the receiver $A$ and possibly other persons as well. In this system recipient $A$ has a public key $P_a = F(a)$ built from his secret $a$.

$B$ wants to send a secret message $M$ to $A$, and of course $B$ wants to be sure that only $A$ can decipher the message. For simplicity, let's assume for a moment that the message is a number in $Z/pZ$.

$B$ chooses a random number $k$ and sends $q^k \bmod p$ and $M * P_a^k \bmod p$ to $A$. Note that in this system the cryptogram is twice as long as the original message.

Anyone intercepting the coded message needs to know the value $P_a^k$ to find $M$. $P_a$ is widely known, so all that is needed is $k$. But, as we have seen above, it would take an enormous amount of calculations to find $k$ from $q^k \bmod p$ if $p$ is large enough. So, unless a government agency with a large budget is really determined to discover $M$, $B$ can be pretty safe.

For $A$, it is very easy to find $M$ from the information sent by $B$. First, we have to remark that $Pa^k = (q^a)^k = (q^k)^a$. $A$ knows $q^k$ and $a$, so can easily calculate the value of $Pa^k$ and deduce $M$ immediately.

Public key encryption is very CPU-intensive, and should never be used when not strictly necessary. It is much more efficient to use it until such a shared secret has been determined and then use a secret key algorithm, such as DES.

(d) RSA

The **RSA algorithm** is based on the difficulty of decomposing a large number into its prime factors when some of these factors are very large primes. The principle is to encrypt a message $m$ using $m^e \bmod n$ as the cryptogram. $e$ is a prime (e.g., $e = 3$). $n$, for instance 15, is the *public key* of the recipient of the message. The public key $n$ is not just a random number, it is also the product of two large primes $p$ and $q$. In our oversimplified example $n = p * q = 3 * 5$. Both $p$ and $q$ are kept secret.

For instance, if '7' is the message to transmit securely, $C(7) = 7^3 \bmod 15 = 343 \bmod 15 = 13$. Therefore, '13' is the corresponding cryptogram.

In order to decipher the message, the recipient seeks a number $d$ with the property $m^{ed} = m$. This is equivalent to saying that $m^{(ed-1)} = 1 \bmod m$.

If the greatest common denominator of $m$ and $n$ is 1 ($\gcd(m, n) = 1$) we know from Euler's generalization of the Fermat theorem that $m^{\phi(n)} = 1 \bmod m$, where $\phi(n)$ is the cardinal of the set of numbers having no common divisors with $n$. When $n$ is a product of primes it is easy to calculate $\phi(n) = (p - 1) * (q - 1)$. Of course, there could be cases where $\gcd(m, n) \neq 1$, but the probability $(p + q - 1)/pq$ is negligible for large primes. Calculating $\phi(n)$ is straightforward when you know $p$ and $q$. But, if you know only the public key $n = pq$, you cannot find $p$ and $q$ in a reasonable period of time.

So in our example the recipient is seeking $d$ such that $ed = 1 \bmod \phi(n)$, with $\phi(n) = (p - 1)(q - 1) = 8$. The problem now reduces to finding $d$ and $k$ in $3d + k8 = 1$. We know that a solution can be found with the Euclidean algorithm because $\gcd(3, 8) = 1$ since $e = 3$ is a prime. Here, the solution is $d = 11(3 * 11 - 8 * 4 = 1)$. $d = 11$ is the *private key* of the recipient and can be used to decipher the message. $M = C^d \bmod n = 13^{11} \bmod 15 = 1792160394037 \bmod 15 = 7$.

Note that the roles of the private key $d$ and the public key $e$ are completely symmetric. So, it is possible to encrypt a message using the private key and then decrypt it with the public key. This is used for digital signatures.

(e) Digital signatures

There are many ways to cryptographically sign a document. In this section we present only one of these methods, based on the ability to cipher with a private key and decipher with a public key in the RSA algorithm.

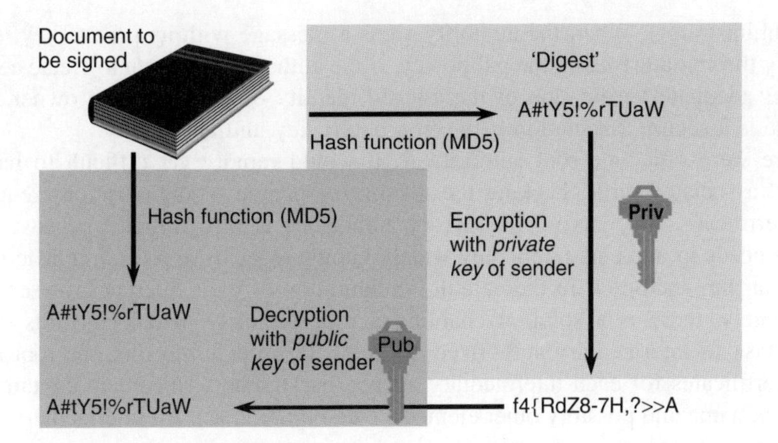


**Figure 2.29** Principle of a digital signature.

First, a hash of the document to be signed is calculated (see Figure 2.29). A hash is a function that takes a long document as input and produces a small string as output. If the initial document changes slightly (e.g., only one bit changes), a good hash function should lead to a completely different and unpredictable result. One of the most popular hash algorithms is called MD5 (Message Digest 5).

Then, the result of the hash function is encrypted with the private key of the person who signs the document; this becomes the signature.

It is easy to check whether a document is original (has not been modified since the signature) and really comes from the alleged author:

- First, a hash of the document is calculated with the same algorithm.
- Then, the signature is deciphered using the public key of the alleged author. If the alleged author is really the author of the document, we obtain the hash code of the original document.
- Finally, both digests are compared, if the document has been modified in any way since the signature they will differ.

#### *2.6.2.1.2.3 Certificates*

Digital signatures are very useful, but only if you are sure that the public keys used by receivers to check the authenticity of the message are associated with the correct identity. The public key can be distributed using a secure method, but this is not very practical. Certificates are a much more efficient way of ensuring that a public key is not a fake.

Certificates usually contain the public key of the presenter, along with some identity information (name and address, corporation name, etc.) and a validity period. In order to avoid any falsification, all the information contained in the certificate is digitally signed by an authority.

An authority is someone owning a widely known public key—so widely known that no one can fake it (e.g., it can be included by default in the operating system, or configured by an administrator). When the authority signs a message with its private key, everyone can verify the signature using the public key. If the authority is known to create certificates only after adequate verification of the alleged identity of the certificate owner, then the certificate is a secure association between a public key and an identity.

If there were only one root authority *R*, it would rapidly get difficult to handle the workload associated with checking the identity of people or organizations requesting a digital certificate. But, having many root authorities is also difficult because the new authority needs to make its public key widely known (e.g., by approaching Microsoft and asking that they include it in the default configuration of their Internet browser).

Fortunately, there is a solution enabling a root authority to delegate this certificate creation task to intermediary authorized agencies. In order to do this, the root authority creates certificates for each intermediary agency by signing a document containing their public key, name and possibly other elements (see Figure 2.30). An intermediary authority *A*, when requested to create a certificate *C* by signing a message which contains the name of the customer, his public key, etc. can just sign this message with its public key *Pa*.

The new certificate *C* is returned to the client, *with the certificate of the intermediary authority A signed by R*.

How does it work? When someone needs to verify the validity of certificate *C*, he first checks that the signature of the certificate document by *A* is valid. He can do this because the public key of *A* is included in the certificate of the intermediary authority *A* signed by *R*.

But the public key of *A* is not necessarily well known, it could just be an untrusted local agency. So, it is also necessary to check that the certificate of the intermediary authority *A* has been properly signed by *R*, certifying that *A* can be trusted. This is easy, because the public key of *R* is well known.

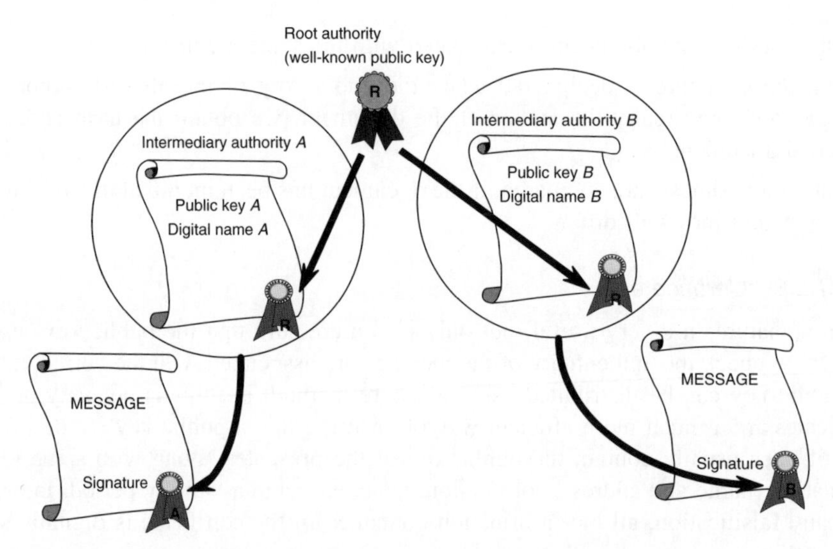


**Figure 2.30** Delegation of authority using a hierarchy of certificates.

It is relatively easy to check the identity of someone once you are in possession of his certificate. You need to check that the person you are communicating with is the legitimate owner of this certificate (i.e., is the person owning the private key corresponding to the public key contained in the certificate). This is not trivial because the presenter of the certificate cannot simply show his private key, as it would become compromised! Fortunately, there are other ways to perform this verification without compromising the secret key: for instance, one can encrypt a random string with the public key, send it to the presenter of the certificate and ask him to decrypt it. If the presenter of the certificate has managed to decrypt the string, then he has access to the private key and must be the owner of the certificate.

#### *2.6.2.2 Securing H.323 with H.235*

During the ITU meetings that led to H.235, because of the general perception that the Internet was not secure and that it would be easy to listen to calls, the initial focus was on securing the media channels. The H.245 procedures were extended to support the encryption of media channels by adding security parameters in the **OpenLogicalChannel** message. If the H.245 channel itself is secure in the first place, then the parameters within the **OpenLogicalChannel** message need no specific protection. This was the main motivation for securing the H.245 channel, but other reasons are equally important (e.g., protecting the DTMF information carried in H.245 **UserInputIndication** messages which may contain sensitive credit card data or passwords).

With a little more experience it became clear that on most IP networks the media channels were hard to access, and therefore already secure enough for the average user. The key requirement was in fact coming from service providers who need to avoid charging the wrong account for a call and also need to be able to prove that someone placed a call in case of a disrupted call. So, the call-signaling channel must be authenticated and optionally encrypted. In the context of H.235, if signaling encryption is used, any server in the network which needs to know the contents of the H.225.0 or H.245 messages needs to be trusted by the communicating endpoints, because it will have access to all confidential information elements: DTMF digits, encryption keys of the media channels, etc. These servers include the gatekeepers in the gatekeeper routed model, the MCUs and gateways otherwise.

Today, apart from specific niche markets (military or financial applications), H.235 is only used to make sure that access to public H.323 networks is restricted to subscribers and to authenticate the calling party in order to prevent any misuse of network call-accounting functions. In many networks, notably when customer premises equipment are owned and managed by the service provider, these goals can be achieved without H.235, using data layer security or source IP address validation. For instance, in a VoIP network providing voice VPN capabilities between corporate PBXs, the VoIP gateway connected to the PBX is probably already using some form of IP or transport-level security, and can be considered a trusted element of the network even without H.235.

#### 2.6.2.2.1 H.235 tools

##### 2.6.2.2.1.1 Tokens

Tokens are parameters transmitted within H.323 messages that are opaque for H.323 itself but can be used by higher level protocols. H.235 uses two types of tokens:

- A **ClearToken** is an ASN-1 sequence of optional parameters, such as timestamp, password, Diffie–Helmann parameters, challenge, random number, certificate,... ClearTokens are appropriate whenever the need is to ensure only the integrity of the transported information.
- A **CryptoToken** contains an object identifier of the encrypted token, followed by a cryptographic algorithm identifier, some parameters used by the algorithm (e.g., initialization vector), and the cryptographic data itself. CryptoTokens can be used to convey hidden tokens, signed token, or hash values. The cryptographic algorithm needs a key of a specific size $N$. For symmetric key algorithms the key is derived from a secret shared between the communicating parties. If the secret is shorter than the required key, the secret is simply padded with zeros, if it is longer than the key then the secret is split into blocks of size $N$ octets (or less for the last chunk) which are XORed. The resulting value is used as the key. When the shared secret is not configured in advance a method to negotiate a common secret is required (Section 2.6.2.2.1.2).

As stated above, the most frequent requirement in real deployments is to make sure the devices registering to the VoIP network cannot forge their identities. For this purpose, it is possible to include in the SETUP message a ClearToken containing a call ID, a timestamp, and a hash value (computed from the call ID and the timestamp—the letter prevents replay attacks). This coupled with the calling and called party information is usually enough.

##### 2.6.2.2.1.2 Generating a shared secret with Diffie–Helmann

Many H.235 procedures require a shared secret. If the communicating endpoints do not already share a secret, they must create one common secret, beginning with a communication that someone can potentially intercept.

The Diffie–Hellman key can be negotiated as described in 2.6.2.1.2.2 by using H.235 tokens.

In Figure 2.31, the *DhA* parameter contains $p$, $q$ and $q^a$, the *DhB* parameter contains $p$, $q$, and $q^b$. The random value passed in $B$'s reply is used for XORing parameters for further exchanges to prevent replay attacks. The CryptoToken is optional and can be used to digitally sign some parameters in order to prove the identity of the sender.

#### 2.6.2.2.2 Securing RAS

RAS messages are exchanged between an endpoint and a gatekeeper prior to any other communication. H.235 does not provide a way to ensure privacy on the RAS link, but it does provide authentication and integrity. If the security mechanism to be used is not known in advance, two parameters are present in the Gatekeeper Request (GRQ) message

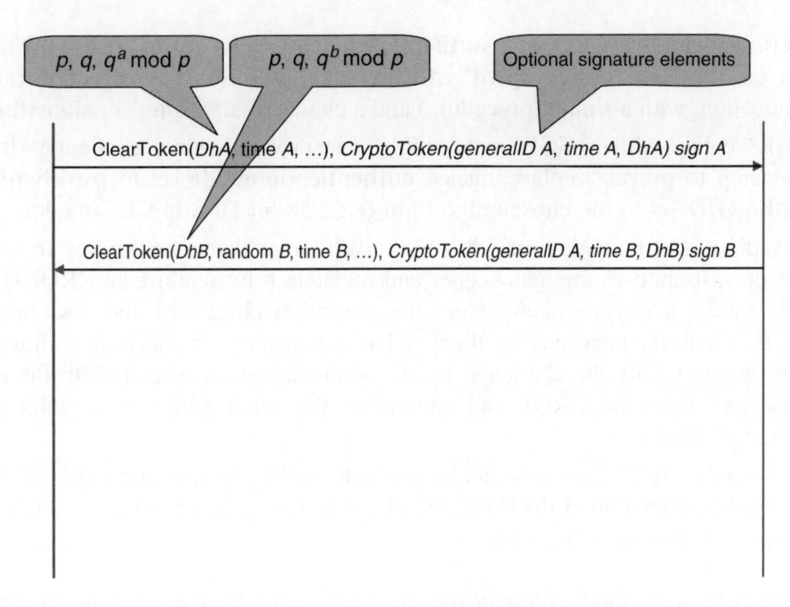


**Figure 2.31** Diffie–Helmann parameters encapsulated in ClearToken and CryptoToken structures.

that allow negotiation of the right mode and algorithms: **authenticationcapability** indicates the authentication mechanism that can be used, and **algorithmOIDs** contains the list of algorithms supported (DES CBC, DES ECB, RC2, ...). Of course, in real networks the service provider will usually preconfigure all equipment with the selected algorithm, so this is a bit theoretical.

There are two modes of operation depending on whether the gatekeeper and the endpoint share a secret or not.

If there has been no previous relationship and no shared secret between the gatekeeper and the endpoint, they need to negotiate one. For this purpose a Diffie–Helmann negotiation occurs during the GRQ, GCF phase using a ClearToken as described in Section 2.6.2.2.1.2. After this, the gatekeeper and the endpoint share a common secret. This secret can be used to authenticate any subsequent RAS message between the gatekeeper and the endpoint, in particular the RRQ and URQ. This is done by including in those messages a CryptoToken (encrypted using the DH secret) containing an XORed combination of the GatekeeperIdentifier, the sequence number of the request, and the last random value received from the gatekeeper (in the RCF or an xCF message). The key used to code the CryptoToken is derived from the Diffie–Hellman secret as described above. The gatekeeper provides new random values in each xCF in a ClearToken.

When the gatekeeper and the endpoint share a common secret, defined at subscription time, then the easiest procedure is to include in each RAS message a ClearToken with a timestamp and a hash code computed on the calling and called party numbers, the call ID and the timestamp. But there are more complex options; for instance, the following procedure can be used:

- The terminal sends a GRQ with **authenticationcapability** set to **pwdSymEnc** (other modes can be used besides **pwdSymEnc**, such as hash-based or certificate-based authentication, with a similar procedure) and a choice of algorithms in **algorithmOIDs**.
- The GK replies with a GCF containing a ClearToken with a challenge string and a timestamp to prevent replay attacks, **authenticationmode** set to **pwdSymEnc** and **algorithmOID** set to the chosen algorithm (e.g., 56-bit DES in CBC mode).
- At this point, the endpoint may have received more than one answer from several gatekeepers. It chooses one gatekeeper and registers it by sending an RRQ. This RRQ should contain a CryptoToken (using the algorithm chosen by the GK, here 56-bit DES CBC) with the encrypted challenge. The gatekeeper can check the validity of this answer by encrypting the challenge locally with the key associated with the endpoint alias (known from the GRQ), and comparing the result with the endpoint-provided encrypted challenge.
- After this, other RAS messages can be authenticated by including a CryptoToken with the XORed combination of the GatekeeperIdentifier, sequence number, and GK random values provided in xCF messages.

Using one of these methods, the gatekeeper can authenticate the RAS messages of each terminal in its zone.

##### *2.6.2.2.3 Securing the call-signaling channel (H.225) and the call control channel (H.245)*

The call-signaling channel can be secured using transport-level mechanisms like TLS or IPsec. An endpoint knows that it needs to secure a channel using TLS if it receives the call on port 1300. This is the more advanced option of H.235, providing confidentiality in addition to integrity and authentication. In practice, such complexity is not required, because only integrity and authentication are required. Most commercial H.323 endpoints implement one of the following methods:

- The SETUP message includes a token with a timestamp and a hash value computed from the most important parameters of the call (at least the calling party number, called party number, and callID). The gatekeeper, in routed mode, can verify this token with the shared secret, ensuring integrity and authentication.
- The RAS Admission Request (ARQ) message includes a token with a timestamp and a hash value computed from the most important parameters of the call (at least the calling party number, called party number, and callID). The gatekeeper performing the RAS function can verify this token from the shared secret. If the token is verified, it sends back in the ACF a token that should be included in the subsequent SETUP message. This token will have to be verified by the call control server receiving the SETUP message. The usefulness of this hierarchical authentication is clearer when considering the scalability of authentication in a large H.323 network (see Section 2.6.2.3.3).

In the SETUP, the caller will indicate which security schemes it supports for the H.245 channel in the **h245SecurityCapability** data structure. **h245SecurityCapability** includes a specific object identifier for each cryptographic algorithm, 56-bit DES CBC, and 56-bit DES OFB (e.g., each has its own identifier). The callee chooses one in the **h245SecurityMode** data structure carried by one of the Q.931 response messages (e.g., CONNECT). If no common security mode can be found, the callee can release the call with the reason code set to **SecurityDenied**. The necessary messages needed to secure the H.245 channel are exchanged before any other H.245 message.

Different methods can be used to initiate the secure channel, depending on whether the communicating endpoints share a secret or not. These procedures are very similar to those described for the RAS channel. Again, if a shared secret does not exist, it can be created using a Diffie–Hellman procedure. To our knowledge there is no commercial implementation of this yet; so, we will not detail the procedure further.

#### *2.6.2.2.4 Encryption of media channels*

Once the H.245 channel is secured, the terminals need to know which security modes can be used for the media channels. This is part of the capabilities exchange (e.g., terminals can signal that they support GSM capability, and/or encrypted GSM capability). A new capability has to be defined for each combination of codec and encryption mode. Since encryption algorithms can use a significant portion of the CPU, it is possible to signal such capabilities as plain GSM + H.263 video *or* Triple DES-encrypted GSM. H.323 is very powerful when it comes to expressing capabilities.

When a new logical channel is opened, selected security mode is specified (chosen by the source) and the key that will be used for logical channel encryption is *provided by the master* (as determined in the master–slave negotiation, see Section 2.2.1.2.2) either in the **OpenLogicalChannel** or in the **OpenLogicalChannelAck** using the **encryptionSync** field. The key is associated with a dynamic payload type, so a receiver which has just received a new key in the **encryptionSync** field will know it must use it as soon as the payload type of the RTP packets it receives matches the payload type associated with the key. The key can be refreshed at any time using the dedicated H.245 commands, **EncryptionUpdateRequest** and **EncryptionUpdate**. If the master decides to update the key (using the H.245 **EncryptionSync** message), then the payload type of the RTP stream must change for the RTP packets that use the new key.

Key negotiation can be made inherently secure using certificate exchange, or can be secured by first securing the H.245 channel. If the H.245 channel is not encrypted for some reason, then H.235 has provisions to open a separate specific LogicalChannel of type **h235Control** to negotiate key parameters for the logical channels. Again, there is no commercial implementation of this.

For multipoint communication, the secured H.245 channel is established with the MCU, and therefore the MCU must be trusted. New endpoints arriving in the conference can retrieve other endpoints' certificates, through **ConferenceRequest/ConferenceResponse** messages. However, they must trust the MCU to check whether the endpoints actually own those certificates.

As already mentioned in Chapter 1, many popular algorithms, such as DES ECB or CBC, are block-oriented. They are designed to code data aligned on the block size (64 bits for DES). The most simple way to cope with this is the RTP padding method described in

RFC 1889 (see Chapter 1 for more details). When it is used, the P bit of the RTP header is set. However, there are other techniques that can be used with DES. In addition to regular RTP padding, H.325 mandates that all implementations support ciphertext stealing for ECB and CBC, and zero pad for CFB and OFB. These techniques are modifications of the regular ECB/CBC/CFB/OFB chain-coding process for the incomplete data block and its predecessor, leading to a cryptogram exactly as long as the original message. When payload length is not a multiple of block size and the P bit is not set, then the decoder must assume that one of these methods is used.

In all cases, when an initialization vector is needed, it is constructed from picking as many octets as needed from the concatenated sequence number and timestamp octets, repeated if needed.

### *2.6.2.3 Scalable and secure H.323 deployments in a multi-vendor environment*

#### *2.6.2.3.1 Split gatekeeper architecture*

In very large deployments of H.323 networks, from 50,000 endpoints up to several hundred thousand, the RAS and routed mode call control functions cannot be performed by the same gatekeeper. A typical network of 100,000 endpoints will generate about 500,000 calls a day, but will generate at least 144,000,000 registration requests during the same time (assuming one refresh registration request per minute per endpoint). This translates to about 1,700 RRQ messages per second. Assuming a single computer were powerful enough to handle this load in case of a temporary network failure, and assuming all endpoints randomly send the first RRQ after failure over a period of 20 s, the average load during network restarts may exceed 5,000 RRQs per second. A single softswitch is unlikely to be able to reliably handle such a load, in addition to the call control functions.

The right solution is to separate the access function, handled by regional 'access gatekeepers', and the call control function, handled by a central call control gatekeeper (Figure 2.32). There are typically several access gatekeepers for each call control gatekeeper in the network. A router-based access gatekeeper can be expected to handle about 30 to 50 registrations per second, representing about 5,000 endpoints. Call control gatekeepers see only active calls, and some scale up to about 15,000 simultaneous calls, sufficient for 300,000 to 500,000 residential users. Note that most vendors publish the calls-per-second limitation of the call control softswitch; this shows 500 active residential calls translate to only about 5 calls per second (the average call duration is about 3 min, and 30% of calls are dropped due to a busy or no answer condition). A softswitch with 15,000 active calls will receive only about 150 calls per second.

#### *2.6.2.3.2 Securing the network edge*

The security of the registration of an endpoint to its access gatekeeper uses one of the H.235 methods explained above. Access gatekeepers may check the identity and password of endpoints (most of the time, the shared password methods of H.235 are used) with Radius or LDAP interfaces to a central database. Only the first RRQ is checked, the

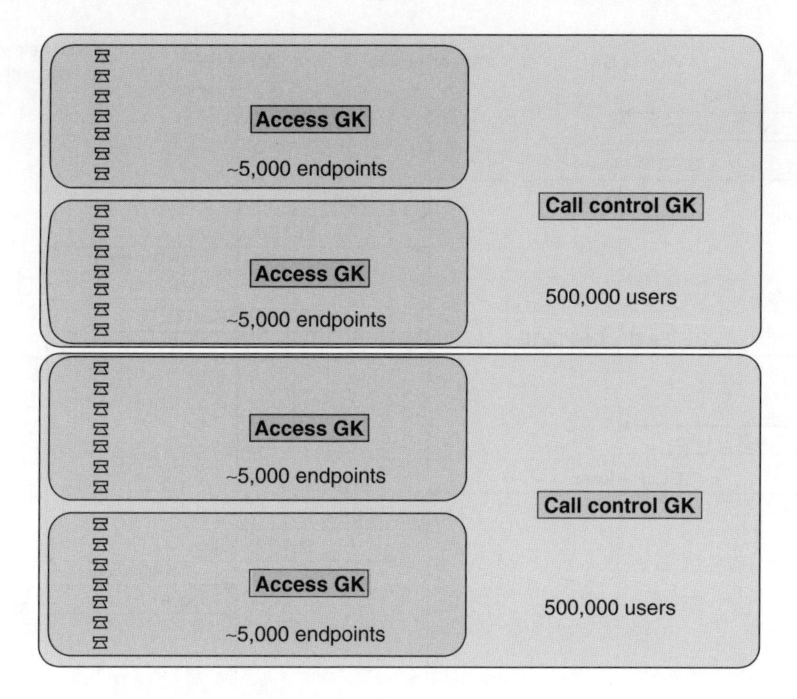


**Figure 2.32** Split gatekeeper architecture.

other keep-alive RRQs do not need an external password database access. A significant advantage of the 'split gatekeeper' architecture described above, besides scalability, is that it facilitates the deployment of multi-vendor solutions. Because of the number of options in H.235, most vendors' security implementations do not interoperate with one another, because the exact information provided in each security token is slightly different, the hash code calculation method is different, etc. With the split architecture it is possible to group endpoints of the same brand together and point them to an access gatekeeper that can understand secure tokens from this brand while keeping call control centralized.

Providing security for the call admission function is more complex. Even if the ARQ/ACF RAS exchange with the access gatekeeper is secure, it is still necessary in most cases to secure the Q.931 call control channel. There is one exception: if the network is from a single vendor and if the PSTN access gateways support the same format of security tokens as the edge endpoints; in this case, it becomes possible to secure the network at the edge. There are several large-scale instances of such networks today.

In the simple case illustrated in Figure 2.33, a token is provided by the access gatekeeper back to the edge endpoint in the admission confirm (ACF) message. The endpoint is required to copy this token in the SETUP message. The call control gatekeeper ignores this token and assumes this call can be forwarded to the destination. The destination is located with a Location Request (LRQ) to the proper access gatekeeper, and the SETUP is forwarded to the destination with its security token. The destination is required to get an authorization from its own access gatekeeper before it can continue processing the call. The token is passed to the access gatekeeper in the Admission Request, and the access

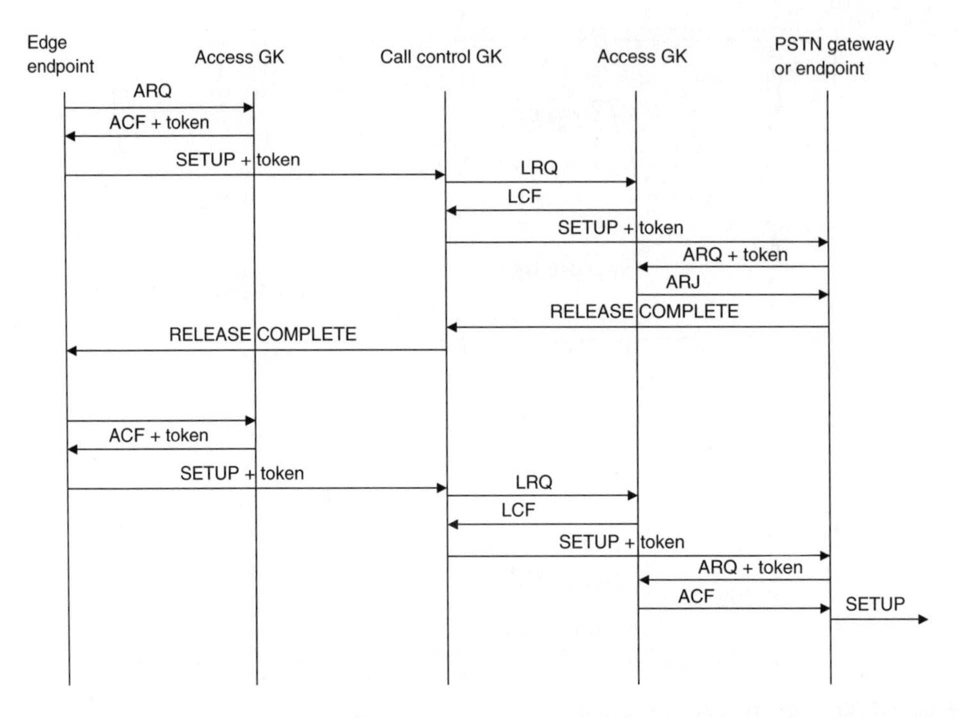


**Figure 2.33** Call secured at the edge by tokens.

gatekeeper verifies it by controlling that the hash code is correct for the given timestamp, callID, and call destination. If it is correct, the call is accepted with an ACF; otherwise, the call is rejected (ARJ) and released by the gateway.

This simple single-vendor case is completely secure for calls to the PSTN, because the gateways are the property of the network provider and can be trusted. It is slightly less secure for calls to endpoints, because potentially an endpoint could be hacked to not do any admission request before it proceeds with a received call. However, this requires both the calling and called endpoints to be hacked, a fairly limited probability.

#### *2.6.2.3.3 Security at the access level and at the call control level*

If the network involves multiple vendors, or if the PSTN gateways do not belong to the same service provider as the calling endpoints (e.g., if a residential network sends international calls to a clearing house), then the simple solution above is not sufficient and the call control gatekeeper needs to enforce the security policy. The simplest way to do this is to have the call control gatekeeper simply validate the token passed in the SETUP message, which was copied from the ACF. But, even this is not trivial. Most vendors simply send a hash code based on the endpoint password and call parameters. This forces the call control gatekeeper to know the password of each endpoint to validate the token, which is time- and memory-consuming for 500,000 users. A more scalable way of doing this is to have the access gatekeeper return a new token, signed with the access gatekeeper password, once the call has been authorized. This makes things much easier for the call control gatekeeper, because now it only needs to know the password of the access gatekeepers, not of the end users.

The assumption for this solution to work is that the access gatekeeper and the call control gatekeeper support the same format for security tokens. In a multi-vendor environment, this is unlikely. A more involved call flow is necessary to implement call control security in this environment.

The call flow of Figure 2.34 also leverages the idea of hierarchical authentication. Once the access gatekeeper has authorized the call, it requests an authorization token from the call control gatekeeper by sending a location request message (LRQ). The call control gatekeeper does not need to perform any specific authentication work, as it only accepts LRQ messages from access gatekeepers, as long as these access gatekeepers have not sent an LRQ if edge authentication failed. For each LRQ received, the call control gatekeeper returns an authorization token (tokenCC) in the LCF, which is copied by the access gatekeeper in the Admission Confirm, and then by the edge device in the SETUP message.

When it receives the SETUP message, the call control gatekeeper simply validates the call control token (tokenCC); it does not even need to look at the access token (tokenA). In fact, the access token is really unnecessary in this case.

This method works even with multiple vendors of CPEs and access gatekeepers, since the call control gatekeeper is not even attempting to understand the format of the security token of the edge device.

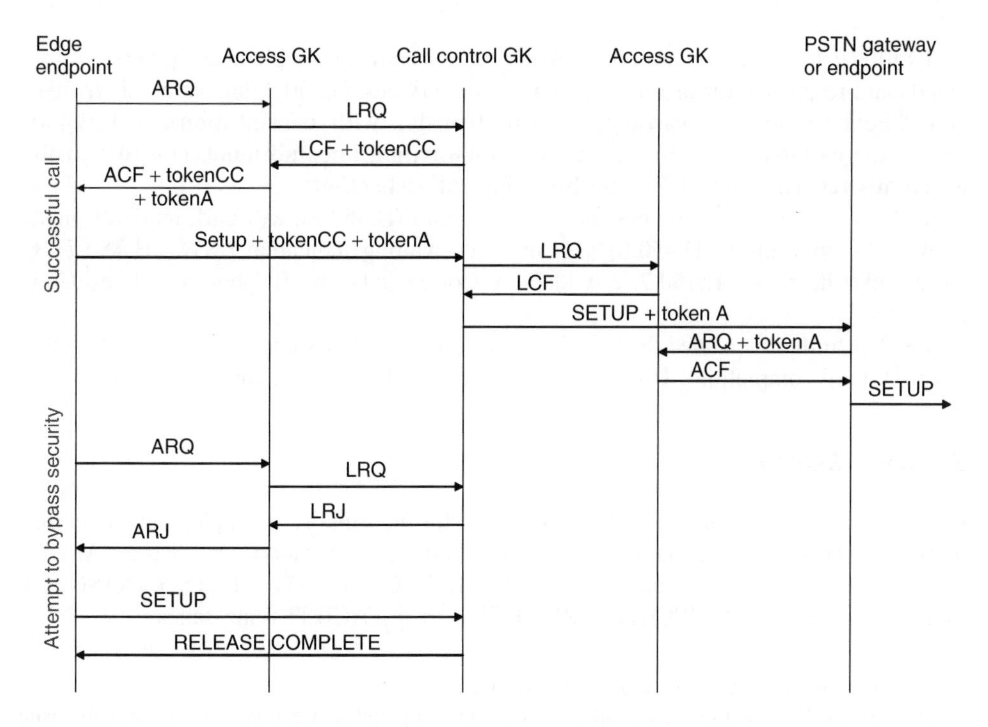


**Figure 2.34** Hierarchical security with token assigned by the call control GK.

## 2.7 SUPPLEMENTARY SERVICES

### 2.7.1 Supplementary services using H.450

H.323v2 introduced the new H.450 standard series, supplementary services for H.323. H.450 is based on the QSIG extensions of ISDN for use by PBXs, and therefore is targeted at private installations. H.450 should be used with caution on public networks, as only a few of the H.450 services (e.g., H.450.7 for message-waiting indication) can be deployed safely in a public environment. H.450 caused a lot of confusion as many perceived it as 'the' way of doing supplementary services in H.323. For clarification, H.323v4 added the following note:

> *Within the H.323 environment, there are several different methods by which services can be provided: the H.450 series of Recommendations, H.248*[10] *in association with its packages, stimulus signaling and Annex K. Although there is commonality of certain design goals for each of these solutions, the emphasis varies and each is more appropriate for certain circumstances. [...] The H.450 series of Recommendations is designed for interoperability of services at a functional level. Its derivation from QSIG ensures interworking with many private networking systems. Services are defined for peer–peer relationships, with feature intelligence typically resident in the endpoint. An H.450 based service must normally be explicitly supported by each affected endpoint in the system.*

H.450.1 defines the general framework for exchanging supplementary service commands and responses for use by supplementary services, H.450.2 defines a **call transfer** procedure[11] (**blind call transfer** and **call transfer with consultation**), and H.450.3 defines the **call diversion** procedures (**call forwarding unconditional**, **call forwarding on no answer**, **call forwarding on busy** and **call deflection**).

H.323v3 added a few services, notably call hold (H.450.4), call park and call pickup (H.450.5), call waiting (H.450.6), and message waiting indication (MWI, H.450.7). Of these, only the MWI, H.450.7, is widely supported today by IP phones and residential gateways.

H.323v4 further expanded the H.450 series with H.450.8: (Name Identification Service), H.450.9: (call completion), H.450.10: (call offer), and H.450.11:(call intrusion).

#### *2.7.1.1 H.450.1*

H.450.1 defines the 'generic functional protocol for the support of supplementary services in H.323'. This recommendation is based on Application Protocol Data Units (APDUs) carried in the call-signaling messages (ALERTING, CALL PROCEEDING, CONNECT, SETUP, RELEASE COMPLETE, PROGRESS) or in FACILITY messages.

[10] H.248 is a stimulus protocol very similar to MGCP.

[11] Another possibility is the use of call control tromboning and redirection of media streams using the Null Capability Set sequence, or TCS=0, as described in Section 2.7.1.2.2.

H.450.1 can be used to convey call-related instructions (e.g., redirecting a call) or call-independent instructions (e.g., program call screening). In the latter case, a special SETUP message with specific bearer capability and conferenceGoal information elements is used to transport the APDU. H.450.1 APDUs have the following structure:

- Optional **Network Facility Extension (NFE)** with the source entity type (endpoint or anyEntity) and address, and the destination entity type and address.
- A description of what to do with unrecognized messages (discard, clear call,...).
- A structure with the actual operation invoked.

The NFE part of the APDU provides a way to route supplementary service messages. The network entity receiving a SETUP message with an H.450.1 APDU may not be the intended recipient of the instructions contained in the APDU. It may have to relay it, or choose to intercept it in the case of a gatekeeper. All H.450 services are built on top of H.450.1.

#### 2.7.1.2 H.450.2 (call transfer)

This recommendation provides a way of transferring calls between H.323 endpoints once the initial call is established (the callee has answered).

##### 2.7.1.2.1 Call transfer between H.450.2-aware endpoints

The scenario shown in Figure 2.35 is an example of call transfer between endpoints. The call could be routed through a gatekeeper, but the gatekeeper would simply relay all H.450.2 APDUs:

- User B calls user A (the transferring user). This is the primary call.
- User A answers the call and uses H.450.2 to transfer the call to user C. User A may previously establish a separate call (secondary call) with user C to announce the transfer, for instance. If this secondary call exists, endpoint A notifies C of the pending call transfer, C returns a temporary identifier I for this secondary call if it can participate in the transfer. Otherwise, the attempt aborts here.
- Endpoint A sends an H.450.2 request to user B to call C (if there is an A–C secondary call, the temporary identifier I is mentioned). The endpoint may handle this directly if it is H.450.2-capable, or A's gatekeeper may choose to do it.
- When the new call request initiated by B arrives at C, C releases the secondary call if it existed. Then, if C answers the call the primary call is also released. B and C can now talk.

In this scenario, A could have called B in the first place, or C could have called A. The next steps of the call transfer would remain the same. The invoke and result APDUs are carried in normal Q.931 messages whenever possible, in FACILITY messages otherwise, as shown in Figure 2.36.

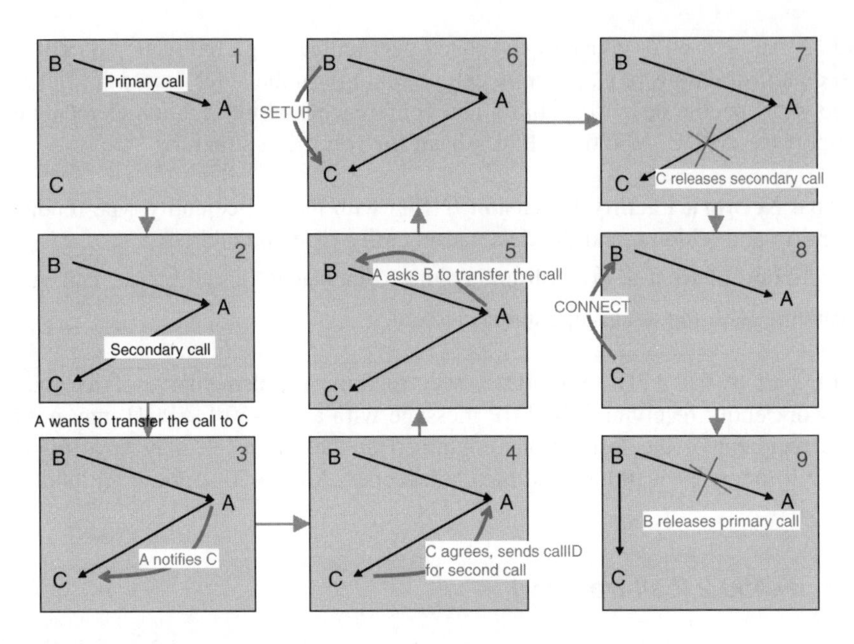


**Figure 2.35** Call transfer with consultation using H.450.2.

***A note on FACILITY REDIRECT***: H.323 mentions another simple way of supporting call transfer. An endpoint can simply send to the transferred endpoint a FACILITY message with the address of the endpoint transferred to. When it receives such a FACILITY message, a terminal should release the current call and restart a new call to the address specified in the FACILITY message. This is a simple way of transferring a call without consultation, but to our knowledge only very few endpoints and gateways support it.

#### *2.7.1.2.2 Transfer using the gatekeeper*

H.450.2 is not very easy to implement, and is unusable end to end in public network environments (see Chapter 6 for more details). The flow chart (Figure 2.36) describes only the normal case, but it would get much more complex if it took into account the many options of H.450.2 and the error conditions. Because of this complexity, many H.323 endpoints, such as stand alone IP phones with stringent memory constraints, may not implement H.450.2.

However, in Section 2.7.1.1 we emphasized the fact that all H.450 APDUs could be routed (using the origin and destination addresses found in the NFE) or intercepted by a gatekeeper. Therefore, if the terminals involved in the primary call were using the gatekeeper routed model (all Q.931 and H.245 messages get relayed by a gatekeeper), then the intermediary gatekeeper can intercept and act on H.450.2 APDUs on behalf of endpoints B and C. This allows H.450.2 to be used even if only terminal A is H.450.2-aware. Terminal A could be a sophisticated secretary terminal, while endpoints B and C could be ordinary simple IP phones. If we go one step further, endpoint A itself could be a simple IP phone, but the gatekeeper would have a web interface allowing the user of

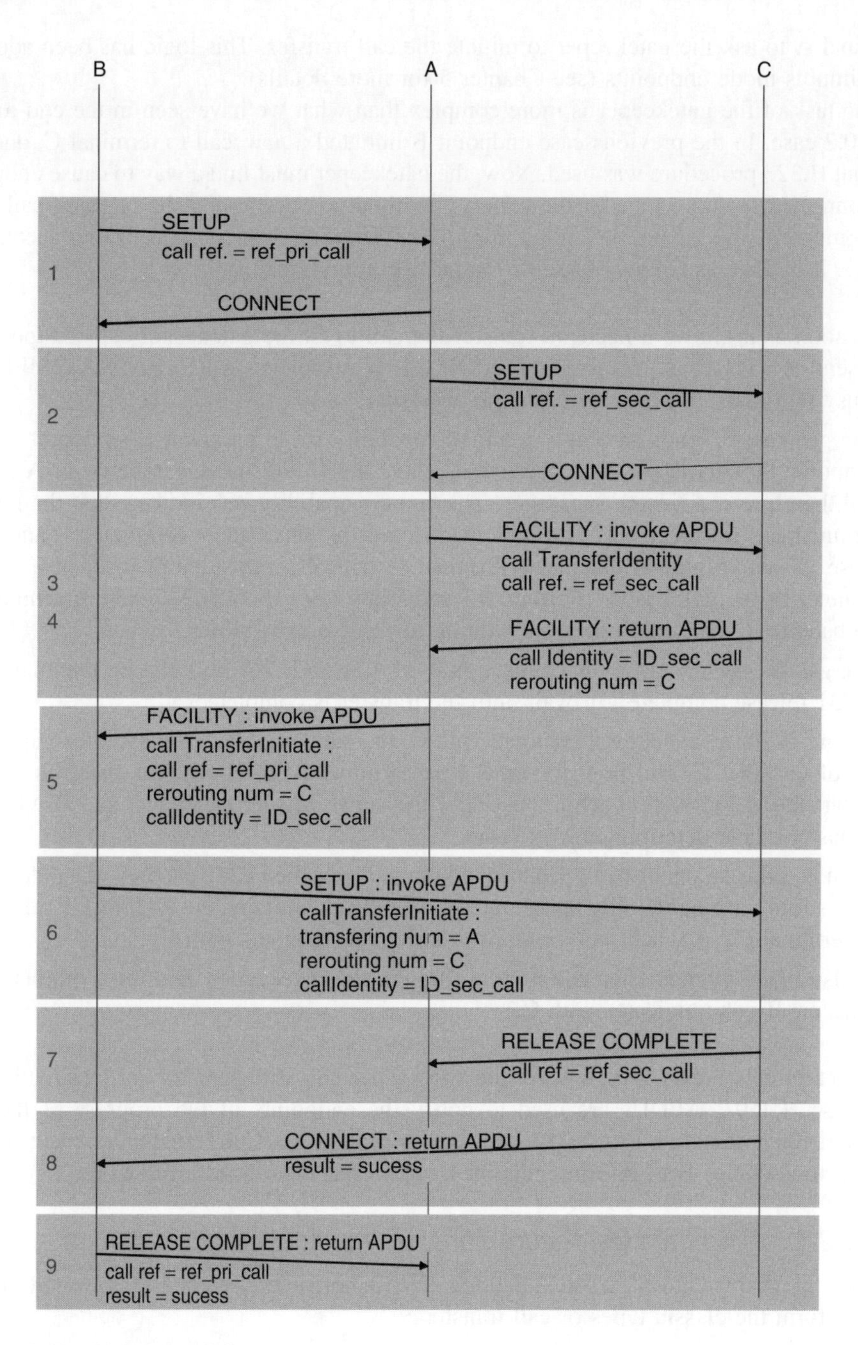


**Figure 2.36** H.450.2 call flow.

terminal A to ask the gatekeeper to initiate the call transfer. This logic has been adopted by stimulus mode endpoints (see Chapter 5 for more details).

The task of the gatekeeper is more complex than what we have seen in the end-to-end H.450.2 case. In the previous case endpoint B initiated a *new* call to terminal C, and the normal H.323 procedure was used. Now, the gatekeeper must find a way to cause endpoint B (connected to A) to transfer the call to C without ever releasing the ongoing call.

Fortunately, this operation, called third-party rerouting, has been taken into account in H.323. It is done as follows (case of a blind call transfer):

- As soon as it knows it needs to transfer the call to C, the gatekeeper calls endpoint C: it send a SETUP and receives a CONNECT. If it receives a RELEASE COMPLETE (busy terminal, ...), then it aborts the operation.
- The gatekeeper sends an empty terminal capability set to endpoint B, endpoint A, and endpoint C. This is possible because it relays the H.245 messages between A and B and therefore can 'insert' messages. The empty capability sets indicate that the remote terminal has no receive capabilities, and, logically, this causes terminals A and B to close all active logical channels. Terminal C will also not attempt to open a logical channel to the gatekeeper. Further, all endpoints reset their H.245 state machine and go back to a state where they are waiting to receive capabilities.
- The gatekeeper can close the connection with A now (H.245 end session command and Q.931 release complete), or wait until the transfer is completed.
- During H.245 channel establishment with C, the gatekeeper has received the capability set of terminal C, and now forwards it to terminal B. This will cause terminal B to restart the H.245 state machine just after the capability set exchange, and B will start a master–slave determination exchange.
- Then B sends an openLogicalChannel command over the H.245 channel, the gatekeeper relays it to terminal C. The openLogicalChannelAck contains the RTP/RTCP addresses of terminal C, so B will now establish the logical channels with C.
- C also opens logical channels with B through the gatekeeper, and this completes the transfer: B communicates with C.

If the endpoints are H.450.2-aware, the gatekeeper can still perform call redirection. In this case H.450.2 APDUs are used to notify the endpoints of the progress of the call transfer. For instance, a FACILITY message is sent with a CallTransferComplete invoke APDU to endpoint B to inform endpoint B that it has been transferred to C.

##### 2.7.1.2.3 Blind transfer, secure transfer, transfer with consultation

To sum up what we have learned in Section 2.7.1.2, here is how an H.450.2-aware terminal can perform the classic types of call transfers:

- Blind transfer: in this type of transfer, A doesn't want to check if C is available before disconnecting from B. As A and B are still in an active call, A sends a FACILITY message to B, the FacilityReason field is of type ***callTransfer*** and contains a

CallTransfertInvoke (CallTransferInitiate) invoke APDU informing B of the address of C. Then, A terminates its conversation with B using the regular H.323 procedure. When it receives the FACILITY message, B initiates a call with C.

- Secure transfer: now, if B cannot connect to C, A wants to remain in conversation with B. A sends the CallTransferInvoke to B but does not disconnect from B immediately. Instead, it waits until it receives the FACILITY message from B with the result of the call transfer (CallTransferResult=success on failure). Depending on the result, A releases the conversation with B (B might also send the RELEASE COMPLETE) or keeps it active.
- Transfer with consultation: now, A might be a secretary who needs to check whether C is available or in a meeting. In a regular private phone system, A can put B on hold. But for some reason the Q.931 message HOLD is forbidden in H.323. So, A can simply stop sending media to B (or send prerecorded music).
- Another solution would be to use the H.450.4 hold supplementary service. Now, A can establish a new call with C. Once A has been allowed to perform the transfer, A can use either the blind transfer or the secure transfer procedure. The procedure shown in the flow chart (Figure 2.36) is a transfer with consultation using secure transfer after the consultation.

#### 2.7.1.3 H.450.3: call diversion. Introduction to H.323 annex K

Recommendation H.450.3 is focused on the redirection of calls between H.323 endpoints before the call is established. This includes call forwarding on busy, call forwarding on no reply, call forwarding unconditional and call deflection. The diversion might be performed by a gatekeeper, or by the endpoint itself. H.450.3 allows the number of successive call diversions to be controlled/limited.

When activating the call forwarding unconditional (CFU) supplementary service for a particular address A, a user can still originate calls, but all calls to A will be redirected to another address. The user can activate/deactivate this service directly on the endpoint associated with A, or remotely on a gatekeeper. H.450.3 also provides ways to interrogate an entity to ascertain whether the supplementary service is activated or not, and for which addresses. The endpoint receiving the diverted call is notified that the call has been diverted, and also where the last diversion point was. The calling endpoint may also optionally be notified that the call has been diverted, with or without the new call destination address.

When activating the call forwarding on busy service, the same operations as in CFU will occur if the line of the user is busy. There might also be more specific conditions (diversion if more than $N$ calls are waiting, ...).

The call forward on no reply is similar, but occurs if the called user using this supplementary service has not answered after a programmable period of time.

The call deflection supplementary service can be invoked by a called user dynamically before the user answers a call. It causes the call to be diverted to the address entered by the called user.

Many types of call redirection can easily be performed by a routed mode gatekeeper. For the call forwarding unconditional service it can simply change the called party information element of the SETUP message and forward it to the new destination. For call forward on busy it will first forward the SETUP to the original destination address, then if it receives a RELEASE COMPLETE (cause busy), send a new modified SETUP to the next destination. Call forwarding on no answer is more involved, because the gatekeeper needs to make sure that audio channels are not established before redirecting the call. Chapter 6 gives more details and call flow examples on call redirections.

The activation of all these H.450.3 supplementary services is performed by exchanging H.450.1 APDUs. For instance, the call deflection supplementary service needs to be triggered by the called endpoint: this is done using the callRerouting invoke APDU. Even this simple task requires many information elements:

- The `reroutingReason`, if needed.
- The new `calledAddress` to use for the redirected call.
- A `diversionCounter` that is useful to avoid loops.
- The `lastReroutingNr` with the address of the last endpoint that performed the rerouting.
- `subscriptionOptions`: does the terminal want to inform the calling party?
- The original `callingNumber` (note that the 'number' here can be any H.323 address).
- More textual information in fields, such as `callingInfo`, `redirectingInfo`,...

At this stage the reader probably wonders whether it is necessary to introduce so much complexity for such a simple feature. This is true of many supplementary services. In fact, when the services are activated by a gatekeeper, a lot can already be done with the ubiquitous web interface for user interaction if you have a web phone and you want to program call forward on busy; this can be done by filling in an HTML form. With a little imagination it is even possible to let the user customize his call control with much more flexibility with a web interface: something like 'if-my-boss-is-calling-then-ring-my-desk-phone-and-try-my-cellular,-otherwise-go-to-the-answering machine,-or-if-it's-my-banker-calling-again-to-say-my-account-is-low-then-sound-as-if-I-wasn't-here...' H.323 annex K describes how an IP phone with a proper Web-capable user interface can use it for the control of supplementary services, which makes it possible to present any user interface for any feature with no impact for the IP phone. But this annex has had no success so far, probably because it provides a lot less functionality than its MGCP equivalent (business phone package, see Section 5.2.2.2.3 for more details).

## 2.7.2 Proper use of H.450 supplementary services, future directions for implementation of supplementary services

H.450, the VoIP version of QSIG, is an appropriate way to convey supplementary services in a private network of PBXs, but end-to-end call transfer supplementary services should not be used in public networks. In Chapter 6 we will describe in more detail the specific requirements of public networks (i.e., networks interconnecting multiple enterprises or residential users, where the service provider must bill for calls originated from each connected device and cannot trust connected devices). Clearly, because of security, call-routing, and accounting issues, only a handful of the H.450 standards can be reasonably deployed in a public network. Just like QSIG is not used today on public networks, H.450 will probably remain only a PBX interconnection protocol.

H.450 can also be implemented to provide supplementary services on business phones, but it has a serious competitor in MGCP. Today, virtually all PBXs use a stimulus protocol to control PBX business phones ('Unistim' for Notel, 'ABC' for Alcatel, etc.). This simplifies the phone design and gives a lot more control to the PBX. For instance, many PBXs offer services as soon as the phone is off-hook, even without dialling any number: you can have notification of voicemail, warnings if the line has been forwarded, etc. Such services cannot be implemented with an H.323 or SIP phone, because these phones do not send any notification to the network when they are off-hook. The rigidity of the H.323 and H.450 standardization process is also a great obstacle to its use in business systems, where the race for new features and differentiation leaves no room for endless discussions in standard bodies. Most PBXs today offer over a hundred features. H.450 took over two years to sort out the first dozen!

MGCP, with its new extensions (business phone event package, BTXML from Cisco, etc.), now has the ability to control virtually any device, see MGCP Chapter, Section 5.2.4 including feature buttons, screens, loudspeaker modes, etc. For the first time a standard protocol gives access to the same power as proprietary stimulus protocols. Many implementations of IP-PBX and Centrex services already exist for MGCP, which goes far beyond the reach of H.450. Automatic off-hook, CTI calls, paging calls are already available! Therefore the prospect of H.450 for business phones also seems very limited, probably restricted to the same niche markets as ISDN business phones today.

## 2.8 FUTURE WORK ON H.323

H.323 is now carrying billions of VoIP minutes per year. Most networks are running H.323v2 with some version 3 and 4 extensions. The current version of the protocol benefits from the experience accumulated by VoIP vendors in hundreds of VoIP networks. However the telephone network is a lot more complex than anticipated in the early days of voice over IP, and the H.323 protocol, despite its maturity, is still in need of improvement and extensions to cover the very specific call flows found in PSTN networks. Among the most problematic call flows, one can cite:

- Pre-connect announcements. Although the basic pre-connect announcement is covered by early H.245 and Fast Connect, the H.323v4 protocol is still not flexible enough to allow multiple media servers to stream media to the calling endpoint (dynamic update of the Fast Connect media information). H.323v5 solves this problem with H.460.6.

- Call release scenarios using Q.931 messages 'forgotten' by the H.323 standard (e.g., the DISCONNECT message). In some instances media is being played while the call is still in a half-released state: one of the most common cases is in-band network announcements played when dialing the wrong number. A consequence is that it is relatively difficult to implement **Advice of Charge (AOC)** on H.323. In ISDN, AOC is sent at the end of the call in one of the messages releasing the call. Since the release messages use three messages, there is always one that is sent by the network to the endpoint. In H.323, when the endpoint releases the call first with a RELEASE COMPLETE message, the network has no chance of sending the AOC information to the endpoint.
- Precise rules on how to transport and use the 'progress indicator', which specifies whether in-band information is present or not (it should be ignored when present). Some complex call flows can be found where locally generated ring-back tones alternate with in-band tones. In H.323, vendors need to be careful to take the progress indicator information into account, as opposed to blindly playing the media they receive from the other party.
- Precise rules on how to interwork with the ISDN network, and in particular the handling of media-type information (3.1-kHz audio, fax, ...). Simply carrying this information transparently causes errors in the ISDN network because it advertises capabilities not yet present in the VoIP network.
- H.323v4 does not handle call loop detection in a robust way (i.e., with hop counters in every message and mapping rules with SS7 ISUP messages). H.323v5 solves this problem.
- More experience and understanding of the implications of call redirection in VoIP networks (see Chapter 6).
- In general, a much better way of mapping SS7 ISUP messages to H.323 messages without loss of information, H.246 and H.323 annex M works only if the ISUP flavor on both sides is the same, because the ISUP information is considered a 'black box' (SIP-T uses a similar approach, and has the same problem). The Global Transparency Descriptor (GTD), a work in progress authored by Cisco Systems and based on a complete decoding and mapping of the ISUP information to a network-independent format, is a significant improvement for both H.323 and SIP. It is not yet standardized though.

Despite all the improvements that are still required, H.323 is, with MGCP and the 3GPP IMS profile of SIP, one of the most mature VoIP protocols, and the most advanced regarding videoconferences. Many international traffic clearing houses use H.323 to exchange calls, and even incumbent carriers increasingly make use of VoIP and H.323 when terminating international traffic. There are H.323 class 4 transit networks in production today that serve millions of end users, and even much more complex class 5 residential networks with multiple vendors, serving over 5,000,000 users (Orange, France), and providing all regulatory features, such as emergency calls, local number portability, lawful call interception, in addition of traditional class 5 services (call forward, call hold, three-way conferencing, etc.).

H.323 has also entered the 3G space: the H.324-M standard used for videoconferencing of recent 3G handsets also uses H.245 for session control and, therefore, interworks seamlessly with H.323. H.323, sometimes represented as a 'legacy' protocol with no future, does seem to keep a significant momentum notably for video applications.

# 3

# The Session Initiation Protocol (SIP)

## 3.1 THE ORIGIN AND PURPOSE OF SIP

The concept of a 'session' was first introduced in RFC 2327 (the Session Description Protocol) as a set of data streams carrying multiple types of media between senders and receivers. A session can be a phone call, a video conference, a user taking remote control of a PC, or two users sharing data, chatting, or exchanging instant messages.

The Session Initiation Protocol (SIP) was originally defined in RFC 2543 by the MMUSIC (Multiparty Multimedia Session Control) working group of the IETF. The MMUSIC working group focused on loosely coupled conferences as they existed on the MBONE (see the companion book, *Beyond VoIP Protocols*, Chapter 6 for additional details on the MBONE) and was working on a complete multimedia framework based on the following protocols:

- The Session Description Protocol (SDP, RFC 2327) and the Session Announcement Protocol (SAP, RFC 2974).
- The Real-Time Stream Protocol (RTSP, RFC 2326) to control real-time, or more precisely isochronous,[1] data servers.
- SIP.

These protocols complement existing IETF protocols, such as RTP (RFC 1889) from the AVT working group (Audio/Video Transport), used for the transfer of isochronous

[1] The data elements of an isochronous data stream, for instance, voice samples, must be played back with the same relative time intervals as when they were recorded.

*IP Telephony: Deploying VoIP Protocols and IMS Infrastructure, Second Edition* O. Hersent

data, or RSVP from the INTSERV (INTegrated SERVices) working group for bandwidth allocation.

SIP now has its own working group within the IETF, which maintains close coordination with the MMUSIC group, as the latter is still working on improving the SDP which is used extensively in SIP.

One of the initial goals of SIP was to remain simple, and to this purpose, 'classic' telecom protocol design principles, such as protocol layers isolation or complete separation of functional blocks (e.g., message syntax, message encoding and serialization, retransmission), were initially left behind as unnecessary heaviness. The initial SIP RFC aimed at defining in a single 150 pages document all the technical details required for session management, covering message reliability, transport, security, and a set of generic primitives for the following functions:

- **User location**: determination of the technical parameters (IP address, etc.) required to reach an end system to be used for communication and association of end users with end systems.
- **User availability**: determination of the reachability of an end user and the willingness of the called party to communicate.
- **Endpoint capabilities**: determination of the media types, media parameters, and end system functions that can be used.
- **Session setup**: 'ringing' a remote device, establishment of media session parameters at both the called and calling parties.
- **Session management**: including transfer and termination of sessions, modifying session parameters, and invoking services. The scope of SIP has been restricted to loose multiparty conferences, i.e., functions such as chair control are out of the scope of the current SIP specification. These conference control functions are left to extensions that can be carried within SIP messages.

It took just about a year for SIP to become surprisingly popular for a telecom protocol, but this can be understood from the context. Just like its contemporaries WAP or UMTS, the development of the SIP occurred at the peak of the Internet bubble, and many start-up companies spent an inordinate amount of marketing resources to promote SIP to omnipotent status. Just as the 'new economy' was being praised as a simple new paradigm vastly superior to the 'old economy', burdened by obsolete conventions and processes, the word began to spread that SIP was a new simple way of designing telecom systems and that the old public network was unnecessarily complex and inefficient. Even the H.323 protocol, only a couple of years older than SIP, was caught in this wave, and began to be criticized for its heaviness and traditional telecom heritage.[2]

[2] Indeed, H.323 is based on the Q.931 protocol used in current telecom networks, and uses the most recent software modelling tool, the Specification and Description Language—SDL—capable of automatic test case generation. H.323 defines and separates many functional software modules, and uses an abstract syntax (ASN.1) to describe its messages, which allows to automatically generate parsing and serialization code.

After the explosion of the Internet bubble, the marketing clouds slowly began to dissipate, and after a few years of experience, the real strengths and weaknesses of SIP are now easier to assess. One strength of the protocol is that the authors constantly tried to abstract it from any specific use. For instance, most of the time, the SIP primitives were used to carry 'opaque' objects required for a specific application or media and not understood by the SIP protocol stack.[3] This did stimulate the imagination of developers, and led to interesting ideas—for instance the use of SIP for IM (see Section 3.5 of this chapter).

The simplicity of Figure 3.1 also explains much of the initial enthusiasm for SIP.[4] From this simple example we can see that SIP is very efficient: the callee to caller media channel can be setup in exactly one round trip and the caller to callee media channel can be setup in one and a half round trips. This was much better than the many round trips that were required by the bootstrap nature of H.323v1. A similar call flow, call fast connect, was only introduced in H.323v2 (see Section 2.3 in Chapter 2).

Unfortunately, the weaknesses of the protocol were also many, and the SIP community had to work hard to solve or improve them. Today, only the IMS profile of

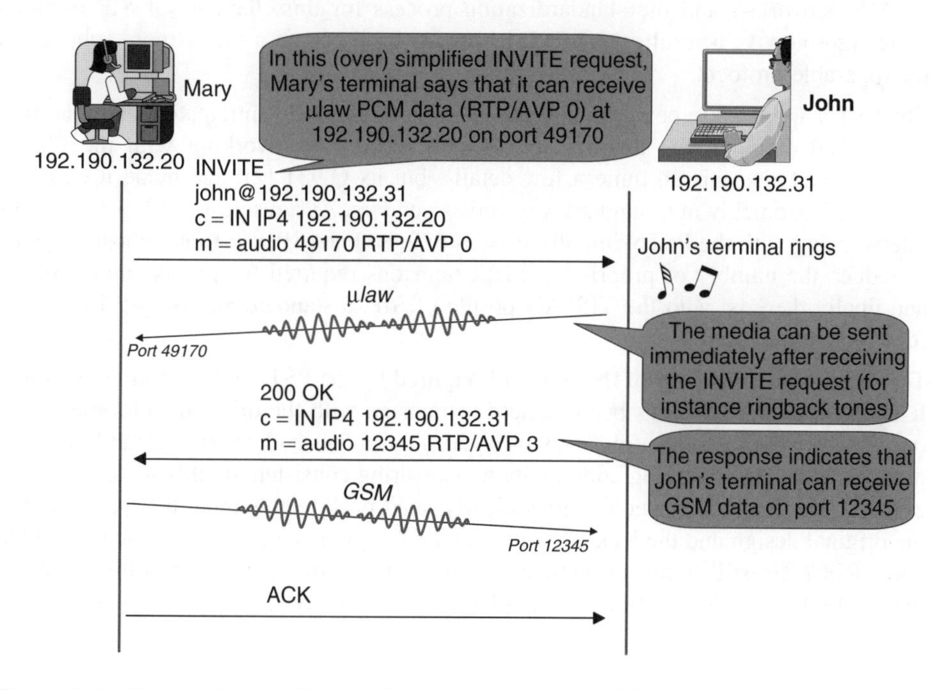


**Figure 3.1** Simple phone call scenario, as per the original SIP RFC.

[3] This ability to transport opaque parameters is also present in most other protocols, notably H.323 using the 'Non-Standard Parameters' that can be freely defined within the framework of the standard. Note also that in SIP the size of opaque parameters is restricted by the fact that no segmentation mechanism has been defined for SIP over UDP.

[4] This figure does not use the offer-answer model introduced by RFC3261; see Section 3.2.2.7.2.

the protocol has really solved most of these issues, which still affect some 'plain SIP' implementations.

- Because SIP 'can potentially' be expanded, it is often believed and touted that SIP 'does' everything. This is the well known 'it's just software' syndrome. Year after year there has been an accumulation of proprietary extensions of SIP—sometimes described in draft documents, while sometimes not even documented—but the lack of a well-defined standardization process has often prevented convergence of implementations to occur. The reality, despite claims to the contrary in 'sponsored' interoperability events, is that only the most common call flows work across vendors, and they are often too trivial to fully address the complexity of real-world applications. Too often, SIP is still only a reassuring name hiding many proprietary extensions. As a result, operational SIP networks today are still built mostly with infrastructure equipment provided or integrated by a single vendor. However, the involvement of ETSI 3GPP and ETSI TISPAN working groups greatly improved the situation—these groups introduced significant modifications to the protocol as they defined a SIP profile for use in IMS networks—and the standardization process for this 'flavour' of SIP is much more rigorous. As a result, the SIP IMS profile has become a much more robust and interoperable protocol.
- The PSTN appeared to be a lot more complex than originally anticipated, and therefore SIP lacked many of the features required for proper interworking with the PSTN. H.323v1 had also missed quite a few details, but its Q.931 heritage made it easier to fix the issues quickly in a standard way across vendors. As a result, most VoIP networks interworking with the PSTN initially used H.323, and not SIP.[5] It took almost 10 years to reduce the number of proprietary SIP extensions required for proper interworking, and finally there is, with the TISPAN profile of SIP, a standard and robust SIP profile for PSTN interworking.
- The increased complexity of the protocol required by the PSTN interworking and other fixes in the initial RFC has become hard to manage with the original 'informal all-in-one design' approach. The 'old' way of layering protocols and defining clean functional modules aimed at managing complexity and ensuring consistent quality as the software evolves. The latest SIP specifications clearly head back to this modular approach, but the original design and the lack of formal methodology make this very difficult, and the latest RFCs are still burdened with exceptions and shortcuts between software layers that make the protocol difficult to implement and test. SIP is certainly not 'simple' any more.

[5] In November 2002, the VASA consortium (BellSouth, Chunghwa Telecom, Equant, France Telecom, SBC, Sprint PCS, Telecom Italia Lab, VeriSign, Verizon, and WorldCom) published an independent study of 'SIP in Carriers Networks' which emphasized that 'some network operators have experienced significant difficulties in interworking different vendors' products. In contrast with the initial objectives of SIP, operators are driven towards single vendor solutions'. The study concluded: 'For existing networks, the arguments against immediate migration from TDM or H.323 to SIP outweigh the potential benefits'.

This section will describe the most common PSTN interworking scenarios, which work without extensions of SIP, and will list the major cases where extensions are still required. When available, the documented extensions of major SIP vendors will be discussed.

One of the applications that has emerged out of the multiple theoretical possibilities of the protocol is Presence and Instant Messaging (IM). The adoption of SIP for IM by Microsoft made it a serious option, as an alternative to the only other open standard for IM : Jabber. The SIP applications for Presence and IM are described in Section 3.5 of this chapter.

### 3.1.1 From RFC 2543 to RFC 3261

SIP remained a draft document for a long time before it was finally published as an RFC in March 1999 (RFC 2543). The first published version of the protocol was SIP 2.0. Unfortunately, this first version of the RFC was trying to embrace too much, contained many errors, and was too vague and ambiguous to be a real specification document. It was more a sort of technical brainstorming document and was taken as such by the many start-up companies that began to implement SIP products. All the first trial SIP networks used their 'flavour' of SIP, with their own corrections and expansions to the original SIP specification, and used only the simplest call flows defined by the RFC.

As the first useful feedback was gathered from these trials, the RFC was updated with nine 'bis' versions, and finally all changes were merged in June 2002 in a new RFC, RFC 3261. Important aspects of the initial specification were split into separate RFCs. Although RFC 3261 does not update the SIP version number, which remains SIP 2.0, it does not only correct errors and clarifies ambiguities but also really makes major changes to RFC 2543. The protocol is now more robust and more clearly documented, although the RFC is still a bit verbose and vague, with expressions like 'modest level of backwards compatibility' or 'almost identical' that can be misleading. RFC 3261 is in reality a major new version of SIP, and is not backward compatible with RFC 2543, although most simple call flows will work across the two RFC versions.

In the process, the SIP protocol lost the apparent simplicity of its early days, and the size of the main RFC nearly doubled to 270 pages. The new RFC is an umbrella document that points to other RFCs for specific details or applications; the complete documentation (see Figure 3.2) includes hundreds of additional pages. Among the most important documents are the following:

- **RFC 3262: Reliability of Provisional Responses in Session Initiation Protocol (SIP).** This RFC is required in all cases where SIP needs to interwork with a telephone network.
- **RFC 3263: Session Initiation Protocol (SIP): Locating SIP Servers.** The location of SIP servers is really an independent module in a SIP implementation, and is now documented separately from the main SIP RFC.
- **RFC 3264: An Offer/Answer Model with Session Description Protocol (SDP).** This was one of the most necessary clarifications of the original SIP RFC, where the exact use of the SDP syntax was ambiguous and led most vendors to implement

| RFC | Titre |
|---|---|
| 2046 | MIME part two,media type |
| 2246 | TLS protocol version 1.0 |
| 2617 | HTTP Authentication: Basic and Digest Access Authentication |
| 2778 | A Model for Presence and Instant Messaging |
| 2779 | Instant Messaging/Presence protocol requirements |
| 2806 | URLs for telephone access (deprecated by RFC 3966) |
| 2848 | PINT |
| 2915 | NAPTR DNS Resource Record |
| 2916 | E164 number and DNS |
| 2976 | SIP INFO Method |
| 3087 | Control of service context using Request-URI |
| 3204 | MIME Types for ISUP and QSIG |
| 3261 | SIP : Session Initiation Protocol |
| 3262 | Reliability of provisional responses in SIP (PRACK) |
| 3263 | SIP : Locating SIP servers |
| 3264 | Offer-Answer model |
| 3265 | Event packages |
| 3266 | Support for IPv6 in SDP |
| 3310 | HTTP digest authentication using authentication and key agreement (AKA) |
| 3311 | UPDATE method |
| 3312 | Integration of resource management and SIP |
| 3313 | Media authorization |
| 3320 | Signaling Compression (SigComp) |
| 3321 | SigComp extended operations |
| 3323 | 'A Privacy Mechanism for the Session Initiation Protocol (SIP)'. |
| 3324 | Short Term Requirements for network asserted Identity |
| 3325 | Private extensions for SIP for asserted identity within trusted networks |
| 3326 | Reason Header |
| 3327 | SIP extension header fied for registering non adjacent contacts (Path) |
| 3329 | Security Mechanism Agreement |
| 3372 | SIP-T |
| 3388 | Grouping of media lines in SDP |
| 3398 | ISDN ISUP in SIP |
| 3420 | Internet media type message/sipfrag |
| 3427 | Change process for the Session Initiation Protocol |
| 3428 | SIP extensions for instant messaging |
| 3455 | Private Header (P-Header) Extensions to SIP for the 3rd-Generation Partnership Project (3GPP) |
| 3485 | SIP and SDP static dictionary for Signaling Compression |

**Figure 3.2** Main RFCs related to the SIP protocol.

| RFC | Titre |
|---|---|
| 3486 | compression of SIP messages |
| 3515 | SIP REFER method |
| 3520 | Session Authorization Policy Element |
| 3550 | RTP: A Transport Protocol for Real-Time Applications (updates RFC 1889) |
| 3578 | Mapping of ISDN ISUP Overlap to SIP |
| 3581 | An Extension to the Session Initiation Protocol (SIP) for Symmetric Response Routing |
| 3608 | Extension Header field for service route discovery during registration |
| 3665 | SIP Basic Call Flow Examples |
| 3666 | SIP PSTN Call Flows |
| 3680 | 'A Session Initiation Protocol (SIP) Event Package for Registrations' |
| 3725 | Best Current Practices for 3 PCC calls flows |
| 3761 | ENUM |
| 3824 | Using E.164 with SIP |
| 3840 | Indicating User Agent Capabilities in SIP |
| 3842 | Message Summary & Waiting Indication Event for SIP |
| 3856 | A presence Event Package for the SIP |
| 3857 | SIP Event Template Package for Watcher Information |
| 3858 | An extensible markup language (XML) based format for Watcher Information |
| 3862 | Common Presence and Instant Messaging : Message Format |
| 3863 | Presence Information Data Format (PIDF) |
| 3891 | SIP Replaces header |
| 3892 | SIP Referred-By mechanism |
| 3903 | SIP extension for event state publication |
| 3966 | Subscriber Number Scheme |
| 4028 | Session Timers |
| 4032 | Update to the Session Initiation Protocol (SIP) Preconditions Framework |
| 4040 | RTP Payload Format for a 64 kbit/s Transparent Call |
| 4353 | A Framework for Conferencing with the Session Initiation Protocol (SIP) |
| 4538 | Request Authorization through Dialog Identification in the Session Initiation Protocol (SIP) |
| 4579 | Call Control—Conferencing for User Agents |
| 4575 | A Session Initiation Protocol (SIP) Event Package for Conference State |
| 4733 | RTP Payload for DTMF Digits, Telephony Tones, and Telephony Signals (updates RFC 2833) |
| 5629 | A Framework for Application Interaction in the Session Initiation Protocol (SIP) |

**Figure 3.2** (*continued*)

a single codec or have their own interpretation of codec negotiation. The new RFC is much clearer and implements a mechanism that is almost identical to the H.323 FastStart and the tunneled H.245 logical Channel Management.

- **RFC 3265: Session Initiation Protocol (SIP)-Specific Event Notification.** This RFC is used by some vendors to transport DTMF tones, but its main use is for IM. Only the main methods are specified, and the content of events is left to application specific profiles (see Section 3.5 later in this chapter).
- **RFC 3266: Support for IPv6 in Session Description Protocol (SDP).** While IPv6 was taken into account since the beginning in H.323, it required some syntax extensions in SIP, now covered by this new RFC. The use of IPv6 for IP telephony remains very questionable,[6] however, because VoIP is the most sensitive application to IP protocol overheads, which are going from bad to worse in IPv6. In addition, the IP address depletion issue, which is the main motivation for the introduction of IPv6, can be overcome by the use of application level proxies and intelligent use of private IP addresses (see Chapter 7 for details). The handling of quality of service is also virtually identical in IPv4 and IPv6.

### 3.1.2 From RFC 3261 to 3GPP, 3GPP2 and TISPAN

In order to manage the transition of GSM networks towards '3G', the European Telecom Standards Institute (ETSI) started in 1998 a joint collaboration effort with several other international[7] standardization bodies, called the 'Third Generation Partnership Project'. The 3GPP project aimed at harmonizing the standards for third generation networks.

One of the major evolutions of these third generation networks was the perspective of a generalized use of packet data. In order to manage packet data communications, release 5 and release 6 of the 3GPP technical specifications introduce the 'IP Multimedia Subsystem' (IMS). The IMS is a complete architecture for the management of services in packet mode, and it selected SIP as the underlying protocol. The IMS architecture uses the loose routing capabilities of SIP (see Section 3.3.2.3) as a fundamental tool for the chaining of services provided to users.

In a similar way, the 3GPP2 partnership aims at harmonizing the transition of ANSI/TIA/EIA-41 and CDMA 2000 networks towards 3G. 3GPP2 regroups the same partners as 3GPP, with the exception of ETSI, and uses, with a few exceptions, the specifications of 3GPP.

In order to take into account the needs of 3GPP (documented in RFC 3113) and of 3GPP2 (RFC 3131), many modifications had to be introduced to RFC 3261, mainly as new headers (RFC 3455) and more precise specifications. The resulting protocol, even if it still appears in signaling messages as SIP 2.0, is in fact a very specific profile, which we will call in the rest of this book SIP/3GPP.

---

[6] The use of IPv6, which was mandatory in IMS Release 5, became optional in IM's release 6 (refer to Chapter 4 for more details).

[7] ARIB and TTC (Japan), CCSA (China), T1 (USA), and TTA (Korea).

The architecture of 3GPP and 3GPP2 is quite complex, because of the constraints of mobility and the extreme diversity of services taken into account. 'Push To Talk' and 'Push to See' were among the first services to be fully defined. In fact, telephony, as expected, proved to be a tough nut to crack. Mobile operators did not plan to use packet mode voice before many years, while fixed operators, experiencing massive deployments of VoIP networks, were in a hurry to extend IMS to fixed networks and fully define the telephony profile.

In order to accelerate the extension of the IMS architecture to fixed networks (Next Generation Networks) and the standardization of the telephony service, ETSI formed in 2003 the TISPAN working group. TISPAN uses 3GPP IMS specifications as its foundation, but simplifies (e.g., makes SIP signaling compression optional) and complements them as needed. For instance, TISPAN specifies network attachment and resource control procedures required by fixed networks. The work of TISPAN focuses on two scenarios:

- the replacement of circuit based PSTN network by a packet network transparently for users ('emulated' PSTN telephony service);
- the replication of existing PSTN services for users of 'intelligent' SIP terminals ('emulated' services).

The IMS architecture and 3GPP/SIP are described in Chapter 4.

## 3.2 OVERVIEW OF A SIMPLE SIP CALL

### 3.2.1 Basic call scenario

In this section, we assume that the initiator of the call knows the IP address of the called endpoint. For instance the caller might be calling the following SIP address, also called a Uniform Resource Locator (URL) or a Uniform Resource Identifier (URI):

sip: john@192.190.132.31

In Section 3.3.1.1 we will see that there are many other types of SIP addresses; this one is just a simple case where the IP address of the called endpoint is directly specified. Note that the syntax looks a bit similar to web addresses e.g., http://www.netcentrex.net. This is because they both use the URI syntax, defined in RFC 2396. A URI begins with a 'scheme' portion before the colon (':'). The scheme portion indicates how the rest of the string is to be interpreted and which syntax to expect. If the scheme is 'http', then the syntax for an http resource is expected after the colon: a double slash ('//'), then a host name, and then an optional path.

SIP uses two `scheme` names: 'sip' for communications over non-secure transport protocols, and 'sips' for communications over secure transport protocols. SIP can also be used with the 'tel' `URI` scheme defined in RFC `2806`. The minimal expected syntax after 'sip:' or 'sips:' is a host name. A host name can be described either directly by its IP address in dotted form (e.g., 10.11.10.13), or using its name in the Domain Name System (e.g., `hostname.subdomainname.domainname.rootdomain`). The DNS system has been described in Chapter 2—DNS based number resolution.

Usually a SIP call is to a specific person, and therefore the SIP URL syntax allows more optional parameters to be used. The generic syntax is:

```
sip:user:password@host:port;uri-parameters?headers
```

Our example uses only the user and host portions.

SIP entities communicate using 'transactions'. SIP calls a TRANSACTION a request requiring a specific action (e.g., the `INVITE` request below) and the response(s) it triggers (200 OK in our example) up to a final response (see the definition below, all 2xx, 3xx, 4xx, 5xx, and 6xx responses are final). The initiator of a SIP request is called a SIP client and the responding entity is called a SIP server for that transaction.

Most communications using SIP need several transactions. If the caller is Mark and the callee is John, then the end systems used by Mark and John (the SIP RFC calls them 'User Agents') will play the role of the client or the server for each transaction, depending on which user agent initiates each transaction (Figure 3.3).

SIP uses several types of request methods: `REGISTER`, `INVITE`, `ACK`, `CANCEL`, `BYE`, `OPTIONS` (defined in the main RFC), `PRACK` (added in RFC 3262, and required for interoperability with the PSTN), `SUBSCRIBE`, `NOTIFY` (both used for IM for instance; these were added in RFC 3265).

The simple communication scenario in Figure 3.3 uses the `INVITE`, `ACK` and `BYE` request methods. A SIP client calls another SIP endpoint by sending an `INVITE` request

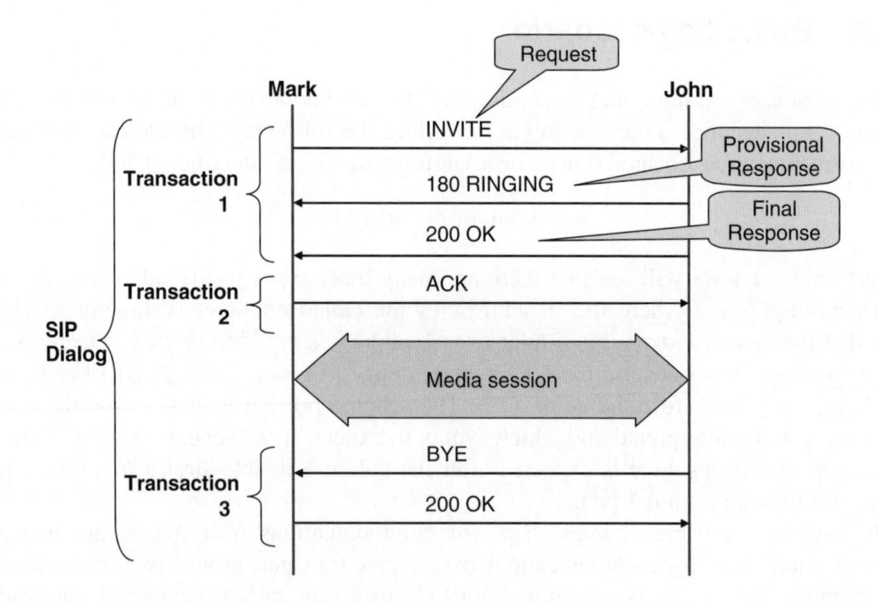


**Figure 3.3** SIP Dialog and SIP transactions.

message. The INVITE message usually contains enough information to allow the called terminal to immediately establish the requested media connection to the calling endpoint.

The called endpoint needs to indicate that it is accepting the request. This is the purpose of the 200 OK response message. Since the request was an invitation, the 200 OK response usually also contains the media capabilities of the called endpoint and where it is expecting to receive the media data.

The other messages are explained in detail below; the ACK is a simple acknowledgement of the 200 OK final response, and the BYE is the request to 'drop the call'.

This exchange of SIP transactions between two user agents during a communication is called a SIP dialog. The combination of the To tag, From tag, and Call-ID completely defines a SIP dialog. These information elements are defined below.

## 3.2.2 Syntax of SIP messages

SIP messages are encoded using HTTP/1.1 message syntax (RFC 2068). The character set is ISO 10,646 with UTF-8 encoding (RFC 2279, see Section 3.5.1.1.1 later in this chapter for more details on UTF-8).

Lines are terminated with CR LF (Carriage Return, Line Feed), but receivers should be able to handle CR or LF as well.

There are two types of SIP messages: REQUESTS and RESPONSES. They share a common format as indicated in Figure 3.4.

### 3.2.2.1 Headers

SIP messages use many header lines. The header types defined by RFC 3261 are listed in Table 3.1.

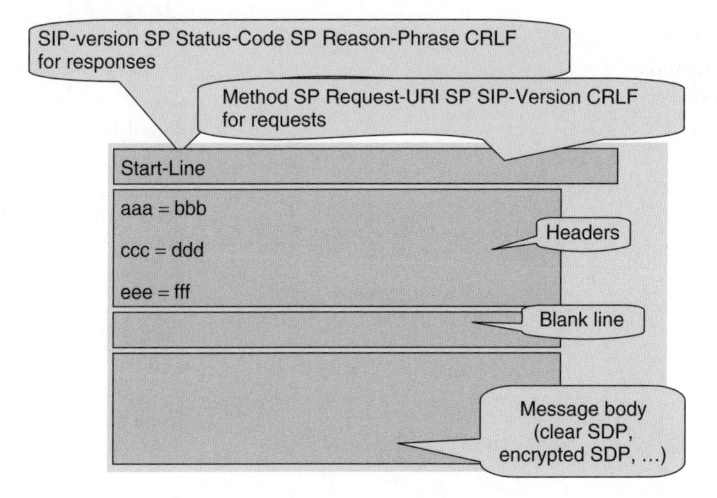


**Figure 3.4** SIP message format: Requests and Responses. (SP is an abbreviation for 'single space').

**Table 3.1** Headers defined in RFC 3261

| | Present in requests | Present in responses |
|---|---|---|
| Accept | ✓ | 2xx, 415 |
| Accept-Encoding | | |
| Accept-Language | | |
| Alert-Info | ✓ | 180 |
| Allow | ✓ | ✓ |
| Authentication-Info | | 2xx |
| Authorization | ✓ | |
| **Call-ID** | ✓ | Copied |
| Call-Info | ✓ | ✓ |
| Contact | ✓ | ✓ (mandatory in 2xx response to an INVITE) |
| Content-Disposition | ✓ | ✓ |
| Content-Encoding | | |
| Content-Language | | |
| Content-Length | | |
| Content-Type | | |
| **CSeq** | ✓ | Copied |
| Date | ✓ | ✓ |
| Error-Info | | 3xx, 4xx, 5xx, 6xx |
| Expires | ✓ | ✓ |
| **From** | ✓ | Copied |
| In-Reply-To | ✓ | |
| **Max-Forwards** | ✓ | |
| Min-Expires | | 423 |
| MIME-Version | ✓ | ✓ |
| Organization | ✓ | ✓ |
| Priority | ✓ | |
| Proxy-Authenticate | | 401, 407 |
| Proxy-Authorization | ✓ | |
| Proxy-Require | ✓ | |
| Record-Route | ✓ | 2xx, 18x |
| Reply-To | ✓ | ✓ |
| Require | ✓ | ✓ |
| Retry-After | | 404, 413,480, 486, 500, 503, 600, 603 |
| Route | ✓ | |
| Server | | ✓ |
| Subject | ✓ | |
| Supported | ✓ | 2xx |
| Timestamp | ✓ | ✓ |
| **To** | ✓ | Copied (+tag) |
| Unsupported | | 420 |
| User-Agent | ✓ | ✓ |
| **Via** | ✓ | Copied |
| Warning | | ✓ |
| WWW-Authenticate | | 401, 407 |

#### 3.2.2.1.1 The general header

The header fields part of the 'General Header' are mandatory in all SIP requests. Most header lines of the general header are also present in SIP responses where, usually, they are just copied from the corresponding request. These header fields appear in bold font in Table 3.1, and their role is explained below:

- **Call-ID**
  Example: `Call-ID: f81d4fae-7dec-11d0-a765-00a0c91e6bf6@foo.bar.com`
  The `Call-ID` header contains a *globally unique* identifier for the current multimedia session. For instance, for a SIP session corresponding to a telephone call, the Call-ID will remain identical in all SIP requests and answers exchanged during the call. In the case of a multiparty conference, the various call legs may contain the same Call-ID.
  In the case of isolated transactions, e.g., a `REGISTER` or `OPTIONS` request and their responses, the Call-ID identifies the various requests belonging to the same session: each `REGISTER` (initial and refresh `REGISTER`) and the corresponding responses will contain the same `Call-ID`.[8] For the `INVITE` and `REGISTRATION` requests the `Call-ID` value also helps to detect duplicates (duplicate INVITE requests can occur when there is a forking proxy in the path).
  The combination of the 'tag' values of the To and From headers and the `Call-ID` value uniquely identifies a given SIP dialog.
  A SIP user agent that initiates a multimedia session may form a new `Call-ID` value by concatenating a locally unique identifier with a globally unique identifier for the user agent, for instance its public IP address, or its DNS name.
- **Cseq**
  Example: `Cseq : 1234 INVITE`
  The `Cseq` header value was used to match requests and the corresponding responses within a SIP transaction in RFC 2543. In RFC 3261, the combined value of the `CSeq` and of the '`branch`' parameter of the `Via` header is used.
  The `Cseq` value consists of a sequence number which may start at any arbitrary value, followed by the name of the request method. The response to a given request must copy its `Cseq` header. As shown in Figure 3.5, SIP user agents must increment the `Cseq` sequence number by one unit for each new request sent within a dialog (not for retransmissions), except for `ACK` and `CANCEL` requests (for these requests the `Cseq` header identifies the request being acknowledged or cancelled). As provisional responses are not transmitted reliably in SIP, the sequence numbers received by the server side of a transaction may not be consecutive.
- **From**
  Example: `From: "MyDisplayName"<sip:myaccount@company.com>; tag=221411414`

[8] However, each SIP request is correlated with the corresponding SIP response by a matching value of the branch parameter, not by the Call ID.

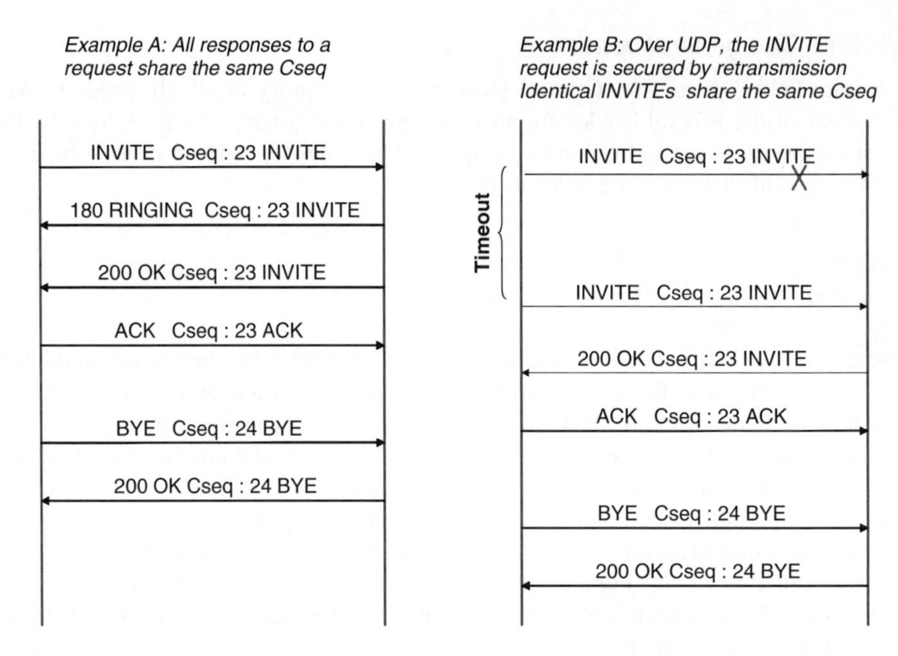


**Figure 3.5** Usage of CSeq header.

The `From` header identifies the sender of the request for SIP implementations according to RFC 3261. This field must be present in all SIP requests and is simply copied in SIP responses. The `From` header may contain an optional display name between double quotes (unless this display name contains no special character), followed by the URI of the sender of the request (between '<''>' signs, unless there is no display name in the header or the URI contains no parameter). This syntax is defined in RFC `2822` and is commonly used by messaging applications. If the sender of the request wishes to remain anonymous, he may use the keyword Anonymous (without double quotes) instead of the display name.
The `tag` parameter became mandatory in RFC 3261 as it is one of the key parameters, together with the `To` tag and the `Call-ID`, identifying a SIP dialog. Other optional tag values may be added. RFC 3261 considers that the default value of a tag, if absent, is null (tag = 0).
In the SIP profile defined by 3GPP and TISPAN, this header becomes informative and is replaced by the '`P-Asserted-ID`' or '`P-Preferred-ID`' headers. See the Section 4.3.1, for more details.

- **Max-Forwards**
  This mandatory header is used to prevent loops when a message is routed by proxies in the network.[9] Each proxy decrements this counter by one as it transmits the request

[9] This very useful feature was only added in H.323v5 (H.225v5HopCounter).

and responds with an error message if the counter value reaches zero. The server side does not copy this header in responses.

- **To**
  Example: `To: Helpdesk <sip:helpdesk@company.com>;tag=287447`
  The `To` header tag value is used, together with the `From` tag value and the `Call-ID`, to uniquely identify a SIP dialog (See Section 3.2.1). The `To` header value is simply copied back in SIP responses, which also add the To-tag value.[10]
  In `REGISTER` requests, the To header value contains the URI that must be recorded and associated to the `Contact` header value, also present in a `REGISTER` request. The URI present in the `To` header then becomes the 'Address of Record' (IETF terminology) or 'Public User Identity' (3GPP terminology) of the user agent, i.e., the address by which it is identified by end users, as opposed to its technical address as a signaling endpoint. In other requests, the `To` field value is only informative and available end to end by user agent implementations. Usually, however, it contains the destination URI of the request.
- **Via**
  Example: `Via: SIP/2.0/UDP PXY1.provider.com; branch = z9hG4bk12345; received=10.0.0.3`
  The only way to distinguish between the two versions of SIP (RFC 2543 or RFC 3261) is to consider the value of the `Via` header. In RFC 3261 it must contain a 'branch' parameter beginning with the 'magic value' 'z9hG4bK'. The value of the `branch` parameter is the identifier of the SIP transaction initiated by the element which inserted the `Via` header. The rather surprising role of this parameter is an example of the modular design violations still present in RFC 3261.
  The main purpose of the `Via` header is to record the route of a request, enabling intermediary SIP proxies to propagate SIP responses back, following the same path. As it forwards a request, each proxy inserts a new `Via` header containing its own address in the topmost position, and adds a specific branch parameter for each downward leg (case of a forking proxy, see Section 3.3.2). This accumulation mechanism is also very useful for the detection of routing loops (see Section 3.3.2.5).
  The server side simply copies back the `Via` headers in its response and may add additional parameters, e.g., a `received` parameter indicating that it received the last request from an IP address which does not correspond to that indicated in the last `Via` header (see RFC `3581` and Section 3.3.2.3).
  If UDP is used as the transport protocol, the address and port to use to respond to a SIP request are contained in the `Via` header: the responses must not be sent back to the source IP address and port of the received UDP datagram, except if the '`rport`' parameter is present (RFC 3581). This mechanism makes it possible to establish and maintain SIP dialogs through Network Address Translation (NAT) routers.

[10] In case the initial request is forked by a proxy, a different tag will be added in the responses on each leg, identifying two or more distinct SIP dialogs initiated by the same request.

#### 3.2.2.1.2 Other headers

These headers appear in Table 3.1, which specifies in which requests and responses they may be used. The most commonly used headers are listed below:

- **Accept**
  Example : `Accept : application/sdp, text/html, application/x-myextension`
  The `Accept` header specifies the message payload types accepted and understood by the sender of the message. If this header is absent then the value '`application/sdp`' is implied. The syntax of this header is specified in RFC1288.
  There are other 'Accept-' headers for more specific information related to the content of messages, for instance, `Accept-Language : fr, en-gb;q=0.8, en;q=0.7` specifies the languages spoken by the sender (with a preference level).
- **Allow**
  Example: `Allow: INVITE, ACK, CANCEL, BYE, PRACK, UPDATE, REFER, MESSAGE`
  This header lists the SIP methods accepted by the sender.
- **Contact**
  Example: `Contact: "John Doe" <sip:doe@phone4.myoffice.net>;q=0.7;expires=3600, "John Doe cell"<sip:doe@phone2.mycellnetwork.net>;q=0.2`
  This is one of the most complex SIP headers.
  When present in a `REGISTER` request, the `Contact` header contains the URI(s) where the 'Address of Record' (IETF terminology)/'Public identity'(3GPP terminology) specified in the `To` header of the same request may be reached. Each contact URI may be qualified by a preference parameter and a validity period for the association. Similarly, when present in a 3xx response, this header indicates alternative addresses where the destination 'Address of Record'/'Public identity' might be reached.
  When present in requests or responses establishing a SIP dialog, this header indicates a URI where the client (resp. server) accepts to receive subsequent requests within the same dialog. This allows the user agents to short-circuit the intermediary proxies. See Section 3.3.2.3.
  The compact form of this header is 'm:'
- **Content-Length, Content-Type**
  Examples: `Content-Type: application/sdp`
  `Content-Type:text/html;charset=ISO-8859-4`
  These headers relate to the payload of the message and form the *entity header*. `Content-Length` indicates the size, in octets, of the payload section of the SIP message (without counting the CR-LF separation between the header and payload section of the SIP message, but counting the CR-LF characters separating the SDP lines if the payload is of type application/sdp). The presence of this header is mandatory when SIP messages are exchanged over a stream oriented protocol (like TCP). `Content-Type` indicates the nature of the message payload. This header is

mandatory if there is a payload. Other headers of type 'Content-' exist to specify other properties of the content, e.g., 'Content-Langage = fr'.

- **Expires**
  Examples: `Expires:Thu,01 Dec 2000 16:00:00 GMT`
  `Expires:5`
  In the case of a `REGISTER` request, the `Expires` header indicates the validity duration of the registration as desired by the user agent. The 'registrar' (see Section 3.3.1.2) may reduce this value in its answer.
  In the case of an `INVITE` request, this header may be used to limit the duration of the request routing and processing (e.g., search algorithms) in the network.
- **Record-Route**
  Example : `Record-Route:`
  `sip:acd.support.com;maddr=192.168.123.234,sip:billing.net`
  `centrex.net;maddr=192.168.0.4`
  Some proxies (see Section 3.3.2.3) may add or update this header if they wish to remain in the path of all messages related to the current SIP dialog. This header is copied in all responses establishing a SIP dialog, enabling the sender to learn that it needs to route the subsequent requests of this dialog through the indicated URIs.
- **Require**
  Example: `Require : sec-agree, 100-rel`
  The user agent may insert a `Require` header within a request (resp. a response) to indicate the SIP extensions that are necessary for the proper functioning of the transaction. Similarly a `Proxy-Require` header may be inserted to indicate that each proxy routing the request must support the specified SIP extension. SIP extensions must be documented in 'standard track' RFCs, e.g., sec-agree (RFC 3329), precondition (RFC 4032), 100-rel (RFC 3262). Some SIP extensions use other extension mechanisms, for instance, the signaling compression mechanism of 3GPP uses a comp = sigcomp parameter in the Route header.
- **Supported**
  Example: `Supported:path, 100-rel`
  When present in a request, this header lists the extensions that the client is ready to accept from the server in its responses. When present in a 2xx response, this header lists the extensions that the client will accept in subsequent requests. These extensions must be listed in a 'standard track' RFC, e.g., RFC 3327 for the 'path' extension.
- **User-Agent**
  Example: `User-Agent: MySoftPhone v1.3.4`
  This header specifies the version of the user agent which generated the request. The format is free. It is very useful to enable servers to adapt to well known limitations or a specific behaviour of a given client.

#### *3.2.2.2 SIP requests*

SIP requests are sent from the client terminal to the server terminal. The following methods are defined:

- "ACK": An ACK Request is sent by a client to confirm that it has received a final response from a server, such as 200 OK, to an INVITE request.
- "BYE": A BYE Request is sent either by the calling agent or by the caller agent to drop a multimedia session.
- "CANCEL": A CANCEL request can be sent to abort a request that was sent previously as long as the server has not yet sent a final response.
- "INFO": defined in RFC 2976, used to carry information which does not change the call state. Some vendors use it for DTMF out-of-band transport (see Section 3.2.2.7.1.3).
- "INVITE": The INVITE request is used to initiate a multimedia session.
- "MESSAGE": This request is defined in RFC 3428 and is used for IM; see Section 3.6.
- "NOTIFY": This request is defined in RFC 3265 and is used to send event notifications; see Section 3.5.2.
- "OPTIONS": A client sends an OPTIONS request to a server to learn its capabilities. The server will send back a list of the methods it supports. It may also in some cases reply with the capability set of the user mentioned in the URL, and how it would have responded to an invitation.
- "PRACK": defined in RFC 3262, is used to implement reliable provisional responses.
- "REFER": this request is defined in RFC 3515 and can be used to redirect sessions; see Sections 3.3.3.1 and 3.3.4.3.
- "REGISTER": This request is used by SIP clients to register their current user agent technical location (one or more Contact URIs) and associate these with a public address ('Address of Record' in IETF terminology, 'Public identity' in 3GPP terminology). A SIP server that can accept a register message is called a registrar.
- "SUBSCRIBE": This request is defined in RFC 3265 and is used to request specific event notifications; see Section 3.5.2 later in this chapter.
- "UPDATE": defined in RFC 3311, is used to change media session parameters in early dialogs, i.e., before the final response to the initial INVITE.

#### *3.2.2.2.1 Start line*

The start line indicates the request method (e.g., INVITE), followed by the request-URI which indicates the user or service to which this request is being addressed. For all requests except REGISTER, it is usually set to the same value as the URI in the To header (this is no longer true in the specific 'strict-routing' mode detailed in Section 3.3.2.3). In the specific case of the REGISTER request, the URI parameter is the one of the registrar server (see Section 3.3.1.2), and the To header line contains the Address of Record (IETF terminology) or Public identity (3GPP terminology) of the user agent.

The last element of the start line is the SIP protocol version, which is SIP/2.0 for endpoints implementing RFC 2543 or 3261.

Figure 3.6 illustrates the SIP request message format.

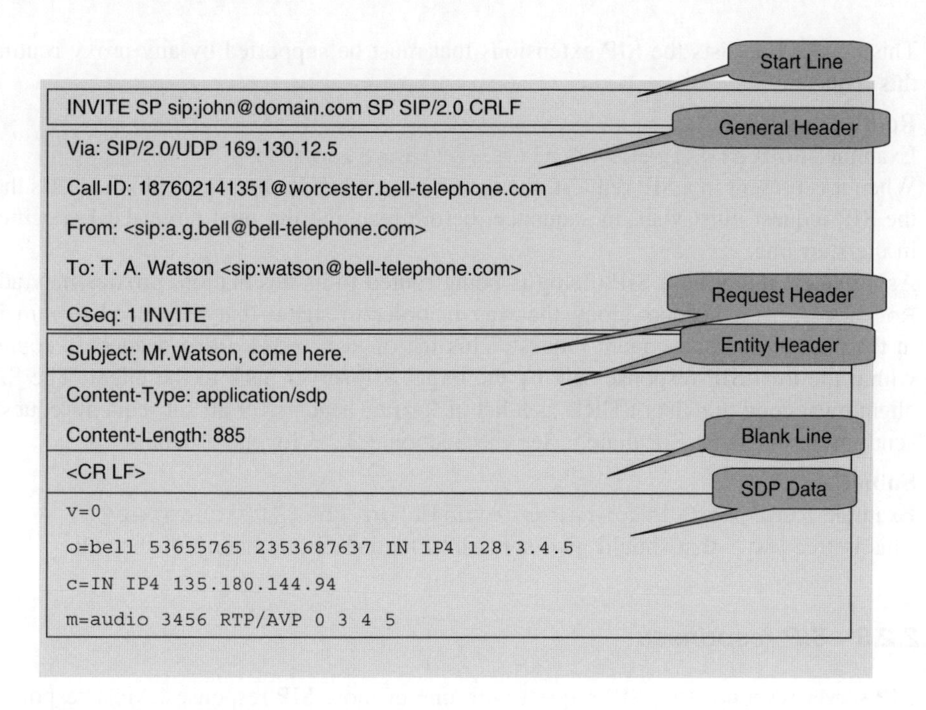
Start Line
General Header
Request Header
Entity Header
Blank Line
SDP Data
INVITE SP sip:john@domain.com SP SIP/2.0 CRLF
Via: SIP/2.0/UDP 169.130.12.5
Call-ID: 187602141351@worcester.bell-telephone.com
From: <sip:a.g.bell@bell-telephone.com>
To: T. A. Watson <sip:watson@bell-telephone.com>
CSeq: 1 INVITE
Subject: Mr.Watson, come here.
Content-Type: application/sdp
Content-Length: 885
<CR LF>
v=0
o=bell 53655765 2353687637 IN IP4 128.3.4.5
c=IN IP4 135.180.144.94
m=audio 3456 RTP/AVP 0 3 4 5

**Figure 3.6** SIP Request message format.

### *3.2.2.2.2 Specific header lines for SIP requests*

In addition to the mandatory fields of the general header, requests can carry additional fields in the *request header*:

- **Authorization**
  Example: `Authorization:Digest username="alice", realm="Atlanta.com", nonce="82a8b34f234ccd345",response="3456ed345ac3434"`
  This header line contains the client authentication data, following a 401 response of the SIP server. See Sections 3.3.1, 3.4.2.1 later in this chapter and Section 4.3.1.1 in Chapter 4 for more details and usage examples.
- **Priority**
  Example : `Priority: emergency`
  This header line specifies the precedence level of the request. Possible values are those of RFC 2076 plus 'emergency'.
- **Proxy-Authorization**
  Just like the Authorization header line, this header line contains the client authentication data, following a 407 response of the SIP proxy. It is rarely used in practice.
- **Proxy-Require**
  Example: `Proxy-Require:sec-agree`

This header line lists the SIP extensions that must be supported by any proxy routing this request.

- **Route**
  Example: `Route:<sip:box3.site3.atlanta.com;lr>`
  When it is present in a SIP request, the list of `Route` header lines indicate the URIs that the SIP request must visit, in sequence, before reaching the final target URI specified in the start line.
  As a request initiating a SIP dialog is being routed to its destination, proxies may add `Record-Route` headers along the way in order to signal that they need to remain in the path of any subsequent request. This list of `Record-Route` headers is copied within the final SIP response sent by the target SIP server back to the client. The SIP client must copy this list of URIs as a list of `Route` headers for all subsequent requests sent within the same SIP dialog. See the Section 3.3.2.3 for more details.
- **Subject**
  Example : `Subject: "Conference call on the SIP chapter"`
  This is free text[11] that should give some information on the nature of the call.

### *3.2.2.3 SIP responses*

A SIP server responds to a SIP request with one or more SIP responses. Most responses (2xx, 3xx, 4xx, 5xx, and 6xx) are 'final responses' and terminate the current SIP transaction. The 1xx responses are 'provisional' and do not terminate the SIP transaction.

The SIP response format is illustrated in Figure 3.7.

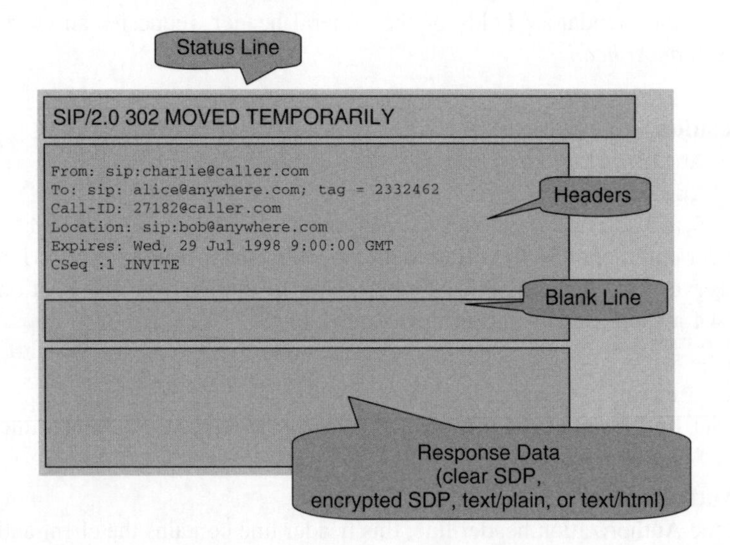


**Figure 3.7** SIP response message format.

[11] It should not contain any character that could cause problems to the parsing of the SIP request.

The first line of a SIP response always contains a status code and a human readable reason phrase. Most of the header section is copied from the original request message. Depending on the status code, there may also be additional header fields, and the response data part may be empty or contain a SDP session description or explanatory text.

#### 3.2.2.3.1 *Response codes*

So far, six categories of status codes have been defined, classified according to the first digit. Common codes are listed in Table 3.2.

This classification makes it easier to add new status codes: in case an old terminal does not understand a new CXX code, it should treat it as a C00 code. Therefore, even old terminals will be able to react 'intelligently' when facing unknown status codes. These terminals can also give some additional information to the user if a reason phrase is present.

RFC 3326 added a new header `Reason` to facilitate interoperability with the PSTN by encapsulating the cause codes defined by Q.850:

```
Reason: Q.850 ;cause=16 ;text="Terminated"
```

#### 3.2.2.3.2 *Specific header lines for SIP responses*

In addition to the mandatory fields of the general header, responses can carry additional fields in the *response header*:

- **Authentication-Info**
  Example: `Authentication-Info: next-nonce=" 2e13164c23473432d0812c 1a5fb2"`
  This header line, specified in RFC 2617 ('HTTP Authentication: Basic and Digest Access Authentication'), may be inserted by a SIP server in any `200 OK` response which has been previously successfully authenticated. It submits a new challenge to the SIP client, which must be used for the next Request requiring authentication.
- **Min-Expires**
  Example : `Min-Expires:60`
  This header line contains the shortest acceptable registration refresh delay, 60 seconds in the example.
- **Proxy-Authenticate**
  Example:
  ```
  Proxy-Authenticate: Digest realm="atlanta.com",
          domain="sip:ss1.carrier.com", qop="auth",
          nonce="f84f1cec41e6cbe5aea9c8e88d359",
          opaque="", stale=FALSE, algorithm=MD5
  ```
  This header line is used within `407 Proxy Authentication Required` responses. In practice 401 responses are more commonly used for authentication and use a WWW-Authenticate header line.
- **Server**
  Example: `Server:NCX CCSv3.3.8`

**Table 3.2** SIP response codes

| | | | |
|---|---|---|---|
| 1xx | Informational | | request received, continuing to process the request |
| | | 100 | Trying |
| | | 180 | Ringing |
| | | 181 | Call is being forwarded |
| | | 182 | Queued |
| | | 183 | Session progress (used to provide inband network announcements, equivalent to ISDN progress message) |
| 2xx | Success | | The action was successfully received, understood, and accepted |
| | | 200 | OK |
| 3xx | Redirection | | Further action must be taken in order to complete the request |
| | | 300 | Multiple choices: several possible locations in contact headers |
| | | 301 | Moved permanently: user can no longer be found at the address specified. New address is in contact header fields |
| | | 302 | Moved temporarily: alternative address in contact header, which may also specify duration of validity |
| | | 305 | Use Proxy: the specified destination must be reached though a proxy |
| | | 380 | For future use: Alternative Service, described in the message body |
| 4xx | Client Error | | The request contains bad syntax or cannot be fulfilled at this server |
| | | 400 | Bad Request |
| | | 401 | Unauthorized |
| | | 402 | Payment Required |
| | | 403 | Forbidden |
| | | 404 | Not Found |
| | | 405 | Method not Allowed |
| | | 406 | Not Acceptable |
| | | 407 | Proxy Authentication Required |
| | | 408 | Request Timeout |
| | | 409 | Conflict |
| | | 410 | Gone |
| | | 411 | Length Required |
| | | 413 | Request Message Body Too Large |
| | | 414 | Request-URI Too Large |
| | | 415 | Unsupported Media Type |
| | | 420 | Bad Extension |
| | | 480 | Temporarily not available |
| | | 481 | Call Leg/Transaction Does Not Exist |

**Table 3.2** (continued)

| | | | |
|---|---|---|---|
| | | 482 | Loop Detected |
| | | 483 | Too Many Hops |
| | | 484 | Address Incomplete |
| | | 485 | Ambiguous |
| | | 486 | Busy Here |
| | | 487 | Request Terminated (by a CANCEL request) |
| | | 488 | Not Acceptable Here. Used if an unacceptable offer is received in an UPDATE |
| | | 491 | Offer received in an UPDATE while an offer was already pending |
| 5xx | Server Error | | The request contains bad syntax or cannot be fulfilled at this server |
| | | 500 | Internal Server Error. Also used when an UPDATE is received before an answer has been generated to a previous UPDATE |
| | | 501 | Not Implemented |
| | | 502 | Bad Gateway |
| | | 503 | Service Unavailable |
| | | 504 | Gateway timeout. Also used if an UPDATE request is not acceptable because the user must be prompted, therefore no immediate response can be generated |
| | | 505 | SIP version not supported |
| 6xx | Global Failure | | The request is invalid at any server |
| | | 600 | Busy Everywhere |
| | | 603 | Decline |
| | | 604 | Does not exist anywhere |
| | | 606 | Not acceptable |

This header line lets the server return information on the vendor, the software release, etc. Clients use the client header line for a similar purpose.

- **Unsupported**
  Example: `Unsupported:100-rel`
  When a server rejects a SIP request because a required SIP extension is not supported, the Unsupported header may specify which extension(s) is not supported using this header line.
- **WWW-authenticate**
  Example: `WWW-Authenticate: Digest realm="homenetwork.com", nonce="ae45abc12c415abc12e5aea12e88d2c41", algorithm=AKAv1-MD5`
  This header line is used with 401 responses, by user agents and registrars as part of the authentication procedures. See Sections 3.3.1 and 3.4.2.1 for more details.

#### 3.2.2.4 Call initiation details

The messages exchanged by a SIP client and a SIP server are independent of the underlying transport protocol (except for some details, for instance, the size of non-TCP messages is limited). The original SIP RFC only required SIP systems to support the UDP transport protocol. RFC 3261 now requires SIP systems to support both the UDP and TCP transport protocols, but UDP is still the most widely used protocol, because it provides better control on retransmission and latency. The only issue of UDP is that it cannot transport large amounts of information without causing packet fragmentation (SIP does not define any application level fragmentation mechanism). RFC3261 recommends the use of TCP or any other reliable congestion controlled protocol (defined in RFC 2914), if the request size is within 200 bytes of the path MTU, or if it is larger than 1300 bytes and the path is unknown. All implementations must still be able to handle packets up to 65,535 bytes (including IP and UDP headers). In practice, this dynamic change of transport protocols from UDP to TCP is cumbersome and does not work well: many implementations simply rely on the IP level standard fragmentation mechanism, which works well.

Connections over UDP or TCP require a port number. If no port is specified in the SIP URI, the connection is made to port 5060 for the transport protocols UDP, TCP, and SCTP (Stream Control Transmission Protocol, RFC 2960). If the transport protocol used is TLS (Transport Layer Security, a secure transport protocol using TCP, defined in RFC 2246), the default port is 5061.

When using TCP, the same connection can be used for all SIP requests and responses of a dialog, or a new TCP connection can be used for each transaction. If UDP is used, the address and port to use for the answers to SIP requests is contained in the 'Via' header parameter of the SIP request. Replies must not be sent to the IP address of the client except when a `rport` parameter is present in the Via line (see description of the Via header in Sections 3.2.2.1.1 and 4.2.3.2).

Since no protocol or port is specified in our sample SIP URI (sip:john@192.190.132.31), Mark's user agent defaults to UDP, and will send its first SIP `INVITE` message over UDP to IP address 192.190.132.31, port 5060. When UDP is used, only one SIP message per UDP datagram may be sent; when a stream-oriented protocol like TCP is used, the message framing uses the `Content-Length` header to determine the end of the message and the beginning of the next: therefore this header must be present when SIP is used over stream-oriented protocol. This violation of the strict transport and presentation protocol layering avoids specifying a transport layer framing mechanism for stream protocols.[12]

The dialog is initiated by the following `INVITE` Message:

```
INVITE sip:john@192.190.132.31 SIP/2.0
Via: SIP/2.0/UDP 10.11.12.13;branch=z9hG4bK776asdhds
Max-Forwards: 70
To: "John" <sip:john@192.190.132.31>
```

[12] H.323 uses a stricter protocol layering and uses RFC 1006 for the purpose of transporting messages over a stream oriented protocol.

```
From: :"Mark" <sip:mark@10.11.12.13>;tag=1928301774
Call-ID: a84b4c76e66710@mydomain.com
CSeq: 314159 INVITE
Content-Type: application/sdp
Content-Length: 228

        v=0
        o=mark 114414141 12214 IN IP4 10.11.12.13
        s=-
        c=IN IP4 10.11.12.13
        t=0 0
        m=audio 49170 RTP/AVP 0
        a=rtpmap:0 PCMU/8000
        m=video 51372 RTP/AVP 31
        a=rtpmap:31 H261/90000
        m=video 53000 RTP/AVP 32
        a=rtpmap:32 MPV/90000
```

The last part of the message, after the blank line, is the media description using the SDP, which will be described in detail in Section 3.2.2.7.2. We will focus first on the Request start line and the most important headers which are mandatory in any SIP request (in bold font). As noted above, SIP is not strictly independent of the transport protocol, and the `Content-Length` header is mandatory only over stream transport protocols, even if there is no payload in the SIP message.

The start line is the first line of the SIP request message. In the example, the SIP method being used is `INVITE` (a request to establish a new session, which will materialize into a new SIP dialog), and the target is URI 'john@192.190.132.31'.

The `Via` header contains the destination address that must be used to send the response to this request: in the example 10.11.12.13, and the transport protocol will be UDP. Also, we note the branch parameter beginning with the magic value `z9hG4bK`: this user agent supports the procedures defined in RFC 3261.

The hop counter (`Max-forwards` header) is initialized to 70: it will be decremented each time the message is forwarded by a proxy.

The From header indicates the URI of the SIP client, and the `tag` parameter is initialized. This is one of the information elements which will uniquely identify the SIP dialog, together with the `Call-ID` and the `To` header tag value which will be set in the response.

The `To` header simply contains informative content with no protocol value. The `To` tag is absent as it will be added by the server side: the SIP request could be forked by intermediary proxies, in which case this single request could initiate multiple SIP dialogs which would each be identified by the To tag of the corresponding response.

The `Call-ID` contains the universally unique identifier for this call. Here, we suppose that Mydomain.com is a universally unique DNS name, and the identifier before the @ sign is locally unique to the client.

The `CSeq` header contains a value which would be used, in RFC 2543, to associate responses to this request. In RFC 3261, however, the combined value of the `CSeq` and the `branch` parameter of the '`Via`' header serves this purpose.

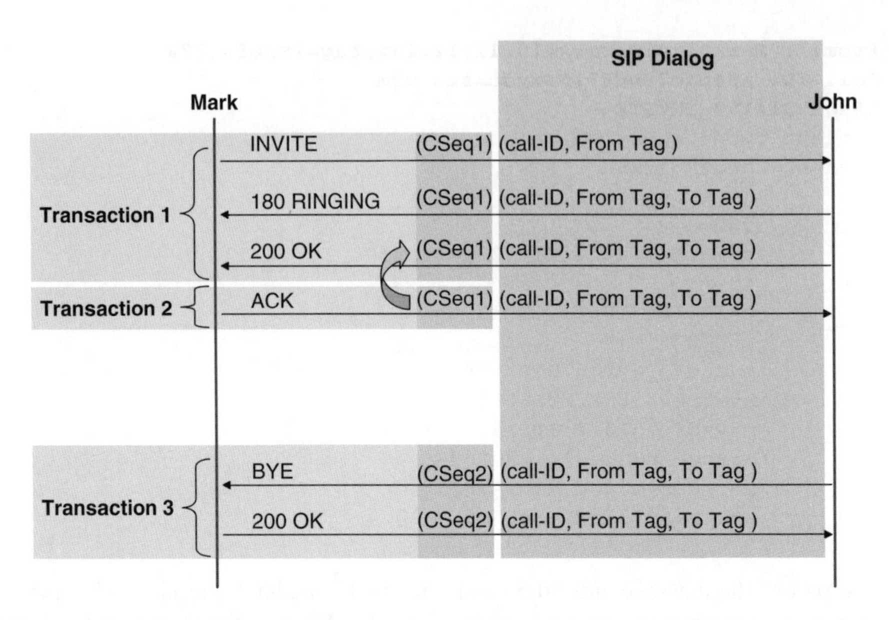


**Figure 3.8** A Dialog is identified by the combination of call-ID, From tag, and To tag.

### *3.2.2.4.1 SIP Dialog and transaction identifiers, message retransmission*

The SIP dialog is identified by the `From` tag, `To` tag, `Call-ID` and `Via Branch` combination (Figure 3.8). Once the dialog is established, all requests and responses of the dialog must include these header values. This makes it easy for a user agent to identify the relevant dialog upon receiving a SIP message.

Each transaction is identified by the common value of its `CSeq` header field (both the method name and sequence number must be identical). The value of the `CSeq` header must be distinct for distinct transactions within the dialog. The only exceptions are the `ACK` after a non 2xx response (see below for details) and `CANCEL` transactions.

The `CANCEL` transaction uses the same `CSeq` identifier as the transaction it is attempting to cancel, but the method name is set to `CANCEL` .

The `ACK` transaction is used only in relation with a prior `INVITE` transaction, which uses a three-way handshake, while all other types of transactions use only a two-way handshake. When it is sent after a 2xx final response, the `ACK` is an isolated, transactionless message. When it is sent after a negative response which terminates the SIP dialog, it is sent as part of the `INVITE` transaction.

#### *3.2.2.4.1.1 Non-INVITE transactions*

When used over non-reliable transport protocols, the reliability of Non-INVITE transactions relies on message retransmission.

The sender of a request will first retransmit the message if it does not receive a provisional or final response within 500 ms[13] (or a better estimation of the network round-trip time if the user agent calculates one using timestamps in SIP messages). It will keep retransmitting the request until it receives a response, doubling the retransmit interval at each occurrence up to an interval of 4 seconds. If a provisional response is received, then the retransmission still occurs until a final response is received, but the retransmission interval is immediately set to 4 seconds. The client will wait up to 64*500 ms, or 32 seconds, for a final response. After this delay it will consider that the transaction has failed. In case the transport layer indicates a failure (for instance, if an ICMP unreachable IP packet has been received), then the transaction fails immediately without trying any retransmission.

If the server response is lost, then the client will retransmit the request. For this reason, the server side of the transaction simply retransmits its last response each time it receives a retransmitted request indicating the response has been lost. Note that because only the *last* response is retransmitted, some *provisional responses* may be lost. SIP does NOT guarantee delivery of provisional responses (this causes problems in certain cases, and RFC 3262 has introduced the new `PRACK` request as an extension to SIP to solve the issue).

Figure 3.9 is an example of retransmission of a `BYE` transaction.

Even after it has sent a final response, the server will keep the response in memory for 32 seconds (64*500 ms) in the case an incoming client request requires a retransmission after the server has sent the final response. This happens if the final response is lost.

It is important to realize that this basic two-way handshake works only if the final response to the request arrives quickly after the request has been sent. SIP expects and requires Non-INVITE transactions to complete within a couple of seconds. If this two-way handshake was used for transactions that take a longer time to complete, then the request would be resent multiple times, which would obviously be very inefficient. Because of this, a different strategy is required for `INVITE` transactions.

##### *3.2.2.4.1.2 INVITE transactions*

(a) A three-way handshake

The handling of the `INVITE` transaction in SIP is completely different from the handling of other transactions. The handling of the `INVITE` is one of the most complex aspects of SIP.

The two-way transaction used for non-INVITE transactions presents a number of issues:

- The initial transaction is resent every 4 seconds until the final response arrives. In a telephony application, the `200 OK` signaling that a user has picked up the phone can

[13] The SIP specification specifies all times in multiples of a default timer T1. In order to simplify reading, we calculated all timer values using the T1 default value of 500 ms recommended in RFC 3261.

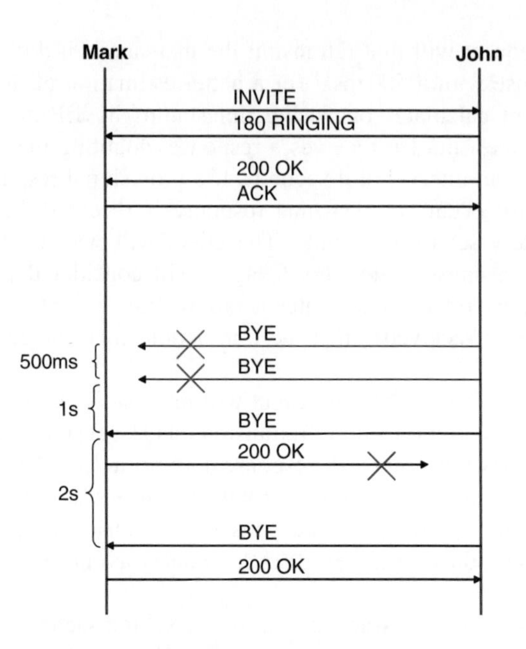


**Figure 3.9** Non-INVITE request retransmission.

arrive up to 3 minutes after the `INVITE` has been sent (this is the amount of time most telephone networks let a phone ring before cancelling the call). In some PSTN interworking applications (prepaid calling card, network prompts), the `200 OK` may never arrive. In order to avoid unnecessary overhead *the retransmission of the* `INVITE` *request stops as soon as a provisional or final response arrives*.

- If the `200 OK` response is lost, it would be retransmitted 4 seconds later in the previously described retransmission scheme. This is obviously unacceptable because no audio path can be established during this delay between the caller and the callee. Instead, if the caller has received a `180 RINGING` provisional response, the status of his call line would continue to appear as 'ringing'. This situation is avoided in an `INVITE` transaction, because the server expects to receive an immediate ACK after it sends a 200 OK. If the `ACK` is not received, then the response is retransmitted after 500 ms.

The three-way handshake works as follows for the client side:

- Over UDP the INVITE request is resent after 500 ms if no provisional or final response is received. The INVITE retransmission continues, but with a retransmission interval that doubles every time, until a provisional or final response arrives.[14] The transaction attempt aborts after seven retransmissions. Note that the server side is required to send a 100 Trying provisional response back within 100 ms, unless it knows that it is going to

[14] RFC 3581 limits the retransmission time to 20 seconds if the user agent is behind a NAT function, in order to keep the NAPT function 'pinhole' open and ready to forward the expected response.

send a final response within 200 ms, so the INVITE retransmission mechanism should normally trigger only if a message gets lost, not if the server side is slow.

- Over TCP or a reliable transport protocol, the `INVITE` request in not retransmitted at application level (it may be retransmitted at the TCP level).

The server side of the `INVITE` transaction over UDP will retransmit the last provisional response if it receives a retransmitted `INVITE` request (this means one or more provisional responses have been lost). The handling of the final response is different:

- In RFC `2543` it is retransmitted after 500 ms if an `ACK` has not been received; then it continues to be retransmitted, with intervals that double every time until an `ACK` is received. The retransmission of the final response is aborted after seven unsuccessful retransmissions or if a `BYE` request is received for this dialog, or, in the case of 3xx, 4xx, and 5xx responses, if a `CANCEL` request is received. If the retransmission that was aborted was for a `200 OK` response, the server should generate a `BYE` request, in case the `200 OK` response did arrive to the client user agent.
- RFC 3261, with its cleaner specification of the transaction layer and modified RFC2543, does not require retransmissions over reliable transports for 3xx responses. However, the `200 OK` response needs to be retransmitted even on reliable transports because it may be routed through several proxies, which may forward it on UDP. Since UDP is unreliable the `200 OK` could get lost, but SIP does not allow proxies to participate in the `200 OK` reliability mechanism, which is ensured end to end only (see the paragraph below on the `ACK` request). Therefore, if the user agent server did not retransmit the `200 OK` response, it could get lost in the network. RFC 2543 extended this behaviour to all final responses, not just `200 OK`, in order to avoid adding exceptions to the already complex specification. In RFC 3261, which has a more formal definition of transactions, only `200 OK` responses are retransmitted over reliable transports.

(b) The ACK request

The `ACK` request that completes the three-way handshake is formed as follows:

- Most of the headers are identical to the original `INVITE` headers; in particular, the `Route` header fields should be identical.
- The `To` header will probably contain a `tag`, which was added by the server side in its response.
- The `CSeq` header serial number is identical to the `CSeq` serial number of the `INVITE` request, which is an exception to the general rule to increase sequentially the `CSeq` header for new requests. This is to have the ability to correlate the `ACK` to the `INVITE` transaction. The `CSeq` method portion is “`ACK`”.

This is an example of an `INVITE` request and its `ACK`:

```
INVITE sip:john@192.190.132.31 SIP/2.0
     Via: SIP/2.0/UDP 10.11.12.13;branch=z9hG4bK776asdhds
```

```
    Max-Forwards: 70
    To: "John" <sip:john@192.190.132.31>
    From: :"Mark" <sip:mark@10.11.12.13>;tag=1928301774
    Call-ID: a84b4c76e66710@mydomain.com
    CSeq: 314159 INVITE

ACK sip:john@192.190.132.31 SIP/2.0
    Via: SIP/2.0/UDP 10.11.12.13;branch=z9hG4bK776asdhds
    Max-Forwards: 70
    To: "John" <sip:john@192.190.132.31>;tag=12344235
    From: :"Mark" <sip:mark@10.11.12.13>;tag=1928301774
    Call-ID: a84b4c76e66710@mydomain.com
    CSeq: 314159 ACK
```

The rules applying to the ACK request in a SIP network are also a bit complex. The handling of 200 OK final responses and 3xx, 4xx, and 5xx final responses is different. If an INVITE transaction is routed through several proxies, then each proxy can acknowledge 3xx, 4xx, 5xx, and 6xx responses by sending its own ACK. This applies only to proxies capable of acknowledging such requests locally (for instance, stateless proxies do not have this intelligence and remain passive, acting as simple message routers). But 200 OK responses are ALWAYS acknowledged by the client user agent. This difference was made because the SIP specification wants SIP clients to know all servers that have accepted an INVITE transaction. Because of forking proxies, more than one server may accept an INVITE if this INVITE has been forked in the network, and a single INVITE request can generate more than one SIP dialog.

In the example of Figure 3.10, one INVITE generates two dialogs. The example also shows that while provisional responses are not transmitted reliably, the final 200 OK response is transmitted reliably. Because multiple 200 OK responses can be received for a single INVITE, in theory, the client could continue to wait for 200 OK responses forever after sending an INVITE request. The SIP specification limits the waiting time to 32 seconds after the first 200 OK was received.

In the second example, Figure 3.11, a forking proxy generates a ACK for a 486 final response, while it forwards the 200 OK response to the client user agent.

The ACK is a very special type of request:

- It is never acknowledged. If an ACK gets lost, the server will retransmit the final response, and the client will retransmit the ACK when receiving the duplicate final response. This mechanism is very similar to the two-way handshake of non-INVITE transactions, with the final response playing the role of the request, and ACK playing the role of the response.
- Because the ACK is never acknowledged, proxies cannot signal any failure back to the client. This requires a specific procedure for authentication. For instance: since the ACK cannot be challenged, it must contain the same credentials as the INVITE. See Section 3.5.2 for further details.

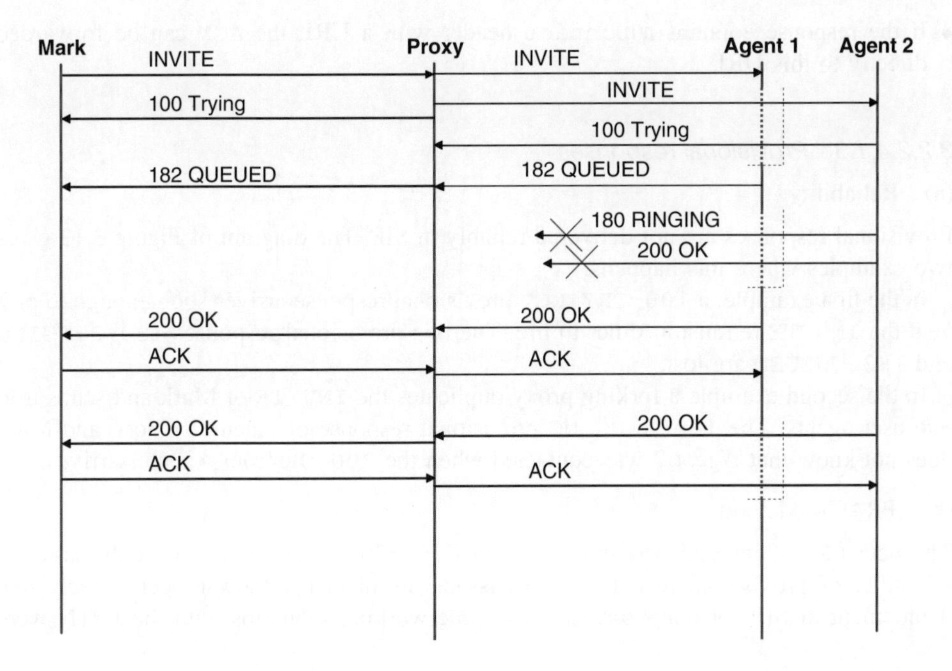


**Figure 3.10** Forking an INVITE request.

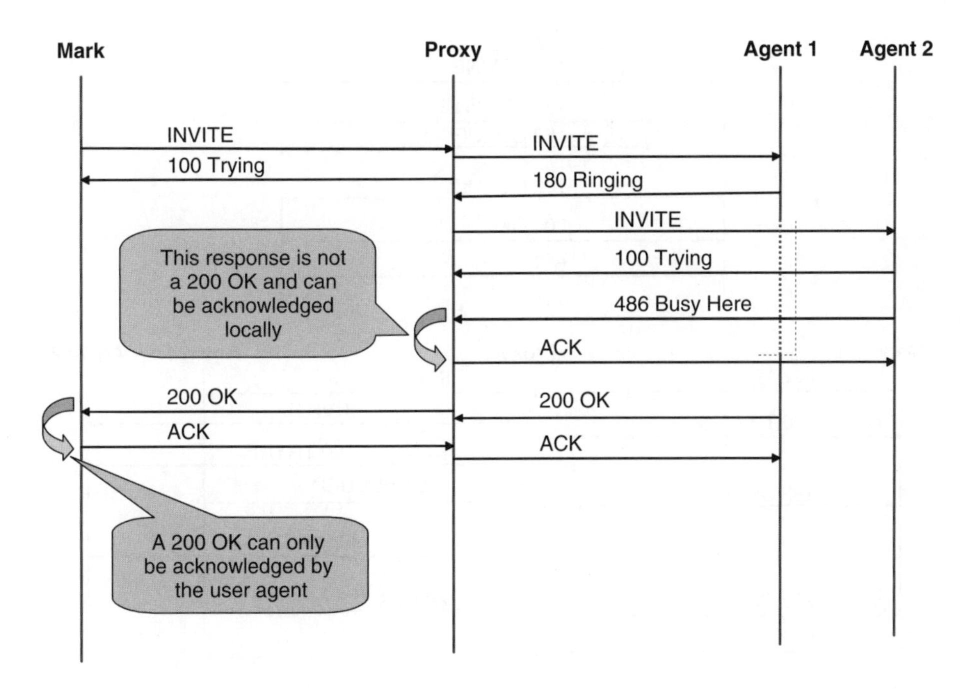


**Figure 3.11** 200 OK responses must be acknowledged end to end.

- If the response contains a `Contact` header with a URI, the `ACK` can be forwarded directly to this URI.

#### *3.2.2.4.1.3 Provisional responses*

(a) Reliability

Provisional responses are not delivered reliably in SIP. The diagram of Figure 3.12 gives two examples where this happens.

In the first example, a `100 TRYING` provisional response arrives soon enough to prevent the `INVITE` retransmit timer to fire. The two provisional responses `180 RINGING` and `182 QUEUED` are lost.

In the second example a forking proxy duplicates the `INVITE` of Mark and sends it to two user agents. The `180 RINGING` provisional response of Agent 2 is lost, and Mark does not know that Agent 2 was contacted when the `200 OK` from Agent 1 arrives.

(b) PRACK Method

The new `PRACK` method was introduced in RFC 3262 (June 2002) to fix the issue of reliability of provisional responses. This issue was in fact a showstopper for any real deployment of SIP for telephony, as many interworking scenarios with the PSTN were

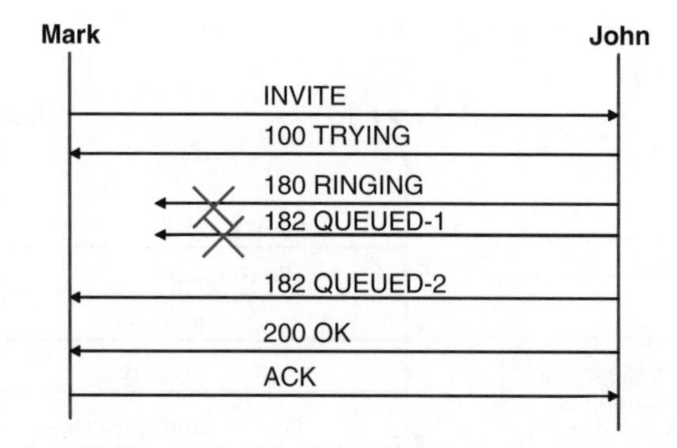


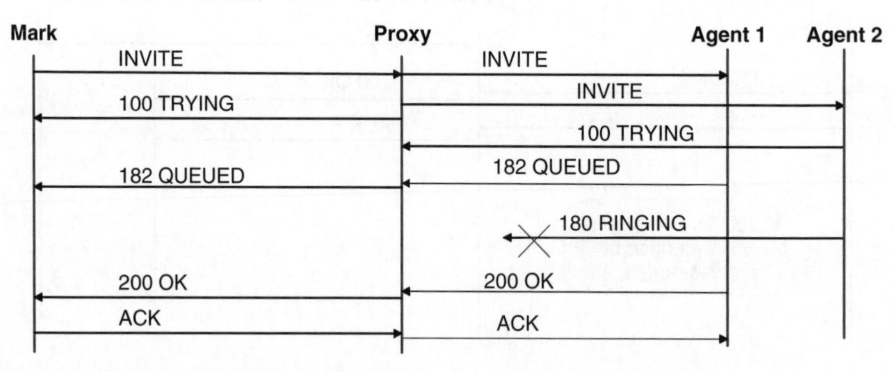


**Figure 3.12** Lack of reliability for provisional responses in baseline SIP.

not supported. Mobile networks, for instance, make an extensive use of pre-call announcements, which could not be delivered reliably in SIP prior to RFC 3262.

The optional `PRACK` extension mimics the end to end reliability algorithm of the `200 OK` response. Provisional responses are retransmitted periodically until the acknowledgment (`PRACK`) arrives. The major difference is that `PRACK`, unlike `ACK`, is itself a normal SIP method, acknowledged hop-by-hop by each stateful proxy, and requiring its own response from the server. The `PRACK` transmission reliability is ensured as for a normal request by the expected 200 OK.... This means that if the client receives a retransmitted provisional response, it should not retransmit the `PRACK`, but rely only on the `PRACK` response to decide if it should retransmit it or not.

Each provisional response contains a serial number in a `RSeq` header, mirrored in the corresponding `PRACK` method `RAck` header. The format of this header is:

```
RAck: <response serial number> <Cseq number> INVITE
```

There should be a `PRACK` for each provisional response, unlike reliability mechanisms such as TCP, also based on serial numbers, which are cumulative (the acknowledgment serial number validates all received messages with lower values).

The sender of the `INVITE` should indicate that it supports the `PRACK` mechanism by including a `100rel` option tag in the `Require` header field. This option can be rejected by the receiver with a `420 Bad Extension response` (unsupported header field with 100rel option tag). If it accepts the `PRACK` option, the receiver may send all 1xx responses (except `100 Trying`), reliably. In reliable provisional responses, it needs to include a `Require` header field with option tag `100rel`, and a `Rseq` header (unique within the transaction).

Unacknowledged provisional responses are simply retransmitted with an exponentially increasing delay, with the same serial number. New provisional responses are sent with the next higher serial number (see Figure 3.13).

If a media offer (see the Section 3.2.2.7.2.3) appears in a reliable provisional response (for instance 183 Session Progress), the `PRACK` should contain an answer to that offer. A `PRACK` can also contain an offer, in which case the corresponding `200 OK` must contain an answer.

###### *3.2.2.4.1.4 Managing the complexity*

The exact procedure of the SIP reliable transmission mechanism therefore depends on the transport protocol used (reliable or not), the method used (special handling of `INVITE` requests), the type of response in the case of an INVITE request, and a special optional mechanism that exists for provisional responses (`PRACK`). This complexity is typical of monolithic designs and unfortunately tends to lead to spaghetti code and subtle bugs. RFC 3261 attempts to reintroduce functional layers more formally in the specification, and isolates a functional block responsible for retransmissions, handling transactions: If the request is an `INVITE`, the transaction includes not only the provisional responses and the final response but also, IF the final response was not a 2xx response, the `ACK` acknowledging the final response. The transaction layer of user agent clients (and some types of proxies) handles `ACKs` of non-200 OK responses for `INVITE` requests. If the

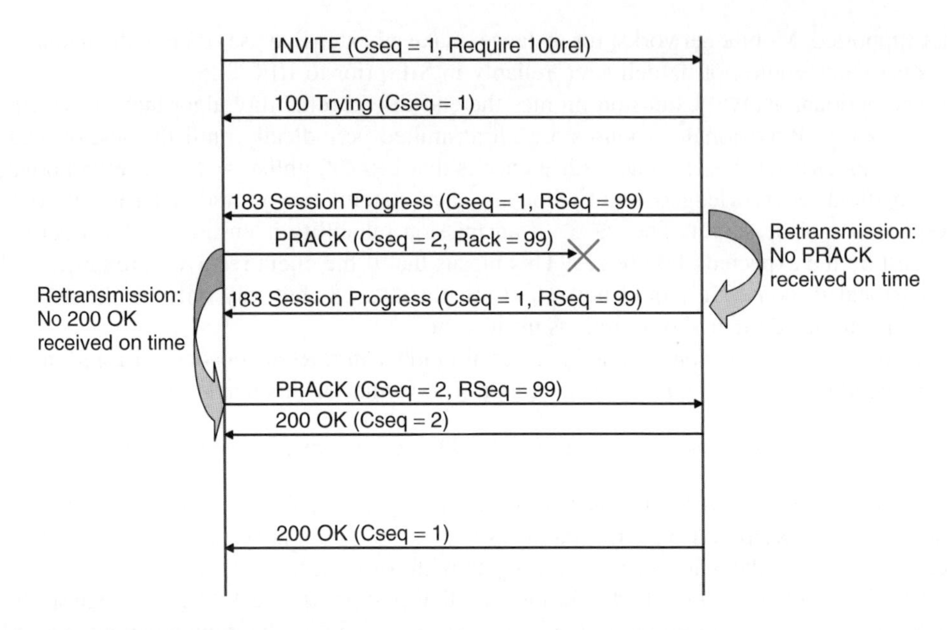


**Figure 3.13** Reliability of provisional responses using the PRACK request.

final response is a 2xx, then the `ACK` is an independent transaction. This is because the SIP specification requires the `200 OK` response to be handled by the application, and not by the transaction functional block. The transaction functional block is not allowed to retransmit `200 OK` responses, and this task must be handled by the user agent application (called the 'user agent core' in the specification).

For all other requests a new transaction is created, which includes all the responses to that request, up to the final response (and its retransmissions if any). There is no `ACK`, and retransmissions are handled completely by the transaction functional block.

This formalism improves the robustness of the specification, but there are still an unusual number of exceptions to the protocol layering principles in the current SIP specification. As SIP evolves, this will be an obstacle to manage the specification complexity. A possible way to simplify the SIP specification would be to drop direct support for UDP and instead use a reliable transport layer on top of UDP, but this would cause backward compatibility problems to SIP implementations. Note that restricting SIP to TCP only would be simpler, but TCP does not always have the right deterministic latency properties required to support high-quality telephony applications. Dropping support for forking proxies would also simplify SIP significantly;[15] there is no significant application using forking proxies yet, and most forking functions can be handled by 'back to back user agents'.

[15] This restriction was adopted by 3GPP, considering the added complexity and the number of potential problems, for little or no substantial benefits.

#### 3.2.2.5 Ending a session

The above example is a simple and successful call setup. Figure 3.14 is a more complete example where Mary calls John, including the call termination by John. If Mary had terminated the call, she would have sent the `BYE` request, and the `From` and `To` headers would be reversed. The media flows are not shown, but the signaling messages include all mandatory headers.

Some SIP headers have abbreviated forms that can help in keeping the total size of a message below the MTU. In this example John's terminal is using the abbreviated form.

#### 3.2.2.6 Rejecting a call

There are occasions when John may be unable to receive a call from Mary. He may not be at home, may not be willing to answer, or he may be already in another conversation. Some of these situations can be expressed in the reply message. SIP provides codes for the usual causes, but also defines more sophisticated replies, such as "`Gone`", "`Payment Required`", or "`Forbidden`".

Figure 3.15 is an example of a simple "`Busy Here`" reply. This reply tells Mary that John cannot be reached at this location (but she might try to reach another location, such as John's mobile phone through a gateway, or a voice mail). Another reply, `600 Busy Everywhere`, can be used to tell that John cannot be reached at any location at this moment.

#### 3.2.2.7 Mid-dialog requests

Once a dialog is established, many situations will require some control information to be transmitted in the middle of the call. For a real-time communication application, the most frequent uses of mid-dialog requests are

- transmission of DTMF information;
- renegotiation of media streams;
- redirection of media streams.

##### 3.2.2.7.1 Transmission of DTMF and flash-hook information

DTMF (Dual Tone Multi-Frequency) signals, are those generated by modern analogue phones when you press one of the keys. Older rotary phones generate a series of small interruptions in the current loop through the phone, corresponding to the digit dialed. Each such small interruption is called a 'flashhook'. They are also frequently used to control some class five features of the phone line, such as three-way calling.

The original SIP specification was focused on PC-based IP telephony and had oversimplified its specification for the transmission of DTMF and Flashhook signals for complex real-world telephony applications. Of course, the problem emerged quickly: without DTMF, you cannot call your answering machine, a prepaid telephony service,

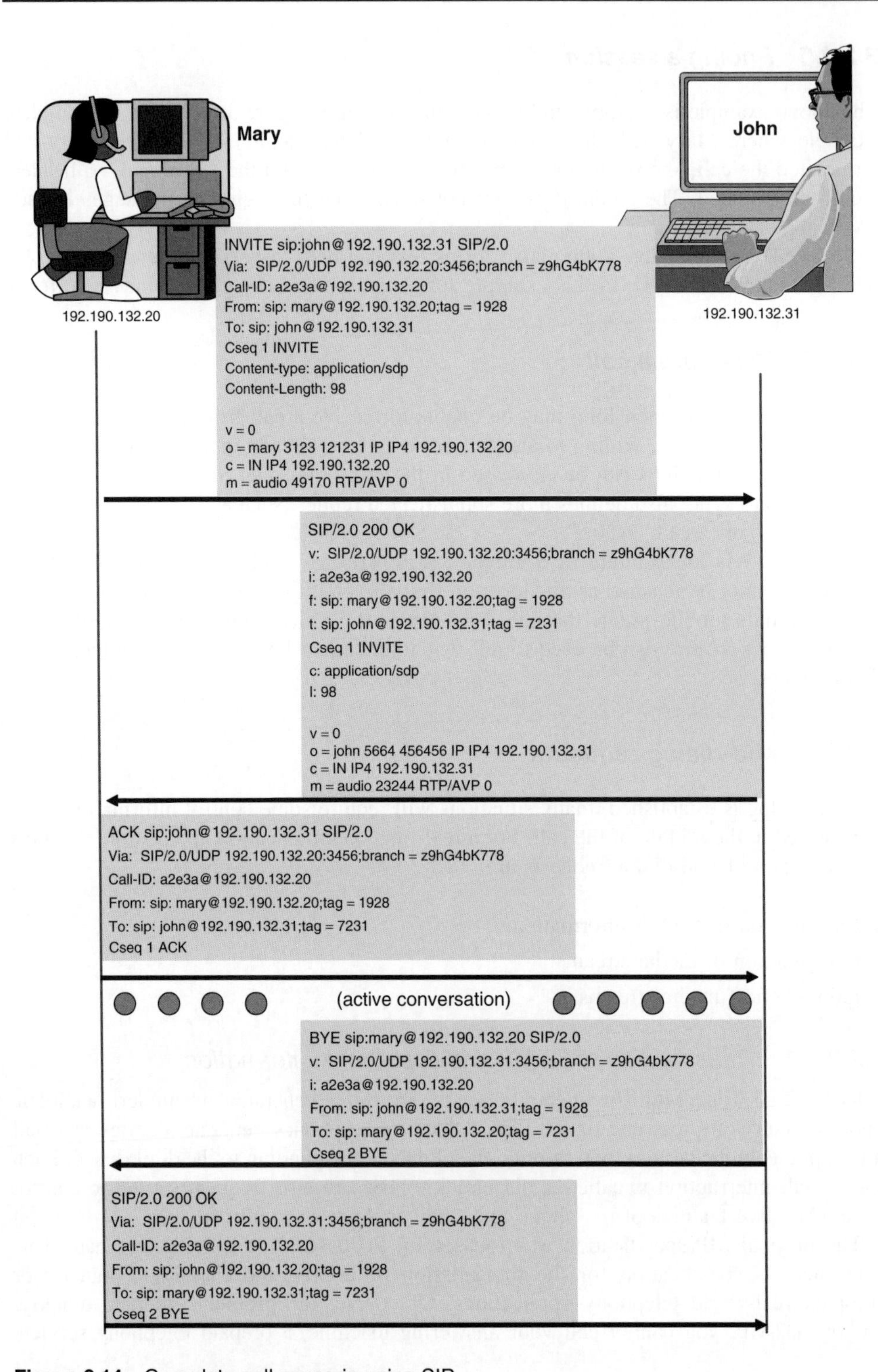


**Figure 3.14** Complete call scenario using SIP.

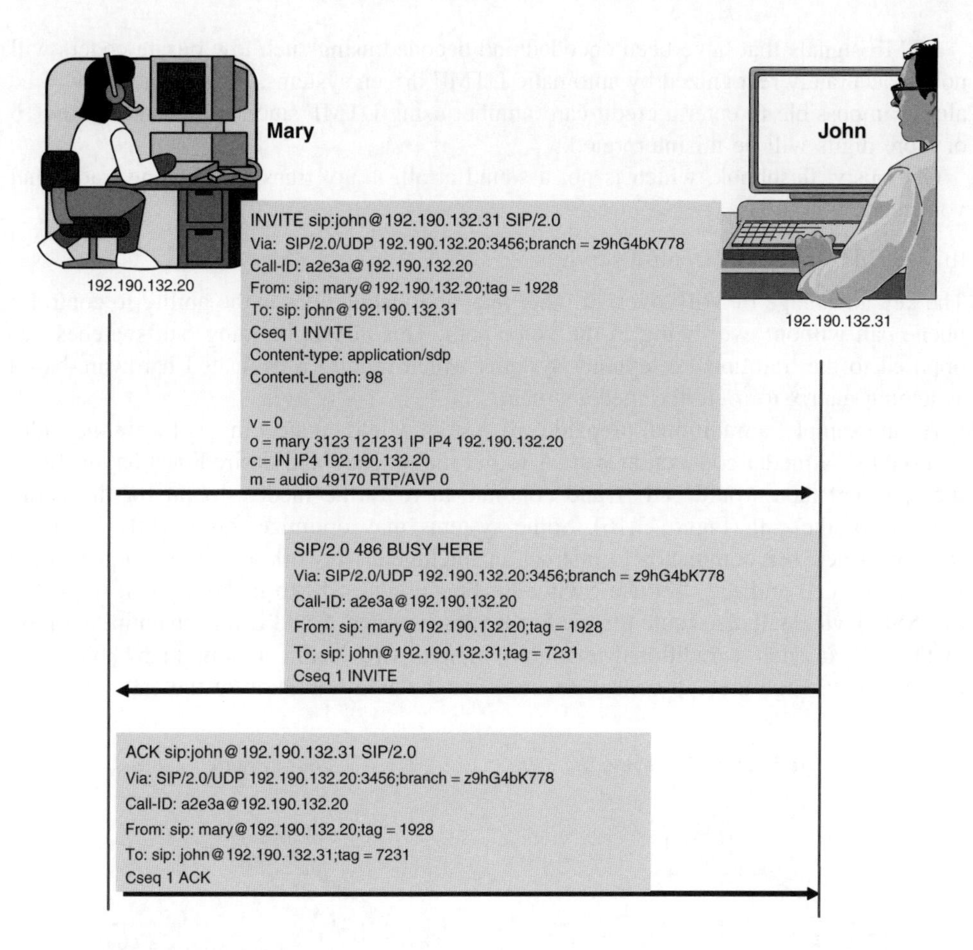


**Figure 3.15** Case of a busy call.

and most call centers, as these information systems frequently use DTMF tones to get some information from you.

### *3.2.2.7.1.1 The issues*

(a) Telephony signals and low bitrate voice coders

Low-bitrate voice coders (in practice anything lower than the 32 kbit/s) usually cannot reliably transport DTMF tones. The reason is that these tones are composed of a mix of two pure frequencies that are almost impossible to find in human voice. Many low bitrate voice coders work by modeling a set of basic human speech components and transmitting only the model parameters on the other side, making it impossible to reproduce exactly pure frequencies. For this reason most of these coders also degrade music significantly, as they are designed for voice only.

DTMF signals that have been encoded and decoded using such low bitrate coders will not be accurately recognized by automatic DTMF-driven systems. For instance, it will be almost impossible to enter a credit card number using DTMF since at least one of the 16 or more digits will be misinterpreted.

Obviously, flashhook, which is not a sound at all, is not transported using traditional voice coding systems.

(b) DTMF-driven call control services

The key advantage of VoIP over all other telephony techniques is the ability to control a phone call without ever being in the voice path. This allows building 'softswitches', as opposed to the traditional telephony systems which require a dedicated hardware-based switching matrix to route the media stream.

As an example, a traditional prepaid card system would connect the call from the caller A, establish a media connection with A to get the pincode and desired destination B for the call, and then would call B and continue to relay the media stream for the entire duration of the call (Figure 3.16). Some systems may optimize this slightly by using Intelligent network commands to instruct an Intelligent Network telephony switch in the path of the call and the 'Service Switching Function' (SSF) to make the call to B, but this SSF device will also route the media streams between A and B for the entire duration of the call. In such a traditional telephony system, two media streams of 64 kbits/s are established through the call control function for the entire duration of the call.

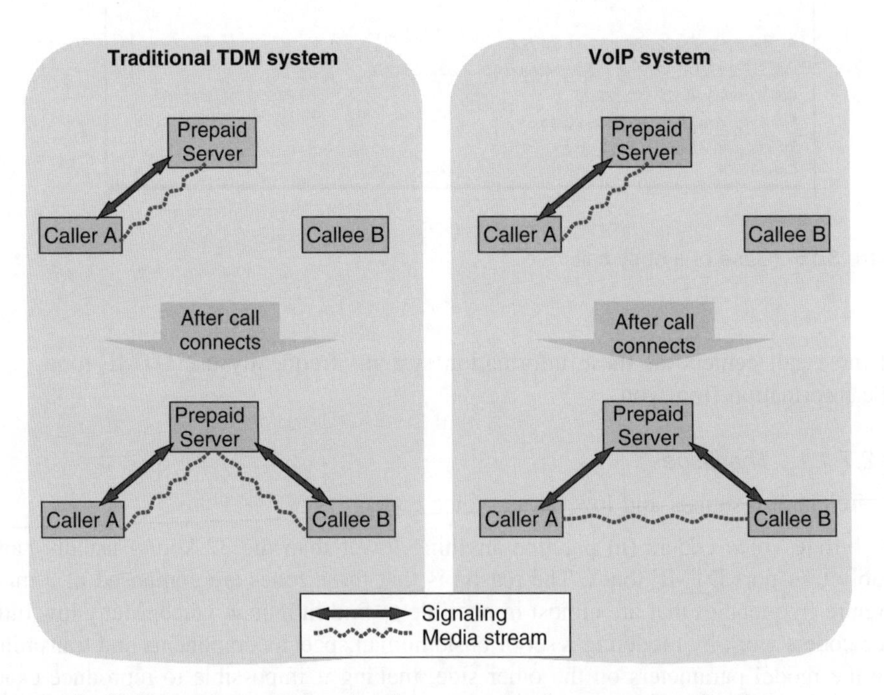


**Figure 3.16** VoIP avoids media tromboning: example of a prepaid application.

A properly designed VoIP network-based prepaid telephony server would establish a media stream with the caller A only during the initial phase of the service, in order to get the pin code, and destination of the call. The server would then call B but instruct A and B to exchange media streams directly over the IP network (see 'Redirection of media streams' in Section 3.2.2.7). It is no longer in the media path of the call.

The issue is that many such DTMF-driven call control services will still need to receive DTMF information, even when the media stream is going directly between the caller A and the callee B. In most prepaid telephony services, for instance, it is possible to stop the current call by pressing the '#' key, and getting the opportunity to make another call to a new destination C without having to re-enter the pin code.

This requires DTMF information to be available on the call control link.

#### *3.2.2.7.1.2 RFC 2833 and 4733*

A quick fix to the first issue, the transmission of DTMF and other events for communications using low-bitrate coders, was presented in RFC 2833 (RTP Payload for DTMF Digits, Telephony Tones, and Telephony Signals), published in May 2000.

RFC 2833 requires edge media devices to implement DTMF detection algorithms for all the media streams they generate. It is trivial for an IP phone as it obviously gets the information of which keypad key is pressed, but for VoIP gateways connected to the PSTN this requires the implementation of DTMF detection algorithms in the G.711 stream received from the PSTN.

The idea is to send the DTMF information in the RTP stream, but as a named event, not as an audio encoded signal. If the resulting RTP stream is received at another PSTN gateway, the PSTN gateway has enough information to regenerate the DTMF information as a waveform. The transmission of DTMF events in the RTP stream, with the same sequence number and timestamp reference as the rest of the RTP stream, allows perfect synchronization of the DTMF and media information, and avoids the possible duplication of DTMF signals at a VoIP gateway (one received directly from the media stream, the other received later in RFC 2833 encoded form).

Figure 3.17 shows the format of a telephony event encoded into an RTP packet. Such events should be generated as soon as a tone of more than 50 ms is detected. Each tone packet is sent three times for redundancy purposes, with the RTP sequence number incremented, while all other fields remain identical. Very short tones can be encoded in a single packet (by setting the 'end' bit). Longer tones can be sent either by continuously sending tone packets with a shorter duration until the tone stops or by forming two packets, one signaling the beginning of the tone and the other signaling the end of the tone. This prevents the sender from having to wait until the end of the tone to send the tone packet, which would obviously require an unacceptable delay.

The volume is a value in negative dBm0. For instance, the value 20 denotes a volume of −20 dBm0. The possible range is between zero and −63 dBm0, but values lower than −55 dBm0 should be rejected. The counter can encode durations up to 8 seconds if the timestamp unit is 1/8000 s, which is more than enough for most uses. A DTMF tone should always be longer than 40 ms in order to be properly recognized by in-band detectors.

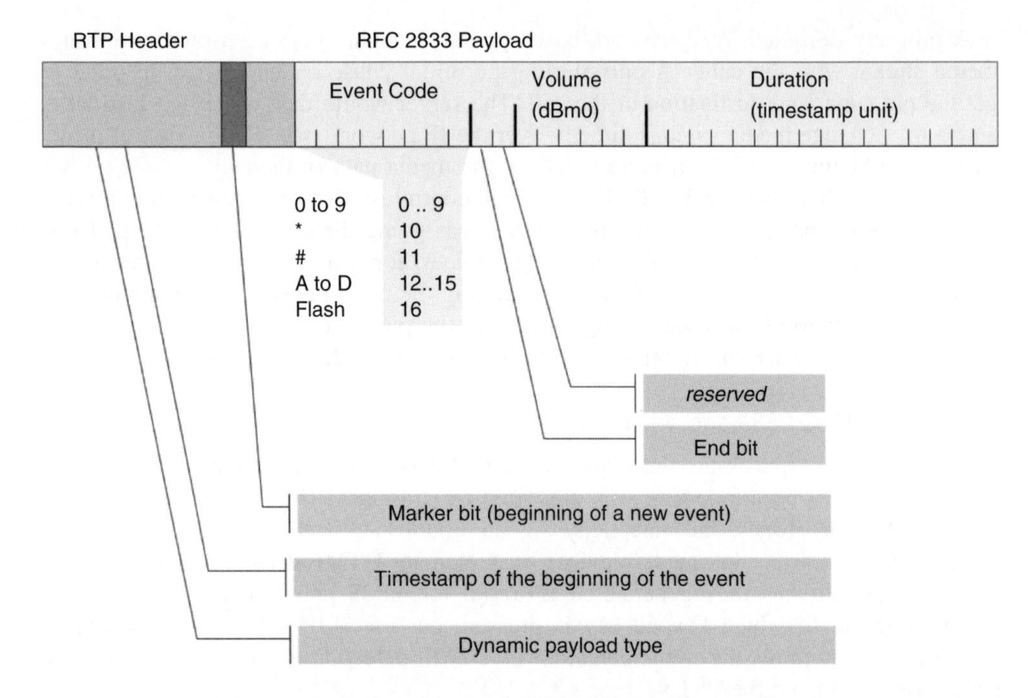


**Figure 3.17** RFC 2833 RTP packet format for DTMF transport.

As SIP uses the SDP to declare the type of media encoding, it was necessary to add a SDP payload format to declare which types of events a receiver can understand.

The following 'm' line can be used for receiving telephone events:

```
m=audio 44143 RTP/AVP 110
a=rtpmap:110 telephone-events/8000
a=recvonly
```

In addition, the fmtp specifier can be used to detail which events can be received. The format is:

a=fmtp:<format> <list of values>

For instance, a receiver understanding all the events in the figure except A,B,C, and D, with dynamic payload type 100, would declare it using

$$a\text{=fmtp: } 1000-11, 16$$

In fact, all implementations are required to handle event 0 to 15, so the fmtp line is optional.

RFC 2833 also describes another format where tones are sent as a series of frequency, amplitude modulation, volume, and duration parameters.

One of the advantages of signaling DTMF information as an event is that all waveform analysis is now performed by edge devices, making IP-based interactive voice response servers much easier to implement.

RFC 2833 is an interesting discussion of telephony signals in a VoIP network, and it solves the problem of transmitting DTMF and other signals in simple class 4 VoIP networks (these networks only route phone calls, without performing any complex service). It is a comprehensive reference of all types of tones and signals found on current networks, including DTMF, modem and fax tones, special information tones, etc.

RFC 2833 can also in principle solve the more complex problem of DTMF and call control because any intermediary proxy can add its own 'm=' line requesting to receive the telephony events to a specific IP address, in which case a c= line must be present in the media section of the SDP right after the m= line,[16] while all other media streams are directed to the target user agent. However, many SIP user agent implementations have overlooked this requirement and are not able to send media and telephony events to different destinations, let alone duplicate the events for transmission to multiple destinations. Right now, in practice, it is not possible to implement a reliable prepaid system with RFC2833 without routing the RTP stream.

RFC 2833 also caused some confusion as it leaves the implementer free of transmitting DTMF simultaneously through the media stream and in event encoded form, or to mute the media stream for the duration of the DTMF signal sent in event form.

Only the latter is safe, as in many complex call flows where the synchronization information may be lost, the simultaneous transmission of RTP in the regular media stream and using an event may cause duplicates.

Note that in H.323 it is mandatory to transmit all DTMF information out-of-band using the H.245 signaling channel. The audio signal is muted for the duration of the event transmitted out of band. However, some vendors can disable this mechanism and use RFC2833 instead. This was introduced to allow some interworking between H.323 networks and SIP networks implementing RFC2833 and lacking any signaling channel DTMF transmission function. This is not a good solution, however, and the use of one of the methods described in Section 3.2.2.7.1.3 for signaling channel DTMF tone transmission in SIP networks is recommended.

The current weaknesses of RFC2833 could be addressed by a more formal specification of how SIP should handle telephony events in complex call control applications, how events should be duplicated and sent to multiple destinations, and when the DTMF tones should be removed from the audio stream. Another issue is feature overlap of SIP application servers, as RFC2833 possibly enables several call control devices on the signaling path to request telephony events, which may cause several of them to take incompatible actions simultaneously.

RFC 4733 was published in December 2006 as an update of RFC 2833. It retains only the most basic signals as part of the main RFC and defines other events (modem, fax, text telephony, and channel-associated events) in a dedicated IANA registry. RFC 4733 also adds support for long duration events. Support for DTMF events is no longer mandatory,

[16] For servers which can handle multiple simultaneous calls, this also requires the allocation of a specific DTMF destination port for each call, which is very cumbersome.

but can be negotiated. RFC 4733 also recognizes the need for DTMF transmission over a reliable control protocol rather than RTP packets 'in circumstances where exact timing alignment between the audio stream and the DTMF digits or other events is not important', and also that 'Depending on the application, it may be desirable to carry the signaling information in more than one form at once'.

This clarification work will probably continue in future revisions of SIP, but for now, most vendors who faced these issues decided to solve the problem by using new SIP messages, as described below.

#### *3.2.2.7.1.3 Alternatives to RFC 2833*

VoIP is a technology breakthrough for the design of value-added services for telephone networks. The possibility to control calls without routing the media stream greatly enhances the density and scalability of application servers. It also decreases the cost of such servers, as many functions now do not require specialized telephony hardware and can run on standard computer platforms. Last, but not least, the services are cheaper to operate because the application servers no longer need to be located close to the end-users, and therefore most services can be implemented using a single point of presence.

The implications of this paradigm shift are just beginning to be fully understood, and as expected, many VoIP devices have been designed with the old TDM model in mind, assuming all servers that control the call are also controlling the media stream.

A properly designed VoIP edge device (gateway, IP phone) should be able to send all information that is possibly of interest to an application server over the signaling link because only this link is guaranteed to reach all application servers. This is mandatory in H.323, using the H.245 channel.

Unfortunately, at the time of writing, there was no agreed standard way of doing this in SIP. Most network VoIP gateway vendors faced the problem and solved it using their own methods, some of which are described below. Interestingly, most SIP phones seem to be using only RFC2833 most of the time without the ability to create separate UDP connections for telephony events. Unfortunately, this is a serious problem when implementing class 5 services using such SIP phones as many services (Interactive voice response, prepaid, call centers, etc.) need access to the DTMF information and would require the application servers to route the media streams if such DTMF information is sent in the RTP packets.

The methods used by VoIP gateway vendors to send events on the signaling link in SIP roughly fall in two categories:

- The use of the new `INFO` mid-dialog request is defined in RFC 2976 to carry the telephone event. Some vendors use one of the mime types defined by RFC2833 (audio/telephone-event and audio/tone mime types.); other vendors use encodings derived from H.323 or MGCP.
- The use of the new `SUBSCRIBE/NOTIFY` mechanism to carry the telephone event.

Unfortunately, since there is no common agreement of the exact encodings, there is no interoperability between the various SIP implementations, and most existing SIP networks use a single gateway vendor to overcome this problem. Some proxies are capable or understanding several formats and convert between them. The following sections describe the encoding used by some popular SIP gateway vendors.

(a) Cisco

Cisco uses a combination of the general `INFO` message defined in RFC 2976 (other methods are available as well, e.g., RFC 2833), and the SUBSCRIBE/NOTIFY method. Cisco implemented DTMF transport according to an Internet Draft: draft-mahy-sip-signaled-digits-00, Signaled Digits in SIP.

A SIP device can instruct a Cisco gateway to send a DTMF tone by sending it an INFO message formatted as follows:

```
INFO sip:1978551212@192.168.20.10 SIP/2.0
Via: SIP/2.0/UDP 192.168.0.1
From: 19785551234@192.168.0.1
To: 19785551212@192.168.20.10
Call-ID: 662606876@192.168.0.1
CSeq: 20 INFO
Content-Type: application/dtmf-relay
Content-Length: 22

Signal=9
Duration=250
```

The duration is in milliseconds. The gateway will confirm the receipt of this indication by responding to the SIP INFO message with a '200 OK' response.

```
SIP/2.0 200 OK
Via: SIP/2.0/UDP 192.168.0.1
From: 19785551234@192.168.0.1
To: 19785551212@192.168.20.10
Call-ID: 662606876@192.168.0.1
CSeq: 20 INFO
```

In order to be notified of DTMF events from a Cisco gateway, a SIP application must first request to receive the DTMF events using the SUBSCRIBE mechanism. The advantage of using the subscribe mechanism is that any SIP application server, in case the call is processed by a chain of proxies, can request DTMF notification at any time during a call. The problem is that many application servers can react simultaneously and take incompatible actions:

```
SUBSCRIBE sip:1010@192.168.110.239 SIP/2.0
Via: SIP/2.0/UDP 213.56.166.173:5060
From: <sip:5500@213.56.166.173;user=phone>
To: <sip:1010@192.168.110.239>
Call-ID: 6CD8C67B-C0A011D3-806DB047-B37E77AF@192.168.110.239
CSeq: 1 SUBSCRIBE
Contact: <sip:5500@213.56.166.173>
Expires: 3600
```

```
Events: telephone-event;duration=2000
User-Agent: NetCentrex IN Stack
Content-Length: 0

SIP/2.0 200 OK
Via: SIP/2.0/UDP 213.56.166.173:5060
From: <sip:5500@213.56.166.173;user=phone>
To: <sip:1010@192.168.110.239>;tag=A1E07C4-694
Date: Sun, 02 Jan 2000 23:08:57 GMT
Call-ID: 6CD8C67B-C0A011D3-806DB047-B37E77AF@192.168.110.239
Server: Cisco-SIPGateway/IOS-12.x
Content-Length: 0
CSeq: 1 SUBSCRIBE
Expires: 3600
Contact: <sip:1010@192.168.110.239:5060;user=phone>
```

If one of the requested events is received from the gateway, a SIP NOTIFY message with a representation of the signalled digits is sent to the requesting application server. In the following sample, the '9' key is pressed:[17]

```
NOTIFY sip:5500@213.56.166.173:5060 SIP/2.0
Via: SIP/2.0/UDP 192.168.110.239:5060
From: <sip:1010@192.168.110.239>;tag=A1E07C4-694
To: <sip:5500@213.56.166.173;user=phone>
Date: Sun, 02 Jan 2000 23:08:57 GMT
Call-ID: 6CD8C67B-C0A011D3-806DB047-B37E77AF@192.168.110.239
User-Agent: Cisco-SIPGateway/IOS-12.x
Max-Forwards: 6
Timestamp: 946854581
CSeq: 102 NOTIFY
Event: telephone-event;rate=1000
Contact: <sip:1010@192.168.110.239:5060;user=phone>
Content-Length: 10
Content-Type: audio/telephone-event

0x0980010E

SIP/2.0 200 OK
Via: SIP/2.0/UDP 192.168.110.239:5060
From: <sip:1010@192.168.110.239>;tag=A1E07C4-694
To: <sip:5500@213.56.166.173;user=phone>
Call-ID: 6CD8C67B-C0A011D3-806DB047-B37E77AF@192.168.110.239
CSeq: 102 NOTIFY
Server: NetCentrex IN Stack
Content-Length: 0
```

[17] Expressed using the format defined by RFC 2833, cf Figure 3.17.

(b) Nuera

Nuera also uses SIP INFO messages, encapsulating an MGCP-like syntax. A message body containing MGCP event information will be formatted as follows:

```
Content-Type: application/mgcp-event
Content-Length: <length of payload>
<MGCP event information>
```

The application requiring the DTMF out of band information must request it using an MGCP notification request, embedded in an INFO message:

```
INFO sip:10.0.0.157 SIP/2.0
Via: SIP/2.0/UDP 10.0.0.168:5060
Route: NUERA-ID<sip:216.188.94.117>
From: 1003<sip:1003@10.0.0.157>
To:
NUERA-
ID<sip:216.188.94.117;user=phone>;tag=216.188.94
.117-eg101483118153
Call-ID:
tac12320020227093301525205-
54444D0000000DD8BC5E753C7D184D10@216.188.94.117
CSeq: 1 INFO
User-Agent: NetCentrex IN Stack
Content-Type: application/mgcp-event
Content-Length: 15

R: [0-9*#](N)

SIP/2.0 200 OK
Via: SIP/2.0/UDP 10.0.0.168:5060
Record-Route: <sip:10.0.0.157;maddr=10.0.0.157>
From: 1003<sip:1003@10.0.0.157>
To:
NUERA-
ID<sip:216.188.94.117;user=phone>;tag=216.188.94
.117-eg101483118153
Call-ID:
tac12320020227093301525205-
54444D0000000DD8BC5E753C7D184D10@216.188.94.117
CSeq: 1 INFO
Content-Length: 0
```

If one of the requested events is received from the gateway, a SIP INFO message with an MGCP event message body containing this observed event is sent to the application server. In the following sample, the '*' key is pressed:

```
INFO sip:1003@10.0.0.157;maddr=10.0.0.157 SIP/2.0
Route: <sip:1003@10.0.0.168>
```

```
To: "1003" <sip:1003@10.0.0.157>
From: "NUERA-ID"
<sip:216.188.94.117;user=phone>;tag=216.188.94
.117-eg101483118153
Via: SIP/2.0/UDP 216.188.94.117:5060
Via: SIP/2.0/UDP 216.188.94.117:5061
Call-ID:
tac123200202270933015252O5-
54444D0000000DD8BC5E753C7D184D10@216.188.94.117
CSeq: 2 INFO
Content-Type: application/mgcp-event; version=1.0
Content-Transfer-Encoding: text
Content-Length: 5

O:*

SIP/2.0 200 OK
Via: SIP/2.0/UDP 216.188.94.117:5060
Via: SIP/2.0/UDP 216.188.94.117:5061
From:
NUERA-
ID<sip:216.188.94.117;user=phone>;tag=216.188.94
.117-eg101483118153
To: 1003<sip:1003@10.0.0.157>
Call-ID:
tac12320020227093301525205-
54444D0000000DD8BC5E753C7D184D10@216.188.94.117
CSeq: 2 INFO
Server: NetCentrex IN Stack
Content-Length: 0
```

(c) Sonus

Sonus offers two mechanisms, DTMF relay and DTMF trigger.

In DTMF relay, a mechanism similar to the signal and signal-update from H.245 is used for precise control of DTMF detection and generation. The signal parameter indicates the detected DTMF tone, the duration parameter indicates the total duration of the tone if known or an initial estimate of the tone duration, and signal-update subsequently updates the estimate of the total duration. The Content-Type header is set to 'application/dtmf-relay'. In the following example a DTMF is pressed for 250 ms.

```
INFO sip:1978551212@192.168.20.10 SIP/2.0
Via: SIP/2.0/UDP 192.168.0.1
From: 19785551234@192.168.0.1
To: 19785551212@192.168.20.10
Call-ID: 662606876@192.168.0.1
CSeq: 20 INFO
Content-Type: application/dtmf-relay
Content-Length: 22
```

```
Signal=9
Duration=250
```

The server is expected to confirm the receipt of this indication by responding to the SIP INFO message with a '200 OK' response.

```
SIP/2.0 200 OK
Via: SIP/2.0/UDP 192.168.0.1
From: 19785551234@192.168.0.1
To: 19785551212@192.168.20.10
Call-ID: 662606876@192.168.0.1
CSeq: 20 INFO
```

The DTMF trigger mechanism is used for well-defined DTMF events not requiring timing information. The Content-Type header is set to 'application/dtmf'. In the following example the DTMF event '#' is pressed.

```
INFO sip:19785512120192.168.20.10 SIP/2.0
Via: SIP/2.0/UDP 192.168.0.1
From: 19785551234@192.168.0.1
To: 19785551212@192.168.20.10
Call-ID: 662606876@192.168.0.1
CSeq: 20 INFO
Content-Type: application/dtmf
Content-Length: 1

#
```

The server confirms the receipt of this indication by responding to the SIP INFO message with a '200 OK' response.

```
SIP/2.0 200 OK
Via: SIP/2.0/UDP 192.168.0.1
From: 19785551234@192.168.0.1
To: 19785551212@192.168.20.10
Call-ID: 662606876@192.168.0.1
CSeq: 20 INFO
```

(d) Perspectives

The problem of out of band DTMF transport was ignored for a long time by the standards community, but the problem was raised finally by operators. Towards the end of 2008, ETSI TISPAN published TR 183 057—a comprehensive study of DTMF transport methods in SIP—trying to identify the best approach for out of band DTMF.

Two approaches seemed to meet the requirements:

- KPML (RFC 4730). With this method, the entity interested in DTMF events initiates a new dialog and sends a SUBSCRIBE to the DTMF sender, specifying in the body of the SUBSCRIBE the targeted sip session (call-id, remote-tag, local-tag). Unfortunately this correlation method poses problems with B2BUA functions, implying that the monitored

SIP session and the SUBSCRIBE session should be routed over the same path. The SUBSCRIBE could be sent in the same dialog, but such dialog reuse is discouraged by the SIP community due to possible interoperability issues.

- draft-ietf-sip-info-events-03 'SIP INFO Event Framework', an extension of RFC 2976 which defines the required mechanisms to support out-of-band DTMF and negotiate the supported events through the introduction of new headers:
  - 'Recv-Info:<event-package>' to indicate supported packages
  - 'Info-Package: :<event-package>' to indicate the package used in the INFO body.

  The specific event package for DTMF is documented in draft-kaplan-sipping-dtmf-package-00: 'DTMF Info-Event Package'.

The TISPAN study concluded that the INFO method was preferable.

### 3.2.2.7.2 Negotiation of media streams

#### 3.2.2.7.2.1 Session description syntax, SDP

SIP uses the Session Description Protocol (SDP) specified in RFC 2237. SDP is also a product of the MMUSIC working group, and was mainly used in the context of the MBONE, the multicast enabled overlay network of the Internet. In order to be able to receive an MBONE session, a receiver needs to know

- which multicast address is going to be used by the session;
- what will be the UDP destination port;
- the audio and/or video coders that will be used (GSM, H.261,...);
- some information on the session (Name, short description);
- contact information;
- activity schedule.

The primary purpose of SDP is to define a standard syntax for this type of information. The SDP session description can be conveyed with various transport methods, depending on the context: the Session Announcement Protocol (SAP) is used on the MBONE, the Real-Time Streaming Protocol(RTSP) is used for streaming applications, and SIP is used to set up point to point and multipoint interactive communications.

SDP is a human readable protocol, consisting of several <type>=<value> lines terminated by `CR/LF` characters. The field names and attributes use US-ASCII characters, but free text fields can be localized since SDP uses the complete ISO 10646 character set. This text encoding, as opposed to a binary encoding like ASN-1 PER used in H.323, facilitates manual programming and analysis of network traces at the expense of a greater bandwidth usage. However, the bandwidth usage of a signaling protocol is negligible compared to the actual media flows, so the trade-off is very good.[18] The only major

[18] Except perhaps for mobile networks. SIP/3GPP introduced the binary SigComp mechanism to compress SIP signaling on mobile networks.

drawback of this method is that the generation of the serialization and parsing code needs to be manual, which is less reliable (and often slower) than the automatic generation allowed by more formal specifications like ASN.1. Manual programming also leads to more interoperability issues.

The session description is structured with one section which applies to the whole session (starting with v = ...), and several media description subsections (each starting with (m = ...). Parameters in the media subsections may override the default parameters of the session level section.

Table 3.3 describes the various field types described by RFC 2237 for each section.

*(i) Dynamic and static payload types*

Under a particular profile, some RTP payload types are static: i.e., their meaning is fully defined in the profile (e.g., RTP/AVP 0 is 64 kbit/s uLAW PCM).

Other RTP payload types only have a meaning in association with a particular session described in SDP. Those are dynamic payload types. The SDP RFC gives the example of the 16 bit linear encoded stereo audio sampled at 16 KHz. There is no static payload type defined that would exactly correspond to this. Instead, we will be using an arbitrary unused number for the payload type, say 98 (`m=audio 49232 RTP/AVP 98`), and describe the format of the transported data in SDP:

```
a=rtpmap:98 L16/16000/2
```

The format is `a=rtpmap:<payload type> <encoding name>/<clock rate> [/<encoding parameters>]`, which in our example translates to 16 linear, 16000 Hz sampling, two channels.

By extension, the term 'dynamic payload type' applies to any RTP format for which the media encoding characteristics are by external signaling (for instance, in H.323, through an H.245 OpenLogicalChannel message).

#### *3.2.2.7.2.2 SDP in the context of interactive communications*

SDP was initially designed in the context of multicast media transmissions over the MBONE, the multicast overlay of the Internet. The problem it solved was simple: SDP simply needed to convey to all potential listeners the Multicast IP address and port of the media transmission. This media session description using the SDP syntax was encapsulated in the higher level SAP and sent in a multicast packet to all potential listeners.

Simply replacing SAP by SIP does not make SDP a suitable protocol for interactive communications:

- The multicast sessions were essentially one way, with the destination IP address and port selected by the sender. An interactive communication uses two-way media streams, with the destination IP address and port selected by the receiver.
- Multicast sessions are fairly static, they are advertised in advance, and all session changes do not have to occur in real time. In an interactive session, audio, video, and other media streams can be established, stopped, or changed at any time.

**Table 3.3** SDP session description parameters

| Session level field type | Sub-section level field type | Usage | Format and example | |
|---|---|---|---|---|
| v= | | Protocol version | V=0 | M |
| o= | | Owner/creator and session identifier | O=<username> <session id> <version> <network type> <address type> <address><br>O=mhandley 2890844526 2890842807 IN IP4 126.16.64.4 | M |
| s= | | Session name | s=<session name><br>s=SDP Seminar | M |
| i= | | Session information | i=<free text session description><br>i=A Seminar on the session description protocol | O |
| u= | | URI of description | u=<Universal Resource Identifier><br>u=http://www.cs.ucl.ac.uk/staff/M.Handley/sdp.03. ps | O |
| e= | | email address | e=<email address> (Optional free Text)<br>or<br>e=<Optional free Text> "<"email address">"<br>e=mjh@isi.edu (Mark Handley)<br>e=Mark Handley<mjh@isi.edu> | O |
| p= | | phone number | p=<phone number> (Optional free Text)<br>Or<br>p=<Optional free Text> “<”phone number“>”<br>p=+44-171-380-7777 | O |
| c= | | connection information - not required if included in all media | c=<network type> <address type> <connection address> TTL must be included for multicast sessions.<br>c=IN IP4 224.2.17.12/127<br>c=IN IP4 224.2.1.1/127 | O |
| b= | | bandwidth information | b=<modifier (CT Conference Total\|AS Application-Specific Maximum>:<bandwidth-value in kilobits/s><br>b=CT:120 | O |
| One or more time description sections | T= | time the session is active | t=<start time> <stop time>, using decimal NTP in seconds<br>t=2873397496 2873404696 | M |
| | r= | zero or more repeat times | r=<repeat interval> <active duration> <list of offsets from start-time>, by default in seconds<br>r=604800 3600 0 90000 | O |

| | | | | |
|---|---|---|---|---|
| | | | means that the repeat interval is 1 week (604800 seconds), active for 1 hour (3600 seconds) after each offset from the start time T. Offsets are here 0 seconds and 90000 seconds (25 hours); i.e., if *** represents active periods and---idle periods,<br>T***T+1h---T+25h***T+26h------T+1week****T+1week+1h----T+1Week+25h***...<br>The repetition is valid until the stop time<br>Unit modifiers can be used for compactness, and the previous record can also be written as follows:<br>r=7d 1h 0 25h | |
| z= | | time zone adjustments | | O |
| k= | | Encryption key | k=<method>:<encryption key><br>or<br>k=<method> | O |
| a= | | zero or more session attribute lines | a=<attribute><br>or<br>a=<attribute>:<value><br>a=recvonly | O |
| Zero or more media descriptions | m= | Media name and transport address | m=<media> <port> <transport> <format list><br>m=audio 49170 RTP/AVP 0 3<br>means that the media is audio, and can be received on port 49170 (RTP only uses even ports, the next odd port being used by RTCP). The transport is protocol RTP/AVT (IETF's Realtime Transport Protocol using the Audio/Video profile), and the format is media payload types 0 or 1 of the AVT profile (0 is u-law PCM coded single channel audio sampled at 8 KHz, 3 is GSM)<br>Other RTP profiles would be coded after the slash, e.g., a hypothetical profile XXX would appear as RTP/XXX | M |
| | i= | Media title | | O |
| | c= | Connection information—optional if included at session-level | | |
| | b= | Bandwidth information | | O |
| | k= | Encryption key | | O |
| | a= | zero or more media attribute lines | | O |

- Multicast sessions, because of the number of participants, do not even attempt to negotiate a common set of media encodings. The sender chooses one coder, and listeners simply have to adapt or fail to join the session. In an interactive communication such as a phone call, the parties expect to be able to communicate, and this may require the negotiation of a common set of media coders.
- Interactive communications may require intermediary servers to charge the communications, and therefore to be aware of the types of media used by each party, when media streams are established or stopped.

The initial version of SIP overlooked many of these issues. The sample SIP call flow consisted of a very simple case where the interactive communication:

- was established between terminals which were assumed to have the same coders available. The issue of which coder to use first, how to maximize the chance of using symmetrical coders (the same coder for each direction of media stream), or how to negotiate a new coder were not really specified;
- established the stream immediately after the stream was proposed. In reality, an endpoint may want to use a proposed media (e.g., video), only some time after it has been proposed;
- was not redirected or re-established to a different endpoint in the middle of a communication. This situation may happen in real communications, e.g., communications to call centers, transferred calls, etc.

The result was a situation where SIP interoperability was limited to basic calls only. A clarification on the exact procedures to use to open a media stream, close it, or renegotiate it was required. This updated specification, called the 'Offer/answer' model, was published in RFC 3264 (under the umbrella of the new 3261 SIP RFC) in June 2002.

#### *3.2.2.7.2.3 The SDP offer/answer model for unicast streams*

RFC 3264, 'An Offer/Answer Model with the SDP', presents all the procedures that were in the previous SIP RFC, as well as new call flow discussions and examples. This complex portion of the interactive communication protocol is the equivalent of H.245 in the H.32x ITU standards.

RFC 3264 first defines the values of some mandatory SDP parameters that become useless or redundant in the context of SIP, where a lot of information is exchanged in the SIP message itself: the subject should be empty (s =<single space> or s=-), the time of the session should be set to 0 (t=0 0), 'e' and 'p' parameters are not used, etc. RFC 3264 also restricts the number of sessions to only one per SDP message.

The media control model is based on 'offers' and 'answers'.

An 'offer' may contain zero and one or more media stream descriptions, each in a line beginning with 'm =', followed by optional attributes. The order in which the media formats appear in the m line is the order of preference. The answerer should choose the first one that is acceptable to it.

For instance, the following offer proposes G.711 Alaw and MuLaw for the audio media, with a preference for Ulaw, and proposes H.261 and MPEG for video, with no preference.

```
v=0
o=john 4898446519 4898446519 IN IP4 johnendpoint.anywhere.com
s=
c=IN IP4 johnendpoint.anywhere.com
t=0 0
m=audio 41732 RTP/AVP 0 1
a=rtpmap:0 PCMU/8000
a=rtpmap:1 PCMA/8000
m=video 43221 RTP/AVP 31
a=rtpmap:31 H261/90000
m=video 49222 RTP/AVP 32
a=rtpmap:32 MPV/90000
```

An offer can contain multiple 'm =' lines for the various media types (e.g., audio, video). This means that the sender of the offer is willing to send or receive these media streams simultaneously. This is valid even if there are multiple 'm =' lines for the same media type. For instance, this can be used in conjunction with RFC 2833 to transmit telephony events to an application server and regular audio to the destination user agent. See Section 3.2.2.7.1.2 for more details. The expression power of this structure is equivalent to a single H.245 capability Descriptor, where each m = line is an `AlternativeCapability-Set`, and the union of all m = lines is a `SimultaneousCapabilitySet`.

In this sample, all proposed media are bidirectional. A unicast stream offer will have the attribute "`a=sendonly`" if the endpoint only wishes to send media to its peer. If it only wants to receive the indicated media, the attribute will be "`a=recvonly`". If it only wants to 'warn' the peer that it may establish a media stream at a later time and provide a hint on the parameters that will be used, it will use "`a=inactive`". The default is "`a=sendrecv`", if the endpoint wishes to establish a media stream to and from its peer.

The usefulness of "`sendrecv`" and "`recvonly`" is obvious. "`inactive`" or "`sendonly`" can be useful in more complex situations, for instance, in a SIP based call center, if the call center application server wishes to ring an agent phone through a gateway, but does not wish the agent to stream back audio immediately, because the call center application server is currently sending a waiting music to the caller. Another common use of the 'inactive' parameter is to allow an appliance which needs to initialize a DSP with some coder before using it to select the coder using an 'inactive' offer/answer, and then activate it through a new offer (see 3.2.2.7.2.3). These parameters can also be used to workaround bugs in IP phones or media gateway implementations:

- Some endpoints do not support receiving SIP INVITE messages without SDP; sending an inactive SDP media stream offer may work around the bug with the same effect.
- Some high-density VoIP gateways use multiple LAN cards for RTP streams, and do not support 'looped back' RTP streams sent and received on the same card. A way to work around this bug is to give a hint to the VoIP gateway of which source IP address

may be used in the future with an inactive session, so it can avoid selecting the wrong LAN card.

The port number indicated in the m line is the UDP port the endpoint wishes the peer to send media to (`a=recvonly` or `a=sendrecv`). For "`a=sendonly`" and "`a=inactive`" streams, the RTCP stream should be sent to the indicated port number plus one, unless a more specific indication is present in the media description.

Depending on the value of the 'a' attribute, the list of media formats indicates the media formats that can be sent (`sendonly`), are expected to be received (`recvonly`), or both (`sendrecv`). The `sendrecv` offer therefore contains enough information to allow the answerer to use symmetrical coders.

When RTP is used, the offer indicates the dynamic payload type that is expected to be used for the media, but in `sendonly` and `sendrecv` offers, however, the payload type may be changed in the answer, in which case the answer value should be used instead of the one originally proposed. This may be required if the answerer is only able to receive certain payload types.

There may be `fmtp` parameters to include further information on media formats, such as supported events for RFC2833/4733.

In the case of RTP streams, all media descriptions SHOULD contain "`a=rtpmap`" mappings from RTP payload types to encodings. If there is no "`a=rtpmap`", the default payload type mapping, as defined by the current profile in use (for example, RFC 1890) is to be used.

The answer will usually be based on the offer, but may change some elements: it may add a reception IP address and port, remove some media formats, etc. In this case the origin line ("`o=`"), must be changed, and must contain a new version number. The time ("`t=`") line must be identical to the one of the offer.

There must be an exact one to one mapping between the number of 'm' lines in the offer and in the answer, which are matched based on the ordering. Rejected streams must contain a port number of 0. Although SDP requires at least one media type to be present in the m line, it will be ignored. A stream must be rejected if there are no acceptable media formats.

This is the answer of Mark to the offer of John, where he selected PCMA and H.261.

```
v=0
o=mark 4898446720 4898446720 IN IP4 markendpoint.anywhere.com
s=
c=IN IP4 markendpoint.anywhere.com
t=0 0
m=audio 51762 RTP/AVP 1
a=rtpmap:1 PCMA/8000
m=video 53221 RTP/AVP 31
a=rtpmap:31 H261/90000
m=video 0 RTP/AVP 32
```

If the media is accepted, the answer should contain a unicast address if the offer was unicast and should not change the media type. The answer should mark the direction of the stream from the point of view of the user agent sending it, e.g., "recvonly" if the offer was "sendonly", and vice-versa. If the offer was "sendrecv", the answer may choose to select only one of the send or receive modes, e.g., "recvonly" if the answerer is not going to generate any media. In all cases, the answerer may want to mark the media as "inactive" if it is not willing to use it right away. This is the only option if the offer was "inactive".

If a type of media has been accepted for a "recvonly" or "sendrecv" offer, it is still necessary to select among the offered media formats the ones that are acceptable. A "recvonly" response should contain at least one media format selected among those present in the offer. When selecting a media format, it is important to remember that the ordering of the media formats in the offer represents the preference of the offerer, and therefore the selected format should be the first that is acceptable to the answerer. Note that once the answer has been sent the answerer must be prepared to receive media in any of the formats listed.

Similarly, for "sendonly" of "sendrecv" answers, it is necessary to select among the proposed media formats those that can be sent by the answerer ("sendonly"), or sent and received("sendrecv"). At least one of the proposed media formats must be in the offer. In "sendrecv" answers, changing the payload type should only affect the payload type of the stream received by the answerer; the answerer should still use the offer payload type to send media to the offerer.

The order of media formats in the answer represents the preference of the answerer, but in order to maximize the chances of using the same coders in both directions, it should in general be the same relative order as the order of the offer. In "sendonly" or "sendrecv" answers, the media stream sent to the offerer should be the preferred format according to the offer among the formats listed in the answer.

Similarly, the offerer should use the first acceptable coder in the "recvonly" or "sendrecv" response. It may temporarily switch to another media type for special conditions, e.g., when it switches back to G711 for modem transmission, when it uses RFC2833/4733 for DTMF transmission, or if based on the RTCP receiver reports it decides to switch to a lower bitrate coder.

In all cases, the IP address and port in the answer indicates where the answerer is expecting to receive media. In the case of a "recvonly" or "sendrecv" answer, the IP address and port will be the sink for the RTP stream. For all types of answers except 'inactive', the IP address and the port+1 will be the sink for the RTCP packets (RTCP sender reports for "recvonly" answers, RTCP receiver reports for "sendonly" answers, or both for "sendrecv" answers).

#### *3.2.2.7.2.4 The case of multicast streams*

In the case of multicast stream offers, the meaning of the sendonly or recvonly attributes is no longer the direction of the media stream from the perspective of the offerer. If the multicast stream is marked 'recvonly', it means that all participants are only allowed to receive media on this multicast address and port, they cannot send to it. If the multicast stream is marked 'sendonly', it means all participants can send to this multicast address, but should not attempt to receive media from it.

Then answer should be identical to the offer, except that some media formats may be removed to indicate that the answerer does not support them.

#### *3.2.2.7.2.5 Redirection or renegotiation of media streams: Re-INVITE and UPDATE methods*

There are many circumstances in which it is necessary to redirect media streams dynamically. A good example is a prepaid service: the caller initially needs to receive and send media to the prepaid service server, but as soon as the prepaid server has established the communication with the desired called party, the server must redirect the media streams to flow directly between the caller and the callee (see Figure 3.18). Note that this cannot be done by using call level redirection, because the prepaid server must remain in the call signaling path in order to monitor the duration of the call and eventually dynamically cut the call if the caller has exhausted its credit.

The reader may think that the server could simply serve as a relay for the media stream so that the caller would never need to dynamically redirect the media streams. This is true... in fact, this is the way traditional TDM prepaid services work using 'service nodes'. But the single most important technological advantage of VoIP over TDM voice is precisely this ability to do better than simply relaying media streams. A VoIP optimized prepaid server will be able to handle many more conversations than its TDM counterpart, and it will also require a lot less bandwidth between the service platform and the rest of the network. It is possible to not take advantage of this, but in this case VoIP presents no advantage. In fact, TDM will work better, because media relaying adds to the voice path delay, and this delay will be far more noticeable in VoIP applications than in TDM.

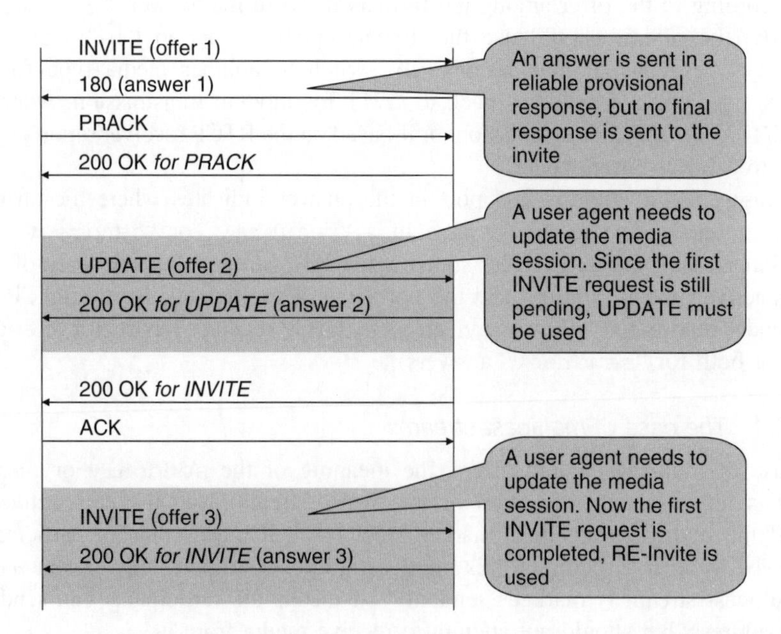


**Figure 3.18** Usage of the UPDATE method.

This ability of all VoIP endpoints to dynamically redirect media streams is a very important requirement. Devices not supporting this feature should be dismissed. It is also important to check that VoIP feature servers, such as prepaid servers or contact center servers, do take advantage of this feature. Many engineers are still 'thinking TDM' when they implement VoIP systems, and media relaying is one of the most frequent bugs found on feature servers. Unfortunately, it does not affect the external functionality of a system, and is detectable in a lab only by taking network traces. But it is very important to not deploy such systems, as in real networks the increased delay is likely to cause echo perception problems, not even considering the scalability issues.

In H.323, the renegotiation of media streams is done using the mandatory H.245 Null Terminal Capability Set procedure, or 'TCS = 0', i.e., a procedure that has no impact on the call control state. In SIP, the re-negotiation of media streams, whether required to change the target of a stream or to use an alternative media format, is done using the offer/answer model. The new offer should be encapsulated in a new `INVITE` message on the same Call-ID, often called a re-Invite. Soon after the release of RFC 3261, a bug was found in this procedure: the `INVITE` message impacts call state; therefore the use of a Re-INVITE does not allow media session changes before the first `INVITE` completes with a final answer. To solve that problem, the new `UPDATE` method was introduced in September 2002 in RFC 3311. The `UPDATE` method does not impact the state of an existing dialog and therefore can be used in an early dialog before the first `INVITE` completes (Figure 3.18). It should be used only in this context. Support for the `UPDATE` method must be specified in the Allow header field.

Either the initial offerer or the answerer can initiate a new offer if they wish to change anything in the existing situation: modify, add, or remove media streams.

The 'o=' line must be identical to the initial offer, except the version number which must be incremented if anything in the SDP has changed. An 'm' line must be present for each m line in the previous offer, but new 'm' lines can be added. Media streams are removed by setting the SDP port to zero.

In the following example, Mark re-sends an offer to change the port for the audio stream to 51780, and indicates that it stops sending video but can still continue to receive it (Mark clicked the 'stop video' button):

```
v=0
o=mark 4898446720 4898446721 IN IP4 markendpoint.anywhere.com
s=
c=IN IP4 markendpoint.anywhere.com
t=0 0
m=audio 51780 RTP/AVP 1
a=rtpmap:1 PCMA/8000
m=video 53221 RTP/AVP 31
a=recvonly
a=rtpmap:31 H261/90000
m=video 0 RTP/AVP 32
```

John accepts the change, and responds with:

```
v=0
o=john 4898446519 4898446520 IN IP4 johnendpoint.anywhere.com
```

```
s=
c=IN IP4 johnendpoint.anywhere.com
t=0 0
m=audio 51780 RTP/AVP 1
a=rtpmap:1 PCMA/8000
m=video 53221 RTP/AVP 31
a=sendvonly
a=rtpmap:31 H261/90000
m=video 0 RTP/AVP 32
```

This procedure is now to be used in all cases, even to put a stream on hold (by changing a "`sendrecv`" stream to "`sendonly`" for instance). This is a change over the recommendation of RFC2543 which used the '0.0.0.0' IP address to hold a stream: this deprecated convention prevented the continuance of receiving RTCP receiver reports when the remote party was on hold but continued to receive media.

#### *3.2.2.7.2.6 Fax*

Just like H.323, SIP either transports the fax G.711 signal transparently (pass through) or uses ITU T.38 fax-relay protocol. While the use of T.38 has been thoroughly documented in H.323, it has been left out of the main specification track in SIP. A draft was published in October 2000 (draft-mule-sip-t38callflows-02.txt) to document the way Clarent Corporation was using SIP to establish T.38 sessions. The call flow was then added to a document listing sample call flows (draft-ietf-sipping-call-flows-00.txt), and is also documented in draft-ietf-sipping-realtimefax-02.txt. T.38 annexe D also discusses SIP call establishment procedures for T.38.

(a) T.38

When a SIP gateway decides, after detecting the fax V.21 preamble flags, that it needs to encode a fax signal using `T.38`, it should use the offer/answer model to include all the T.38 parameters in SDP format. The T.38 transmission replaces the normal audio transmission over RTP (Only the CNG signal may be sent inband if the originating gateway did not detect it). Once the fax transmission is complete, the normal RTP audio transmission should resume. During fax transmission, the normal audio transmission can either be stopped completely or be put on hold.

The call flow presented in Figure 3.19 is a simple case of a gateway with a dedicated fax port; therefore it knows that the media session is going to be T.38 in the first `INVITE`.

The response shown in Figure 3.20 also contains T.38 parameters, as well as the reception port for IFT packets. Note that the selected mode is 'UDP redundancy', not 'FEC' because the parameter a=T38FaxUdpEC:t38UDPFEC has been removed in the answer.

If the fax communication is detected in the middle of a voice call, a SIP re-INVITE should be used with the T.38 parameters in the new SDP session. Once the fax communication terminates, another Re-INVITE is used to re-establish the normal RTP audio session.

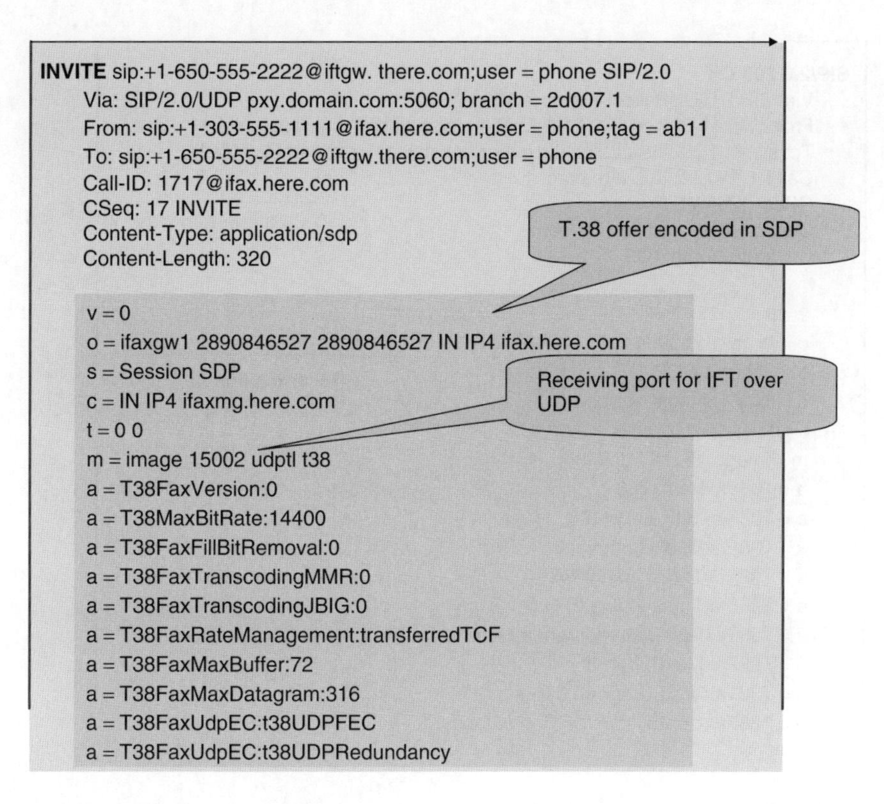

INVITE sip:+1-650-555-2222@iftgw. there.com;user = phone SIP/2.0
Via: SIP/2.0/UDP pxy.domain.com:5060; branch = 2d007.1
From: sip:+1-303-555-1111@ifax.here.com;user = phone;tag = ab11
To: sip:+1-650-555-2222@iftgw.there.com;user = phone
Call-ID: 1717@ifax.here.com
CSeq: 17 INVITE
Content-Type: application/sdp
Content-Length: 320
T.38 offer encoded in SDP
v = 0
o = ifaxgw1 2890846527 2890846527 IN IP4 ifax.here.com
s = Session SDP
c = IN IP4 ifaxmg.here.com
Receiving port for IFT over UDP
t = 0 0
m = image 15002 udptl t38
a = T38FaxVersion:0
a = T38MaxBitRate:14400
a = T38FaxFillBitRemoval:0
a = T38FaxTranscodingMMR:0
a = T38FaxTranscodingJBIG:0
a = T38FaxRateManagement:transferredTCF
a = T38FaxMaxBuffer:72
a = T38FaxMaxDatagram:316
a = T38FaxUdpEC:t38UDPFEC
a = T38FaxUdpEC:t38UDPRedundancy


**Figure 3.19** INVITE from a dedicated fax port.

(b) Fax Pass-Through

Fax `pass-through` is simple, and is the fall-back mode if T.38 is not supported by the two gateways (`488 Not Acceptable Here` or `606 Not Acceptable` response). If a gateway attempted to initiate a T.38 media session in a Re-Invite and this was rejected, it should initiate a new Re-Invite with the pass-through session parameters. Fax pass-through only requires to dynamically switch the RTP payload type to G.711 Ulaw or Alaw to disable any `silence suppression` and to keep the echo `cancellation` active. Specific SDP parameters have been defined in `RFC 3108`. For G.711 ULaw, the following SDP description can be used in the re-INVITE:

```
Content-Type: application/sdp
      Content-Length: 181

      v=0
      o=faxgw1 2890844527 171091 IN IP4 iftgw.there.com
      s=Session SDP
      c=IN IP4 iftmg.there.com
      t=0 0
```

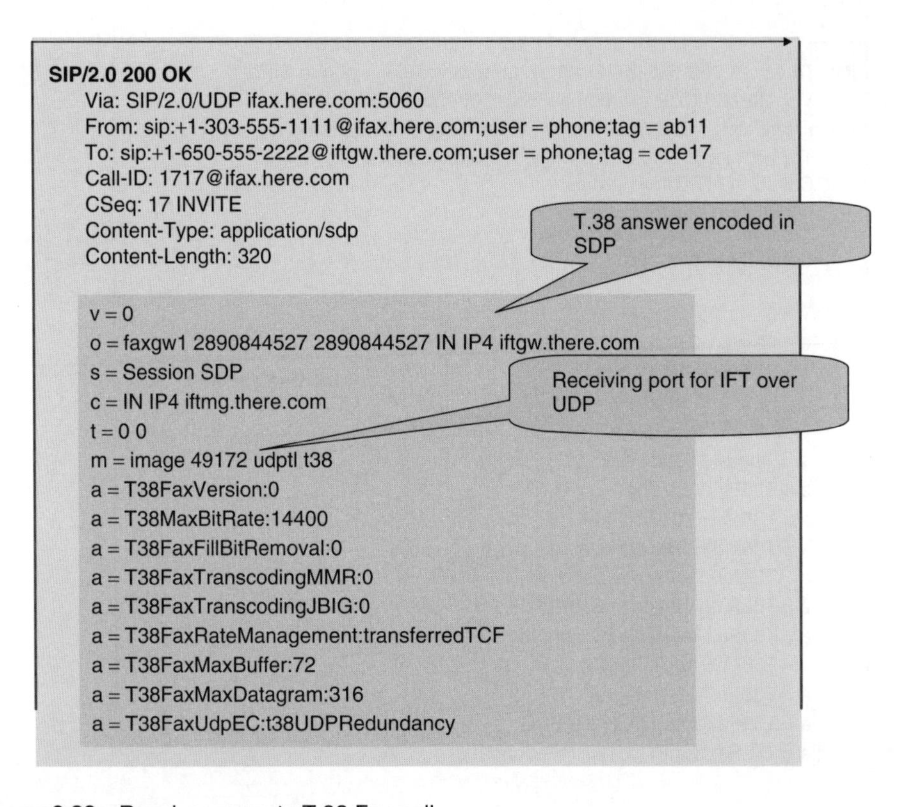
SIP/2.0 200 OK
Via: SIP/2.0/UDP ifax.here.com:5060
From: sip:+1-303-555-1111@ifax.here.com;user = phone;tag = ab11
To: sip:+1-650-555-2222@iftgw.there.com;user = phone;tag = cde17
Call-ID: 1717@ifax.here.com
CSeq: 17 INVITE
Content-Type: application/sdp
Content-Length: 320
T.38 answer encoded in SDP
v = 0
o = faxgw1 2890844527 2890844527 IN IP4 iftgw.there.com
s = Session SDP
c = IN IP4 iftmg.there.com
Receiving port for IFT over UDP
t = 0 0
m = image 49172 udptl t38
a = T38FaxVersion:0
a = T38MaxBitRate:14400
a = T38FaxFillBitRemoval:0
a = T38FaxTranscodingMMR:0
a = T38FaxTranscodingJBIG:0
a = T38FaxRateManagement:transferredTCF
a = T38FaxMaxBuffer:72
a = T38FaxMaxDatagram:316
a = T38FaxUdpEC:t38UDPRedundancy

**Figure 3.20** Receiver accepts T.38 Fax call.

```
m=audio 12322 RTP/AVP 0
a=rtpmap:0 PCMU/8000
a=ecan:fb on -
a=silenceSupp:off - - - -
```

### *3.2.2.7.2.7 Exchange of capabilities using SDP, RFC 3407*

H.323 uses a specific set of messages to negotiate capabilities of endpoints. These capabilities are very flexible and can take into account processing or bandwidth constraints which make only certain combinations of coders possible.

Prior to RFC 3407, SIP did not have a specific syntax to express capabilities, and therefore also used SDP for this purpose ("e =" and "p =" lines could be omitted). This method rapidly proved to be inadequate and specific SDP syntax extensions for capabilities were defined in RFC 3407.

In addition to the SDP sections described above for offers and answers, RFC 3407 defines a capability data section beginning

```
a=sqn : <sequence number of this capability set>
```

Followed by one or more lines containing capability descriptors:

```
a=cdsc : <incremental sequence number of this capability set>
<type of media> <type of transport> <list of payload types>
```

Each capability line may be complemented by parameter lines:

```
a=cpar : < encapsulated 'a=...' lines
```

The following example is given in RFC 3407:

```
v=0
o=- 25678 753849 IN IP4 192.0.0.1
s=
c=IN IP4 192.0.0.1
t=0 0
m=audio 3456 RTP/AVP 18 96
a=rtpmap:96 telephone-event
a=fmtp:96 0-15,32-35
a=sqn: 0
a=cdsc: 1 audio RTP/AVP 0 18 96
a=cpar: a=fmtp:96 0-16,32-35
a=cdsc: 4 image udptl t38
a=cdsc: 5 image tcp t38
```

We can see that the section corresponding to the SDP offer (the streams that this user agent is prepared to receive now) indicates only G.729 (dynamic payload type 18) and telephony events (dynamic payload type 96). But, with the special RFC 3407 capability section, the terminal indicates that it is also capable of receiving G.711, T.38 fax on both UDP and TCP, (and of course G.729 and telephony events). However, it is not prepared to receive these media types right now—for instance, in order to receive a fax, an actual fax signal should be detected before the user agent would be prepared to receive T.38 data.

Unlike H.323, there is no way to specify that some coders of different media types cannot be used simultaneously; for instance, a processor intensive coder such as G.729 cannot be used with H.263. In H.323, it is possible to specify that the H.261 coder can be used with either G.711 or G.729, but H.263 can only be used with G.711.

Similarly, it may be impossible for bandwidth reasons to use video coders and data-sharing simultaneously, but this time a bandwidth constraint may be expressed in SIP by using the SDP session level "b =" parameter.

## 3.3 CALL HANDLING SERVICES WITH SIP

SIP defines many functional names for SIP servers, such as:

- proxy (stateless, stateful, forking . . .) server;
- registrar server;

- redirect server;
- location server;
- back-to-back user agent.

In addition to these names, the industry also frequently uses terms like 'application server' or 'feature server'.

Despite the fact that most of these functions are called 'servers', they really describe functions, and do not necessarily refer to separate servers. In fact, most of the time, a given SIP server will implement the features of many (if not all) of the entities listed above. For instance a server could:

- register SIP user agents in a certain area (registrar behaviour);
- reply to other SIP servers location requests (location server behaviour);
- handle outgoing calls of the locally registered SIP devices (stateful proxy);
- propagate SIMPLE IM messages without modification (stateless proxy);
- implement certain complex applications such as contact center call distribution for calls to certain numbers, using a back-to-back user agent behaviour.

The descriptions below refer to each of the above-listed behaviours and uses the SIP 'server' terminology for them, but the reader should keep in mind that all these functions can coexist in a single box. Typically, it is only in very large deployments that call control and registration features may be separated, requiring, for instance, separate registrars and stateful proxy boxes. The advanced message routing capabilities of SIP (refer to Section 3.3.2.3) facilitate the design of such distributed architectures: it is even possible to implement load balancing servers that handle only the first message in a dialog then drop out of the signaling message path. The 3GPP IMS (see Chapter 4 for details) is such a distributed architecture that makes extensive use of SIP message routing features.

## 3.3.1 Location and registration

### 3.3.1.1 *Locating users from SIP addresses*

SIP addresses are called URIs (Uniform Resource Identifiers). URIs are really names (except SIP addresses using an IP host address, as the address used in our simple call example) and do not refer directly to the transport address to be called but to an abstract entity that can reach the user directly or indirectly.

SIP URIs have two major forms, an 'email-like' form and a telephone number form:

- The general format of email form SIP URIs is `user@host`, where `host` is usually a fully qualified domain name, that can be resolved to an IP address using the DNS system. In many cases, the SIP address of a user will be the same as his email address.

- The general format of a telephone number as a SIP URI is `phone-number @host;user=phone`. The host part is optional and may indicate a server that can reach this phone number, which can be used to specify a preferred service provider. In the IMS architecture (see Chapter 4), the host name is mandatory and indicates the home domain of the user agent. Many telephony systems, however, can decide where to route the phone call based on the phone number only (through prefix analysis or a local number portability query), so the domain name part is not present in many telephony applications.
- A 'Tel URI' format has been defined in RFC 3966 (which replaces RFC 2806). This URI format may be used to write local or global phone numbers. When beginning with a '+' symbol, the number is an E164 global scope phone number, e.g., +33 158713333. Otherwise, the URI contains a local scope or private phone number, and the numbering plan context must be specified in a '`phone-context`' parameter. For instance `Tel:112;phone-context=+33` may be used to express an emergency phone number in France. Clearly the call will reach a different destination depending on the country, and, as it is not a global scope number (cannot be reached from abroad), it cannot be written as an E164 number. The Tel URI was defined not just for SIP: it may also be used in other contexts, e.g., on web pages.

There is also another type of URI which serves a different purpose; it is a sort of command line requiring some action from the user agent, described by a METHOD attribute. See the examples below.

Optionally a SIP URI may also specify a port number and a transport mode if the SIP default transport (UDP) and port (5060) are not to be used.

Table 3.4 lists examples of SIP URIs:

**Table 3.4** Common SIP URL formats

| | |
|---|---|
| `John@netcentrex.net:1234` | Vanilla SIP URI with a custom UDP port |
| `Userdomain.com` | No user part, default port will be 5060 |
| `support@company.fr:2345;transport=UDP` | Wants to be contacted using UDP |
| `192.190.234.3:8001` | Contact the server at this IP address and port |
| `support@netcentrex.net;maddr=239.255.255.1;ttl=32` | Override normal host name to transport address mechanism : use multicast to 239.255.255.1 with a TTL of 32 instead |
| `+33-231759329@cybercall.com;user=phone` | Global phone number |
| `0231759329;isub=10;postd=w11p11@cybercall.com;user=phone` | Local phone number with isdn subaddress, wait for dial tone, then dial 11 (pause) 11 using DTMF |
| `ACD@netcentrex.net?priority=high&customercode=1234` | Using proprietary extension headers to control priority in an ACD system... |
| `Newcomer@reg.usergroup.com;METHOD=REGISTER` | Previous URIs would trigger a SIP INVITE request, this one initiates a registration to the registrar of usergroup: reg.usergroup.com |

Most of the extensions (headers, maddr, etc...) are not allowed in the To, From parameters of SIP requests and responses, but can be used in the contact parameters.

#### 3.3.1.1.1 The original RFC 2543

RFC 2543 described how to locate the physical endpoint from its SIP URI. This was done in two stages:

- Obtaining a destination address for the initial `INVITE`: the destination SIP URI can be resolved into a SIP server address (see below for details). This SIP server will be the destination of the initial `INVITE` message, unless an 'outbound proxy' has been configured, in which case all calls should be directed to the outbound proxy. The SIP server may be the final destination of the call, or just a SIP proxy which will route the request to its final destination.
- Forwarding the `INVITE` to the final destination: The SIP server which receives the initial `INVITE`, if it is not the final destination of the call, becomes responsible for routing the `INVITE` request to the called endpoint. This can be done in two ways: either by instructing the calling endpoint to send a new `INVITE` request to another location using the 302 moved reply (redirect server behaviour) or by transparently relaying the `INVITE` message to the appropriate transport address (proxy behaviour). The first model is similar to the H.323 direct call model and the second is similar to the H.323 gatekeeper routed call model.

In order to locate the first hop SIP server from the URI, a SIP terminal will use `DNS`. A SIP URI domain name must have an SRV record, a MX record, a CNAME or an A record. The resolution algorithm is represented on Figure 3.21.

First, the terminal will retrieve the SRV resource records for the considered domain name. Then it will keep only the records of type `_sip._udp` or `_sip._tcp`. If there is a `_sip._udp` record, the terminal will contact the SIP server using UDP at the specified transport address. It will use the port specified in the SIP URL or default to the port specified in the `_sip._udp` record. If there is a `_sip._tcp` record, the same method will be used, but over TCP.

If no SRV record is found, the user agent will try to retrieve the IP address of a SIP server by looking at the MX records first (normally used to point to a mail server), then CNAME records (pointing to an alias name), and finally an A record (pointing to an IP address).

Pointing to a SIP server instead of the called endpoint directly allows the called endpoint to move (the transport address changes), while enabling the use of DNS caching. If the address of the called endpoint was stored directly in DNS, there could be a lot of trouble with DNS caching. Normally, all DNS records can be cached by the DNS resolver. The cached record expires after a certain period after it has been first retrieved by a DNS query; this period is called the Time To Live (TTL, in seconds). The value of the TTL is stored in the DNS record. Therefore, when the terminal moves, the caller could still have a wrong address in the DNS resolver cache, and the call could fail. The only solution is to set the TTL to zero and update the primary DNS record as the terminal moves... not very easy, and not cache friendly, and therefore not scalable.

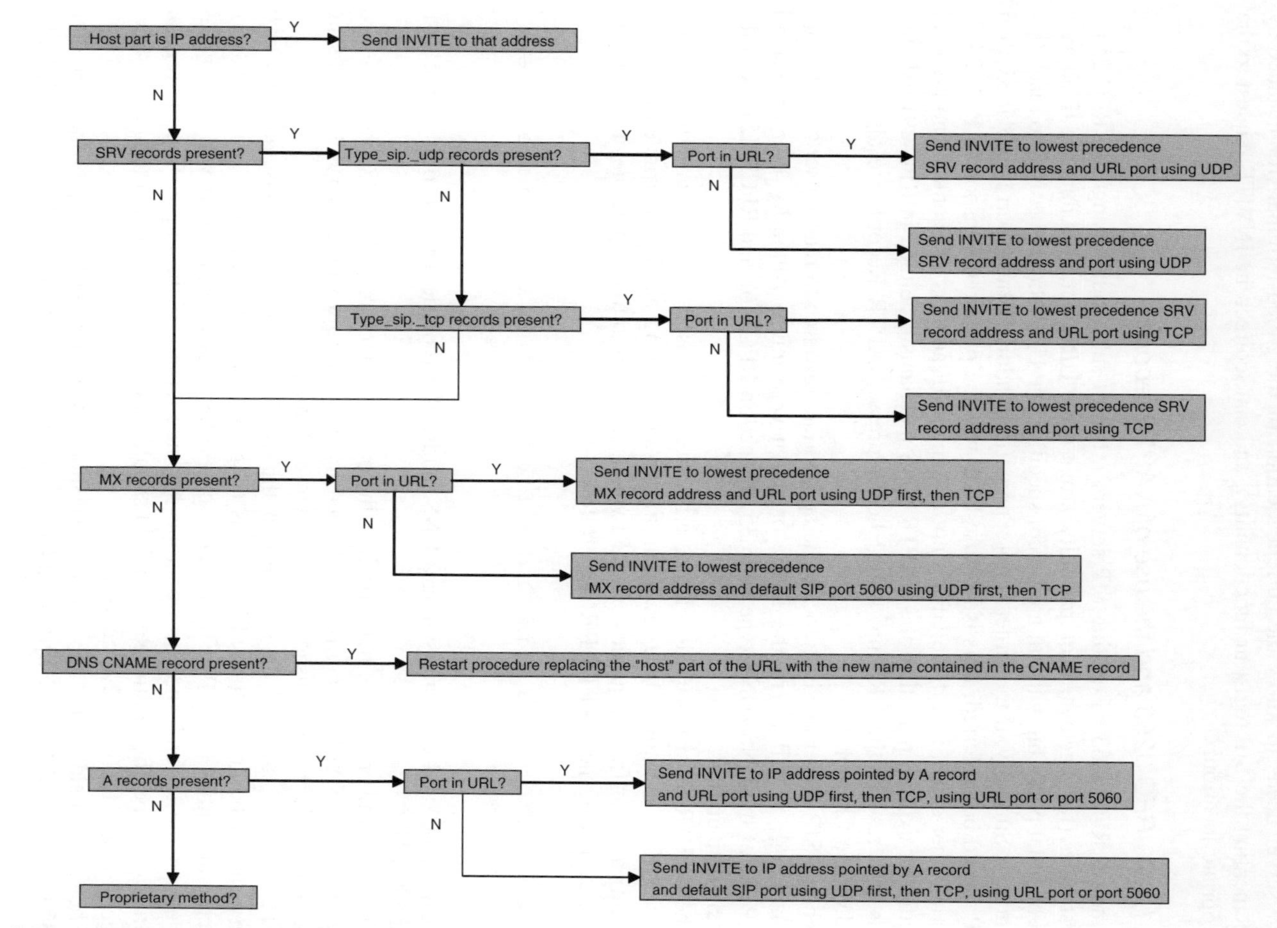


**Figure 3.21** Location of a SIP server using DNS.

On the other hand, a SIP proxy server is not likely to move very often, and storing its address in a DNS SRV, MX, or A record does not cause any trouble. This SIP proxy server needs to know the current location of the called terminal (for instance by implementing the SIP registrar functionality), and can redirect the INVITE request to the appropriate location.

#### *3.3.1.1.2 RFC 3263 and the use of NAPTR records*

The newer RFC 3263 ('Locating SIP servers'), brings substantial changes to RFC 2543. It first states (for 'backward compatibility'), that if the URI contains a numeric IP address (and optional port), but without a protocol specifier, the UDP should be used to reach this IP address. Similarly, if the target is not a numeric IP address, but a port is provided, then UDP should be preferred. This is because UDP was the preferred transport in RFC 2543.

In all other cases, i.e., if the URI is not a numeric IP address and contains no protocol specifier or explicit port, then the NAPTR DNS mechanism defined in RFC 2915 should be used to resolve the URI into a next hop address. NAPTR Records are also used by ENUM (see Chapter 2).

When a SIP server needs to locate the appropriate resource to reach 'user@subdomain .domain.org', it will query the DNS for the new Naming Authority Pointer Record (NAPTR) of the DNS (DNS type code 35, defined in RFC2168 and RFC2915) for 'sub-domain.domain.org'.

The NAPTR record is used to attach a rewrite rule (based on a regular expression) to the DNS domain name. Once rewritten, the resulting string can be interpreted as a new domain name for further queries or a URI which can be used to delegate the name lookup.

The syntax of the NAPTR RR is as follows:

```
Domain TTL Class Type Order Preference Flags Service Regexp
  Replacement
```

Domain, TTL, and class are standard DNS fields. The type field is set to 35 in the case of the NAPTR.

The order and preference field specifies the order in which records MUST be processed when multiple NAPTR records are returned in response to a single query. The ordering is lexicographic; order is used first, and then preference.

The "S", "A" and "P" flags mean that this NAPTR record is the last one; the next query should be done using SRV records (flag 'S'), an A record (flag 'A'), a protocol specific algorithm (flag 'P'). In all these cases the 'replacement' field will be used as the new name to fetch the corresponding resource record. If the flag is 'U', the regular expression[19] specified in the RegExp field should be applied to the domain name in order to get a new URI (recursive query).

The service field defines the protocol that should be used after this step of the resolution (H323, LDAP, SIP, TEL, SIPS), and the type of service that will be provided,'D2U'

[19] An expression composed of a series of symbols each defining a specific modifications to a string, and defined in POSIX.

(UDP transport), 'D2T' (TCP transport), or 'D2S' (SCTP transport). For memory, ENUM uses 'E2U' for this type of service.

A SIP server looking for a next hop protocol for a SIP call will therefore look for NAPTR resource records with the service field set to:

- SIP+D2U for a list of next hops that must be reached via SIP/UDP;
- SIP+D2T for a list of next hops that must be reached via SIP/TCP;
- SIP+D2S for a list of next hops that must be reached via SIP/SCTP;
- SIPS+D2T for a list of next hops that must be reached via SIPS/TCP.

All other service fields are discarded along with the options that are not supported by the requesting server (e.g., SCTP); the responses are sorted according to the preference value (lower values have a higher priority), and the server must try them in order.

Resource records with the SIP+D2U, SIP+D2T, SIP+D2S, or SIPS+D2T service codes also have the 'S' flag : the full resolution requires a request for SRV resource records for the resource indicated in the 'Replacement'. The regular expression field will be empty.

For instance, we have the following NAPTR records for subdomain.domain.org:

```
;        order  pref  flags  service     regexp   replacement
 IN NAPTR  50     50   "s"   "SIPS+D2T"    ""    _sips._tcp.subdomain.domain.org.
 IN NAPTR  90     50   "s"   "SIP+D2T"     ""    _sip._tcp.subdomain.domain.org.
 IN NAPTR  100    50   "s"   "SIP+D2U"     ""    _sip._udp.subdomain.domain.org.
```

This indicates that the server supports TLS (over TCP), TCP, and UDP, in that order of preference. If the client supports TCP and UDP, TCP will be used, targeted to a host determined by an SRV lookup of `_sip._tcp.subdomain.domain.org`. The lookup will, for instance, return this list of SRV records:

```
;;          Priority Weight Port  Target
   IN SRV   0        1      5060  proxy1.domain.com
   IN SRV   0        2      5060  proxy2.domain.com
```

In theory, the NAPTR record could modify subdomain.domain.org into anything. RFC 3263 mentions that if this is the case, then SRV records with the original name must also be present, for backward compatibility with RFC 2543.

The whole procedure requires at least three DNS queries for each transaction: one to resolve the URI into NAPTR records, one to resolve the new resource name specified by the NAPTR resource field into a set of SRV resource records, and at least one to resolve the selected SRV record target into an IP address. In fact, most of the time, more requests will be required if the URI contains many subdomain components because, in the more general case, these domains will not all be in cache. For subdomain.domain.org, this is five DNS requests, each request lasting about 100 ms!

The procedure introduced by RFC 3263 requires one more DNS lookup than the previous RFC 2543 mechanism which looked for SRV records directly . . . but does not bring anything new because it does not make use of any of the possibilities of NAPTR records that go beyond what can be done with SRV records. Overall, this is an overly complicated and verbose RFC, with many exceptions to the general rule... for a very simple address resolution problem. It could evolve into something more interesting if the full power of NAPTR records, with the use of regular expressions, was used. It would then become very similar to ENUM. As it stands now, it seems to be chasing the same concepts as ENUM, keeping the complexity and leaving the features out...

#### 3.3.1.2 The registrar function

A registrar is a server that accepts REGISTER requests. Registrars are needed to keep track of the current 'location' of a user-agent, i.e., what transport protocol address and port (expressed as a URL or URI) should be used to send SIP signaling messages to the user agent. The IP address of a user agent may change under a number of situations: it may be connected via an ISP providing dynamic addresses, it may be a LAN that provides dynamic IP addresses via DHCP, or it may be a roaming user.

In order to be able to reach this user-agent, it receives a stable name called an Address of Record (AoR) in IETF terminology and a Public identity in 3GPP terminology. The registrar maintains the mapping between SIP AoR(s), and the currently valid SIP protocol URI(s) for each user agent.

The `Registrar` can be contacted by unicast if the address of the registrar is known. In this case, the procedure is the same as for any other SIP request. An example registration is given in Figure 3.22 (only the headers with a specific usage during registration are shown).

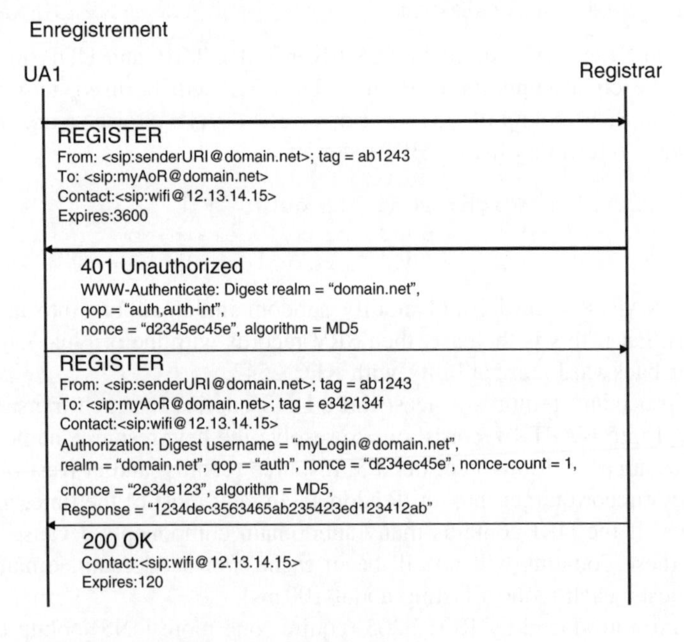


**Figure 3.22** Unicast registration with authentication.

The REGISTER request uses the following headers:

- From: the From header should contain the URI of the sender of the request. In general, it is the same URI as the AoR of the user agent and the From header will be identical to the To header. However, SIP allows third-party registrations; in this case the From header would contain the URI of the server performing the third-party registration
- To: contains the AoR (Public identity) of the user agent. The user agent indicates, through this registration, that it is ready to receive SIP signaling messages sent to this AoR (i.e., containing this AoR in the top line) on the contact URI(s) specified in the Contact header.
- Contact: contains one or several URIs (transport addresses and ports) where the user agent is listening for SIP messages.
- Expires: indicates the desired validity of the association between the AoR and contact URIs.

In the example of Figure 3.22 the registrar authenticates the client by rejecting the initial REGISTER request with a 401 response containing a WWW-authenticate header. This response indicates the following:

- The registrar wants to use HTTP digest authentication (RFC 2617 : HTTP authentication: Basic and Digest Access Authentication).
- The authentication domain (realm or name space) to use is 'domain.net'. This is a hint helping the user agent (and possibly the end user) to select the right password.
- The message protection mechanisms supported ("qop": Quality of protection). 'qop=auth' indicates that the entire REGISTER message value is taken into account in the authentication hash algorithm, protecting the whole message from alterations.
- The supported hash algorithm(s), usually MD5
- A challenge (value of the 'nonce'), i.e., a pseudo random character string that the user agent will need to combine with the password in order to generate a hashcode. The hashcode allows the registrar to check that the user agent possesses the expected secret, without exposing it in clear on the communication link.

The client then sends a new Register request, containing an 'Authorization' header which indicates:

- the selected message protection mechanism (optional, by default qop=auth);
- the number of times the 'nonce' value has been used ('nonce-count');
- a pseudo random string ('cnonce');
- the hashcode proving that the user agent knows the shared password. In the example, it contains 128 bits computed from the username, realm, password, nonce (challenge), 'nonce-count', c-nonce, qop value, method, and URI. If qop = auth, the entire REGISTER message value is taken into account in the hash algorithm.

In the 200 OK response, the Registrar may, as in the example of Figure 3.22, simply confirm the registration and communicate back to the user agent all the URIs currently configured as contact points for the registered AoR/public identity. The registrar may also restrict the validity duration of the registration. In the example, the Expires header indicates a maximum validity duration of the registration of 120 seconds—the intent is to maintain open a 'pin-hole' in NAPT routers which may be in the path of SIP messages (see Chapter 7), by sending refresh REGISTER messages periodically. The expiration delay may also be specified for each contact precisely - in this case expires is used as a parameter for URIs present in the Contact header.

Mutual authentication is also supported: the registrar must insert its own 'Authentication-Info' header, which contains a hash code computed using the cnonce pseudo random value transmitted by the user agent and the user agent secret. This proves that the registrar also knows the secret of the user agent. This header may also be used by the registrar to convey to the client a new challenge to be used for the next request ('nextnonce'). If this is not used, the registrar will reject the request when the nonce-count is too high (including the stale=true parameter in its response), triggering a new authentication process.

If the user agent moves and needs to update the URI of the registration, it may cancel the previous registrations by positioning the contact header to special value '*', and then register the new location URI.

The detailed syntax of the headers of registration requests and responses headers is discussed in Section 3.2.2.2 (method REGISTER, headers Contact, Expires, To) and Section 3.2.2.3 (2xx, 3xx). The reliability mechanism is the normal SIP two-way handshake (see Section 3.2.2.4.1.1). The specific registration procedure used by 3GPP IMS user agents is described in detail in Section 4.2.1.

In order to facilitate user mobility and avoid manual configuration as much as possible, SIP defines a well known 'all SIP servers' address (sip.mcast.net : 224.0.1.75). A client can therefore, in theory, register his current IP address with a multicast register message. For some unclear reason SIP restricts the TTL (time to live) of this message to one, limiting the discovery method to the local subnet.

Currently SIP servers cannot reply to a multicast "REGISTER" message; therefore the client does not have a chance to learn the address of the appropriate SIP server, or even to know if there is a SIP server that accepted the registration.[20] Owing to this limitation and the lack of multicast support in most IP networks, this multicast procedure is not used in practice.

### 3.3.1.3 Redirect server

A redirect server responds to an INVITE request with a 3xx reply (or rejects the call with a client error or server error).

[20] This feature is roughly equivalent to the Gatekeeper discovery method described in H.323. However, in H.323, the gatekeepers that are willing to handle the request can reply, allowing the client to select the appropriate gatekeeper and contact it directly later on.

- The 300 'Multiple choices' reply can be used when the SIP URL of the request can be contacted at several alternative addresses. The choices are listed as contact fields. This can be used as a simple form of load balancing, or more interestingly to let a caller know all the available means or media that can be used to communicate with the destination user. For instance the returned contact field could be:
  `m: sip:John_gsm@company.com,sip:John_home@family.org`
- The 301 'moved permanently' reply indicates that the SIP URL of the request can no longer be contacted at this location. The client should try to contact the new location given by the Contact header field of the reply. This change is permanent and can be memorized by the client. The contact header can also mention several possible destinations.
- The 302 'moved temporarily' reply redirects the client to a new location, as above, but for a limited duration, as indicated by the Expires field.
- The 305 'use proxy' indicates that the specified location should be reached via the indicated proxy.
- The 380 'Alternative Service' is really for future use, not fully defined in the current SIP RFCs. This reply is more complex, and may seem a bit redundant with the previous replies: in addition to providing a new destination in the contact field, the reply can also contain a session description in the message body that represents the sending capacities of the new destination. The caller is expected to send an INVITE request to this new destination, and offer in its SDP session description the appropriate capabilities (which can be a copy of the SDP parameters of the 380 reply, except for the receiving RTP ports).

Other replies (e.g., 303) were defined in early SIP drafts, but have become obsolete.

A redirect server can be used in conjunction with a registrar to redirect calls to the current location(s) of the caller. It can also act as a basic form of call distribution system, as shown in Figure 3.23.

Redirect servers can be useful tools to improve the scalability of complex call management systems. Inserted as a front end, it can distribute calls among a pool of secondary servers, achieving load balancing. This is permitted by the `maddr` parameter of the `Contact` field:

```
<sip:originaladdress@callcenter.com:9999;maddr=sophisticated
ACD3.callcenter.com>
```

By returning this, the redirect server indicates that the caller should send an `INVITE` with the same destination URI (originaladdress@callcenter.com), but send it to the third ACD server of the pool (ACD3.callcenter.com). The `maddr` parameter instructs the caller to bypass the normal procedure to find the appropriate SIP server from the domain part of the URL, and to use the domain name provided instead.

One of the most interesting uses of the redirect server in conjunction with a registrar is forthe deployment of large scale residential networks. A network serving hundreds of thousands of SIP endpoints cannot be realistically realized with a single server. The reason is that if N endpoint is sending a registration message every S seconds, the number of

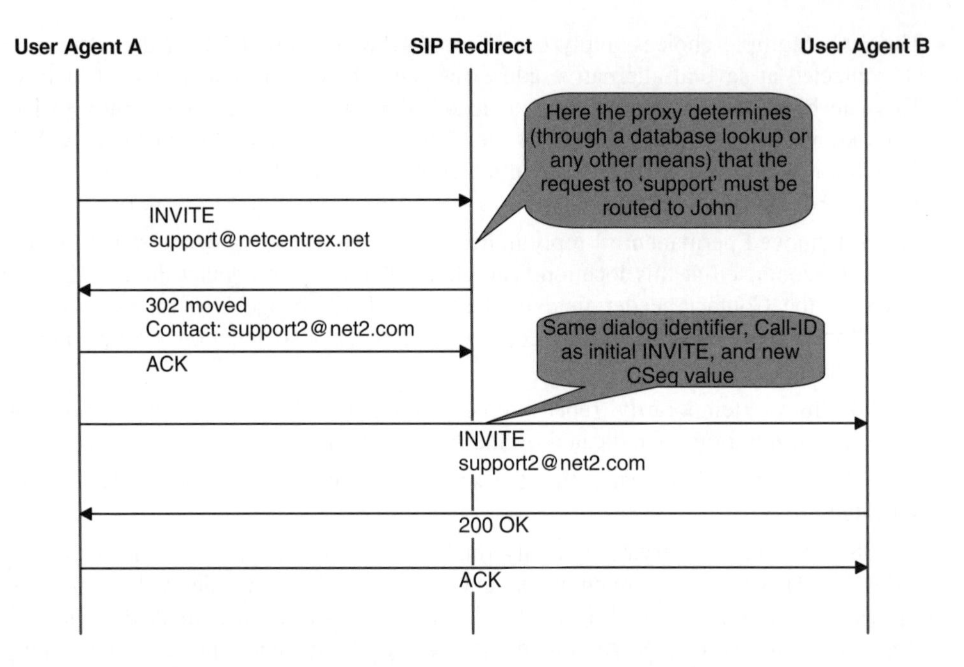


**Figure 3.23** SIP redirect server distributes calls, in a very basic contact center.

messages per second that would need to be processed by the central server would be on average N/S, and much worse when the network starts or re-starts. With $N = 300000$ and $S = 60$, the central server should process over 5000 Registration messages per second. Obviously, in this case it is necessary to use a number of separate registrar servers—for instance, one server per block of 100000 users. The call control function can still be centralized because there are a lot less calls than registration messages (if each user makes 1 call every 5 hours, this is $300000/(5 * 3600) = 16$ calls per second). But in order to route incoming calls to the right user agent, the central call control function will need to query the registrars, which can be done using an INVITE/REDIRECT transaction. This type of strategy is currently used with success in large residential networks.

## 3.3.2 The proxy function, back to back user agents

### *3.3.2.1 Definitions*

A proxy server acts as a server on one side (receiving requests), and as a client on the other side (possibly sending requests).

Strictly speaking, a 'proxy' should be mostly transparent to user agents' messages, simply passing messages and changing them in very limited ways. A proxy can forward a request without any change to its final destination, it can decide to validate requests, authenticate users, fork requests, resolve addresses, cancel pending calls, etc.

Depending on the level of control the proxy has over the SIP messages it processes it can be a stateless proxy, a stateful proxy, or even a back-to-back user agent:

- A stateless proxy simply chooses the next hop destination for an incoming SIP message using the "`To`" header information, but otherwise keeps no state for the call or even the transaction (it will not handle retransmissions, but simply pass them transparently). For instance, a stateless proxy will not do any local processing for a CANCEL request other than forward it, and will not even acknowledge locally any response, but simply pass it transparently to the original sender of the request. This behaviour is made possible because SIP allows a proxy to store some state in the messages, e.g., in the "`Via`" header. This state is copied in the response, and therefore the stateless proxy server does not need to keep in memory any call parameter to be able to forward a response; it simply finds the information it needs in the response itself (e.g., the next hop is at the bottom of the pile of via headers, after discarding the via header corresponding to the proxy itself). Stateless proxies have often been presented as a technology breakthrough that would make SIP networks considerably more scalable than any other network. The reality is that a stateless proxy can serve only very limited purposes (for instance it cannot do billing), and therefore can be used only in simple infrastructure call flows, for instance, performing load balancing or basic message routing in core networks. Even in this role it cannot do very much, as even a simple load balancing function usually needs to keep in memory the number of calls it has sent to each destination, and discard any destination that appears to fail frequently.
- Stateful proxies are much more useful, as they can keep any state relative to the call and all transactions involved in the call. Stateful proxies also manage locally some aspects of the transactions, e.g., they will handle retransmissions locally and acknowledge the final responses (except `200 OK`) and `CANCEL` requests. Stateful proxies can serve most call control purposes required in a SIP network, such as choosing an egress route for a phone call among multiple gateways by offering the call in sequence to multiple gateways and analyzing the responses to eventually try another gateway if the current attempt failed due to congestion or any other reason. Since a stateful proxy memorizes when a call has begun, it can generate call detail records when the call ends with the duration of the call. The IMS architecture makes an intensive use of stateful proxies, in particular the S-CSCF (see Chapter 4 for more details).
- Most applications need so much control on the call that they cannot be implemented within the restrictions set on simple proxies. For instance the requirement to transparently forward any `200 OK` response received from a destination may not be compatible with applications which need to filter the responses for security or any other purpose. Many of the most sophisticated applications, such as business telephony applications and contact centers, need the complete range of possibilities of a user agent. Therefore they act as a full user-agent receiving a call, and reinitiate a call as a user agent. Strictly speaking, such servers are no longer proxies, but should be called "back-to-back user agents".

The names 'feature server' or 'application server' first emerged in marketing presentations and are now widely used by the industry, but they have no precise meaning. A 'feature

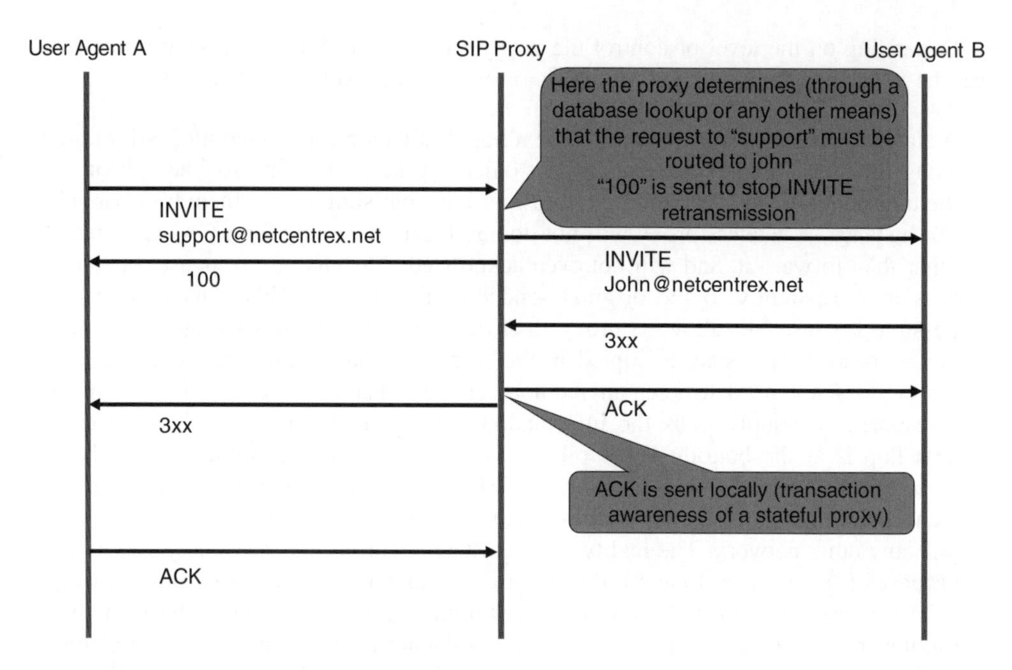


**Figure 3.24** Simple call through a stateful proxy.

server' is any server that implements an application! It could be a stateful proxy, a back-to-back user agent, or even an interactive voice response server that can receive, generate, and bridge calls. It is only in the 3GPP IMS architecture that 'Application server' has a precise meaning: it is a server that implements the ISC SIP interface (see Chapter 4 for more details).

There is a frequent confusion leading to the idea that a SIP application server 'replaces' the 'old' Intelligent Network' model on telephone networks. In fact, the intelligent network model refers to the ability of a 'Service Control Point (SCP)', to use an abstract protocol-independent model of a call to remote control a 'Service Switching Function (SSF)', which is the only one to be part of the telephony network and to implement telephony protocol stacks. The function implemented by an SSF with SIP stacks and the SCP together can be described as an 'application server'. But a simple monolithic SIP-only application server does not replace an IN architecture, which has been designed precisely to facilitate the programming of protocol-independent application.

In Figure 3.24, the proxy is at least a stateful proxy because it locally sends a `100` reply and generates ACKs (transaction awareness).

Note that a stateful proxy is not allowed to send ACKs locally for `200 OK` responses (Figure 3.25). This message and its reliability must be handled end to end ensuring that the call is established and media can start flowing only when the end-to-end handshake is complete. Only a back-to-back user agent can send an ACK locally to a 200 OK response (and it must understand the consequences).

Most useful functions (e.g., the ability to drop a call from the proxy) go beyond the strict definition of a 'proxy': most commercial server implementations are 'back-to-back

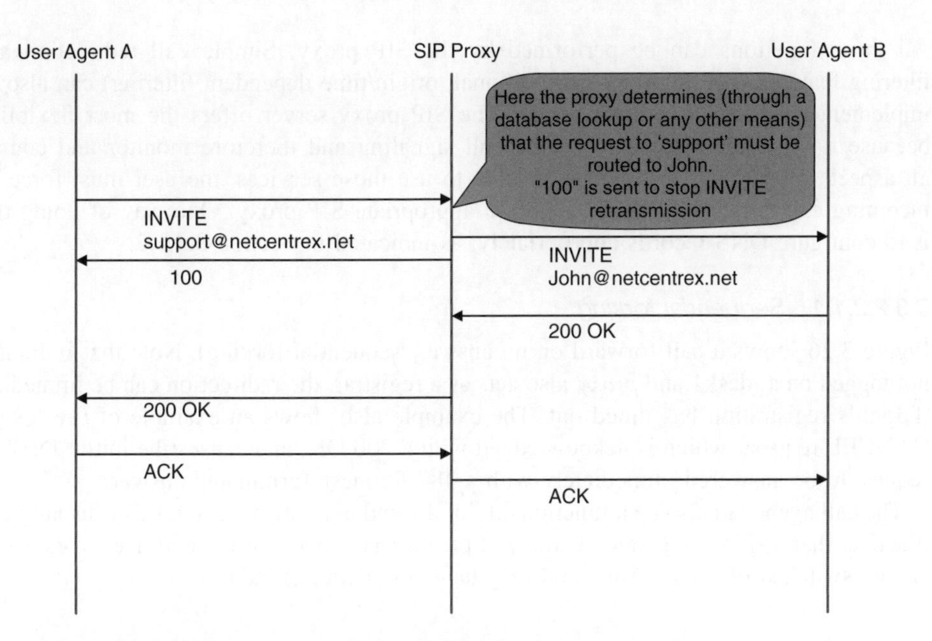


**Figure 3.25** 200 OK response is acknowledged end to end.

user agents', according to the SIP terminology. In the following text, we do not restrict ourselves to the strict 'proxy' terminology, and describe the various functions of a server that has control over SIP signaling during the duration of the call (that encompasses the 'proxy', and 'back-to-back user agent' features).

### *3.3.2.2 Examples of proxies*

#### *3.3.2.2.1 Call agent function*

A call agent is a service that handles incoming and/or outgoing calls on behalf of a user. In traditional telephony this type of function is performed by the Intelligent Network infrastructure of the operator, or by the PBX of the company. The concept of call agent was introduced in the IP telephony area in Scott Petrack's description of a Call Management Agent (CMA).

A call agent can perform the following tasks:

- Try to find the user by redirecting the call setup messages (SIP INVITE or H.323 SETUP) to the proper location, or several possible locations simultaneously.
- Implement call redirection rules such as call forward on busy, call forward on no answer, call forward unconditional.
- Implement call filtering with origin/time dependent rules.
- Record unsuccessful call attempts for future reference.

All these functions can be performed by the SIP proxy. Simple call redirection and filtering features (call forward unconditional, origin/time dependent filtering) can also be implemented on a SIP redirect server. The SIP proxy server offers the most flexibility because it can choose to relay all the call signaling and therefore monitor and control all aspects of the call. In order to be able to use those services, the user must force all incoming call attempts to go through the appropriate SIP proxy. One way of doing this is to configure DNS records appropriately, as indicated in Section 3.3.1.2.

#### *3.3.2.2.1.1 Sequential forking*

Figure 3.26 shows a call forward on no answer (sequential forking). Note that if John is not logged on at desk1 and *proxy* also acts as a registrar, the redirection can be immediate if John's registration has timed out. The example also shows an example of use for the CANCEL request, which is acknowledged with a 200 OK, and causes the initial INVITE request to be answered immediately with a 487 Request Terminated answer.

The call agent can also be a functionality of the end user software, but this is usually less practical than using a separate centralized proxy server because the end user workstation can be switched off at any time and may have a dynamic IP address.

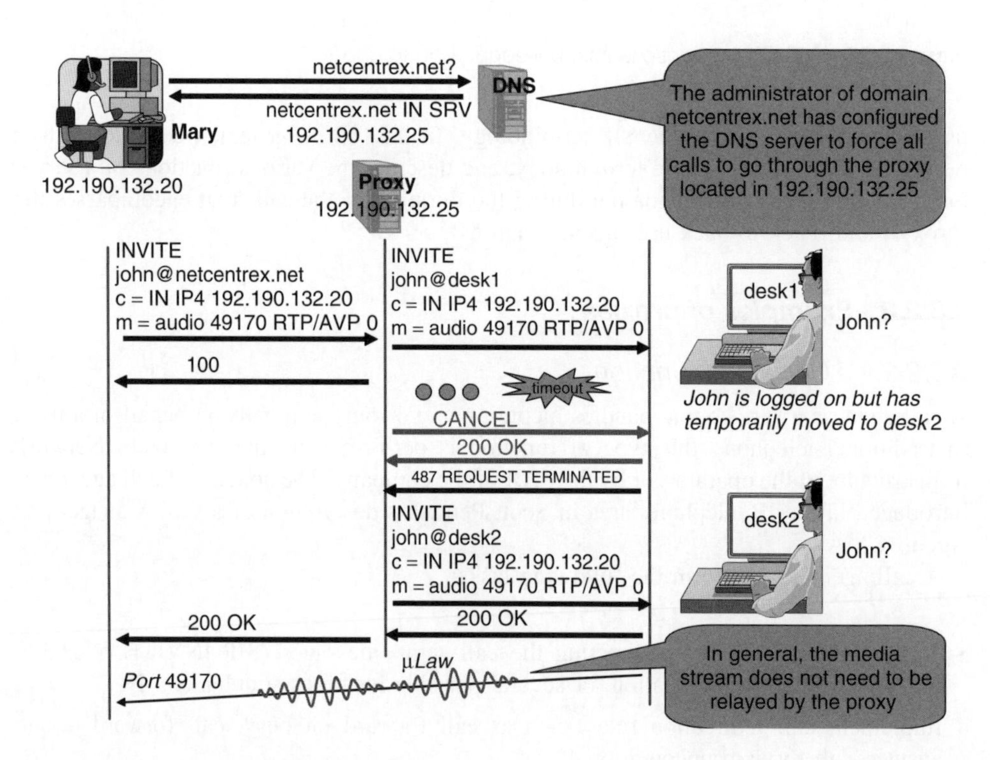


**Figure 3.26** Sequential forking example.

By accessing the database of a registrar, a SIP proxy can solve most user mobility/address change issues of the end user terminal. For instance, each time a user connects to the Internet via an ISP, he gets a new IP address. But if his SIP software registers this new IP address, the proxy will be able to relay all calls to the new IP address.

##### *3.3.2.2.1.2 Parallel forking*

Forking proxies can duplicate a request and send copies of it to several hosts, each with a specific 'branch' parameter. This is called parallel forking. Parallel forking proxies are not necessarily transparent to responses and may filter out some replies before forwarding them back to the client. Parallel forking proxies may cause some hosts to receive duplicates of the same request with the same call ID (but different branch parameters), but SIP clients must reply to each request.

Parallel forking proxies can simultaneously contact several endpoints belonging to the same person. Some manufacturers call this the 'simultaneous ringing' function. Although this call flow will work in a demo lab, it unfortunately cannot be implemented in a real telephony network because in a real network each INVITE can cause the network to send back a one-way announcement (using a 183 session progress response) which can be an error message (network busy), some information ('please type a pin-code', 'the cell phone you are calling is being located') or sometimes even advertising ('welcome to the X network'). The forking proxy has no way of merging multiple audio sources to provide feedback to the user in the unlikely, but possible, case of multiple in-band messages, and therefore can only be used if such a possibility does not exist, for instance, if all called numbers are private extensions directly controlled by the proxy.

There can be other potential applications of parallel forking proxies. A possible use is to handle `NOTIFY` messages when SIP is used for presence or alert applications (SUBSCRIBE/NOTIFY methods). The forking proxy is then acting as a concentration point for notify messages. A forking proxy can also be used at the edge of a VoIP network to try to send a call simultaneously to multiple VoIP peer networks, which are expected to reject the call or redirect it using a 302 moved response (This is identical to the LRQ-blast procedure used in H.323 networks).

### *3.3.2.3 Routing of requests and responses in SIP networks*

The routing of requests and responses is a key feature of SIP networks, and one of the areas where SIP introduced many innovations. The routing mechanisms have evolved from RFC 2543 to RFC 3261, and again in IMS. Each time, new possibilities were introduced to allow proxies to drop out of the path of session signaling as soon as possible. These 'short circuits' increase the scalability of SIP networks.

While the original SIP RFC had defined the 'short circuit' mechanisms, the evolutions of the protocol mainly consisted in adding new ways of controlling requests and responses paths.

#### *3.3.2.3.1 Routing of requests within a SIP dialog. Path optimization mechanisms in SIP: the Contact header*

SIP user-agents should[21] add a 'Contact' header to all SIP requests which initiate a dialog (e.g., INVITE, or SUBSCRIBE). This 'Contact' header indicates a URI where the user agent is ready to receive the responses to the request. When such a "Contact" header is present in a SIP request, the SIP proxies in the path of the request will be bypassed by the SIP responses: the destination user-agent will respond directly to the Contact URI of the Request.

Similarly, when responding to a SIP request establishing a dialog, a SIP server should add a "Contact" header including where the server wants to receive subsequent requests within the dialog. For instance, it may receive the ACK following an INVITE/200 OK sequence directly on this contact URI. Again, any intermediary proxy which forwarded the INVITE request may be bypassed by the ACK.

#### *3.3.2.3.2 Routing of responses to a SIP request, the Via header*

A request from A to B can be routed through several proxies. In many cases, it is desirable to force the response(s) to such a request to follow the same path as the request: for instance, a proxy might be billing the call or controlling a firewall and needs to have access to all the information regarding the call.

When a TCP connection is used for a SIP transaction, this is not generally an issue: the reply to a request automatically gets back to the other end of the TCP 'pipe' because TCP maintains a context throughout the connection.[22] On the other hand, when UDP is used, some information must be present in the request datagram in order to allow the receiver to know where to send the reply.

Since SIP is protocol independent, all SIP requests and replies contain "Via" headers for exactly this purpose. This also helps avoid routing loops (each proxy checks if it is already in the Via list). Each time a SIP proxy forwards a request, it appends its name to the list of forwarding proxies recorded in the Via headers. When a proxy forwards a reply, it reverses the process and removes its name from the list. Additional details on the use of Via headers can be found in Section 3.2.2.1.

If a proxy ('P') no longer needs to be in the path of SIP responses, it will not append its own 'Via' header to the list. When handling the response, the 'P+1' proxy (which was the next hop after proxy 'P' on the request path) will therefore use the 'Via' header inserted by proxy or user agent 'P-1', and forward the response directly to it. Proxy 'P' will be skipped.

When a proxy or user agent is connected to the next-hop Sip server via one or more NAPT IP routers (see Chapter 7), the URI indicated in the topmost "Via" header should not be used as a destination URI for the SIP response. Behind an NAPT function, com-

[21] RFC 3261 says that this is mandatory for all requests initiating a dialog. In practice many implementations omit the Contact header when not needed (considering that SIP is already verbose enough). Regarding the scope of the Contact URI, it should not be limited to the SIP dialog.

[22] For this reason many SIP stacks do not support Loose Routing with TCP transport... which makes the 'UDP fall back to TCP for large messages' strategy impossible.

munication is only possible if the proxy and user agent sends and receives SIP messages on the same IP address and port: the source IP address and port of the Requests it initiates must be used as destination IP address and port of the responses. In order to indicate that it supports sending and receiving on the same IP address and port and that it may be situated behind a NAPT router, a proxy or user agent may use the 'rport' extension defined by RFC 3581 ('Symmetric Response Routing'). A proxy that receives a message with a "Via" header comprising 'rport' should forward responses to the source IP address and port of the last SIP request sent by this user agent or proxy, ignoring the "Via" URI. Before propagating the SIP request, proxies should add to the "Via" line a 'received' parameter with the request apparent source address, and initialize the "rport" parameter to the apparent source port of the SIP request (see Figure 3.27). Since these parameters will be copied in responses, stateless proxies can implement symmetric routing without having to memorize the IP address and port information: all required information is present in the "Via" header line.

The mechanism introduced by RFC 3581 not only allows the client to receive SIP responses behind a NAT function but also enables SIP user agents to find out whether they are connected to the rest of the network via a NAT function: the user agent just needs to use the "rport" parameter in Register request, and examine the "rport" and "received" values in the SIP response.

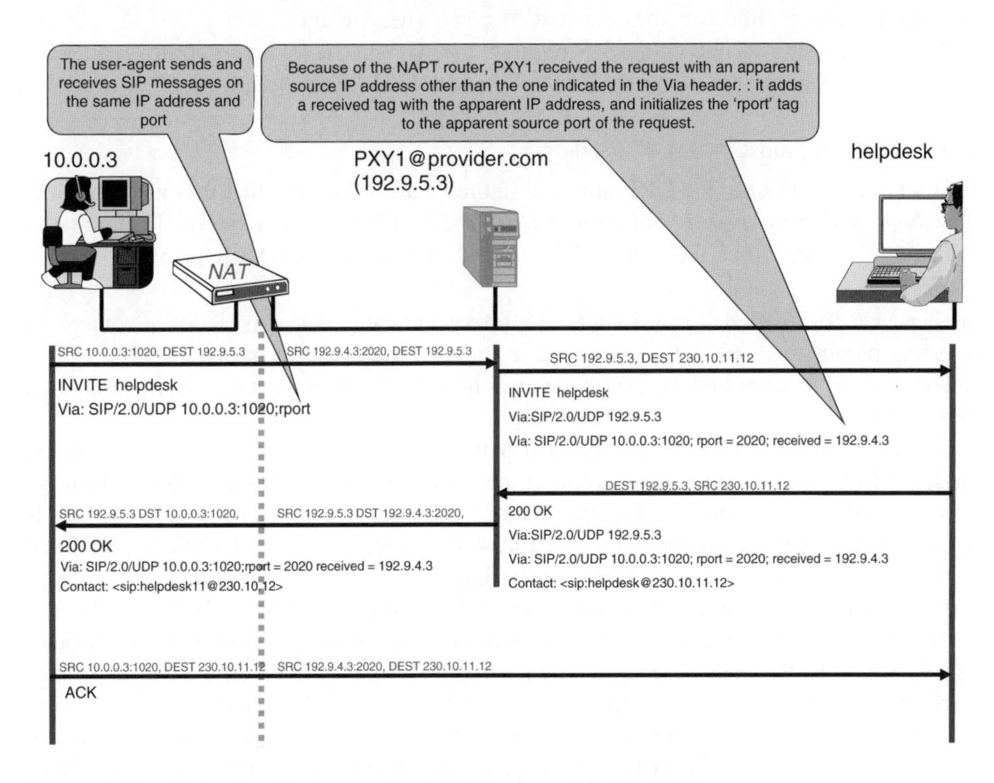


**Figure 3.27** Using the Via header for NAPT traversal (RFC 3581).

### 3.3.2.3.3 Record-route and route headers. Loose routing (RFC 3261), Strict routing (RFC 2543)

If not only the requests and the associated replies (transaction) but also all requests within a SIP dialog (e.g., ACK, NOTIFY) must be routed along the same path, the Via header is not sufficient and proxies must use the Record Route header. This is because SIP servers can add a Contact header field that enables SIP user agents to send them requests (e.g., BYE requests) directly, and therefore proxies are not guaranteed to be on the path of all requests in a SIP dialog.

When a proxy needs to remain in the path of all SIP requests within a SIP dialog, it adds a "`Record-route`" header including its SIP URI and an optional maddr parameter, in the first position of the list of "`Record-route`" headers (if any).

The SIP server that responds to a SIP request must copy all "`Record-route`" headers in the response. In addition, for all SIP requests that it will initiate as part of this SIP dialog, it must copy the content of these "`Record-route`" headers in as many "`Route`" headers.

The SIP client which initiated the SIP request creating the SIP dialog therefore receives, in the final SIP response, a copy of all "`Record-route`" headers: in all subsequent SIP requests within the SIP dialog, it must copy the content of these "`Record-route`" headers in as many "`Route`" headers, but inverting the order of these headers (the last "`Record-route`" line becomes the first "`Route`" header line).

When processing an incoming SIP request, the SIP proxies which comply with RFC 3261 should examine the topmost "`Route`" header URI (if any), and

- remove the topmost "`Route`" header;
- forward the request to the URI indicated in the "`Route`" header that was just removed, *without modifying the original request URI* (the first line of the request). If there is no "`Route`" header, then the request URI should be used as the next hop.

Such proxies are called 'loose routers'. A proxy indicates that it is a 'loose router' by inserting parameter '`lr`' in the '`Record-route`' headers that it inserts.

The use of `Route` and `Record-Route` headers by loose routers is illustrated in Figure 3.28.

Requests can be routed on a predefined path by using the Route header. The routing model of RFC 3261 is called 'loose routing' because it allows proxies to route the message to additional hops not indicated in the 'Route' list (Figure 3.29). The only constraint is that all proxies indicated in the 'Route' list must be visited before the request is forwarded to the target indicated in the original request URI.

This handling of 'Route' headers is new, and was introduced after RFC 2543 bis 05. In previous specifications, proxies had to route messages strictly according to the 'Route' list. In addition, *the original Request URI was overwritten and could not be recovered*. Such 'old' proxies are called 'strict routers'. This issue in the specification was fixed by 'loose routing', and a work-around strategy was specified to enable loose routers to prevent loss of information when a message is routed through an old 'strict router' (see Figure 3.29).

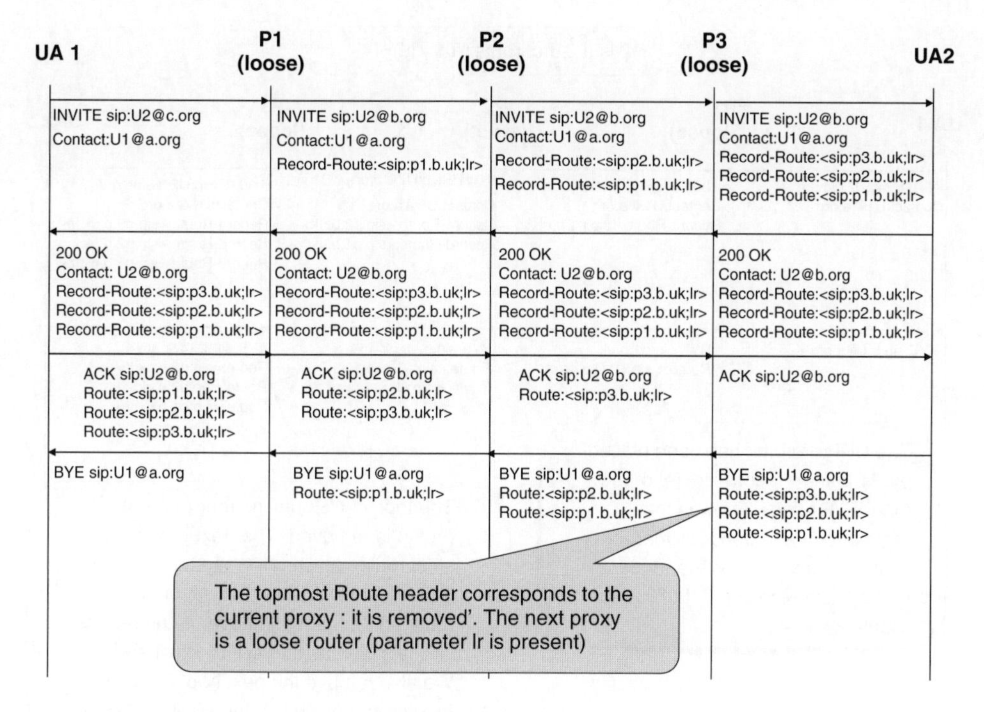


**Figure 3.28** Use of "Record-route" and "Route" headers by loose routers.

It is interesting to note that the "To" header normally contains the original destination of the request, but is not actually used by the request routing mechanism.

In RFC 3261, the behaviour of the last hop proxy is a special case: if the proxy knows that it is responsible for the last hop routing of inbound sessions to the user agent (e.g., it is responsible for the domain of the AoR of the user-agent), then it replaces the Request URI by the URI provided by the localization service (e.g., the URI registered by the user agent in the "Contact" header of a REGISTER request). This operation is called a 'retarget'. This special case has no real justification, and causes many issues because the URI containing the public identity is lost. 3GPP IMS had to introduce a new header, P-Called-Party (RFC 3455) as a workaround for this problem.

### 3.3.2.4 Routing imposed at registration time: Path and Service-Route headers

#### 3.3.2.4.1 Requests from the user agent

In many networks, e.g., in IMS networks, a service proxy ('Home proxy') is identified for the user agent during the registration procedure, with the requirement that all subsequent requests by the user agent will be routed to this service proxy.

Unfortunately, the "Contact" header of the REGISTER response is useless for this purpose because its scope is the REGISTER URI request, which is set to the registrar

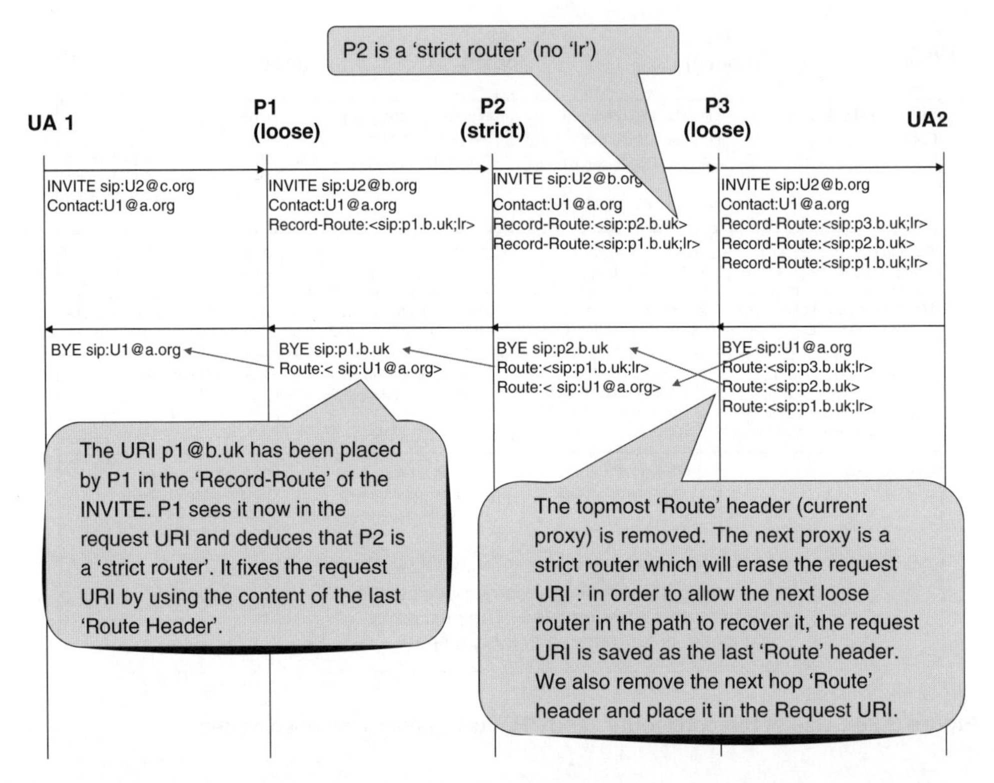


**Figure 3.29** Loose routing workaround for older 'strict' routers that erase original Request URI.

URI. We want *all* future request of this user agent, to any URI, to be routed through the service proxy.

A new header "`Service-Route`" was introduced by RFC 3608 ('Extension Header field for service route discovery during registration') in order to solve this problem. When it is present in a `REGISTER` response, then the URI specified by the '`Service-Route`' header must be copied by the user agent in a "`Route`" header for all subsequent requests, causing these request to be routed to the URI specified in the 'Service-Route' header.

### *3.3.2.4.2 Requests to the user-agent*

Any proxy situated in the path of the REGISTER message may indicate that it needs to remain in the path of all future requests sent towards the AoR of the user agent (the URI indicated in the "`To`" header of the `REGISTER` message).

In order to make this possible, a new header "`Path`" was introduced by RFC 3327 ('Path Extension Header Field for SIP'). Proxies which need to remain in the path of all future requests sent towards the AoR of the user agent insert a `Path` header entry containing their URI in the topmost position (if there are other Path entries).

The registrar will associate this ordered list of URIs to the user agent AoR and store them. It will also copy all "Route" headers in its response. In all networks where the 'Home proxy' is responsible for routing all requests towards the user agent it manages (e.g., an IMS S-CSCF, see Chapter 4) and has access to the registrar database, this stored path of URIs is inserted in any incoming request forwarded to the user agent by the 'Home proxy', as 'Route' headers. This ensures that these requests are routed through the proxy URIs that were listed in the path headers.

#### 3.3.2.5 Loops and spirals

SIP implements detection of loops via two mechanisms:

- Max-forwards: this feature is mandatory in RFC 3261. The Max-forwards header contained in every request is decremented at each hop (requests are initially sent with a default value of 70). If this counter reaches 0, the request should be rejected with a 483 'Too Many Hops' response.
- Loop detection: proxies can detect that they have already processed a request by analyzing the Via header list. If the Request-URI, From or To header fields have changed, this is not a loop: it only indicates that the call has been processed by an application which changed header elements that influence routing, for instance the destination, and that the new destination is also routed by this proxy. This normal situation is called a 'spiral'. On the other hand, if the headers listed above have not changed, this is a loop and the request should be rejected with a 482 'Loop Detected' response.

Note that the loop detection mechanism is more complex but detects loops sooner than the Max-forwards mechanism.

Both mechanisms do not prevent loops involving a segment of the PSTN.

#### 3.3.2.6 Billing for SIP calls

By definition, all participants invited by a common source for a given session are in the same SIP 'call'. This call is identified by a globally unique call ID. Within a call, each leg can be identified by a unique combination of the To, From, and Call ID fields. A proxy performing the call accounting function should be able to distinguish different legs and create a CDR for each call leg. It should also be able to recognize re-Invite messages that change only the media description (see 3.2.2.7.2.3), not the participants, and in this case it should not create a new call leg.

In the PSTN, a call is usually paid by the person who initiates it. A proxy relaying all signaling from the terminal of a user can create appropriate accounting records by logging the `INVITE` (re-Invite are ignored), `CANCEL`, and `BYE` requests, as well as the replies (Figure 3.30). The duration of each leg can be derived from the first accepted `INVITE` request (`200 OK`) up to the first `BYE` request.

One way to force the user to go through the proxy to make calls is to control a firewall in the network from the proxy, as illustrated in Figure 3.30. This prevents the user from

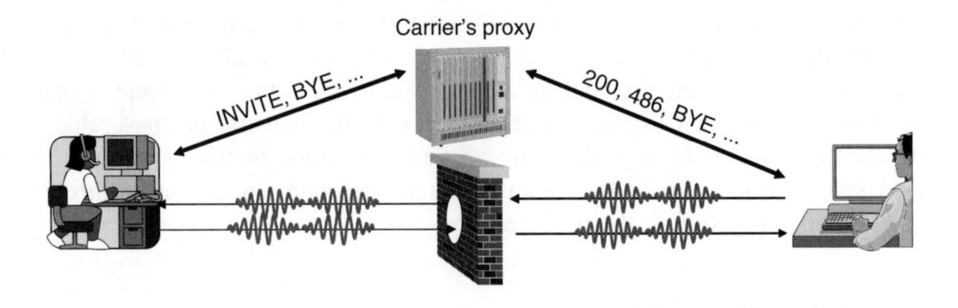


| Call-ID | From | To | Operation | Time Stamp | |
|---|---|---|---|---|---|
| 4321@192.190.12.32 | Mary@domain1 | John@domain2 | INVITE | 11/11/1998 | 11:11:11 |
| 4321@192.190.12.32 | Mary@domain1 | John@domain2 | 200 | 11/11/1998 | 11:11:30 |
| 4321@192.190.12.32 | Mary@domain1 | John@domain2 | BYE | 11/11/1998 | 11:21:12 |
| 4321@192.190.12.32 | Mary@domain1 | Mark@domain3 | INVITE | 11/11/1998 | 11:14:13 |
| 4321@192.190.12.32 | Mary@domain1 | Mark@domain3 | 200 | 11/11/1998 | 11:14:40 |
| 4321@192:190:12:32 | Mary@domain1 | Mark@domain3 | BYE | 11/11/1998 | 11:21:13 |
| 4321@192.190.12.32 | Mary@domain1 | July@domain3 | INVITE | 11/11/1998 | 11:14:13 |
| 4321@192.190.12.32 | Mary@domain1 | July@domain3 | 486 | 11/11/1998 | 11:14:40 |
| 4444@192.190.12.32 | Mary@domain1 | Dilan@domain4 | INVITE | 11/11/1998 | 15:25:25 |
| 4444@192.190.12.32 | Mary@domain1 | Dilan@domain4 | 200 | 11/11/1998 | 15:25:40 |
| 4444@192:190:12:32 | Mary@domain1 | Dilan@domain4 | BYE | 11/11/1998 | 15:31:43 |

**Figure 3.30** A proxy could control a firewall to ensure that all communications are logged.

trying to bypass the call accounting feature of the proxy. In the PacketCable® architecture for cable networks, the call management proxy dynamically sets 'gates' on the cable end CMTS (a sort of router with cable specific features) for media channels using reserved quality of service.

In reality many networks do not need this, as all VoIP devices in the network are configured to accept calls only if the INVITE comes from the service provider proxy (this can be done by simple access control lists—ACLs restricting SIP signaling traffic on the routers connected to these resources). This way, if a user tries to bypass the network proxy, it will not be able to establish a call to PSTN gateways, for instance. Without some form of dynamic firewall control, direct VoIP user-to-user calls on the IP network will be allowed on a best effort basis (if user devices, for instance softphones, are not under control of the service provider). This is usually not a problem, as there are virtually countless number of ways to communicate without control in best effort mode.

## 3.3.3 Some common services

### *3.3.3.1 Call transfer*

In a business environment, telephony networks must implement many types of call transfer (e.g., supervised call transfer, blind call transfer) as well as many related services (e.g., notification of redirections).

The REFER method, introduced in RFC 3515 ('The SIP Refer Method'), allows the user agent which initiates the transfer ('the transferor') to instruct another user agent ('transferee') to initiate an INVITE request towards a new target.

Unfortunately, RFC 3515 does not define all the use cases of the REFER method, and in practice the requirements of a telephony network cannot be implemented unambiguously with only that specification. draft-ietf-sipping-cc-transfer-12, a work that has been ongoing for over 7 years now, aims at providing a comprehensive study of all common use cases and introduces new headers that were necessary to fully cover the requirements for call transfer, such as 'Replaces' (RFC 3891), 'Referred-By' (RFC 3892), or 'Target-Dialog' (RFC 4538).

A simple case of blind transfer is presented in Figure 3.31, illustrating the use of REFER, SUBSCRIBE, and NOTIFY methods to implement the required features.

The REFER request and the NOTIFY requests which follow may be part of the existing dialog established by the initial INVITE (this is the case of most implementations), or use a separate dialog as in Figure 3.31 (the messages part of the new dialog initiated by the REFER appear in bold). The reuse of an existing dialog may present some drawbacks (see draft-sparks-sipping-dialogusage for an overview), but creating a separate dialog is also problematic: the new dialog could be routed to another user agent (multiple user agents can register under the same AoR!). Some of these issues are solved by

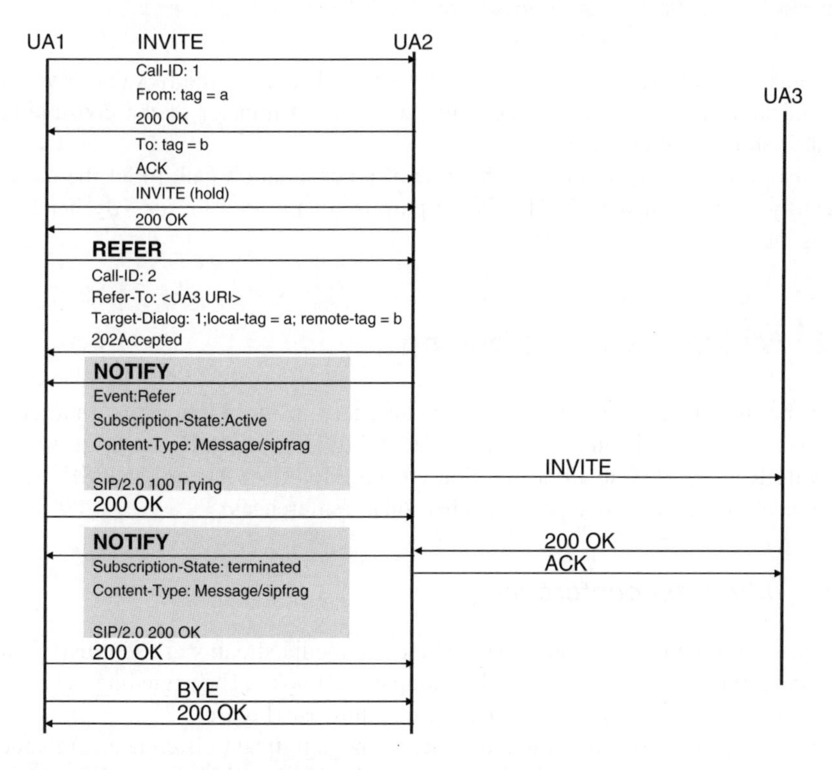


**Figure 3.31** SIP based blind call transfer.

associating the REFER dialog to the original INVITE dialog, by means of the new header Target-Dialog.

In Figure 3.31, note how UA1 uses a Re-INVITE (see Section 3.2.2.7.2.3) with a sendonly or inactive SDP parameter to put the call on hold before sending the REFER to the 'transferee' UA2. This releases media resources which UA2 may need to initiate a new session towards UA3.

UA1 remains informed of the progress of the call transfer by means of NOTIFY requests (light grey on Figure 3.31): the REFER method implicitly subscribes to the session state events of the new session towards UA3. When the 'transferor' has received confirmation of the success of all transfer towards the new target, it terminates its session with UA2. If the new session towards UA3 had failed, UA1 would have been informed in a NOTIFY request, and could have reactivated the media streams of its pending session with UA2.

The call flow discussed above was a call transfer without a consultation call. In the case of a consultation call, UA1 would have established a session with UA3 before transferring the call. This introduces new issues, as UA3 may not have enough resources to maintain two sessions simultaneously, and in any case it should not ring twice. In order to solve this problem, the REFER method would include in this case a "Replaces" header, encapsulated in the Refer-To header:

```
REFER
Refer-To: sip:<UA3 URI>?Replaces=<CallID of consultation
call>, to-tag=...;from-tag=...
```

This syntax forces UA2 to also include a 'Replaces' header in the INVITE it will send to UA3, indicating to UA3 that the incoming session is a transfer of the consultation call, and that it should not ring again.

It is also important, especially in the case of a consultation call, to let the target know the identity of the 'transferor'. This is the purpose of the Referred-By header, which contains the URI of the 'transferor'.

## 3.3.4 Multiparty conferencing

SIP can be used to establish multipoint conferences in multi-unicast or multicast mode (remember this protocol came from the MMUSIC IETF group!). However, it was not until 2006, with RFC 4579 (Call Control - Conferencing for User Agents) that SIP defined the floor control primitives required for professional conferences.

### *3.3.4.1 Multicast conferencing*

A multicast conference is a conference where the media streams are sent using multicast (for more details on multicast, see the companion book). The signaling related to this conference can be sent using multi-unicast or multicast (Figure 3.32).

In the case of multi-unicast signaling, there is no significant difference compared to the point-to-point case, except that the SDP session descriptions indicate multicast addresses,

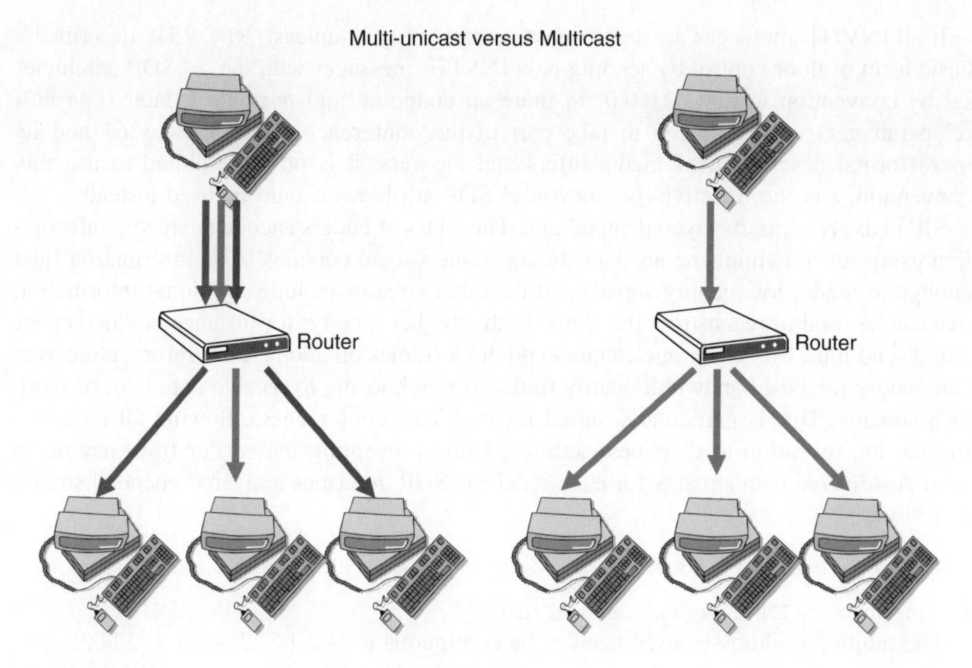


**Figure 3.32** Multi-unicast and multicast signaling.

and the offer/answer model is also a bit modified compared to the unicast media case (see Section 3.2.2.7.2.4 for details).

When multicast signaling is used to establish multiparty conferences, the SIP requests are carried using UDP, since this is the only transport protocol that can be multicast over IP. Multicast requests are expected to be used mostly to set up conference calls, and therefore the destination URL will generally be a conference name rather than an individual's. However, the theory also allows using a multicast request with the URL of an individual, for instance, for multicast searches. The replies to a SIP request are then sent back to the sending UDP port on the same multicast address. In order to reduce network traffic and avoid a possible storm of synchronized replies, there are some modifications compared to the multi-unicast invitation procedure, including the following:

- 2xx are not sent.
- 6xx replies are sent only if the destination URL matches the name of a user on the host (i.e., the request is a multicast search rather than an invitation to a multiparty conference).
- Replies are sent after a 0–1 second random delay.

This form of multicast signaling was described in the first SIP RFC, but is not recommended in RFC 3261. It works in simple cases, but becomes very complex to manage if the full generality of SIP call flows is considered. Therefore, it seems SIP is headed more towards the use of multi-unicast to control multicast media sessions.

If all INVITE messages are sent from a central entity in unicast, RFC 2543 described a basic form of floor control by sending new INVITE messages with the 'c' SDP parameter set by convention to null '0.0.0.0' to mute an endpoint, and re-invite it later (non null 'c' parameter) when allowed to take part in the conference. Since RFC 3261 and its more formal description of media offers and answers, it is now prohibited to use this convention, and the 'inactive' or 'recvonly' SDP attributes should be used instead.

SIP natively supports layered encodings. This class of coders encode the media information using several simultaneous data streams. One stream contains basic information (just enough to render low quality signal) and the other streams include additional information that can be used to reconstruct the signal with a higher quality: for instance, a video coder could send intra frames on one channel and delta frames on another. Therefore, a receiver can choose the best bandwidth/quality trade-off by choosing to receive one, two, or more data streams. This is particularly suited for multicast conferences, allowing all receivers to tune the reception to their best settings, while preventing the sender from having to send customized data streams for each receiver. SDP describes a layered encoded stream as follows:

```
c=<base multicast address>/<ttl>/<number of addresses>
```

For instance: `c=IN IP4 224.2.1.1/127/3`

The multicast addresses used need to be contiguous (224.2.1.1, 224.2.1.2, 224.2.1.3).

Unfortunately, there is no known commercial implementation yet using this facility.

### 3.3.4.2 Multi-unicast conferencing

The support of SIP for multi-unicast media conferences is limited in RFC 3261. A central entity can be set up to act as an MCU to either mix or switch the incoming media streams. The central bridge could implement very simple floor control by using re-Invites with the "`inactive`", "`recvonly`", or "`sendrecv`" SDP attributes. In practice, this is sufficient as most conferencing services use external, application level, user interfaces for floor control, and require the VoIP protocol only to implement basic mute/active/redirect functions, which can be readily provided by SIP and SDP.

However, SIP is still lacking some messages for full support of video transmission control. An example is the request for full frames, present in H.245. Most video coders, e.g., H.261 or H.263, send full frames only from time to time, and deltas in between. Most of the time, the instant at which a participant decides to speak will not coincide with the sending of a full frame. Therefore, if the MCU simply copies the incoming video stream to the output stream, the receivers will have to wait for the next full frame to get an image (Figure 3.33). So the MCU needs to completely recode the stream in order to be able to send a full frame when the video switches. In a similar case, H.323 can mute the video stream of non-active speakers, and request a full frame when it switches to an active speaker (VideoFastUpdate message). Such a message is also useful to quickly recover from video packet loss. Some RTCP messages (Full Intra Request) or new SIP

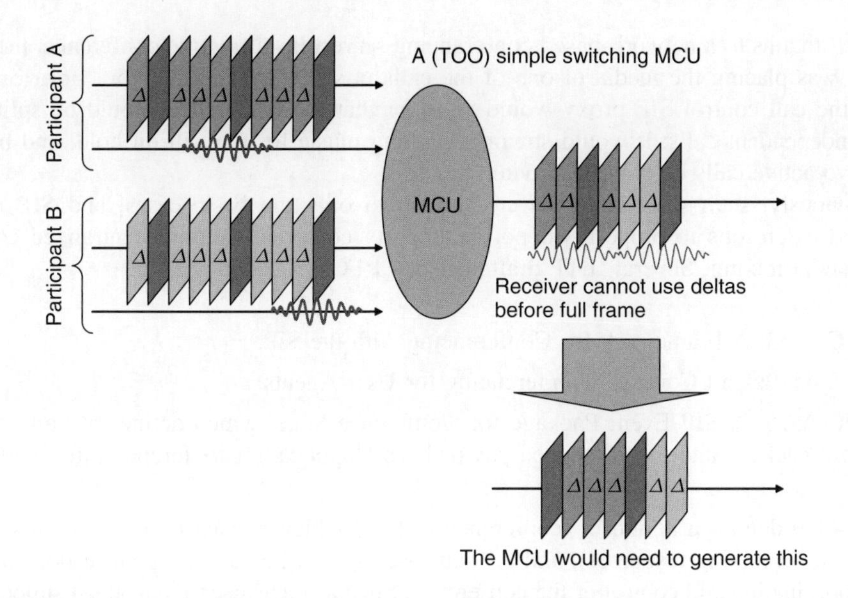


**Figure 3.33** Full intra requests are helpful in all situations where the video source changes.

messages could be used for the same purpose, but video control is still not documented enough to allow for seamless high-quality interoperability across vendors.[23]

### 3.3.4.3 Ad hoc conferencing, RFCs 4353, 4579, 4575

An ad hoc conference starts from a point-to-point session, and dynamically turns it into a multiparty conference. Most SIP phones support this feature locally (three-way conferencing), by establishing two or more point-to-point sessions and mixing all media streams internally (at least for low complexity vocoders such as G711). In fact, it is still difficult to implement network based ad hoc conferencing in the network as SIP because no standard conference activation message has been defined in RFC3261 to instruct the call control SIP proxy to perform the required session changes. This was a significant obstacle for the deployment of SIP endpoints at the edge of public networks, because phone embedded three-way functions are usually limited (G.711 only in most cases) and use twice the bandwidth of normal calls.

While some vendors turned to proprietary extensions for network based ad hoc conferencing, other vendors used a simple workaround: a user agent handling multiple sessions in which SDP offer/answer state is `sendrcv` (see Section 3.2.2.7.2.3) is assumed to implicitly want to conference the sessions. The call control SIP proxy can then redirect the

[23] The most advanced initiative is the IETF work in progress draft-levin-mmusic-xml-schema-media-control-13 'XML Schema for Media Control'.

media streams to a network-based conferencing server. If, during the conference, the user agent was placing the media of one of the calls in state "inactive" or "sendonly", then the call control SIP proxy would consider that the conference should be split into two independent calls: it would stream a waiting music for the call on hold, and bridge the two active calls by using re-Invite requests.

Obviously, such workarounds were limited to only the basic cases, and SIP really needed extensions to properly support multiparty conferencing with centralized control and media mixing. Several IETF drafts became RFCs in 2006:

- RFC 4353: A Framework for Conferencing with the SIP.
- RFC 4579: Call Control—Conferencing for User Agents.
- RFC 4575: A SIP Event Package for Conference State, which defines a 'conference' event package and the associated payload type 'application/conference-info+xml'.

RFC 4579 defines a notion of 'conference factory' which manages conference resources (e.g., media mixing resources), and a 'focus user agent', i.e., a user agent responsible for the coordination and control of the conference. Such a 'focus user agent' must support the event package 'conference' (see Section 3.5). This package name must be present in the Allow-Events header of the SIP requests and responses it sends. The 'focus user agent' uses header 'Replaces' to establish ad hoc conferences.

Figure 3.34 and Figure 3.35 describe a call scenario where focus user agent UA1, under the control of a SIP proxy used as an application server, initiates an ad hoc conference.

The first step is to send a INVITE request to an 'ad hoc conference factory' server CF. The URI of this server must be preconfigured in the user agent.

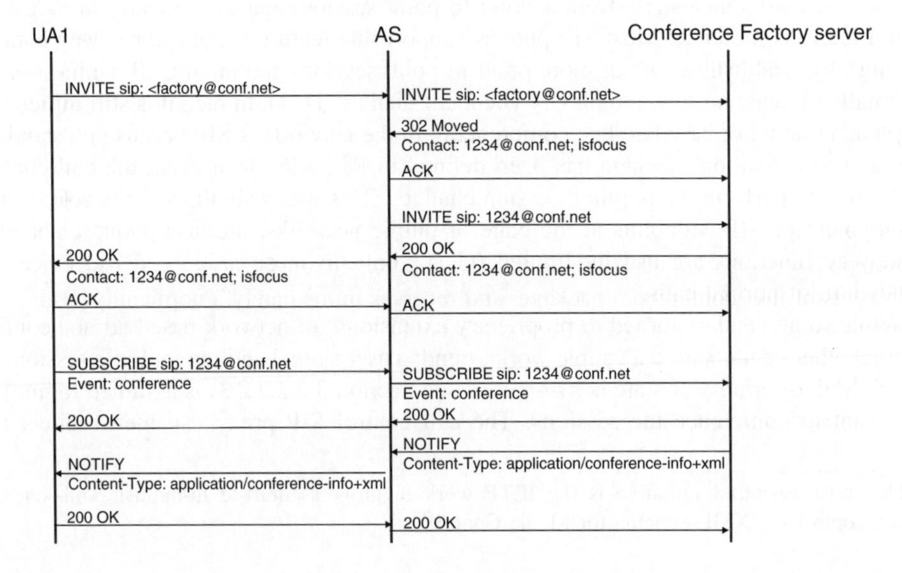


**Figure 3.34** Initializing an ad hoc conference bridge.

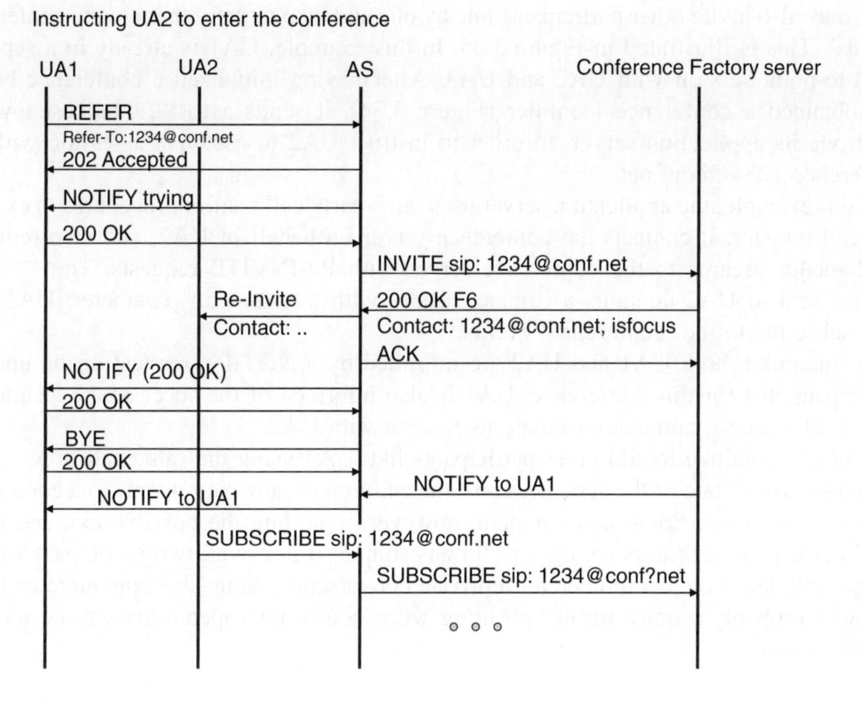


**Figure 3.35** Instructing user agent UA2 to enter the conference.

CF allocates conferencing resources as needed, and assigns a conference identifier which is sent back towards the initiator of the conference in the Contact header of a `302 moved` response. In our example, this redirection is handled directly by the application server, which only returns the final `200 OK` response to UA1. As the Contact header of the response contains a `isFocus` parameter, the user agent understands that this is the Focus of the conference, i.e., the SIP server that will synchronize all signaling activities related to this tightly coupled conference. This mechanism conforms to RFC 3840 (Indicating User Agent Capabilities in SIP). Note that the URI indicated by the Contact header could have been a different server, in the case of a distributed implementation of the Conference factory function.

A user agent supporting this more advanced conferencing profile must subscribe to the events of the 'conference' package, and receives immediately a notification listing all current participants and their status. Initially, only UA1 is present in the list, but the list will be updated each time a new participant enters the conference or if the status of a participant changes.

Now that conference bridge resources have been allocated, UA1 can instruct the 'Focus' to invite other participants in the conference. To this purpose it uses a specific `REFER` request in which the `Refer-To` header indicates the URI of the participant being invited:

```
:
REFER sip:<conferenceID>
Refer-To: <URI of the new participant>
```

UA1 may also invite other participants one by one and then transfer them to the conference 'Focus'. This is illustrated in Figure 3.35. In this example, UA1 is already in a separate point-to-point session with UA2 and UA3. After having initialized a conference bridge and obtained a conference identifier (Figure 3.34), it sends a `REFER` request towards UA2, via its application server, in order to instruct UA2 to establish a session with the conference 1234@conf.net.

In our example, the application server uses third-party call control procedures to execute this call transfer: it contacts the conference server on behalf of UA2, and then redirects UA2 media streams to the conference server with Re-INVITE requests. The `INVITE` request sent to UA2 contains a Contact header with a `isFocus` parameter: UA2 may then subscribe to the 'conference' events.

At this point, both UA1 and UA2 are informed by a NOTIFY request of the updated participants list for this conference. UA1 is also informed of the successful execution of its `REFER` request, and can terminate its session with UA2.

UA1 can continue to add other participants like UA3 using the same call flow.

In the current state of the specifications, the addition of new participants to a conference is clearly specified. Some holes remain, however, regarding the possible exit scenarios. If all participants exit at once, the call flow is simple. If however, two participant wish to temporarily leave the conference for a private conversation, things become more complex and will probably require further profiling work before interoperability can be ensured across vendors.

## 3.4 SIP SECURITY

### 3.4.1 Media security

`Media encryption` is specified by SDP. The k parameter of SDP stores the security algorithm in use as well as the key. The following formats are defined in `RFC 2327` (SDP):

- `k=clear:<encryption key>`. This format refers to the encryption algorithms described in RFC 1890 ("RTP Profile for Audio and Video Conferences with Minimal Control", January 1996). RFC 1890 first describes how to extract a key from a pass phrase in a standard way. The pass phrase is put in a canonical form (leading and trailing white spaces removed, characters put to lowercase, etc.), then hashed into 16 octets by the MD5 algorithm. Keys shorter than 128 bits are formed by truncating the MD5 digest.
  The name of the algorithm in use is concatenated before the key, separated from the key using a single slash. Standard identifiers for the most common algorithms can be found in RFC1423: "DES-CBC", "DES-ECB",... the default is DES-CBC. RFC 1423 also describes how to store additional parameters needed for the particular algorithms, such as the 64 bit initialization vector of DES-CBC. For example, this line can be used to initiate a DES-CBC encrypted session:
  `k=clear:DES-CBC/aZ25rYg7/12eR5t6y`

- `k=base64:<encoded encryption key>`. The format is the same as above, but base64 encoded to hide characters not allowed by SDP.
- `k=prompt`. Prompt the user for a key. The default algorithm is DES-CBC.

## 3.4.2 Message exchange security

### 3.4.2.1 Authentication

Most vendors support the mechanisms defined by RFC 2617 (HTTP Authentication: Basic and Digest Access Authentication) for basic authentication (clear password) or digest authentication (hashcode derived from the message content and a challenge sent by the server or proxy).

User agents, Registrars or Redirect servers should use response code 401 to indicate that they cannot accept the request without further authentication information. Proxys should use response code 407.

#### 3.4.2.1.1 Basic authentication

The use of basic authentication has been deprecated in RFC 3261, as the password is sent in clear form over the network. An RFC 2543 server user agent willing to authenticate a client user agent using the 'basic' method will respond with the following header:

```
WWW-Authenticate: Basic realm="realm_information_here"
```

The realm is simply a context information that should be presented to the user in order to allow him to select the proper username and password. SIP requires that it be globally unique.

The client must re-issue the request, sending back the user ID and password, separated by a single colon and encoded as a base64 string, using the Authorization header.

```
Authorization: Basic QWxhZGRpbjpvcGVuIHNlc2FtZQ==
```

Note that since the ACK accepts no response, any authentication information that was accepted for an INVITE must be accepted also for the corresponding ACK (same Authorization and Proxy-Authorization headers). This is true for all authentication methods.

#### 3.4.2.1.2 HTTP Digest for user agent authentication, registrars and redirect servers

In order to avoid sending the password in clear form, many user agents support the HTTP digest method. By default, the `Authorization` header field contains the MD5 digest of:

- the Request URI;
- the user name;
- the nonce value;
- optionally, the message body.

The use of the 'digest method' is also specified in the `WWW-authenticate` header of the `401 Unauthorized` response.

```
WWW-Authenticate = Digest realm="realm_information_here",
qop="auth,auth-int",
nonce="0d1128f1806872deac4e01029b7c96b3",
stale=FALSE, algo-
rithm=MD5, opaque="5ccc069c403ebaf9f0171e9517f40e41"
```

The nonce should be generated randomly for each 401 response, and must not contain any double quote. If an 'opaque' string is included by the server, it should be passed back by the client in the response. The stale flag, when set, indicates that the previous authentication data was correct, but was rejected because the nonce information was stale. The quality of protection parameter indicates that the server supports authentication only (auth) and authentication with integrity protection (auth-int), in which case the message body can be included in the hash value calculation.

The client then re-issues the request (keeping the same Call-ID), including the requested authentication information in the Authorization header. The MD5 hash value is in the response parameter.

```
Authorization = Digest username="81@realm_information_here",
realm="realm_information_here",
                        nonce="0d1128f1806872deac4e01029b7c96b3",
                        uri="destination.test.org",
                        qop=auth,
                        response="2923fb70ddfdf57f7ffe5cc436ab4889"
                        opaque="5ccc069c403ebaf9f0171e9517f40e41"
                        algorithm=MD5
```

In response to the resubmitted request, the server can provide some feedback regarding the successful authentication in the Authentication-Info header. In addition, it may provide a new nonce parameter (nextnonce parameter), and may even include a hash value proving it also knows the client secret (rspauth parameter).

Authentication-Info: nextnonce="47364c23432d2e131a5fb210812c",
rspauth="29364c52832d2e131a545211212c"

Note that another authentication method based on PGP, now deprecated, was defined in RFC 2543. This semantic allowed to exclude some variable fields (such as the Via field) from the signed data. Figure 3.36 shows that the PGP signature could protect both the clear part and the encrypted part of the SIP message.

#### *3.4.2.1.3 HTTP digest for proxy servers*

A proxy may decide to authenticate a request by using the `407 Proxy Authentication Required` response, which contains a `Proxy-Authenticate` header containing a challenge. The client must then resend the request with a Proxy-Authorization header providing the credentials matching the challenge. The content of these headers is similar to that of WWW-Authenticate and Authorization.

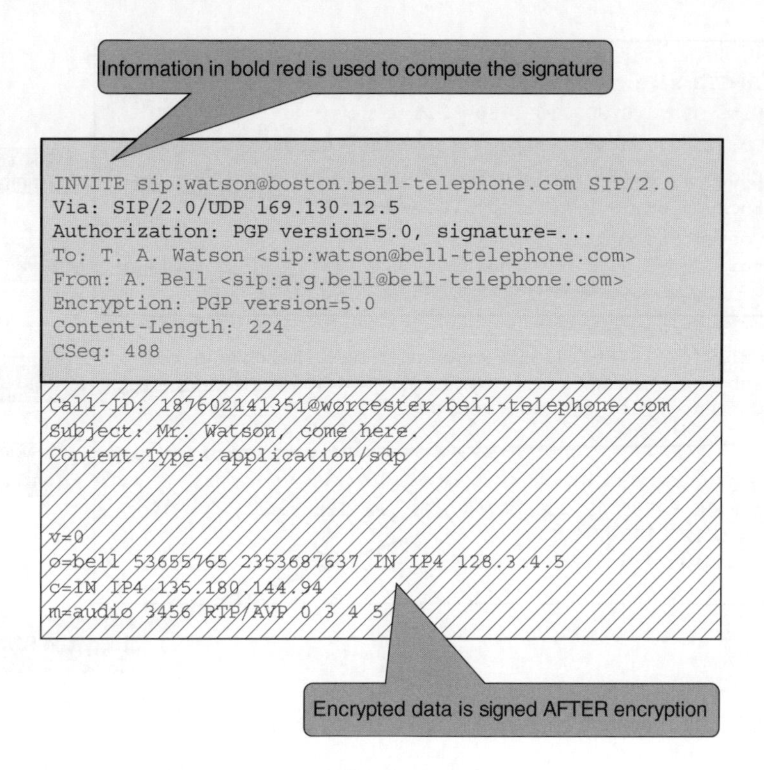

Information in bold red is used to compute the signature
INVITE sip:watson@boston.bell-telephone.com SIP/2.0
Via: SIP/2.0/UDP 169.130.12.5
Authorization: PGP version=5.0, signature=...
To: T. A. Watson <sip:watson@bell-telephone.com>
From: A. Bell <sip:a.g.bell@bell-telephone.com>
Encryption: PGP version=5.0
Content-Length: 224
CSeq: 488
Call-ID: 187602141351@worcester.bell-telephone.com
Subject: Mr. Watson, come here.
Content-Type: application/sdp
v=0
o=bell 53655765 2353687637 IN IP4 128.3.4.5
c=IN IP4 135.180.144.94
m=audio 3456 RTP/AVP 0 3 4 5
Encrypted data is signed AFTER encryption


**Figure 3.36** Scope of PGP signature in RFC 2543.

In order to avoid this round trip, the client can of course provide the credentials in the first message, if the authentication mechanism replay protection mechanism allows it.

On subsequent responses, the server sends a `Proxy-Authentication-Info` header, with the same parameters as those of the Authentication-Info header field.

Proxies must be completely transparent to the WWW-Authenticate, Authentication-Info and Authorization headers, and must forward them without any change.

#### *3.4.2.2 Encryption of messages*

If the media encryption key must be protected, then the SDP requests and replies must be encrypted. There are many other reasons for protecting the SIP messages, for instance, to hide the origin or destination of calls and the related information fields (Subject,...). In general, however, SIP messages only need to be authenticated, which is useful not only to prevent call spoofing but also for accounting and billing.

SIP messages can be encrypted hop by hop, for instance, using IPSEC. They can also be transported over a secure transport layer such as TLS (in this case 'sips:' URI are used). SIP also describes an end-to-end encryption strategy based on a shared secret key between the sender and the receiver or on a public key mechanism. If a common secret key is used, then the receiver of the message is able to decrypt a message encrypted by the

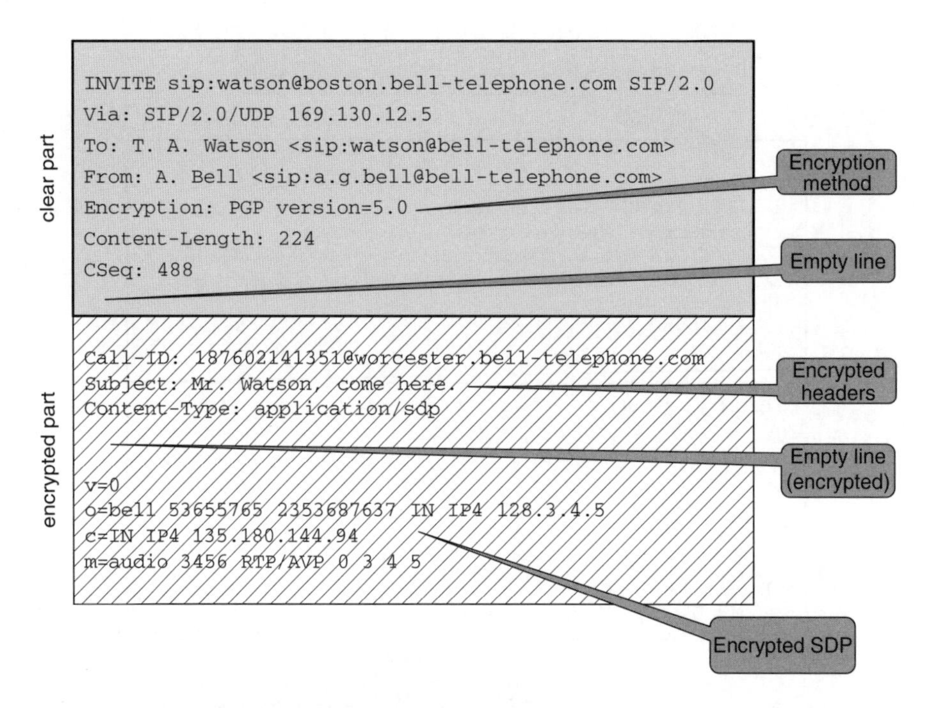


**Figure 3.37** SIP message encryption in RFC 2543, using PGP.

sender by using the shared password. If a public key scheme is used, the sender encrypts the message using the public key of the receiver. This encryption can be performed by the sender of the request or by an intermediary security proxy.

RFC 2543 also defined an encryption mechanism based on PGP, which has been deprecated. The request line and unencrypted headers were sent first, followed by an 'Encryption:' header field, which indicates the encryption method in use, for instance:

Encryption: PGP version = 2.6.2, encoding = ascii

The encrypted part began after the first empty line (CRLF of the previous line immediately followed by CRLF). The example of Figure 3.37 is taken from RFC 2543.

If only the message body had to be encrypted, an extra empty line had to be inserted in the body before encryption to prevent the receiver from mixing up the message body data with encrypted headers. There were specific issues with the Via header, since it is used by proxies to route the request back to the source.

## 3.5 INSTANT MESSAGING (IM) AND PRESENCE

Instant Messaging (IM), and more generally 'Presence' applications, are only the most popular applications of a more general use of SIP for the subscription and exchange of stateless event messages. These capabilities which can be used in any SIP or IMS network, will likely be the backbone of many future innovative applications, e.g., machine-to-machine communications.

IM and presence capabilities have been added to SIP by RFC 3265 (SIP- Specific Event Notification) in a very general way that was not particularly targeted at IM. In fact, it was first used by some VoIP vendors to implement out-of-band DTMF transmission during a call (as described in Section 3.2.2.7.1.3). With Windows XP, Microsoft introduced a new version of its Messenger® client, with included support for voice and video, but also instant messages. The client works primarily with proprietary protocols using a specific Microsoft central server as a message reflector, but can be configured to use SIP as well. Microsoft chose to use the mechanism defined in RFC 3265 for the subscription of presence information and notification of state information. This instantly made this extension of SIP a 'de facto' standard.

For users strongly in favour of a convergence of unified messaging protocols, the use of RFC 3265 was a step in the right direction, but still did not define the exact message content. The convergence of message contents is a more difficult step, as it may depend on the capabilities of each client, and has enormous implications for the large IM systems that currently 'own' their users. The convergence of the IM format is the main task of the IMPP working group of the IETF.

## 3.5.1 Common profile for instant messaging (CPIM)

This specification of the IMPP (Instant Messaging and Presence Protocol) working group of the IETF defines a number of operations and features to be supported by IM systems. The profile aims at facilitating the interworking between various IM systems, by providing an intermediary canonical format which facilitates the design of transcoding gateways. Obviously, this format can also be used to format instant messages, not just as a conversion intermediary format. Today, two popular IM protocols follow the CPIM guidelines: XMPP (eXtensible Messaging and Presence Protocol) used by the Jabber IM client, and SIMPLE (SIP for Instant Messaging and Presence Leveraging Extensions).

### *3.5.1.1 Common presence and instant messaging message format*

The CPIM canonical message format specification is described in RFC 3862 (Common Presence and Instant Messaging : message format). RFC 3862 defines a new MIME format 'message/CPIM', intended to be a common format for CPIM compliant messaging protocols. MIME, the Multipurpose Internet Mail Extensions, are defined in RFC 2045, 2046, and 2048.

One of the key reasons to encourage instant message systems to support the CPIM message format natively is to allow a gateway between two IM systems to preserve the electronic signatures that can be added to a CPIM message. Signatures are lost if any transcoding has to occur.

Although the defined format complies with MIME, it does not allow for all the options of MIME. This simplification aims at suppressing or restricting all the options of MIME that can present an obstacle for interoperability or the verification of electronic signatures (e.g., suppression or addition of headers, extensibility of header formats, weak internationalization, etc.).

One of the key requirements for an IM system is to be able to support all character sets. CPIM uses the UTF-8 encoding.

###### 3.5.1.1.1 The Universal Character Set and the UTF-8 Format

Computer science discovered very late that the world was not using only the well known US-ASCII characters, which can be encoded on only 7 bits. As a consequence, many operating systems could only handle 8 bit character sets, which led to a system where each language required its own code page (Latin-1, Hebrew, Arabic, Greek) encoded on 8 bits, and where a given character could have multiple encodings depending on the character page. For instance, the Euro € sign is coded as 0xA4 in Latin-9 (ISO 8859-15), 0x80 in Latin-2 (CP1250), and 0x88 in Cyrillic (CP1251). Any program that needed to use characters from multiple pages simultaneously had to have a very cumbersome design....

The Universal Character Set, defined by ISO/IEC 10646-1 contains in a single character set all (almost all) the symbols used by all known writing systems on earth. This is a multi-octet character set: UCS-2 contains the first 64.000 characters and is encoded on two octets; it is also called the Basic Multilingual Plane (BMP); UCS-4 is encoded on four octets and can contain potentially many more characters beyond the first 64,000 (although there is currently no character defined beyond those already contained in the BMP). The UCS character set is identical to the Unicode character set defined by the Unicode consortium, but Unicode defines more character properties, semantic conventions, and more character-rendering options. UCS and Unicode maintain a close coordination and so far use the same code points for each character.

Multi-byte character sets are not compatible with many current applications or systems which are byte oriented. Many systems are also only able to handle correctly seven bit US-ASCII characters. For instance, in any C program '0' means 'end of the string', but this sequence can be found in the middle of a multi-byte character stream, and therefore 'printf' cannot be used with UCS-four character streams. Even recent systems understanding two octet characters cannot handle UCS-four characters. In order to facilitate the use of UCS in such systems, UCS Transformation Formats (UTF) have been defined:

- UTF-7 encodes all BMP characters using only octets with the first bit set to '0', and therefore is transparent even to older seven bit mail systems.
- UTF-8, defined in RFC 2279, uses variable length (one to six octets) encodings for the UCS-2 or UCS-4 characters, but preserves all seven bit US-ASCII characters, which are encoded on one single octet, with the usual seven bit ASCII value. ASCII character values are encoded in UCS-4 as 0000 0000 to 0000 007F, and are encoded in UTF-8 as 00 to 7F. Because of this, UTF-8 is 'file system safe' (it was originally called UTF-FSS). For multi-octet sequences, the first octet indicates the number of n octets in the sequence with n high-order bits set to '1'. All the following octets have the first two bits set to '10', and six variable bits. The remaining (8-1-n) bits of the first octet,

**Table 3.5** UTF-8 encoding of 4 octet characters preserves 7 bit ASCII values

| UCS-4 range (hex.) | UTF-8 octet sequence (binary) |
|---|---|
| 0000 0000-0000 007F | 0xxxxxxx |
| 0000 0080-0000 07 FF | 110xxxxx 10xxxxxx |
| 0000 0800-0000 FFFF | 1110xxxx 10xxxxxx 10xxxxxx |
| 0001 0000-001F FFFF | 11110xxx 10xxxxxx 10xxxxxx 10xxxxxx |
| 0020 0000-03 FF FFFF | 111110xx 10xxxxxx 10xxxxxx 10xxxxxx 10xxxxxx |
| 0400 0000-7FFF FFFF | 1111110x 10xxxxxx ... 10xxxxxx |

and the 6*(n-1) bits of the following octets, are used to encode the UCS character, as shown in Table 3.5.

Some examples:

- The copyright sign © (Unicode character U + 00A9 = 1010 1001) is encoded in UTF-8 as 11000010 10101001 = 0xC2 0xA9.
- The Euro sign € (U + 20AC) is encoded in UTF 8 as 0xE2 0x82 0xAC.
- A good list of the Unicode Fonts for the Microsoft Windows® operating system can be found at http://www.alanwood.net/unicode/fonts.html.

#### *3.5.1.1.2 Message format*

The message/CPIM format is a multipart MIME format which encapsulates

- content and message related metadata;
- the message itself in the form of any MIME content;
- optionally an electronic signature according to S/MIME, RFC 2633.

Figure 3.38 shows an example without electronic signature.

The end of the message body is defined by the framing mechanism of the transport protocol used.

##### *3.5.1.1.2.1 MIME headers part*

In the example of Figure 3.38 it is composed only of the mandatory Message/CPIM content type header, but other headers can be added before the blank line if necessary. Each line is ended with the CR-LF characters.

##### *3.5.1.1.2.2 Message headers part*

This part must remain intact end to end. The headers and their values must not be changed in any way or even reordered.

Each line has a key: value form (with a single space after the ':'). The key must contain only US-ASCII characters (some control characters like or '''must be escaped),

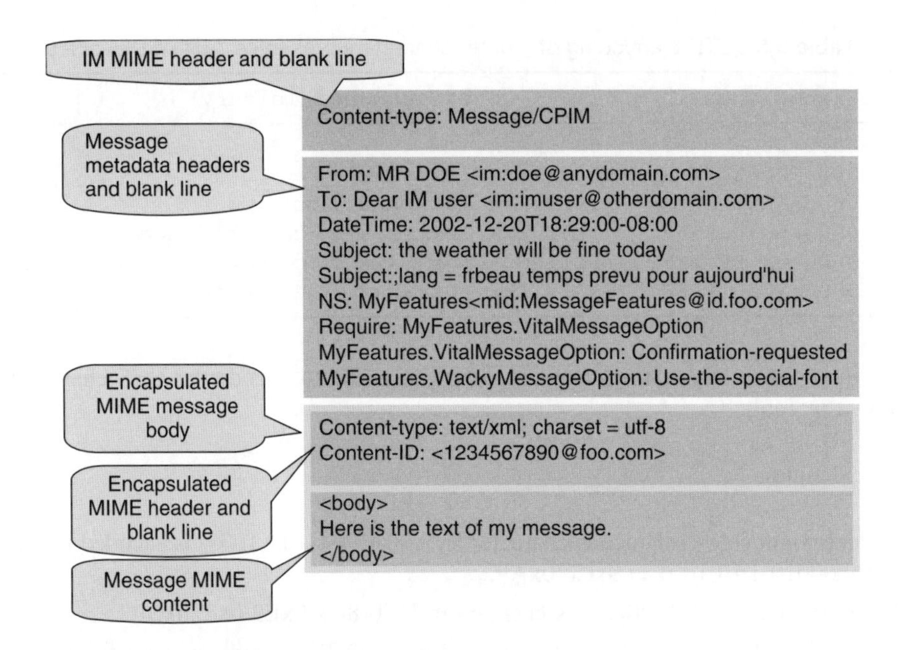
IM MIME header and blank line
Content-type: Message/CPIM
Message metadata headers and blank line
From: MR DOE <im:doe@anydomain.com>
To: Dear IM user <im:imuser@otherdomain.com>
DateTime: 2002-12-20T18:29:00-08:00
Subject: the weather will be fine today
Subject:;lang = frbeau temps prevu pour aujourd'hui
NS: MyFeatures<mid:MessageFeatures@id.foo.com>
Require: MyFeatures.VitalMessageOption
MyFeatures.VitalMessageOption: Confirmation-requested
MyFeatures.WackyMessageOption: Use-the-special-font
Encapsulated MIME message body
Content-type: text/xml; charset = utf-8
Content-ID: <1234567890@foo.com>
Encapsulated MIME header and blank line
<body>
Here is the text of my message.
</body>
Message MIME content

**Figure 3.38** CPIM multipart MIME format.

while any UTF-8 character (with the same escaped control characters) is allowed in the value portion. A header can be tagged to indicate that it contains a specific language by using the ';lang = tag' after the header name and colon, where 'tag' is a language identifying token (defined in RFC 3066).

From, To, Subject, DateTime (RFC 3339: date/UTC time/time offset) are headers defined in CPIM. The example of Figure 3.38 also shows the extension mechanism for the CPIM format. A developer can define his own extension namespace (here MyFeatures) by using the NS (namespace) header. New header keys beginning with 'MyFeatures' can then be used. They will be ignored if not understood. It is possible to indicate to the receiving system that it needs to support an extended header in order to understand the message by using the 'Require' header followed by the header key that must be supported.

### *3.5.1.1.2.3 Encapsulated MIME object*

This is the message itself; any MIME type can be encapsulated. Like any MIME encoded object, it is composed of a header part and a content part, separated by a blank line. For simple text only IM systems, the text/xml MIME type using the UTF-8 encoding can convey any written symbol from any language.

### *3.5.1.1.2.4 MIME security multipart message wrapper*

The message can be secured and signed using Multipart MIME, as shown in the example of Figure 3.39.

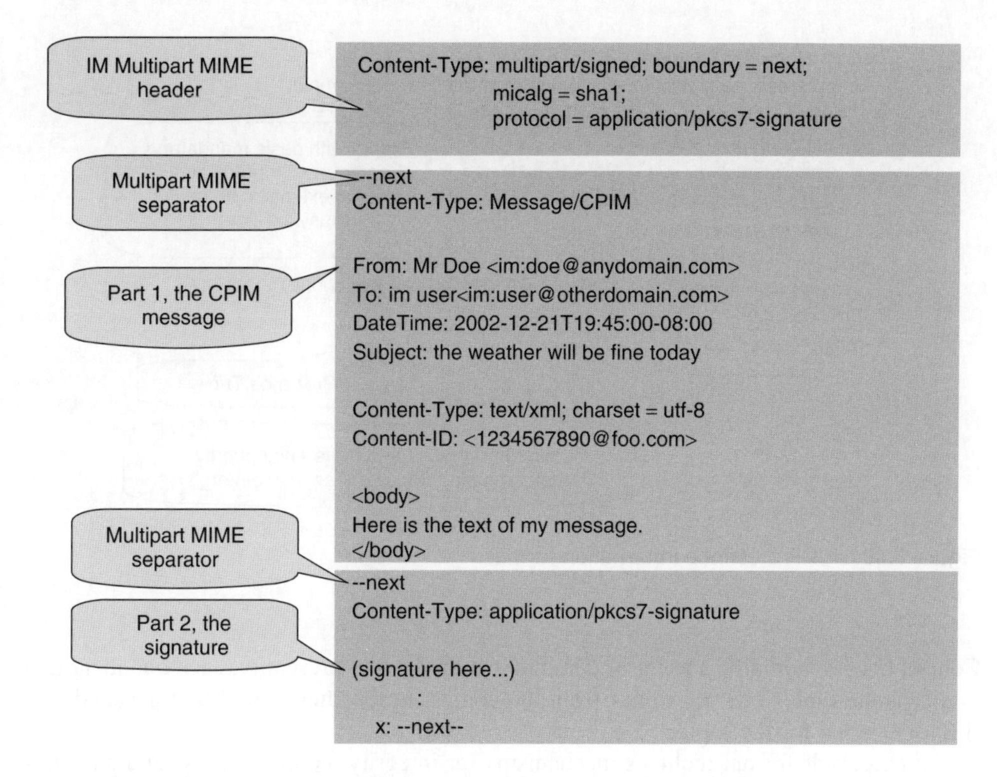


**Figure 3.39** CPIM message electronic signature using multipart MIME.

### *3.5.1.1.3 Common Presence and Instant Messaging (CPIM) Presence Information Data Format (PIDF)*

The IMPP working group of the IETF worked to define a standard format for presence information sent from a *presentity* (the entity about which presence information is generated), to a *watcher*. Note that multiple devices may send presence information for a given presentity.

RFC 3863 ('Presence Information Data Format') defines the new XML MIME media type 'application/cpim-pidf+xml', with an optional charset parameter.

The presence information of a *presentity* consists of one or more tuples (Figure 3.40) with STATUS, optional COMMUNICATION URI, and optional other presence markup information (relative priority, timestamp, human readable comment). Status may contain one or multiple values; the 'OPEN' and 'CLOSED' values mean the entity is 'ready' and 'not ready' (respectively) to receive Instant Messages, but does not imply anything for other communication means. Other status values may be defined in extensions (busy, off-line, away, on the phone...). There may be more than one tuple for a presentity if multiple devices/applications can reach the presentity and each one creates a presence component in the form of a tuple. For instance, in a SIP REGISTER message, the To header field (address of record) would be considered the presentity, while each URI in the

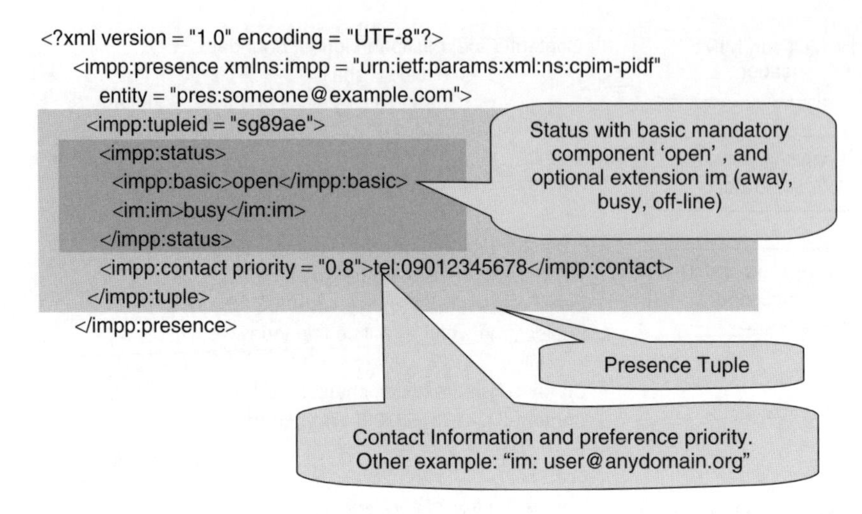


**Figure 3.40** PIDF presence information format.

Contact header would be a point of communication for that presentity, each one identified in a separate tuple. The 'q' values from the contact header field could be translated into 'priority' value for the tuple.

The cpim-pidf format includes mechanisms for integrity, confidentiality, and authentication, independently of SIP.

## 3.5.2 RFC 3265, Specific Event Notification

RFC 3265 specifies SIP based transport mechanisms for watchers to subscribe to presence information, and for 'presentities' (entities sending presence information) to send presence information updates to the watchers.

### *3.5.2.1 Subscribe and notify requests*

RFC 3265 defines two new optional requests: SUBSCRIBE and NOTIFY. These requests are generic in nature and must be further specified by 'event packages' (the name was taken from MGCP, but the events defined are different from MGCP events).

The SUBSCRIBE and NOTIFY requests are normal SIP requests that can be routed by proxies using the 'from' and 'to' headers, and can be acknowledged by a 200 OK or a 202 response. 200 implies that the request has been accepted, while 202 only acknowledges that the SUBSCRIBE message was received and the syntax was correct. The response to a subscribe request must be immediate, and this makes it impossible to ask for any form of user authorization before sending the response. 202 would be used in the case of a buddy list request to publish presence information, to respond immediately before the user has accepted or rejected the request. All other responses defined for other SIP requests are also valid and imply that the SUBSCRIPTION has not been accepted as is.

The SUBSCRIBE request can be sent by a SIP client willing to receive certain events (the 'subscriber') to a SIP server generating these events, or already receiving these events (the 'notifier'). The SUBSCRIBE REQUEST contains an 'Expires' header limiting the duration of the subscription, which can be shortened if the response contains a shorter period in its own 'Expires' header. In order to improve scalability for heavy load notifiers (e.g., voicemail systems), longer periods can be requested by rejecting the SUBSCRIBE with a '423 Interval too small', but convention intervals above 1 hour must be accepted). With the 'Expires' header, the subscription is a 'soft state', which is a very common approach of Internet protocols, also used by RSVP. Soft states are more tolerant to protocol errors or network instability, avoiding any undesired accumulation of state in any network entity.

When a subscriber wishes to stop subscribing to a certain set of events, it can do so by setting the 'Expires' header value to zero.

The 'Event' header in the SUBSCRIBE request specifies the category or set of events that are requested. The exact syntax of the 'Event' header is free and must be defined by specific Event packages. Optionally, the body of the SUBSCRIBE request can also be used to further specify the subscription, but again it must be specified by the Event package.

Both SUBSCRIBE and NOTIFY can create a SIP dialog, as defined above for INVITE requests. Therefore, these requests do not require any prior INVITE request and can be sent asynchronously at any time. Alternatively, they can be sent within an existing dialog; in this case the Event header must contain an 'id' parameter to distinguish between the various subscriptions. Sending multiple SUBSCRIBE requests with identical 'id' parameters within an existing dialog can be used to refresh subscriptions (if it does not correspond to an active subscription, it will be rejected with a 481 response).

Because the SUBSCRIBE/NOTIFY mechanism was primarily defined to handle state change notifications, the first SUBSCRIBE will trigger an immediate NOTIFY, within the same dialog, to synchronize the initial state status of the subscriber. This is also true even if the Expires header has a value of 0. This allows for simple state polling, with SUBSCRIBE requests having an 'Expires' value of 0. Examples of commonly used state information include voicemail box status, busy state of a user (for call completion on busy), buddy lists with 'presence' status...

The notifier can decide to terminate a subscription at any time by sending a NOTIFY message with a 'Subscription state' header with a value of 'terminated' and a reason parameter. One useful reason parameter is 'rejected', which can be used when a user has decided to not accept a subscription. This mechanism is often used, as shown in Figure 3.41, because the initial SUBSCRIBE request has been acknowledged by a 202 response.

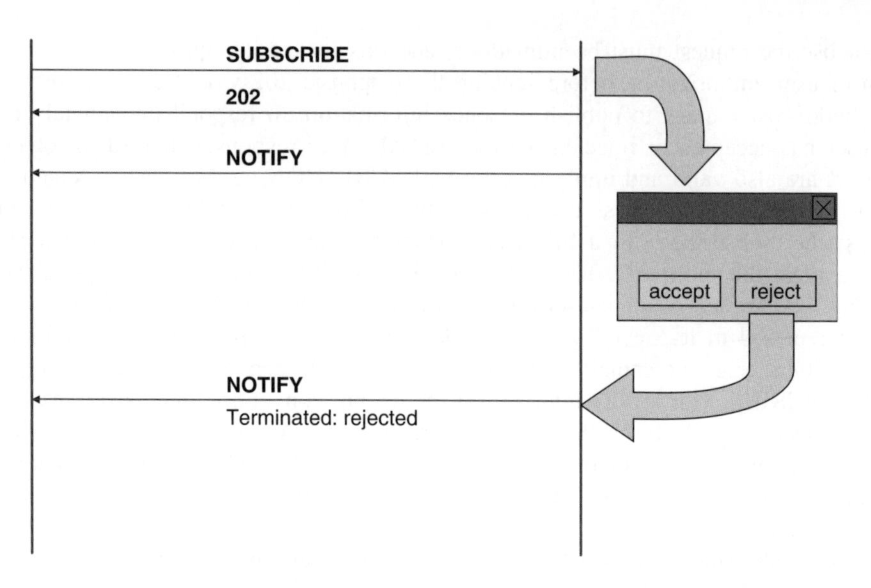


**Figure 3.41** Case of a rejected Subscription.

### *3.5.2.2 Use of RFC 3265 for presence*

The SIMPLE working group (SIP for IM and Presence Leveraging Extensions) is attempting to get the various IM and Presence implementations to converge on an interoperable standard based on SIP. SIMPLE works in close coordination with the IMPP working group.

RFC 3856 ('A presence Event Package for the SIP') specifies how to use RFC 3265 for presence. It defines 'Presence Agents' as SIP devices able to receive presence subscription requests, and to send presence information for a given presentity. Presence is handled by creating a specific 'presence' event package. In the future, other specific types of events may be created to handle the requirements for a 'buddy list' (a party is typing a message, message delivery confirmation, typical party states).

According to RFC3265 the name of the event package 'presence' must be in the event header field of SUBSCRIBE and NOTIFY requests. No SUBSCRIBE body is yet defined and therefore should normally be empty (Figure 3.42).

In the example, the subscription has been accepted immediately with a 200 OK. As soon as the presence user agent receives the subscription, it must, according to RFC 3265, immediately send back the current presence state in a NOTIFY message. The notification data should use the CPIM body type defined by IMPP: application/cpim-pidf+xml. In the example of Figure 3.43 the presentity is CLOSED and therefore not ready to receive Instant Messages. A non-standard extension explains the cause: the user is busy.

As soon as the presence information changes, the presence user agent sends a new NOTIFY message. In the example of Figure 3.44 the presentity is now OPEN and can receive instant messages. The rate at which presence notification updates can be sent is limited to at most one every 5 seconds.

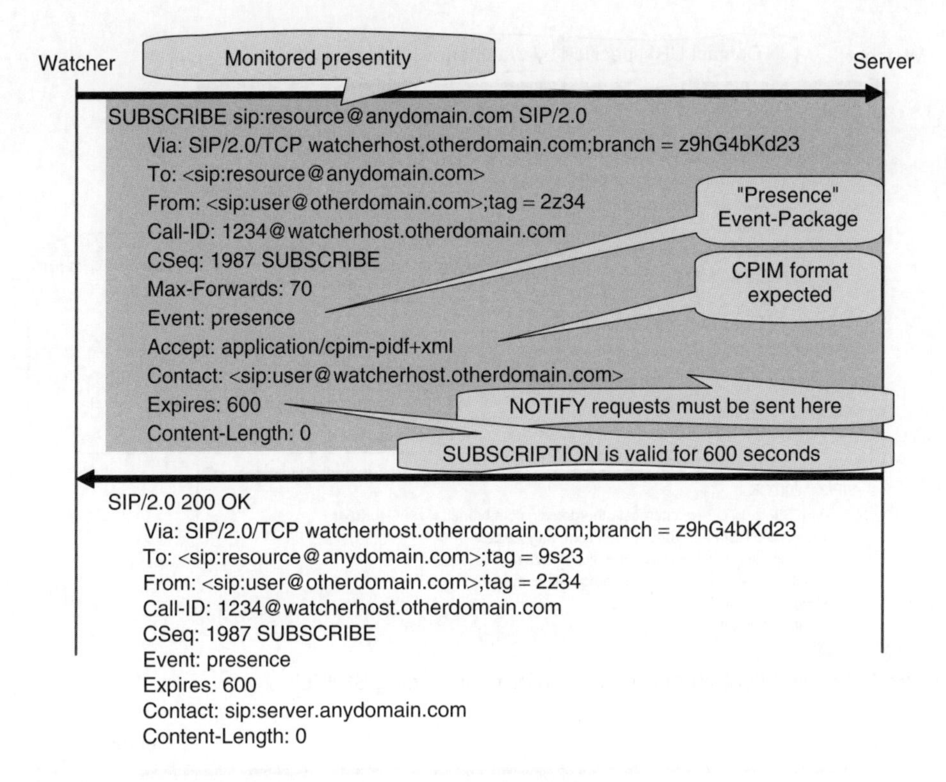


**Figure 3.42** Subscribing to presence information with SIMPLE.

#### *3.5.2.2.1 Watcher Information*

RFC 3858 ('An XML Based Format for Watcher Informartion') defines an XML format for the 'watcher information', and defines a new payload type for it:application/watcherinfo+xml. The watcher information is the list of all active and pending requests to receive event notifications (subscriptions) for a specific resource. The watcher information (Figure 3.45) includes the URIs of the watchers, an id, the current status of the subscription, the event that caused transition to that status, and optionally other parameters such as the duration of the subscription.

In order to receive the watcher information, a normal SUBSCRIBE request can be sent to the presence server, as illustrated in Figure 3.46.

If the subscription is accepted, the updates to the watcher information are reported in NOTIFY requests (Figure 3.47).

#### *3.5.2.2.2 Procedure for new subscriptions, presence authorization*

RFC 3857 ('A Watcher Information Event Template Package For SIP') defines a framework for the authorization of presence subscriptions. When a request for presence information arrives on a presence server, the presence server will usually require the user to authorize the new subscription. This is possible only, of course, if the user is first made

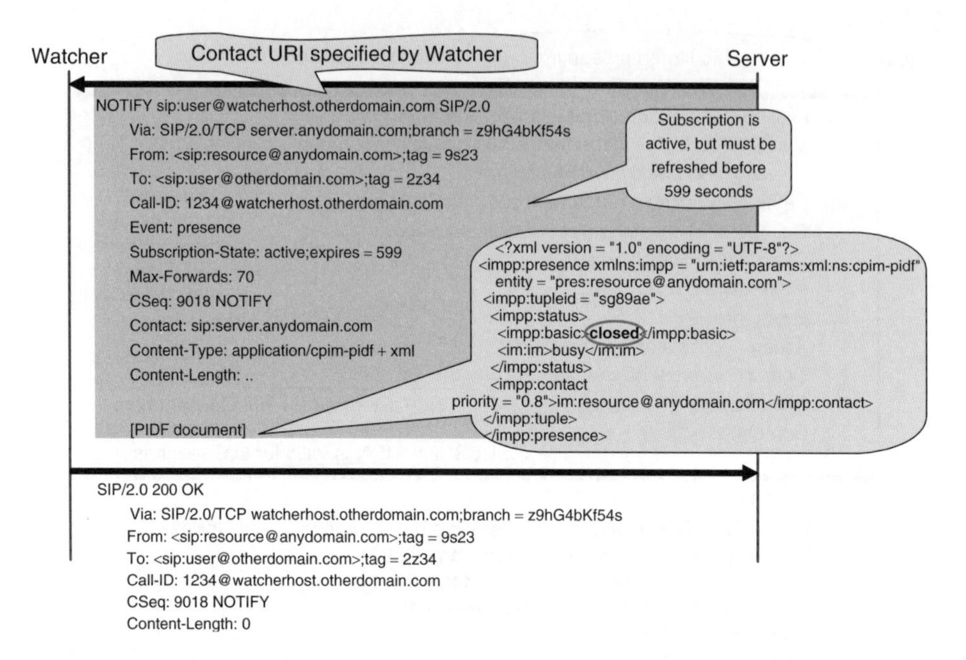


**Figure 3.43** Notification of presence information using SIMPLE.

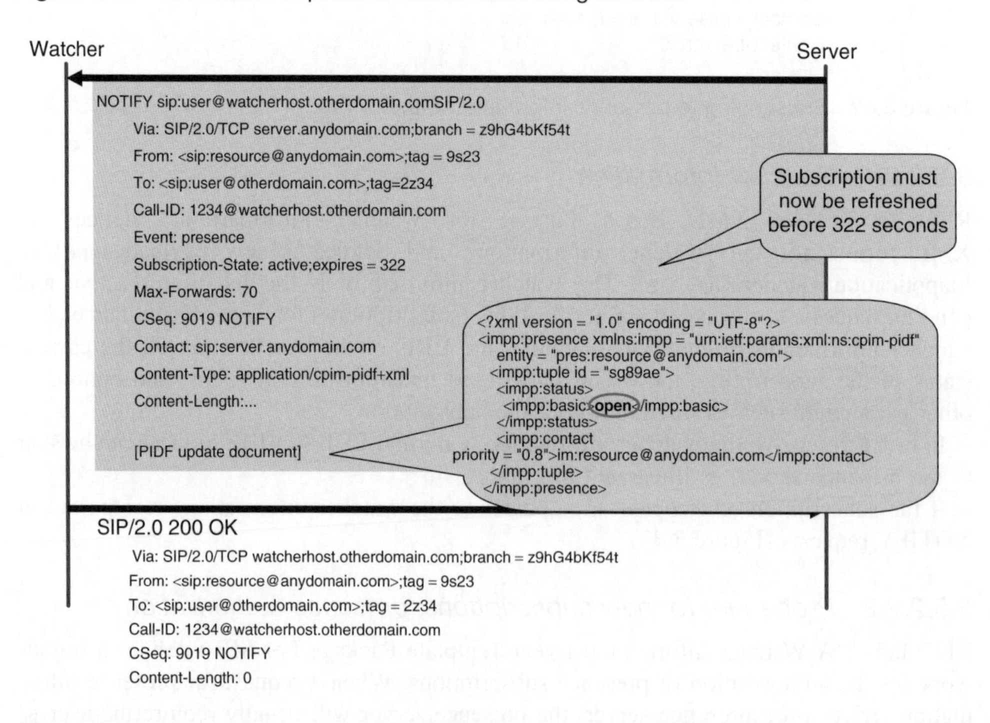


**Figure 3.44** Updated presence information sent in a new NOTIFY request.

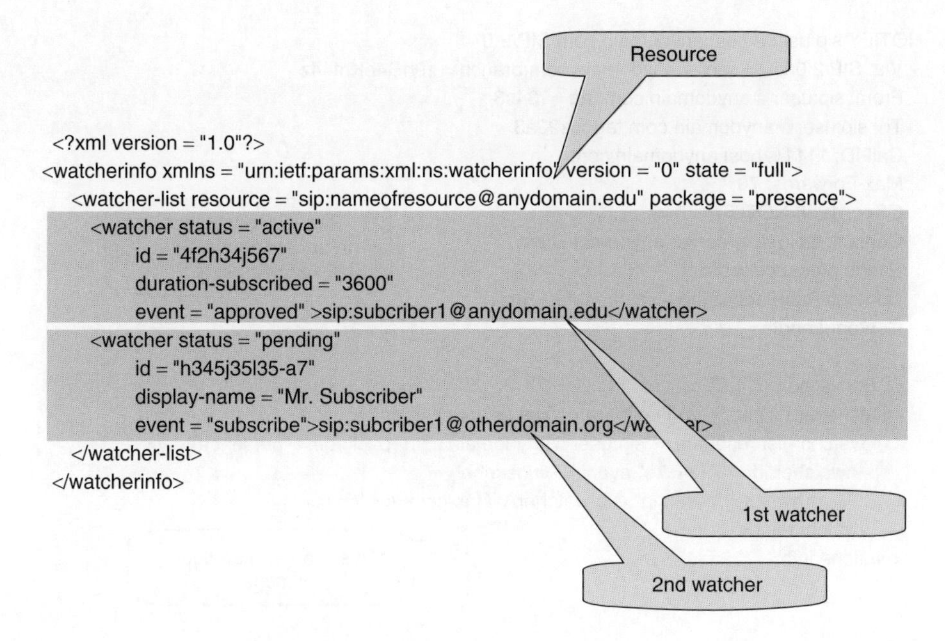


**Figure 3.45** Watcher information format.

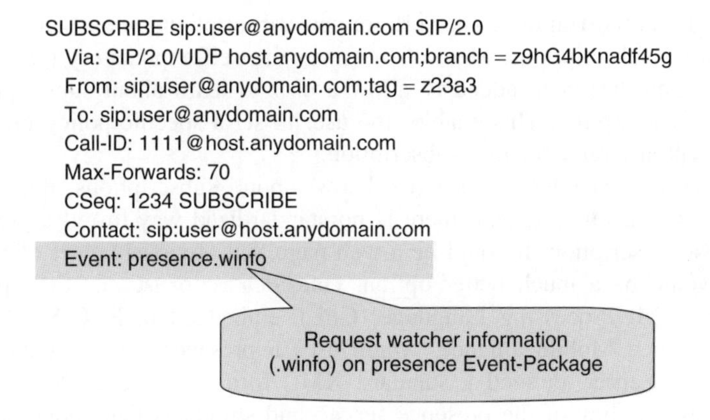


**Figure 3.46** A SUBSCRIBE request for watcher information.

aware of the new subscription, which is relatively easy if the presence server is the user agent of the user, but becomes more complex if the presence server is a network-based device. The idea of the RFC is to always allow a user to subscribe to any modification of the watcher information relative to his own presence.

New subscriptions will update the watcher information, and therefore the user will receive a NOTIFY with the update for watcher XML information (only the changes are included, therefore the user needs to cumulate these changes to get a complete up-to-date view of the watcher information).

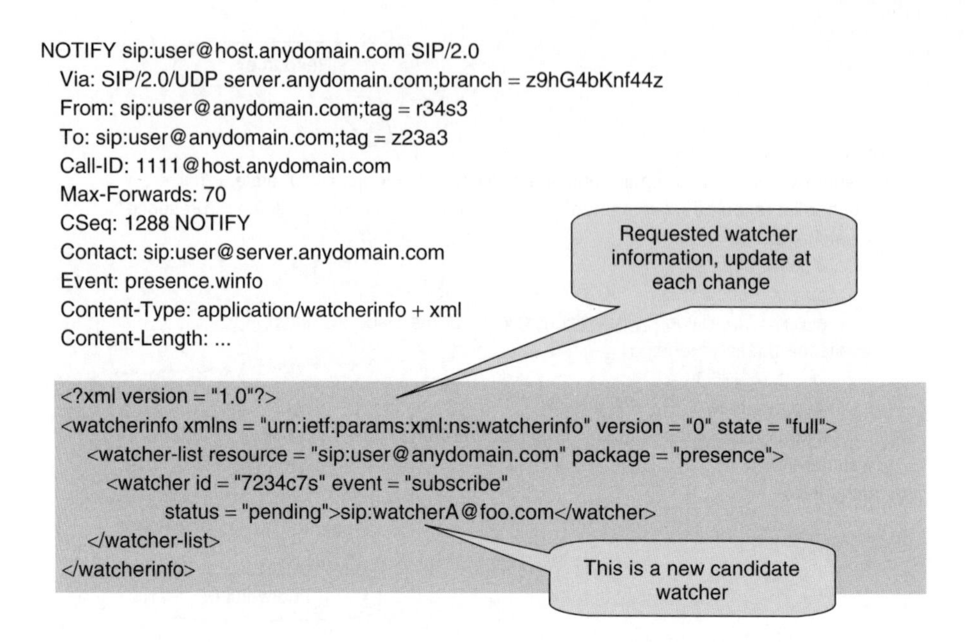


**Figure 3.47** Watcher information update sent through a NOTIFY request.

As part of this authorization framework for new subscriptions, some standard states have been defined for subscriptions. Most of the states are self-explanatory (see Figure 3.48); the 'waiting' state has been added to allow a user to learn about subscription requests even if they have expired. This enables the user to set a specific policy on the presence server to accept any retry for that subscription.

This framework enables a user to know about subscriptions that need to be authorized... but at this moment, there is no standardized way to tell a presence server to authorize a subscription. It could be a web page, but obviously a set of SIP messages to do this would be a much better option. One idea is to use an XML policy syntax similar to the Call Processing Language (CPL), embedded in REGISTER requests to do this. The Open Mobile alliance, which uses a presence server in its specification of the presence server, defined a standard XML format (Presence XDM) to describe the authorization policy of the presence server, and stipulates that endpoints should use XCAP (RFC 4827) to manipulate this document.

## 3.5.3 RFC 3428: SIP extensions for instant messaging

RFC 3428 defines a new `MESSAGE` request which carries MIME body parts representing the content of an Instant Message. The message request is usually sent outside the context of an existing dialog, and does not create its own dialog. It can also be sent as

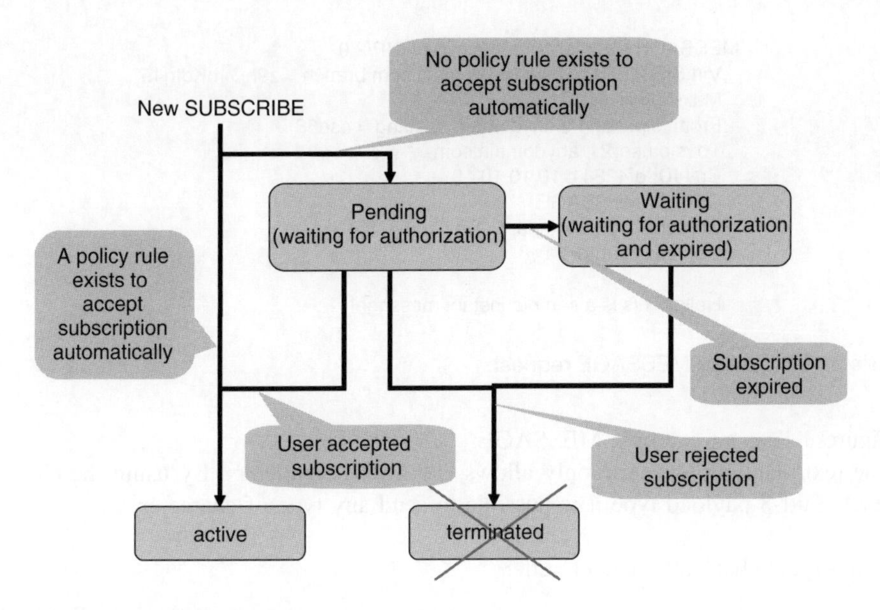


**Figure 3.48** Subscription states.

part of an existing dialog in some circumstances, e.g., if an instant message is sent as part of an existing voice call. The response can be a provisional or a final one; usually it will simply be 200 OK if the message has been received or 202 if it has been stored and will be presented to the target user as soon as possible. Neither the `MESSAGE` request nor the 200 OK reply are allowed to have a 'Contact' header (they do not create a dialog). A MESSAGE request can have an 'expires' field, which is a simple indication of its validity for proxies that may try to store it if the target user is not immediately available.

RFC 3428 defines an instant message URI, im:user@domain, which is independent of the underlying instant message transport protocol. For all practical purposes, however, it is translated into a SIP URI immediately and placed in the Request-URI of the message request before it is sent. An instant message can be sent to an Instant Message URI (im:someone@domain.org), or to a SIP URI. If the IM URI is used, the next hop server and SIP transport method can be found by performing a SRV DNS request for_im._sip.domain.org which should return a resource record of SIP proxy that can route the message (This is described in RFC 3861 'Address Resolution for IM and Presence').

The size of the message payload is limited to 1300 bytes or 200 bytes less than the path MTU, if known (this is usually not the case), in order to avoid the message segmentation problems of SIP. A basic form of congestion control is ensured by requiring clients to never send a MESSAGE until the previous MESSAGE request has been acknowledged with a response.

```
MESSAGE sip:user2@domain.com SIP/2.0
  Via:SIP/2.0/TCP user1pc.domain.com;branch = z9hG4bKdfg45
  Max-Forwards: 70
  From: sip:user1@anydomain.com;tag = 4sd83
  To: sip:user2@anydomain.com
  Call-ID: ef4234@10.10.10.10
  CSeq: 1 MESSAGE
  Content-Type: text/plain
  Content-Length: 41

  Hello. This is a sample instant message!
```

**Figure 3.49** Sample MESSAGE request.

Figure 3.49 is an example MESSAGE:

The text/plain content type only allows US-ASCII characters. By using the char/xml; charset = utf-8 payload type it is possible to send any type of character.

Content-Type: char/xml; charset=utf-8

This is an utf-8 message, it can contain all non US-ASCII characters like € or © !

RFC 3428 are also required to support the 'message/CPIM' content type, and therefore instant messages can carry any type of MIME content.

The framework defined by RFC 3428 is still quite basic compared to the GSM SMS specifications; for instance, it does not cover confirmation of message reception. However, RFC 3428, together with the CPIM format, provides a good foundation to be able to send instant messages across SIP-based IM systems from various service providers, and in the future also, across non-SIP-based IM systems like the popular GSM Short Message and Multimedia Message Services.

# 4 The 3GPP IP Multimedia Subsystem (IMS) Architecture

## 4.1 INTRODUCTION

### 4.1.1 Centralized value added services platforms on switched telephone networks: the 'tromboning' issue

In today's switched telephone networks, deploying centralized value-added services (VAS) platforms remains problematic: if a given call must be handled by a VAS switch, then not only the signaling but also the voice media stream must be relayed by that switch. For services provided to a majority of end-users, the obvious solution is to deploy the VAS platforms regionally, or even in each local telephone exchange. Since such massive services are already deployed by all operators, today, it is the accumulation of innovative niche VAS services which plays a key role in the telecom operator's 'long tail' service strategies. Unfortunately, such a distributed services architecture is too expensive for niche VAS services, which target only a small fraction of telephone subscribers. Ideally, a single centralized VAS switch should provide the service to all subscribers, but then potentially a VAS call between two subscribers living in the same city would be routed through a VAS switch located thousands of miles away. Not only does this situation waste network resources but it can also lead to delay and echo problems (see Chapter 1).

Often, telephony engineers believe that the SS7 signaling system solves this problem, because transport links carrying signaling (e.g., SS7/ISUP) and transport links carrying media streams (G711 data) are separate. This is true; however, each telephone exchange or VAS switch must have access to both media streams and signaling information, creating

*IP Telephony: Deploying VoIP Protocols and IMS Infrastructure, Second Edition* O. Hersent

this 'tromboning' problem: a traditional telephony switch cannot route calls by only handling the signaling messages.

### 4.1.2 The 'Intelligent Network' (IN)

The Intelligent Network (IN) technology has been developed to solve some of the challenges associated with the tromboning issue. IN is based on an abstract call model, which enables an external Service Control Point (SCP) to interact with a remote programmable switch implementing this abstract call model (see Figures 4.1 and 4.2), the Service Switching Function (SSF), using only standardized control primitives. The tromboning issue is partially solved, because SSFs can be distributed and shared by niche applications. However, there is still a need to deploy SSFs regionally.

The INAP protocol, standardized by the International Telecommunication Union (ITU-T Q1224), is one of the most popular and standardized call models for fixed networks, and CAP (CAMEL Application Part) is the equivalent for mobile networks. There are also many proprietary variants introduced by individual switch vendors.

The IN model has been extended by ITU in order to handle VoIP calls (IN Capability set 4, Q1244).

However, it does not seem that the market is likely to adopt this approach. The main problem with INAP is that it relies on a standardized call model and set of control primitives, which obviously restrains the possible features of VAS applications. As we will see, the IP Multimedia Subsystem (IMS) fully solves the tromboning issue in a very elegant way, and without the complexity and heaviness associated with a standardized call model.

### 4.1.3 How VoIP solves the 'tromboning' issue. The value added services architecture of 3GPP IMS

Unlike TDM networks, VoIP totally separates the media streams plane and the signaling plane. In order to fully control a call, a VoIP server only needs to have access to the signaling. It can then switch media streams without relaying them. This fundamental improvement is valid for all VoIP protocols.

For instance, a VoIP prepaid telephony application is able to bridge the media streams of users A and B without accessing any media packet of A or B. VoIP solves the root cause of the 'tromboning' issue.

Realizing that this was a revolution for the VAS architecture of telecom networks, the 3GPP[1] (3rd Generation Partnership Project) led by ETSI (European Telecom Standards Institute) defined a new VAS architecture for telecom networks optimized for IP networks: the IMS. The IMS introduces the concept of an Application Server (AS) and specifies how

[1] For more information on 3GPP and 3GPP2, see Section 3.1.2 in Chapter 3.

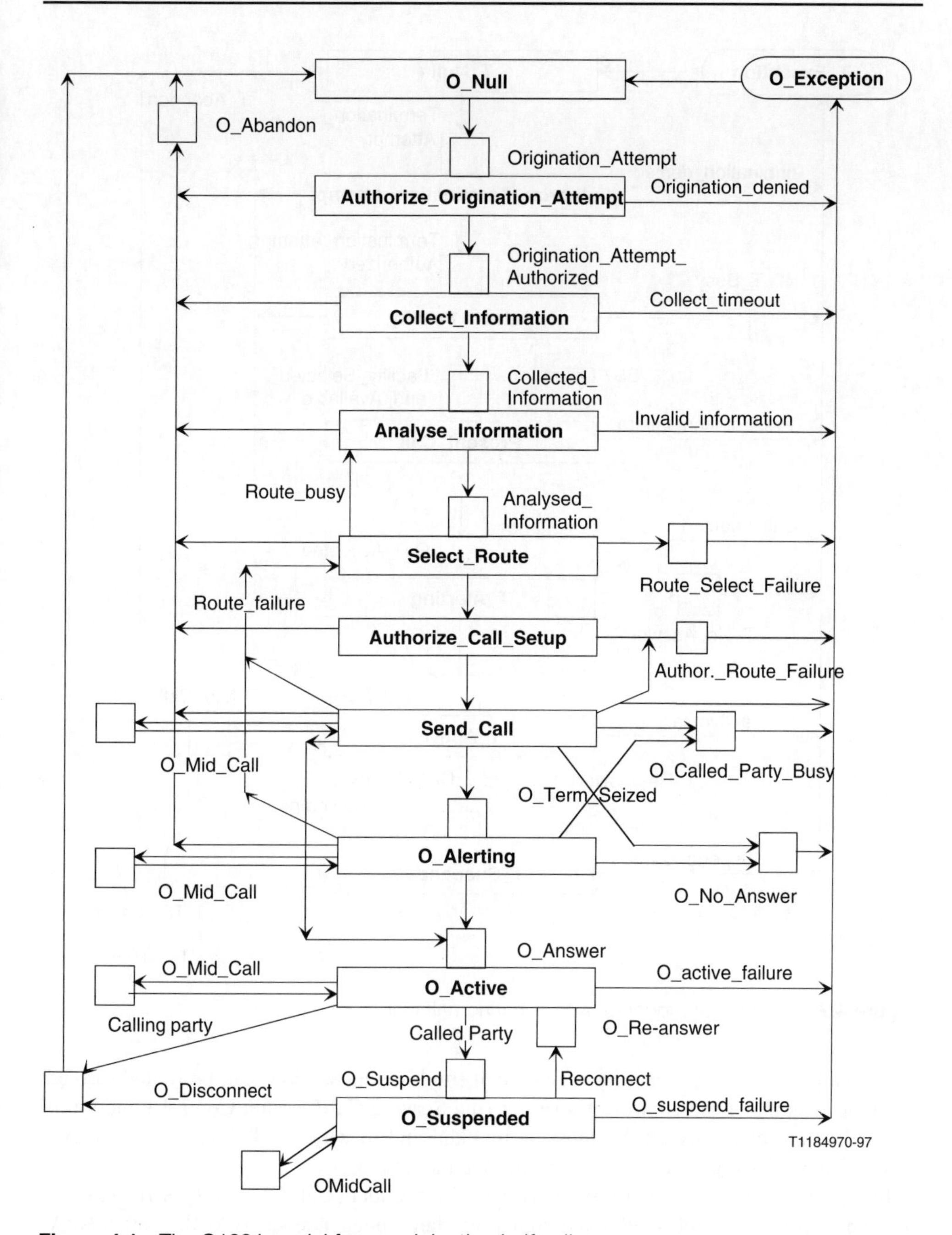


**Figure 4.1** The Q1224 model for an originating half call.

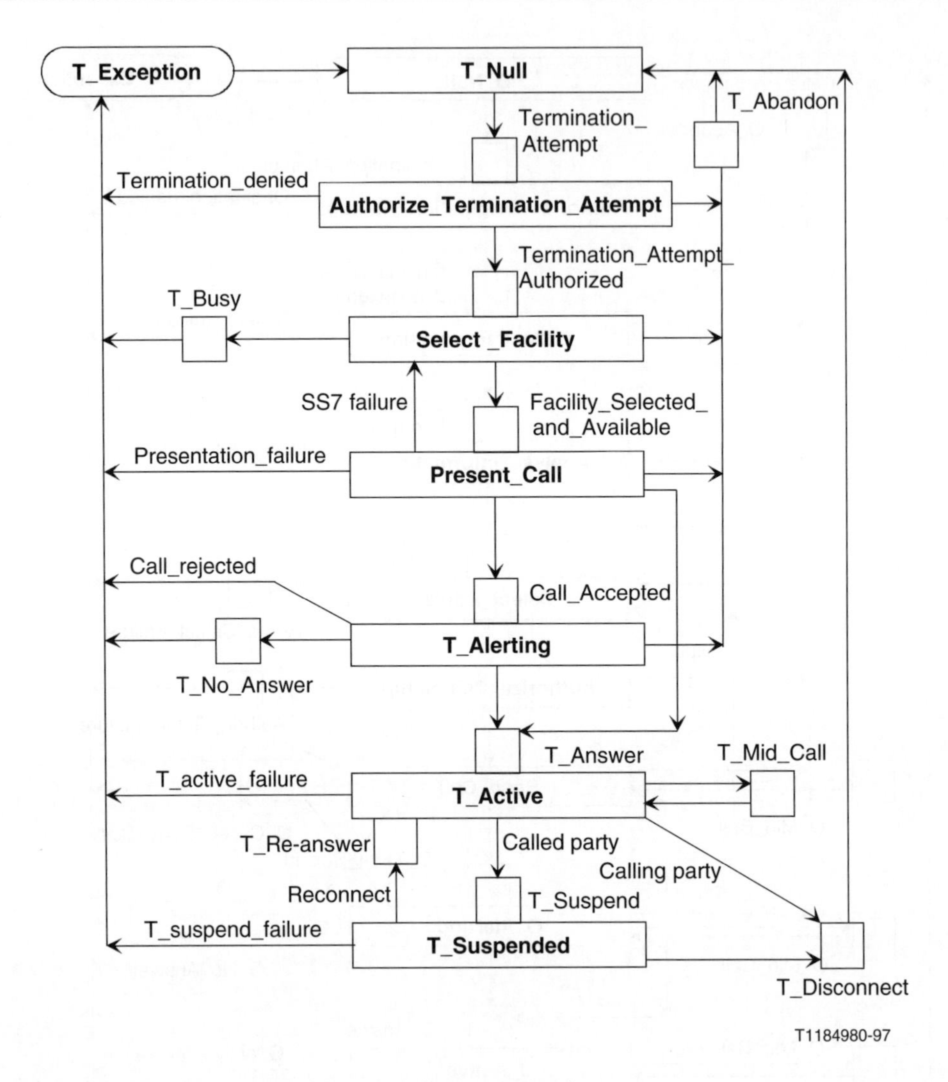


**Figure 4.2** The Q1224 model for a terminating half call.

the signaling messages related to a given call (multimedia session) can be routed through multiple ASs under the control of a server (the Serving Call/Session Control Function or S-CSCF) that selects which AS must be invoked and in which order. This chain of ASs can be invoked without any impact on the media streams.

In fact, a recent ITU protocol, BICC (Bearer Independent Call Control, Q1901), exhibits the same property of full media and signaling plane independence as VoIP protocols. As a successor of ISUP, BICC was also very compatible with all telephony applications. Despite this, when the time came to select a protocol for the 3GPP VAS architecture, the

Session Initiation Protocol (SIP) was selected for no other fundamental reason than the apparent simplicity of this text-based protocol.

### 4.1.4 The IMS architecture is ideal for mobile networks... but not only

The use of VoIP has an important consequence for the IMS architecture. In all traditional telephony networks, it is necessary to define a standard call model in order to enable applications to control the network resources (e.g., Q1224), but IMS ASs have access to the SIP signaling directly, and therefore can implement any feature made possible by the protocol without a need for any standard call model.

This gives application developers much more flexibility and reduces, significantly, the complexity of the network (mainly by simplifying the interactions between servers). It opens new VAS possibilities which were previously impossible.

The most striking example, perhaps, is the support of VAS applications in a roaming context for mobile networks.

On current mobile networks, most VAS services, for instance, short numbering services for virtual private networks, stop working when users are roaming. The reason is that routing all calls dialed by the user to his home network would cause a tromboning problem. In theory, IN services could be executed by the visited network under control of the home network, using CAMEL, but this is rarely done in practice as this requires a complete alignment of the call models, and poses serious security problems.

In an IMS network, since the 'tromboning' problem no longer exists, the routing of all outgoing and incoming calls through the home network becomes the standard behaviour. As a consequence, all services which work in the home country also work in a roaming situation. This is the concept of the 'virtual home environment'. No wonder, mobile operators were the first promoters of the IMS model for their 3G networks, even though, in the end, fixed operators were the first to fully deploy operational IMS networks.

In this introduction, we insisted, voluntarily, on the impact of IMS on the design of VAS. The IMS is much more than a profile of SIP. The IMS not only adds a lot of rigor to the SIP specification but it defines, for the first time, a telecom network *architecture*, which takes fully into account the fact that the 'tromboning' is no longer a problem in VoIP networks.

This has a profound impact on the way services will be implemented and deployed, and on the relationships between operators. It is of course still possible to implement services as in traditional circuit switched networks, for instance, using an IMS AS acting as an IN SSF,[2] but this amounts to importing into IMS the now unnecessary complexity of TDM telecom networks.

[2] Such an AS implementing the IN SSF call model is defined in the standard and called an IM-SSF (IP Multimedia Service Switching Function) or OSA-SCS (Open Service Access-Service Capability Server).

## 4.2 OVERVIEW OF THE IMS ARCHITECTURE

### 4.2.1 Registration

When it registers to the network, an IMS User Equipment (UE) may not be in its 'home domain'. During the attachment procedure to the 3GPP network, an IMS UE receives the URI of its first contact point in the visited network, the Proxy-CSCF (P-CSCF).

The P-CSCF analyzes the URI of the `REGISTER` request which contains the name of the UE home network and executes a DNS request to locate the entry point of the home network, as per RFC 3263 (see Section 3.3.1.1.2). In general, this entry point is an Interrogating-CSCF (I-CSCF).

Before forwarding the `REGISTER` request, the P-CSCF inserts IMS-specific headers, which are used for call accounting and user localization, for instance, a 'Path' header with its own URI (see Section 3.3.2.4 in Chapter 3).

The I-CSCF, after receiving the `REGISTER` request, interrogates the Home Subscriber Server (HSS) database, providing the user's public (header 'To') and private (username of the Authorization header) identifiers of the user. In its response, the HSS provides the URI of the Serving-CSCF (S-CSCF) of the user (if it is already allocated), or a set of capabilities which allow the I-CSCF to select an S-CSCF. With the information provided by the HSS, the I-CSCF forwards the `REGISTER` request to the appropriate S-CSCF (Figures 4.3 and 4.4).

For the first `REGISTER` request of the UE, the S-CSCF interrogates the HSS, which provides several 'authentication vectors' (a set of challenge strings, and the expected response for each challenge). The HSS also memorizes the URI of the S-CSCF now handling the specified public identity.

The S-CSCF responds to the `INVITE` with a `401 Unauthorized` response, which contains one of the challenges proposed by the HSS in the `WWW-Authenticate` header (Figure 4.4, message 4).

The UE sends a new `REGISTER` request, now including a 'response' parameter in the Authorization header (Figure 4.4, message 5). This new `REGISTER` request reaches the same S-CSCF as the initial `REGISTER` because the HSS now indicates to the I-CSCF the URI of the previously allocated S-CSCF.

The challenge response is calculated according to RFC 3310 ('Digest Authentication using Authentication and Key Agreement'). This method uses the authentication parameters stored in the terminal's UICC (Universal Integrated Circuit Card, see Figure 4.5). The authentication parameters are accessible, in read-only mode, only after the user has provided the correct PIN code.

After receiving a *`REGISTER`* request with the correct challenge response, the S-CSCF creates a 'binding' which associates:

- the public identity of the user indicated in the *`REGISTER`* request (a user can have multiple public IDs);
- the 'Contact' header indicated in the *`REGISTER`* request;
- the 'Path' header indicated in the *`REGISTER`* request.

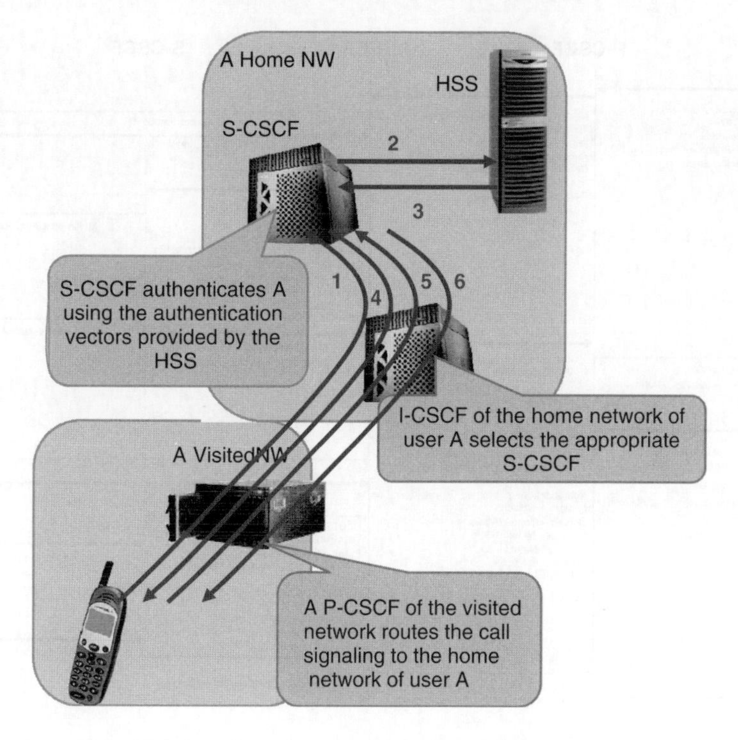


**Figure 4.3** Overview of the registration of an IMS User Equipment.

The S-CSCF then again interrogates the HSS to load the profile information of the user. The user service profile contains the 'Initial Filter Criteria' (IFC), a list of filters based on SIP requests header information which define, for each matching filter, an ordered list of ASs which must be invoked in the SIP session. The user profile lists the public identities that this UE is allowed to use, and may also specify other public identities which must be administratively associated to this UE, even if not registered by the UE. This list is included in the '`P-Associated-URI`' header of the `200 OK` response to the second `REGISTER` (Figure 4.4, message 6).

## 4.2.2 SIP session establishment in an IMS environment

In the more general situation, both the calling user, A, and the called user, B, are currently in a roaming situation. Their IMS terminals are connected to the local networks ('Visited networks'), but need to receive services provided by their respective 'Home networks'.

Figure 4.6 illustrates the call flow of a session establishment in IMS, and the role of the IMS servers directly involved in the SIP session routing.

UE A sends the initial SIP request for the new session to the P-CSCF of the visited network (currently providing IP connectivity to user A). UE A discovered the address

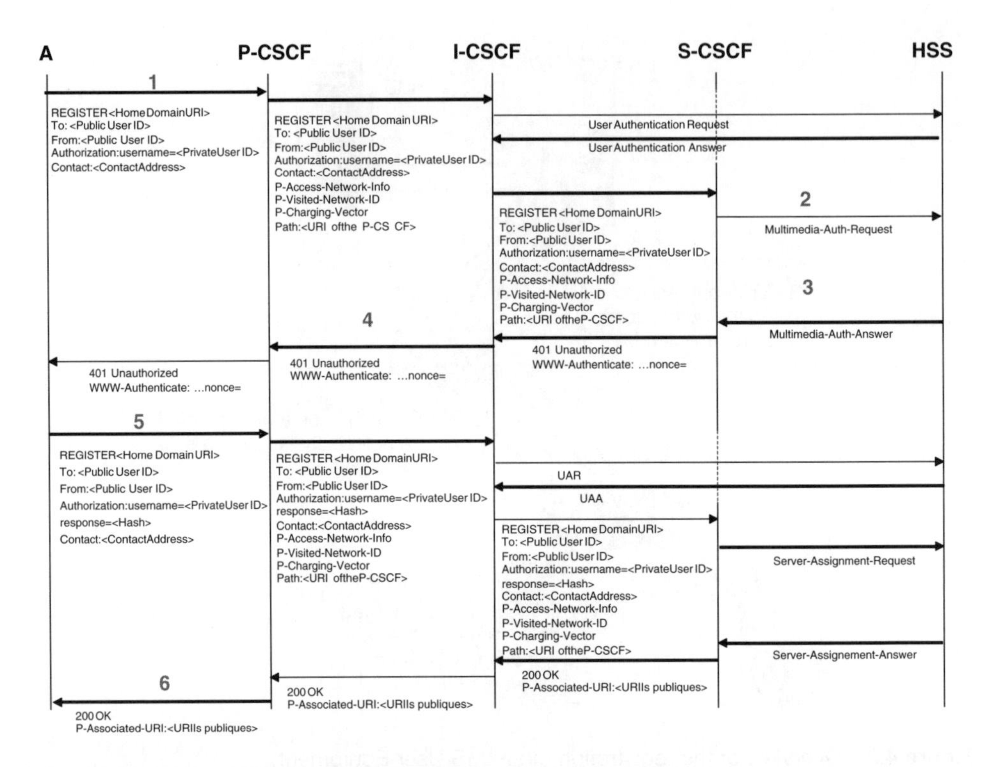


**Figure 4.4** The main SIP headers used during the registration process.

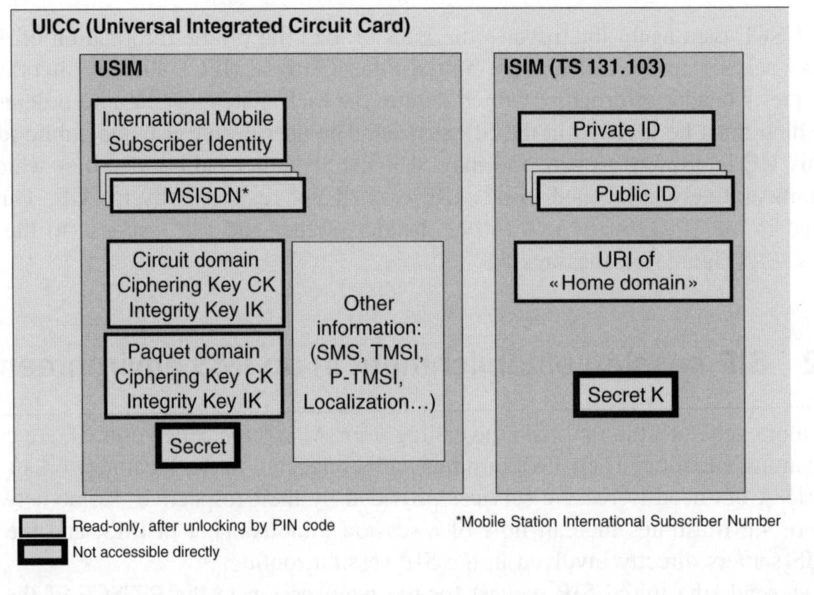


**Figure 4.5** simplified content of a UMTS UICC with the ISIM parameters used for IMS authentication.

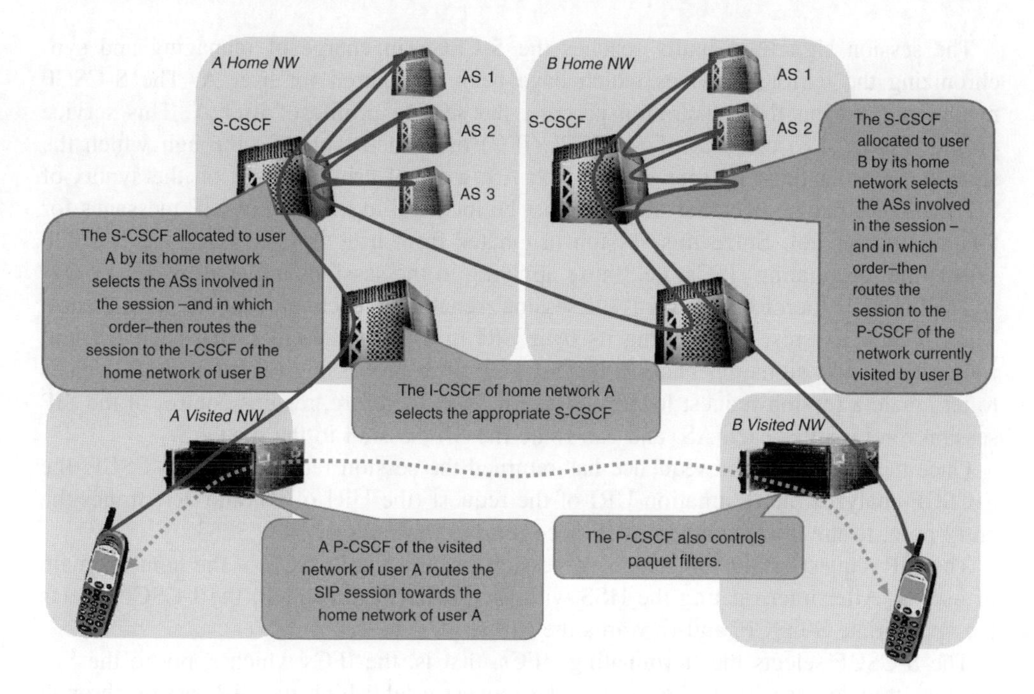


**Figure 4.6** Simplified overview of an IMS session routing, in a roaming situation.

of this server during the network attachment procedure, which is not detailed here. The P-CSCF performs three main functions when handling the new session:

- It ensures the security of the visited IMS network, by checking the syntax of SIP messages, and also optionally by using packet relays to check the media streams (packet format and stream characteristics).
- It analyzes the URI of A and determines from this URI the appropriate entry point to the home network of A.
- It relays the session signaling to the entry point of the home network of A, so that the services of the 'Virtual Home Environment' of A can be performed.

The session signaling messages, therefore, reach the entry point defined by the home network of A, usually an I-CSCF. The I-CSCF interrogates the HSS, a database which stores the service profile of user A, in order to learn the URI of the S-CSCF which was allocated to the terminal of A during the registration process (see Section 4.2.1), then it propagates the session signaling to this S-CSCF.

Depending on the configuration selected by the operator, the I-CSCF may also secure the network by filtering the session signaling, in which case it will include a 'Path' header to remain in the path of the SIP messages for this session (see Section 3.3.2.4 in Chapter 3), but it is not mandatory. In a simpler configuration, the I-CSCF is simply a stateless load balancer distributing sessions to the S-CSCFs, under control of the HSS.

The session signaling finally reaches the S-CSCF in charge of managing and synchronizing the various services which have been configured for user A. The S-CSCF memorized, during the registration process, the service profile of user A. This service profile (Figure 4.8) consists of an 'IFC' (IFC) ordered list of ASs through which the session signaling must be routed. For each AS, a set of criteria based on the syntax of SIP messages defines whether this AS must be included in the path of SIP messages for this session, or not. Since this session originates from user A's terminal, the S-CSCF selects the 'originating' IFCs, i.e., those applying to the sessions initiated by A.

The S-CSCF, therefore, forwards the session signaling to each relevant AS, in sequence. The S-CSCF ensures, by inserting its own URI in a 'Route' header, that each AS will route the session signaling back to the S-CSCF after processing (unless an AS decides to terminate a session request locally). The S-CSCF, therefore, regains control of the SIP session routing after each AS, and can route the SIP session to the next AS.

Once the last AS in the sequence has returned the session request to the S-CSCF, the S-CSCF analyzes the destination URI of the request (the URI of B), and determines the entry point to the home domain of B, according to the rules of RFC 3263.

The SIP request establishing the new session reaches the I-CSCF of the home domain of user B. After interrogating the HSS with the public identity of B, the I-CSCF selects the appropriate S-CSCF, and forwards the SIP request to it.

The S-CSCF selects the 'terminating' IFCs, that is, the IFCs which apply to the SIP sessions towards user B. It then routes the request establishing the SIP session through the entire ASs listed in the IFCs, which match the criteria.

When the last terminating AS sends back the session request to the S-CSCF, the last task of the S-CSCF is to route the request to the UE which has registered the public identity of user B. The S-CSCF scans its 'bindings' table: If a UE is registered with this public identity, the binding entry stores the 'Contact' and 'Path' headers for this UE. The S-CSCF replaces the SIP request URI with the contact of the 'Contact' header (this operation is the 'retarget'), and inserts all the URIs listed in the binding Path headers into as many 'Route' headers, thereby, forcing the session path to reach all SIP servers which indicated at registration time that they needed to see all SIP requests establishing a session to B.

One such server is the P-CSCF handling the UE of B, and therefore the SIP request reaches that P-CSCF, which forwards it, finally, to the UE of user B.

### 4.2.3 A few remarks on the IMS architecture

- Unlike the IN architecture (based on a standard call model), which can involve specific ASs each time the call reaches a defined state (a 'Detection Point'), the IMS service architecture involves all ASs at the beginning of the SIP session, even though some of these ASs may only be useful if the session reaches certain states (e.g., forward on no-answer service). This reflects the general idea that each signaling loop through an AS 'costs nothing', i.e., does not impact media streams and therefore is acceptable.

  In practice, however, the difference between the IN and the IMS models is negligible: virtually, all real life IN applications use triggers for the majority of call states, and virtually all real life IMS ASs handling telephony also need to see most SIP session messages.

- The amount of signaling 'ping pong' between the P-CSCFs, I-CSCFs, S-CSCFs and the various ASs does impact the session set-up time, which can become significant and negatively impact the user experience. In practice, however, each 'type' of application, e.g., telephony, is usually handled by a single AS, and the latency of messages from the I-CSCF to the S-CSCF and the various ASs is low, since it remains within the home network service platform. In a full mobility situation, where the home domains of A and B are distinct, the latency will be much higher, but not higher than what cellular phone users already experience.
- The IMS architecture is particularly well adapted to mobility situations, because it natively handles roaming. However, IMS in itself is only a service platform, not a mobility platform. IMS terminals cannot change the IP address without consequence: at a minimum, they need to terminate all SIP sessions and re-register. The underlying IP transport layer (IP Connectivity access network or IPCan) is responsible for handling terminal mobility: 3GPP uses UTRAN and GPRS tunnels, 3GPP2 and WiMax use mobile IP.

## 4.3 THE IMS CSCFs

### 4.3.1 The Proxy-CSCF

The P-CSCF plays the role of a SIP 'outbound proxy' for the IMS terminal, i.e., it is the first contact point of the IMS network. For this reason, the P-CSCF is a very important server in the IMS architecture:

- It forwards the UE requests to the entry point of its home domain, which is responsible for originating services.
- It ensures the safety of the IMS core network by filtering SIP messages.
- It ensures the confidentiality and security of communications between the IMS UUE and the IMS core, by using IPSec tunnels.
- It has access to all SIP signaling and therefore knows the characteristics of the media streams being set up. It can interface with packet filters to ensure media level policy enforcement (e.g., compliance with token bucket parameters see companion book, QoS chapter), and media related QoS measurements.
- It can duplicate signaling messages and forward them to lawful interception authorities and also instruct packet filters to duplicate media streams.
- It can locally perform the routing of special calls, such as emergency calls.

The P-CSCF may also handle the compression of SIP messages on the P-CSCF/UE link. This mechanism is called 'SigComp' and is defined in RFC 3581 using a dictionary defined in RFC 3485. This compression mechanism is not the best thing IMS has specified: it is a complex state machine-based binary compression scheme, which decreases the robustness of the SIP message transmission as well as defeats the original 'simplicity' of the SIP text based, human readable encoding. Using ASN.1 PER, an extremely efficient binary encoding mechanism already widely used over telecom networks, would have been a much better choice.

#### 4.3.1.1 Handling of REGISTER requests

After receiving the first REGISTER request from a UE:

- It finds the home domain by applying the procedures of RFC 3263 (DNS request, see Section 3.3.1.1.2) on the URI of the REGISTER request.
- It adds a Path header to the REGISTER request. The Path header is an IMS extension defined in RFC 3327 'SIP extension header field for registering non-adjacent contacts' (see Section 3.3.2.3). By putting its own URI in a Path header, the P-CSCF indicates to the S-CSCF handling this UE registration that any future SIP request to the UE must be routed through this P-CSCF URI.
- If the UE did not include a P-Access-Network-Info header in the request (it should), the P-CSCF adds one. The P-Access-Network-Info header is documented in RFC 3455 'Private Header extensions to SIP for 3GPP'. It contains information, specific to the access network technology, which helps locating the UE. For the 3GPP-UTRAN-TDD access network, for instance, it is a cell identifier (utran-cell-id-3gpp=<identifier>). This information is used by emergency services.
- It adds a P-Visited-Network-ID header, defined in RFC 3455. It is a character string identifying the visited network, which allows the home domain to check the existence of roaming relationships between the 'home' domain and the visited domain.
- It adds a P-Charging-Vector (RFC 3455), which contains a globally unique identifier (icid-value=<UID>) for a given SIP dialog or isolated transaction. This identifier is used to correlate the charging records generated by proxies.
- It analyzes the Security-Client header (specified in RFC 3329 'Security Mechanism Agreement') which contains the UE side information[3] necessary to establish two IPSec tunnels: one for the P-CSCF to terminal messages and one for the UE to P-CSCF messages. When it forwards the 401 Unauthorized response from the S-CSCF, the P-CSCF adds a Security-Server header (RFC 3329), which contains the server side information necessary for the set up of IPsec associations.

The second REGISTER request (following the 401 response to the first REGISTER) must be received by the P-CSCF *through the IPSec security association established after the first registration attempt*. The processing of this second request by the P-CSCF is identical to the first one, except for security aspects: the P-CSCF only checks that the

[3] List of supported algorithms, index of the security association, ports which should be secured.

content of the `Security-Verify` header is identical to the content of the `Security-Server` header which was sent in the previous `401` response. This check will detect attacks based on modifications of the Security-Server header which would remove the options corresponding to the most secure algorithms.

When receiving the 200 OK response from the S-CSCF, the P-CSCF:

- Records the content of the `P-Associated-URI` header (RFC 3455). This will be used by the P-CSCF to check that the UE uses only identities authorized by its home domain.
- Records the content of the `Service-Route` header (defined by RFC 3608 'Extension Header field for service route discovery during registration', see also Section 3.3.2.4 in Chapter 3). This will be used by the P-CSCF to check that the UE respects these routing constraints in all subsequent requests, and if needed, the P-CSCF may add the missing `Route` headers on the fly.

#### *4.3.1.2 Handling INVITE requests*

After receiving an `INVITE` request from the UE, the P-CSCF:

- Checks that the `P-Preferred-Identity` header contains one of the public identities authorized by the home domain for this UE, by comparing the URI with those listed in the previously recorded `P-Associated-URI`. If the public identity is allowed, the P-CSCF copies in a `P-Asserted-Identity header`. If the URI is incorrect, the P-CSCF may, according to its local policy, reject the session or place one of the allowed public identities of the UE in the `P-Asserted-Identity` header. These headers are defined in RFC 3325 ('Extensions to SIP for Asserted Identity within Trusted Networks').
- Checks that the `Via` header contains a transport address, which corresponds to the reception port of the IPSec security association configured during registration.

If this INVITE request receives a `183 Session Progress` response, the P-CSCF adds a `P-Media-Authorization` header in the 183 Session Progress. This header defined in RFC 3313 ('Private Session Initiation Protocol (SIP) Extensions') contains the authorization token for the UE media streams, which has been obtained by the P-CSCF from the network bandwidth controller (PDF or Policy Decision Function) through the IMS Gq interface. This token, the format of which is documented in RFC 3520 ('Session Authorization Policy Element'), will be transmitted by the UE to the network GGSN when activating the GPRS link, for the activation of the PDP context. The GGSN will check the authenticity of the token directly from the PDF, using the COPS protocol. This model, called Service-Based Local Policy, is documented in 3GPP document TS 23.207.

After receiving an `INVITE` request towards the UE, the P-CSCF:

- Adds a `P-Media-Authorization` header (see above).
- Removes the `P-Access-Network-Info` header, if present, as this header contains potentially sensitive information on the location of the caller.

### 4.3.2 The Serving-CSCF (S-CSCF) and Application Servers (AS)

#### *4.3.2.1 Role of the S-CSCF in the registration process*

The S-CSCF is responsible for the authentication of registration requests sent by the UE. When it receives such a registration request, the S-CSCF provides the IP Multimedia Private Identity (IMPI, the username of the Authorization header) to the HSS, and retrieves from the HSS a set of single use authentication vectors (see Figures 4.3 and 4.4). Each authentication vector contains:

- a random challenge, RAND;
- a network authentication token, AUTN—AUTN is generated from the shared secret K corresponding to the private ID contained in the ISIM (see Figure 4.5), and a sequence number SQN synchronized between the terminal and the HSS; a resynchronization procedure is described in RFC 3310;
- the expected response to the challenge, which should be sent by the legitimate UE, XRES;
- an integrity key to be used for the session, IK;
- a ciphering key to be used for the session, CK.

The S-CSCF picks the first authentication vector which has not yet been used to challenge the UE, builds a string containing RAND and AUTH (base 64 encoded), and places this string in the 'nonce' parameter of the '`WWW-authenticate`' header of the `401` response, together with the IK and CK keys.

The P-CSCF receives the `401` response, but removes CK and IK before forwarding the response to the UE.

The UE asks the ISIM to check the validity of AUTN. Once AUTN has been verified, the UE returns the response to the challenge, calculated from the challenge RAND, and secret K, in the '`response`' parameter of the `Authorization` header. The UE also calculates keys IK and CK from RAND and the shared secret K.

After comparing the challenge response sent by the UE in the second REGISTER, and the expected response XRES, the S-CSCF confirms the registration by sending a 200 OK response. The routing of the responses sent by the S-CSCF, as well as binding record which results from the new registration, has been described in Section 4.2.1.

#### *4.3.2.2 Role of the S-CSCF in the processing of new sessions*

During the registration process, the S-CSCF retrieves from the HSS the list of service profiles linked to the private identity being registered (Figure 4.7).

In this profile, the HSS will find all the public identities (IP multimedia public identity or IMPU) associated to the private identity (for telephony applications, there will be at

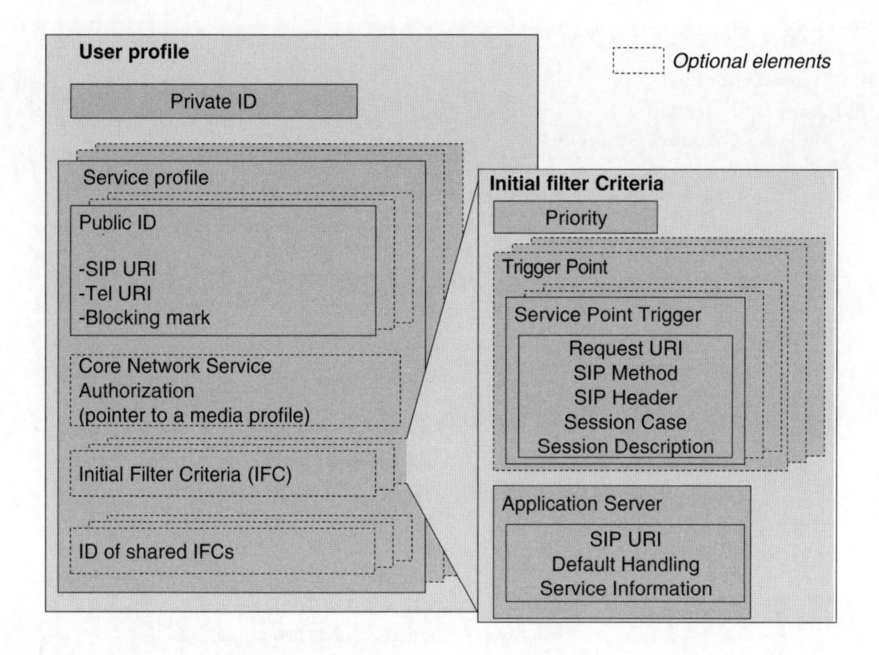


**Figure 4.7** User profile stored in the HSS.

least a SIP URI and a Tel URI), and will return this list in the `Contact` header of the 200 OK response which confirms the registration of the UE.

Each public identity (IMPU) is linked to a service profile, which contains:

- A pointer to a media profile (which defines the media types that the UE is allowed to negotiate in the SDP offer answer).
- The XML encoded 'IFC',. IFCs are specified in 3GPP TS 29.228.

An example IFC XML encoded service profile structure is given in Figure 4.8, also showing IFCs.

Each 'IFC' specifies a set of conditions to be met by a SIP request initiating a new SIP dialog to be routed to a given AS. These conditions are expressed as a set of filters on the Request-URI, the request method, the presence or absence of a given header, the direction of the request (session case: from or towards the UE) and SDP properties. IFCs are only evaluated at the beginning of a new SIP dialog.

The XML form of IFCs is extremely verbose and hard to read. In specification documents, IFCs are generally written as Boolean structures. For instance, the following expression would be a valid IFC associated to a presence AS:

(Method=REGISTER) OR (Method=SUBSCRIBE) AND Header=Event:presence) OR (Method=PUBLISH AND Header=Event:presence) OR (Method=SUBSCRIBE AND Header=Event.presence.winfo)

```
<?xml version=''1.0'' ?>
<IMSSubscription>
     <PrivateID>user</PrivateID>
     <ServiceProfile>
                    <PublicIdentity>
                           <Identity>sip:user@netcentrex.net</Identity>
                    </PublicIdentity>
                    <InitialFilterCriteria>
                           <Priority>0</Priority>
                           <TriggerPoint>
                                  <ConditionTypeCNF>1</ConditionTypeCNF>
                                  <SPT>
                                     <ConditionNegated>1</ConditionNegated>
                                     <Group>0</Group>
                                     <SessionCase>0</SessionCase>
                                  </SPT>
                                  <SPT>
                                     <ConditionNegated>0</ConditionNegated>
                                     <Group>1</Group>
                                     <Method>INVITE</Method>
                                     </SPT>
                           </TriggerPoint>
                           <ApplicationServer>
                               <ServerName>sip:127.0.0.1:5080;lr</ServerName>
                                  <DefaultHandling>0</DefaultHandling>
                           </ApplicationServer>
                    </InitialFilterCriteria>
                    <InitialFilterCriteria>
                          <Priority>1</Priority>
                          <TriggerPoint>
                                  <ConditionTypeCNF>1</ConditionTypeCNF>
                                  <SPT>
                                     <ConditionNegated>0</ConditionNegated>
                                     <Group>0</Group>
                                     <SessionCase>0</SessionCase>
                                  </SPT>
                                  <SPT>
                                     <ConditionNegated>0</ConditionNegated>
                                     <Group>1</Group>
                                     <Method>REGISTER</Method>
                                  </SPT>
                                  <SPT>
                                     <ConditionNegated>0</ConditionNegated>
                                     <Group>1</Group>
```

**Figure 4.8** Example user profile stored in the HSS (XML encoded).

```
                                        <Method>INVITE</Method>
                                    </SPT>
                                </TriggerPoint>
                                <ApplicationServer>
                                <ServerName>sip:127.0.0.1:5080;lr;orig</ServerName>
                                    <DefaultHandling>0</DefaultHandling>
                                    <ServiceInfo>Hello world</ServiceInfo>
                                </ApplicationServer>
                        </InitialFilterCriteria>
        </ServiceProfile>
        <ServiceProfile>
                        <PublicIdentity>
                                <Identity>sip:ims@netcentrex.net<Identity>
                        </PublicIdentity>
        </ServiceProfile>
    </IMSSubscription>
```

**Figure 4.8** *(continued)*

IFC filters are evaluated from highest priority (higher value of the `priority` XML property) to lowest priority. If the criteria conditions are met, then the request is forwarded to the corresponding AS, by inserting in the SIP request a `Route` header containing the value of the IFC `ServerName` property. If the request is a REGISTER, then the IFC `ServiceInfo` parameters are also added to the URI. The IFC `DefaultHandling` parameter defines whether the request processing should continue or not, if the defined AS cannot be reached or if it returns an error.

### *4.3.2.3 Example S-CSCF–AS interaction (number translation AS)*

In this example, the S-CSCF 'scscf.home.com' handles a SIP session matching an originating IFC filter, and routes the session to the associated AS 'as1.home.com'. The example shows how the S-CSCF inserts a `Route` header listing URI 'as1.home.com' (in bold font in Figure 4.9) as first hop, and 'scscf.home.com' as second hop. This instructs the AS to route back the SIP request to the S-CSCF after processing. Both the S-CSCF and the P-CSCF have requested to remain in the path of SIP messages for the rest of this session (see the `Record-Route` headers, in bold font). See Section 3.3.2.3.3 for more details on the usage of `Record-Route` and `Route` headers.

The original Request-URI is '0231463596@home.com'. The AS will reformat it as an E164 number.

After processing the SIP request, AS1 returns the INVITE request to the S-CSCF (Figure 4.10). This AS behaves as a back-to-back user agent (B2BUA, see Section 3.3.2) and translates the URI of the request, as well as the 'To' and 'From' headers (a pure SIP proxy would not be allowed to translate these headers). This AS is aware of the local dialing conventions of the user, and knows how to translate the source and destination

```
    INVITE sip:0231463596@home.com SIP/2.0
        Via: SIP/2.0/UDP
219.133.94.149:10548;branch=z9hG4bK2b4dc5f81;Role=2;Si=44;
    lcmsid=44;LCMID=S20,SIP/2.0/UDP
219.133.94.148:10226;branch=z9hG4bK907ad0e53;Role=2,SIP/2.0/UDP
62.161.167.198:5060;branch=z9hG4bK1222949536
        Route:<sip:as1.home.com;lr>,<sip:scscf.home.com:10548;lr;
    lcmsid=44;ORGDLGID=58-15>
        Record-Route:<sip:scscf.home.com:10548;lr;Role=2;Si=44;
    lcmsid=44;LCMID=S20;X-HwCsfCookie=45>,<sip:pcscf.home.com:
    10226;lr;Role=2;X-HwCsfCookie=22>
        Call-ID: f43847d9-94e57660@62.161.167.198
        From:<sip:+33231463592@home.com>;tag=ed6dc796-94e57660
        To:<sip:0231463596@home.com>
        CSeq: 12 INVITE
        Allow: ACK,INVITE,CANCEL,BYE,REFER,SUBSCRIBE,NOTIFY,
    PRACK,INFO,UPDATE
        Contact:<sip:+33231463592@62.161.167.198;transport=udp>
        Max-Forwards: 68
        Supported: timer,100rel,replaces
        User-Agent: DAVOLINK 0212002 050318
        P-Asserted-Identity:<sip:+33231463592@home.com>
        P-Charging-Vector: icid-value="pcscf.home.com.
    3233994664.95.14";orig-ioi=scscf.home.com
        P-Charging-Function-Addresses: ccf=CA1;ecf=ecf1
        Content-Length: 202
        Content-Type: application/sdp
    Message body
```

**Figure 4.9** SIP request from S-CSCF to originating AS.

numbers to the E164 international format. AS1 signals that it needs to remain in the path of SIP messages for the rest of this session by inserting a `Record-Route` header.

## 4.3.3 The Media Resource Function (MRF)

The Media Resource Function (MRF) is an important server in the IMS service architecture, as it is a centralized media processing resource available for all ASs. In principle, any AS may have its own media processing resources (in practice, this is still often the case), however, the IMS paradigm is that media resources should rather be shared, and that all ASs requiring such media resources should dynamically invoke an MRF.

As defined by 3GPP, the MRF can be decomposed into two subfunctions:

- the Media Resource Function Controller (MRFC), which behaves as a SIP user agent;
- the Multimedia Resource Function Processor (MRFP), which performs media processing.

```
INVITE sip:+33231463596@home.com SIP/2.0
        Via: SIP/2.0/UDP
62.161.167.195:5060;maddr=62.161.167.195;branch=z9hG4bKae3e43ef
    7db668c3fd7f4bc12e23649a
        Record-Route:<sip:62.161.167.195;lr>
        Route:<sip:scscf.home.com:10548;lr;lcmsid=44;ORGDLGID=58-
15>
        From: "DOE John"<sip:+33231463592@home.com>;tag=ed6dc796-
94e57660
        To:<sip:+33231463596@home.com>
        Call-ID: 1e0a712c-00000000@62.161.167.195
        P-Charging-Vector: icid-value="pcscf.home.com.
    3233994664.95.14";orig-ioi=scscf.home.com
        CSeq: 12 INVITE
        Contact:<sip:+33231463592@62.161.167.195>
        Supported: timer,100rel,replaces
        P-Charging-Function-Addresses: ccf=CA1;ecf=ecf1
        P-Asserted-Identity:<sip:+33231463592@home.com>
        User-Agent: NetCentrex CCS Softswitch/4.16.24
        Content-Type: application/sdp
        Max-Forwards: 67
        Content-Length: 202
   Message body
```

**Figure 4.10** Request returned after processing by AS1 to the S-CSCF.

The interface defined by the standard between the MRFC and the MRFP is H.248. In practice however, MRFC and MRFP functions are sold by the same vendors and it does not always make a lot of sense to force the communication between the MRFC and the MRFP to be fully standardized. The added value of MRF vendors often resides precisely in the ability to define richer media processing primitives going well beyond those defined in H.248.

The MRF can be used also in a pure IETF/SIP environment, without requiring the rest of the IMS environment (P-CSCF, S-CSCF, HSS, etc.). As such, it is probably one of the first IMS components to be sourced and installed while preparing SIP network migrations to an IMS architecture. In a UMTS R99 network, for instance, an MRF can be used to handle audio and video services (the latter either natively if the MRF supports the 3G-324M multiplex format, or via a 3G-324M gateway). An example application, illustrated in Figure 4.11, is a portal application using a video interface, which can be used by any 3G phone owner as well as video capable SIP terminals. Such portal applications can be used to create video-based interfaces to other services, e.g., to add video message capabilities to twitter for any 3G phone. Because these services use the standard 3G-324M capabilities present in most 3G phones, they are easier to deploy than services based on add-on applets to smartphones, and can address more users.

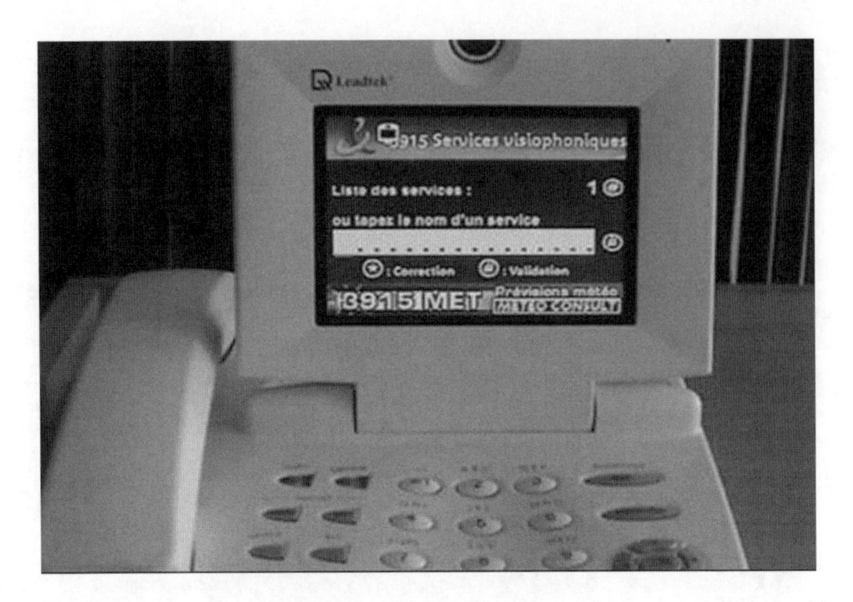


**Figure 4.11** Example video service performed by an MRF.

## 4.4 THE FULL PICTURE: 3GPP RELEASE 8, TISPAN

All the network elements described in our Overview of the IMS architecture (Section 4.3), were already present in the 3GPP release 5, which focused on future mobile networks as part of the LTE evolution. But many operators had short-term needs not covered by 3GPP R5: Packetcable focused on the requirements of cable network operators, ETSI TISPAN defined the extensions required for fixed networks, 3GPP2 had their own additions. In some cases, the requirements of 3GPP R5 were not realistic, e.g., the need for USIM-based authentication would have been too expensive for fixed networks.

3GPP releases 6, 7 and 8 merge and integrate the requirements of all these groups, and this version is also named the 'Common IMS'. For instance, the Access Gateway Control Function (AGCF), which had been defined in TISPAN, was added to the 3GPP IMS R7 architecture. The SIP digest authentication method, by far the most widely deployed in practice, was added as one of the possible authentication mechanisms.

Another major addition of 3GPP release 8 is the Evolved Packet System (EPS) architecture. The EPS was the result of the work of the Service Architecture Evolution working group of 3GPP and achieves the following objectives:

- It defines an evolution path from the GSM/GPRS and WCDMA/HSPA (3G) access networks to an all IP radio access network (RAN).
- It takes into account non 3GPP access networks (e.g., CDMA 2000, WLAN and WIMAX), so that they can be used as part of the EPS and by all applications on top of the EPS like the IMS. It defines handover procedures between 3GPP RANs

and non 3GPP RANs, and introduces the Access Network Discovery and Selection Function (ANDSF), which provides information on available access networks and operator preferences to the terminal.

- It accepts mobility models based not only on the 3GPP Generic Tunneling Protocol (GTP, TS 23.401) but also on IETF Proxy Mobile IP (PMIP, RFC 5213 and TS 23.402) and even based on the terminal (RFC 4877, Client Mobile IP using Dual Stack Mobile IPv6 DSMIP+) user data which is routed directly from the visited network.

## 4.4.1 The packet core domain: the evolved packet system

3GPP R8 defines the architecture of the Packet Core Domain and of the IMS domain, and standardizes all interfaces between the two domains. In this book, our focus is on the IMS domain, however, it becomes impossible to understand the IMS R8 architecture without a minimal knowledge of the EPS.

### *4.4.1.1 Evolution of the Radio Access Network (RAN): the Evolved-UTRAN*

Figure 4.12 shows the evolution of the RAN. With GSM and UTRAN, voice was handled separately in circuit mode by the Mobile Switching Center (MSC). With E-UTRAN (LTE radio access), that option disappears:[4] everything is packet based.

- The eNodeB hosts the radio transceiver and manages the radio resources (radio admission control, radio bearer control), and is capable of forwarding datagrams to another eNodeB for handovers.
- The GGSN functionality has been decomposed into a Mobility Management Entity (MME) which handles only signaling, and a Serving-Gateway (S-GW) which handles user traffic. The MME manages access control (using interface S6a to the HSS), bearer control and local mobility procedures (manages context transfer).
- The S-GW is the mobility anchor point for intra-3GPP handovers. When using PMIP, the S-GW creates a single IP tunnel per UE to the Packet Data Network Gateway (PDN-GW): all datagrams to and from the UE transit over this tunnel, regardless of the radio bearer that they will use. The S-GW routes the datagrams received from the IP tunnel to the appropriate radio bearer over the S1 interface to the eNodeB, based on dynamic packet filters configured by the Policy and Charging Control Function (PCRF). It filters downlink IP packets to the PDN and sets the appropriate QoS level (DiffServ codepoint). When GTP is used, the S-GW creates a GTP tunnel to the PDN for each Traffic class, and its functionality may be limited to mapping GTP-based S1 bearer to the corresponding GTP S5/S8 bearer tunnel: filtering and policing will be performed by the PDN.

[4] Except in handover situations when the terminal must fall back to UTRAN or GSM.

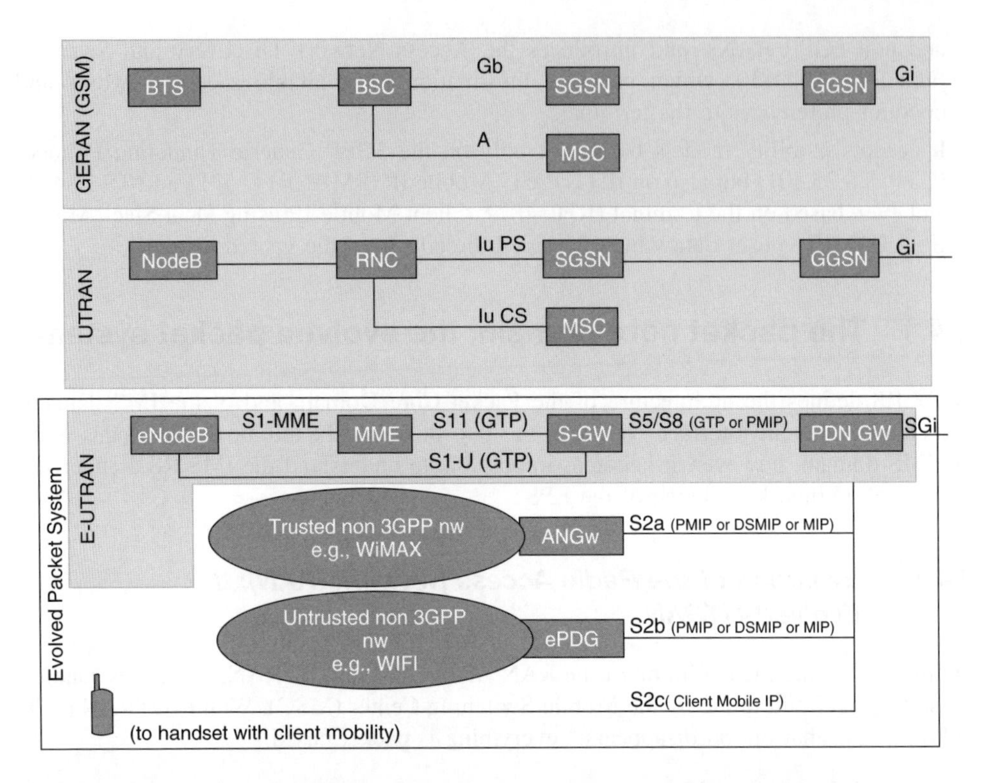


**Figure 4.12** Evolution of the Radio Access Network (GSM, UTRAN, E-UTRAN).

- The PDN is the anchor point (mobile IP home agent) for 3GPP/non 3GPP mobility situations. It is responsible for IP address allocation to the UE. In addition, it performs packet filtering and policing, and QoS marking, based on preconfigured rules, or rules configured dynamically by the PCRF. If the S-GW performs these filtering and policing functions (typically when PMIP is used as the protocol to the S-GW), the functionality of the PDN may be limited to forwarding packets to the PMIP tunnel.
- The Access Network Gateway (AN-GW) is used to connect trusted non 3GPP RANs to the EPC. UEs do not need to establish an IPsec tunnel to the AN-GW, and may use MIP or PMIP directly to communicate with the EPC. The AN-GW may act as a local mobility anchor point (PMIP foreign agent), the PDN acts as a PMIP home agent. Mobility tunnels may be cascaded.
- The evolved Packet Data Gateway (ePDG) ensures connectivity with untrusted non 3GPP RANs, like WiFi. The UE must establish an IPsec tunnel to the ePDG to access operator's services and to connect to the PMIP, acting as a PMIP home agent. The ePDG may implement mobility protocols like PMIPv6, in which case it plays the role of the PMIP foreign agent, the PDN acts as a PMIP home agent. Mobility tunnels may be cascaded.

### 4.4.1.2 The Policy and Charging Control (PCC)

The 3GPP Policy and Charging Control (PCC) architecture has become increasingly sophisticated with each new release.

- Release 6 used a simple architecture performing QoS and packet filtering (gate control), and targeted for IMS only. A PDF controls a Policy Enforcement Point (PEP), via the COPS protocol (Go).
- Release 7 addresses the issue, that not all services will be IMS based: it introduces a new component, the Subscriber Profile Repository (SPR), where the Policy and Charging Rule Function (PCRF, the new name for the PDF) can fetch the subscriber's policy and charging rules, independently of the IMS. In addition, it defines the online and offline charging mechanisms for the user data, and replaces the COPS protocol by a Diameter-based equivalent (RFC 3588, RFC 3589).
- Release 8 (Figure 4.13) takes into consideration the fact that two entities in the RAN are legitimate candidates for policy enforcement and accounting: the S-GW has access to radio bearer level parameters and can block packets closest to the source, and the PDN-GW. From the point of view of PCC, the former is named Bearer Binding and Event Reporting Function (BBERF, using interface Gxx to the PCRF) and the latter the Policy and Charging Enforcement Function (PCEF, using interface Gx to the PCRF.

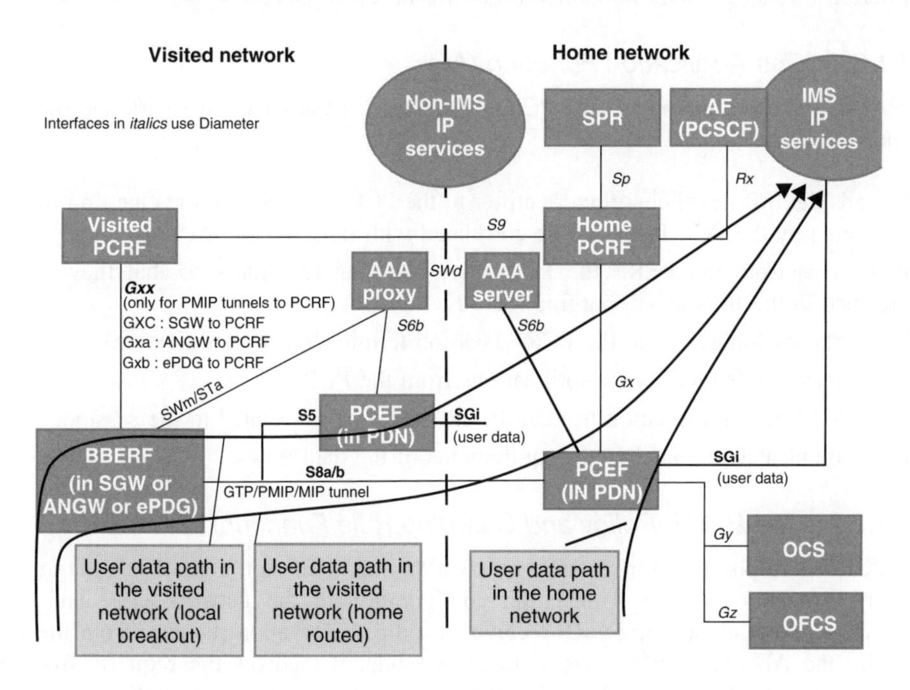


**Figure 4.13** EPS PCC infrastructure overview (UE data flows to the IMS).

The BBERF is specific to each trusted RAN technology (LTE uses interface Gxc for mobile IP-based networks, WiMax uses Gxa).
On networks using GTP as the protocol between the S-GW and the PDN, as GTP supports PCC related signaling, both the S-GW and the PDN can perform packet filtering and policing. On networks using PMIP as the protocol between the S-GW and the PDN, only the S-GW perform packet filtering and policing. Since PMIP does not define access control primitives, an out-of-band mechanism is used from the ANGW or ePDG to an AAA server for authentication (STa, SWa, SWm diameter interfaces to an AAA server, which itself may interface with the home network AAA server or directly to the HSS using interface Wx).

Two PCC mechanisms are defined:

- A pull mode, IMS independent, where the BBERF and PCEF inform the PCRF upon bearer establishment, and request policy and charging parameters. This mechanism enables QoS and charging control in an EPS for any IP application.
- A push mode, where a specific IMS function, the Application Function (AF) provides the PCRF with policy and charging parameters upon IMS session establishment. The AF defined in the context of the EPS is not a stand-alone entity (it usually is implemented by a P-CSCF), and should not be confused with the AS defined in the context of IMS, where it refers to a node implementing a SIP-based application.

#### *4.4.1.2.1 The Application Function (AF)*

The AF is used to interconnect the PCC architecture of the packet network and the IMS application layer:

- It sends the IMS level charging identifier to the PCC layer so that the charging records generated by the PCC layer can be correlated with those of the IMS layer.
- It receives from the PCRF the PCC layer charging identifiers so that they can be included in the IMS layer charging records.
- It reports session events to the PCRF (session termination or modification).
- It receives bearer level event notifications from the PCRF.
- It sends IP filter information to identify the data flow associated to the session.
- It communicates the bandwidth requirements of the data flows.

#### *4.4.1.2.2 The PCRF (Policy and Charging Rule Function)*

The PCRF is responsible for resource reservation and admission control in the transport network, Network Address Translation configuration at the network edge and NAT traversal. In push mode, the PCRF receives resource reservation requests from the IMS (sent by the AF over interface Rx). In push mode, it receives the requests from the network, typically for non-IMS applications.

The PCRF verifies whether the resource requests match the policy rules defined for the requestor. The PCRF accesses the policy rule database stored in the HSS

(Subscription profile repository, not yet standardized as of March 2010) through interface Sp.

In addition, it coordinates the admission requests sent to the PCEF (through interface Gx) and BBERF (interface Gxx), when applicable.

The PCRF and the overall 3GPP charging and policy control architecture are specified in TS 23.203.

The standard interfaces of the PCRF:

| Interface name | Connected subsystems | Purpose |
|---|---|---|
| Gx | PCRF/PCEF | Diameter. PCEF (e.g., GGSN for GPRS communications) requests policy and charging parameters to PCRF and PCRF sends policy parameters (PCC decision) to PCEF. Event reporting from PCEF to PCRF |
| Gxx | PCRF/BBERF | Diameter. Binding and event reporting function of the IP CAN in the home network (BBERF) requests policy and charging parameters to PCRF and PCRF sends policy parameters (PCC decision) to BBERF. Event reporting from BBERF to PCRF |
| Rx | PCRF/AF (PCSCF) | Diameter. Exposure of PCRF services to the service layer, enabling the service layer (e.g., P-CSCF) to request dynamic QoS requirements. It replaced Gq and is equivalent to TISPAN Gq′ |
| S9 | PCRF/PCRF | Diameter, Communication between PCRF in home network and PCFR in visited network of the user, enabling the home PCRF to have dynamic PCC over the visited PCEF/BBERF, and to serve Rx authorizations and subscriptions from an AF in the visited network |
| Sp | PCRF/HSS SPR | Retrieval of policy parameters for a given subscriber ID (e.g., an IMSI) from the HSS SPR. The HSS SPR can also notify the PCRF of policy parameter changes |
| Go (obsolete) | PDF/GGSN | Diameter, Charging correlation between IMS and GPRS |
| Gq(obsolete) | PDF/PCSCF | Diameter, Exchange of policy related queries and responses. Replaced by Rx |

#### 4.4.1.2.3 Policy and Charging Enforcement Function (PCEF)

The PCEF implements resource reservation, packet filtering (gates), policing and marking, under control of the PCRF through interface Gx.

The standardized interfaces of the PCEF:

| Interface name | Connected subsystems | Purpose |
|---|---|---|
| Gx | PCEF/PCRX | Diameter. PCEF requests policy and charging parameters to PCRF and PCRF sends policy parameters (PCC decision) to PCEF |
| Gy | PCEF/OCS | Diameter. Event reporting to the Online Charging System |
| Gz | PCEF/OFCS | Diameter. Event reporting to the Offline Charging System |

The PCC rules can be pre-provisioned into the PCEF and only referenced by the PCRF over interface Gx, or can be set dynamically. A PCC rule is composed of:

- a rule identifier;
- a Service data flow template (a list of data flow filters);
- a precedence level, which is used to order the list of service data flow templates used to identify a given service data flow;
- a charging key, which will be used by the online or offline billing system to identify a tariff;
- a service identifier: the PCEF will accumulate the measurements of all PCC rules with identical charging key and service identifier;
- a charging method (online, offline, both or no charging);
- a measurement method (volume, duration, event based upon identification of preset PCC rules);
- an identifier provided by the AF for correlation purposes;
- a gate status for the service data flow matching the service data flow template: open or closed;
- a QoS class identifier;
- the maximum and guaranteed bit rates for the uplink, and for the downlink, data flows.

#### 4.4.1.2.4 The Bearer Binding and Event Reporting Function (BBERF)

The BBERF offers the same capabilities as the PCEF; in addition, it can associate data flows to specific radio bearers. The PCRF need to configure its policy parameters on the BBERF, as opposed to the PCEF, when it uses PMIP tunneling.

The BBERF is controlled by the PCRF through interface Gxx, and more specifically through:

- Gxc, when the BBERF is an SGW using PMIP tunneling to the PCRF.
- Gxa, when the BBERF is an ANGW (WiMAX IP CAN) using PMIP tunneling to the PCRF.
- Gxb, when the BBERF is an ePDG (WiFi IP CAN) using PMIP tunneling to the PCRF.

### *4.4.1.3 The TISPAN transport layer*

The EPS architecture outlined above focuses on wireless mobile network access. The ETSI TISPAN committee focuses, instead, on fixed networks. The TISPAN reference architecture is described in ETSI ES 282 001. As illustrated in Figure 4.14, the TISPAN architecture isolates two functional layers, each comprising a number of functional subsystems:

- The Service Layer, comprising the Core IMS as described in the previous section, and subsystems dedicated to particular applications, such as the PSTN/ISDN Emulation Subsystem (PES) for telephony, and common components used by several subsystems (charging, user profile management, ENUM routing, etc.)

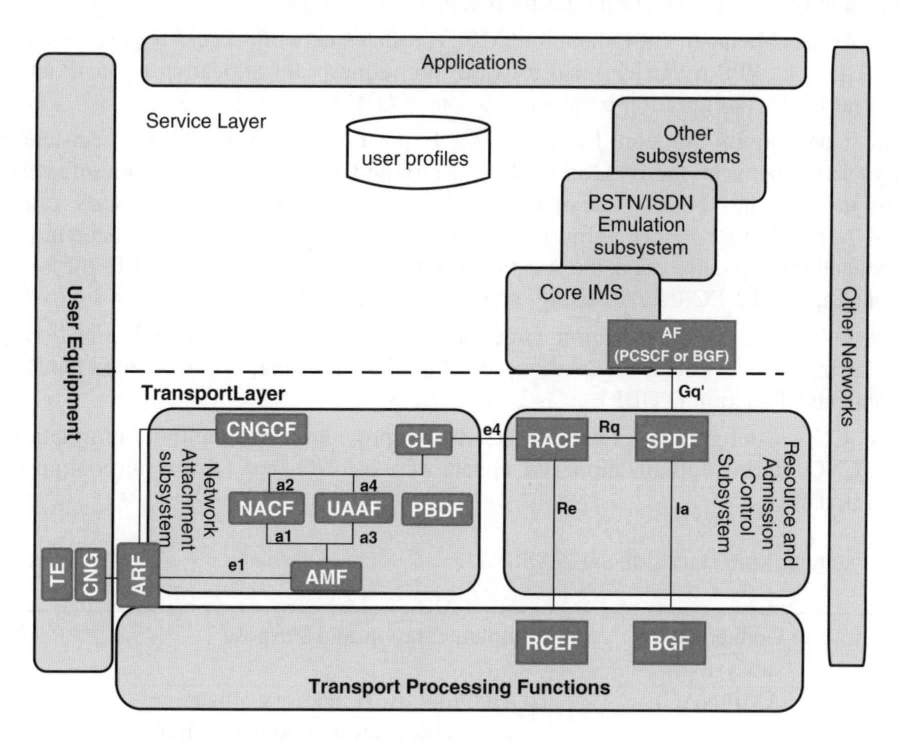


**Figure 4.14** The functional layers of the ETSI TISPAN reference architecture.

- The IP transport layer, comprising the Network Attachment subsystem (NASS), which controls the access to the IP network through the various physical transport layers suitable for IP communications, and the Resource and Admission Control Subsystem (RACS), which controls the use of network resources.

The TISPAN interfaces are not yet fully aligned with those of 3GPP. Probably, in a future release interfaces, Gq′ and Rx will be aligned, and the RACS will appear as a specific flavour of 3GPP PCRF. The Ro and Gy interfaces (offline charging), and Rf and Gz interfaces (online charging) are also likely to align over time.

#### *4.4.1.3.1 The Network Attachment Subsystem (NASS)*

The Network Attachment Subsystem (NASS) is accessed by Customer Network Gateways (CNG) or UE through the Access Relay Function (ARF, typically a DHCP relay or PPPoE relay).

The NASS is responsible for IP address allocation to user devices (including any authentication required during the address allocation and association of the allocated IP address with a physical location) and access network configuration according to the user profile. It is fully defined in ETSI ES 282 004 and is composed of the following functions:

- The Network Access Configuration Function (NACF) is responsible for IP address allocation and network configuration parameters.
- The Access Management Function (AMF), translates network access requests issued by the UE (e.g., PPP to Radius) and forwards the requests for allocation of an IP address and network configuration parameters to the NACF.
- The Connectivity Session Location and Repository Function (CLF), registers the association between the IP address allocated to the UE and related network information provided by the NACF (location, uplink and downlink bandwidth, QoS profile), and responds to network information queries by applications (e.g., emergency call applications typically need the location information, which is inserted in the session signaling by the PCSCF/AS using interface e2 to the CLF).
- The User Access Authorization Function (UAAF) performs user authentication and authorization checking for network access, based on information contained in the Profile Data Base Function (PDBF).
- The CNG Configuration Function (CNGCF) provides configuration information to the CNG, during initialization and update of the CNG (e.g., firmware configuration and updates).

The standard interfaces of the NASS:

| Interface name | Connected subsystems | Implementation and Purpose |
|---|---|---|
| a1 | AMF/NACF | DHCP/radius. IP address allocation, authentication and authorization |
| a3 | AMF/UAAF | Radius/diameter. Authentication and authorization |

| | | |
|---|---|---|
| e1 | UE-NASS | DHCP or PPP. IP address allocation and network configuration parameters |
| e2 | NASS-service layer subsystems NASS (visited NW)/NASS (home network) | Diameter. Export of information on access sessions and notification services. In a roaming configuration, e2 enables information on active IP connectivity sessions to be exchanged across networks |
| e3 | NASS-UE | HTTP, FTP, TFTP. Configuration of the UE by NASS |
| e4 | NASS-RACS | Diameter. See RACS. Export of subscriber access profile information |
| e5 | NASS (visited NW)/NASS (home network) | Diameter or Radius. e5 enables distributed execution of user authorization and authentication procedures to take place between a visited network and a home network |

#### *4.4.1.3.2 The Resource and Admission Control Subsystem (RACS)*

In TISPAN networks, the RACS performs admission control (for the access and aggregation segments of the network), resource reservation (for the access, aggregation and core network), policy control, remote NAT traversal control and NAT control at the edges of the core network. In order to perform these functions, the RACS has access to a topology map of the access and aggregation network, either through local configuration or by accessing external databases.

Both the push (AF requests a given policy to be applied for data streams related to an IMS session) and pull models (traffic policies are requested by the transport network nodes) are supported by the RACS.

The RACS subsystem is defined in TISPAN ES 282 003 and is composed of the following entities (Figure 4.14):

- **The AF**: this function is typically hosted in the P-CSCF or in the Interconnection Border Control Function (IBCF). It maps the session level QoS parameters (e.g., SDP offer answer information) to policy requests (push mode), and sends those requests to the SPDF over interface Gq′.
- **The Service PDF (SPDF)**: the SPDF implements the service-based policy (authorization of requests based on administrative rules) based on the information provided by the AF over Gq′ or other SPDFs over Rd′ (intradomain) or Ri’ (interdomain): requestor name, service class, service priority (e.g., emergency), reservation class, bandwidth. It hides the network topology from the AF and acts as a resource reservation mediation function, coordinating the admission requests to the appropriate Resource and Admission Control Functions (RACFs) (using interface Rq to specify the Subscriber Id or IP address requesting the resource, the Requestor Name/Service Class, the media description and service priority), BGFs (using interface Ia) or remote SPDFs (interface Rd′ and Ri’). The SPDF correlates all Resource Bundle-Ids identifying the reserved resources

from RACFs, BGFs and remote SPDFs to the original resource request received from the AS, so that the resources can be released when the IMS session terminates.

- **The RACF**: an RACF managing the access network is called the Access-RACF (A-RACF), and an RACF managing the core network resources is called the Core-RACF (C-RACF). The RACF is responsible for admission control, i.e., decides if it can grant resource requests received from the SPDF from interface Rq (push mode) or from network elements over interface Re (pull mode) based on its knowledge of current resource usage and existing reservations. Each reservation is identified by a Resource Bundle-Id.

  In the case of the A-RACF, the required subscriber QoS profile is retrieved from the Connectivity session Location and repository Function (CLF) of the NASS over interface e4. This profile contains subscriber attachment information (Subscriber ID, Physical Access ID, Logical Access ID, Access Network Type and Globally Unique IP Address) and QoS Profile Information (Transport Service Class, uplink Subscribed Bandwidth, downlink Subscribed Bandwidth, Maximum Priority, Media Type and Requestor Name, Initial Gate Settings). RACFs can be deployed in hierarchical more, each lower level RACF managing the resources delegated by the higher level RACF: the interface between RACFs is Rr (TR 183 070).

- **The policy enforcement nodes**. Generic transport network nodes which implement packet filtering, QoS marking and policing play the role of the Resource Control Enforcement Function (RCEF). The C-BGF and I-BGF also perform the functions of an RCEF, plus usage metering, NAT and remote NAT traversal for the C-BGF.

For charging purposes, the RACS can generate offline records comprising Charging Correlation Information (CCI, generate locally if not provided by AF), Request Type (Resource reservation, modification or release), Requestor Information, Subscriber Information (Subscriber ID), Service Priority, Media Description, Commit ID, Time Stamp and Reason.

The standard interfaces of the RACS:

| Interface name | Connected subsystems | Purpose |
|---|---|---|
| e4 | NASS/A-RACF | Diameter. Export of subscriber access profile information |
| Gq′ | SPDF-AF | Diameter. Exposure of RACS services to the service layer, enabling the service layer (e.g., P-CSCF) to request dynamic QoS requirements. It replaced Gq and is equivalent to 3GPP Rx |
| Ia | SPDF/BGF | H248. Control of Nat, hosted NAT traversal and gating. Retrieval of media flow statistics |
| Rd′ | SPDF/SPDF (intradomain) | Resource management services within a single domain |

| | | |
|---|---|---|
| Re | RACF/RCEF | Diameter. Control of gating and policing parameters of transport nodes. Event reporting |
| Rf/Gz | RACF, SPDF/ CCF, OFCS | Diameter. Offline charging |
| Ro/Gy | RACF, SPDF/ CCF, OFCS | Diameter. Online charging |
| Ri' | SPDF (domain1)/ SPDF (domain2) | Resource management services between interconnected domains |
| Rq | SPDF/RACF | Diameter. Admission request from SPDF |
| Go (obsolete) | PDF/GGSN | Charging correlation between IMS and GPRS |
| Gq(obsolete) | PDF/PCSCF | Exchange of policy related queries and responses. Replaced by Gq′ (TISPAN) Rx (3GPP) |

#### *4.4.1.3.3 The Border Gateway Function (BGF)*

The Border Gateway Function (BGF) interfaces two IP transport domains (Core-BGF or C-BGF between access and core network, and I-BGF between two core networks). It provides the same functionalities as an RCEF, plus (optionally) usage metering, NAPT, NAT traversal, transcoding. In pre-IMS terms, it is a session border controller sitting between two networks. The BGF is defined in 3GPP TS 23 228 (ETSI TS 123 228).

The standard interfaces of the BGF are:

| Interface name | Connected subsystems | Implementation and purpose |
|---|---|---|
| Di | C-BGF/access network | IP. |
| Ia | BGF/RACS | H.248 configuration of the BGF by the RACS. Control of NAT, hosted NAT traversal and gating. Retrieval of media flow statistics. |
| Ds | C-BGF, I-BGF/transport network entities | IP. |
| Iz | I-BGF/other core network | |

## 4.4.2 The IMS domain

The 3GPP R8 IMS domain reference architecture is described in document 3GPP TS 23.002/ETSI TS 123002 and 3GPP TS 23 228.

In this section, we consider only the packet communication domain (the full 3GPP architecture also covers circuit mode communications).

Figure 4.15, which does not include the functions and interfaces of the packet layer, provides a general overview of the IMS service layer functional architecture. This intimidating diagram gives a feel for the complexity of 3GPP R8. However, the IMS service layer is not more complex than the service layer of most current generation networks, it is only more standardized: most of the interfaces named and represented explicitly in the IMS architecture also exist in current non-IMS networks, but many of them are vendor specific and proprietary ('load balancers', 'front ends', 'OSS systems', etc.).

The functional subsystems defined by 3GPP and TISPAN do not necessarily correspond to physical servers: a single server can implement several functions (in which case the interfaces between these functions remain internal), and a given function may be implemented on several servers if the scale of the network justifies it.

The paragraphs below briefly explain the purpose of each function and describe the key interfaces. The 'Core IMS' is a subset of the 3GPP IMS defined in ETSI TS 123 002 and ES 282 007, comprising the EPS, HSS, SLF and CSCFs.

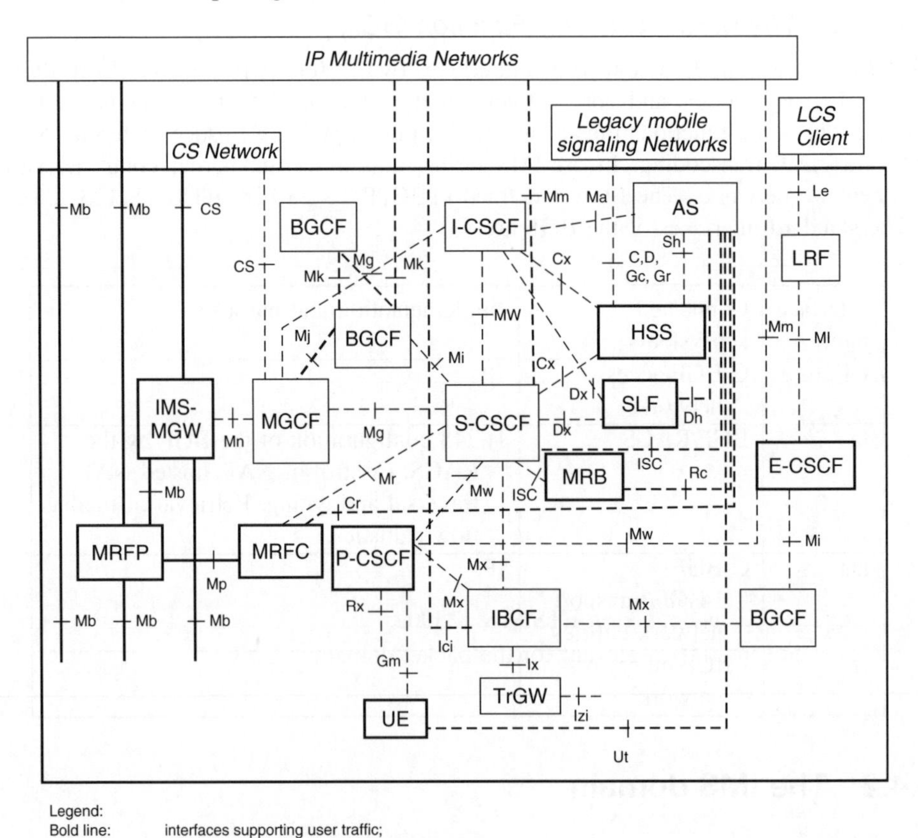


**Figure 4.15** Functional architecture of the IMS service layer (from 3GPP TS 23.002).

#### 4.4.2.1 The Home Subscriber Server (3GPP) or User Profile Server Function (TISPAN)

The HSS and the User Profile Server Function (UPSF), are user profile databases. The HSS defined by 3GPP is a superset of the HLR server used in current GSM cellular networks: it stores the user profile and location related information required by the HLR function and implements the Authentication Center (AUC) function which generates the authentication vectors required to authenticate terminals during the network attachment phase (see Sections 4.2.1 and 3.1.2.1). The user profile data format is described in TS 23.008.

The TISPAN UPSF also stores the user profile information related in particular to authentication, numbering and addressing, security and location. However, it does not implement the HLR/AUC functionality.

In order to ensure scalability, several HSS/UPSF can be deployed. The Subscription Locator Function (SLF) is used to find the appropriate HSS/UPSF for a given user of service.

HSS/UPSF standard interfaces relevant for the packet domain:

| Interface name | Connected subsystems | Implementation and purpose |
|---|---|---|
| Cx | HSS/I-CSCF, S-CSCF | Diameter, TS 23.228. Identification of correct SCSCF, authorization and authentication, retrieval of user profile |
| Sh | AS/HSS | Diameter-based interface. Exchange of application level user or service profile |

#### 4.4.2.2 The Subscription Locator Function (SLF)

The SLF is used by ASs and service control subsystems to retrieve the identity of the UPSF or HSS, where the service-level profile of a particular user or public service is available. The SLF is defined in ETSI TS 123002.

SLF standard interfaces relevant for the packet domain:

| Interface name | Connected subsystems | Implementation and purpose |
|---|---|---|
| Dh | AS/SLF | Diameter-based interface. Query to get the correct HSS for a given subscriber |
| Dx | CSCF/SLF | Idem |

#### 4.4.2.3 P-CSCF, S-CSCF, I-CSCF

The P-CSCF, I-CSCF and S-CSCF are described in TS 23.228 and have been discussed in Section 4.3. 3GPP also introduces an Emergency-CSCF (E-CSCF) described in TS 23.167.

Standard CSCF interfaces in the IMS core:

| Interface name | Connected subsystems | Implementation and purpose |
|---|---|---|
| Cx | I-CSCF, S-CSCF/HSS | Diameter-based interface, retrieval of the user or service location from the HSS/UPSF |
| Dx | I-CSCF, S-CSCF/SLF | Diameter-based interface, retrieval of the HSS/UPSF location for a user or service |
| Mm | I-CSCF, S-CSCF/external network | SIP. Exchange of signaling with other IP networks |
| Mw | CSCFs | SIP, interface between P-CSCF, I-CSCF and S-CSCF |
| Rf/Gz | CSCF/CCF, OFCS | Diameter. Offline charging |
| Ro/Gy | S-CSCF/ECF, OCS | Diameter. Online charging |

#### 4.4.2.4 The Application Server Function (ASF or AS)

The ASF offers VAS.[5] It may be located in the user's home network or hosted by an application provider. The ASF is specified in ETSI TS 123 002.

Some ASs have a specific name, e.g.,

- IM SSF: An AS implementing the IN call model (see Section 4.1.2), which can be interfaced with legacy IN SCP() through CAMEL or INAP.
- The Service Centralization and Continuity AS (SCC AS) defined in TS 23 292 and TS 23.237, formerly called the Voice Call Continuity server (VCC) is an AS managing the handover of sessions between a circuit switched access network (e.g., GSM) and an IMS access network. It is typically used to manage hybrid WiFi/Cellular handsets.
- OSA AS, which implements the IN like OSA interface (defined in TS 22 127).

ASF standard IMS interfaces:

| Interface name | Connected subsystems | Implementation and purpose |
|---|---|---|
| Dh | ASF to SLF | Diameter. Identification of relevant HSS for a user or service profile |

[5] it should not be confused with the AF defined in the Enhanced Packet Core architecture, which is usually implemented in the P-CSCF.

| | | |
|---|---|---|
| ISC | ASF to SCSCF | SIP, see Section 4.3.2 AS. See TS 23.228 |
| Ma | ISCSF/AS | SIP. Used by ICSCF to reach Public Service Identities managed by the AS |
| Rf/Gz | AS/CCF, OFCS | Diameter. Offline charging |
| Ro/Gy | AS/ECF, OCS | Diameter. Online charging |
| Sh | ASF to HSS | Diameter. Retrieval of application specific profile information |

#### 4.4.2.5 The Media Resource Function Controller (MRFC)

The MRF has been discussed in Section 4.3.3. The MRF can be decomposed into:

- an MRF Controller, which interfaces at SIP level with the S-CSCF;
- an MRF Processor (MRFP), which implements media processing functions. If the architecture is decomposed, the MRFC controls the MRFP via interface Mp (H.248).

A pool of MRFCs can be managed by a Media Resource Broker (MRB).

MRFC standard interfaces:

| Interface name | Connected subsystems | Implementation and purpose |
|---|---|---|
| Cr, Sr | MRFC/AS | HTTP-based interface used by the MRFC to retrieve media files or scripts (e.g., VoiceXML) from the AS. Defined in TS 23.218 |
| Mp | MRFC/MRFP | H.248. Control of MRFP by MRFC in split implementations |
| Mr | S-CSCF/MRFC | SIP |
| Rf/Gz | MRFC/CCF, OFCS | Diameter. Offline charging |
| Ro/Gy | MRFC/ECF, OCS | Diameter. Online charging |

#### 4.4.2.6 The Multimedia Resource Function Processor (MRFP)

The MRFP may support announcement server, conferencing or other media processing functions. It is defined in ETSI TS 123 002.

The MRFP is controlled by an MRFC through interface Mp (H.248 protocol).

MRFP standard interfaces

| Interface name | Connected subsystems | Implementation and purpose |
|---|---|---|
| Mb | MRFP/user data plane | SIP |
| Mp | MRFC/MRFP | H.248. Control of MRFP by MRFC in split implementations |

### 4.4.2.7 The PSTN/ISDN Emulation Subsystem (PES)

The PES supports the emulation of PSTN/ISDN services for legacy terminals connected to the IMS network through residential gateways or access gateways. This service is called 'PSTN emulation'. The PES is defined in ES 282 002 and TS 182 012, the PES architecture is illustrated in Figure 4.16.

The PES uses SIP with encapsulated ISUP to ensure interworking between circuit switched telephony networks and ISDN and analogue telephony accesses. The intent is to standardize all the telephony features required in a public electronic communications network.

In the initial release of the specification (v1.1.1, march 2006), the PES was designed as a stand-alone system, a peer to an IMS system. However, it was clear from the beginning that the internal architecture of the PES could be realized as part of a standard IMS architecture, and this implementation is further described in TS 182012.

The only specific component of the PES is the AGCF, which acts as a SIP User agent towards the IMS (generates Registrations on behalf of the connected gateways), performs the role of the P-CSCF (generates asserted identities and charging identifiers, interfaces with the RACS) and controls the residential media gateways via H.248, MGCP or any other suitable protocol (P1 interface). The AGCF performs all functions required to adapt stimulus analogue telephone interfaces to the functional model of SIP networks: it generates dial tone, detects events such as hook flash, it may provide local conferencing services. For basic calls, the mapping to SIP is easy and digit accumulation is provided by the AGCF based on digit maps. However, many countries require overlapped signaling

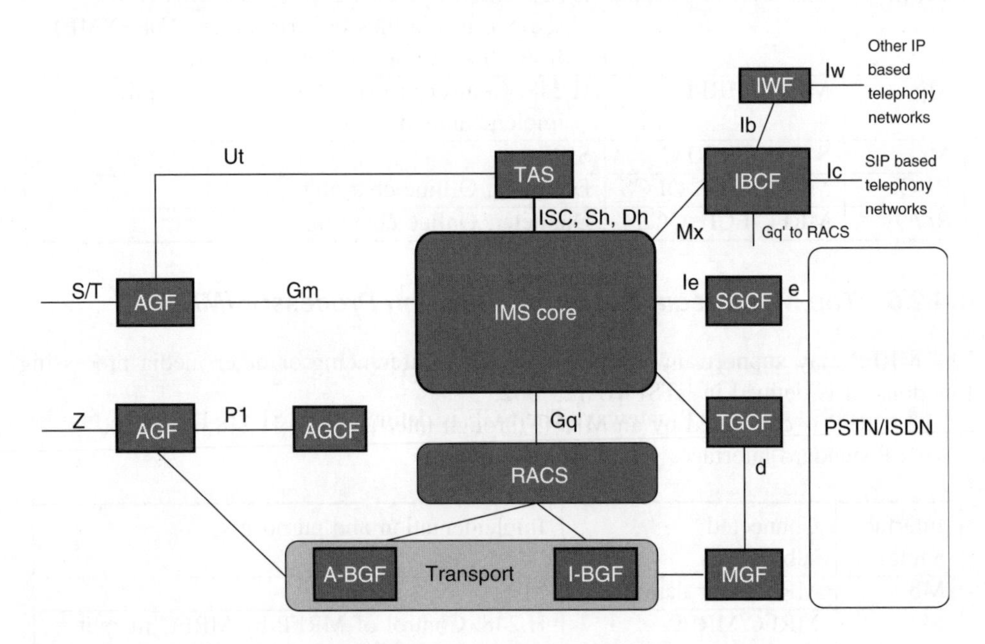


**Figure 4.16** TISPAN PES architecture.

procedures, and TISPAN had to study the mapping of overlapped signaling to SIP (TR 183 056). Also, there is no 'natural' way of mapping some stimulus events to SIP (e.g., hook flash to SIP): TISPAN standardized a set of 'tricks' (usually based on creating a SIP new session) to allow an AS to have control over the user experience of the analogue line user.

The other functional blocks used in the PES are described below.

#### 4.4.2.8 The Media Gateway Function (MGF, AGF, A-MGF, R-MGF, RG, TGF, T-MGF)

The MGF is a media processing VoIP gateway interfacing circuits and the packet domain. Signaling is handled by the MGCF, which controls the MGF through interface Mn (H.248).

Gateways in the core operator network are Trunking Media Gateways Functions (TGF or T-MGF). Gateways located in the access network are Access-Media Gateway Functions (AGF or A-MGF) or Residential-Media Gateway Functions (R-MGF or RG).

#### 4.4.2.9 The Multimedia Gateway Control Function (MGCF, TGCF, AGCF)

The MGCF interfaces a circuit switched network and the CSCFs and Breakout Gateway Control Function (BGCF). It controls one or more MGFs using H.248. The MGCF may also be called TGCF when it controls a T-MGF, and Analogue Gateway Control Function (AGCF), when it controls analogue gateways (AGFs). An AGCF is significantly more complex than a TGCF, because it needs to implement all the logic required to adapt stimulus signaling to SIP.

| Interface name | Connected subsystems | Implementation and purpose |
|---|---|---|
| Mg | CSCF/MGCF | SIP. Mapping of ISUP signaling to SIP |
| Mn | MGF/MGCF | H.248 |

#### 4.4.2.10 The Signaling Gateway Function (SGF, SGCF)

An SS7 to SS7 over SCTP gateway. The SGF is defined in ETSI TS 123 002.. In other documents, e.g., ES 282 002 it is named SGCF (Signaling Gateway Control Function).

The SGF conveys the tunneled ISUP signaling to the MGCF

#### 4.4.2.11 The Breakout Gateway Control Function (BGCF)

The BGCF routes the SIP session when it leaves the IMS operator domain. If the next hop is a circuit network, it selects the MGCF.

The CSCFs interface with the BGCF through interface Mi (SIP).

| Interface name | Connected subsystems | Implementation and purpose |
|---|---|---|
| Mi | CSCF/BGCF | SIP |
| Mj | BGCF/MGCF | SIP |
| Mr | S-CSCF/MRFC | SIP |
| Rf/Gz | MRFC/CCF, OFCS | Diameter. Offline charging |
| Ro/Gy | MRFC/ECF, OCS | Diameter. Online charging |

#### *4.4.2.12 The Interconnection Border Control Function (IBCF)*

The IBCF is used as a gateway to external packet networks, and optionally provides NAT and Firewall functions. Since 3GPP R7, it takes the Topology Hiding Inter-network Gateway (THIG) role which was previously an optional function of the I-CSCF.

IBCF standard interfaces:

| Interface name | Connected subsystems | Implementation and purpose |
|---|---|---|
| Ic | IBCF to 3GPP/TISPAN network | SIP. Interface to a 3GPP or TISPAN network |
| Ib | IBCF/IWF | SIP. Interface to an interworking function which itself interfaces with other SIP profiles or protocols, e.g., H.323 |
| Rf/Gz | IBCF/CCF, OFCS | Diameter. Offline charging |
| Ro/Gy | IBCF/ECF, OCS | Diameter. Online charging |

#### *4.4.2.13 The Interworking Function (IWF)*

The Interworking Function (IWF) interfaces the IMS network with networks implementing other protocols (e.g., H.323) or other SIP profiles.

IWF standard interfaces:

| Interface name | Connected subsystems | Implementation and purpose |
|---|---|---|
| Iw | IWF to non 3GPP/TISPAN networks | Interface with networks implementing other SIP profiles or protocols e.g., H.323 |
| Ib | IBCF/IWF | Interface between IBCF and IWF |
| Rf/Gz | IWF/CCF, OFCS | Diameter. Offline charging |
| Ro/Gy | IWF/ECF, OCS | Diameter. Online charging |

#### 4.4.2.14 Charging: Charging Collector Function (CCF), Session Charging Function (SCF), Event Charging Function (ECF), Offline Charging System (OFCS), Online Charging System (OCS)

The TISPAN Charging Collector Function (CCF) and 3GPP Offline Charging System (OFCS, TS 32 240) collect offline charging records from the P-CSCF, I-CSCF, S-CSCF, BGCF, MRFC, MGCF, or AS. The interface used is Diameter Rf for the CCF and diameter Gz for the OFCS. The IMS Charging Identifier (ICID) is used as a unique identifier for the session by the charging records. In roaming situations, the origin and destination operators are identified by their respective Inter Operator Identifier (IOI).

Online charging is implemented via two interfaces:

- The SCSCF can route the session to a Session Charging Function (SCF). Since the SCF is in control of the session, it can terminate the session at any moment.
- ASs and MRFCs can implement diameter-based Ro interface to a TISPAN Event Charging Function (ECF), or the Gy interface to a 3GPP Online Charging System (OCS, TS 32 296).

#### 4.4.2.15 The user equipment

The UE may be:

- a monolithic terminal equipment unit (TE),
- or in Customer Premises Network (CPN) is composed of one or more Customer Network Devices (CND) connected to a Customer Network Gateway (CNG). The GNG may also include a Residential MGF (R-MGF) or other gateway functions. The GNG is defined in TS 185 003 and TS 185 006, and TR 185 007 provides a comprehensive study of the protocols applicable to a CNG and CND.

The standard interfaces for the UE:

| Interface name | Connected subsystems | purpose |
|---|---|---|
| Dj | UE/Transport processing | Exchange of media and media control flows with the IMS network |
| e1 | UE/NASS | DHCP or PPP, IP address allocation and network configuration parameters |
| e3 | UE/NASS | HTTP, FTP, TFTP. Configuration of the UE by NASS |
| Gm | UE/PCSCF and service layer | SIP, used for registration and session control |
| Ut | UE/Service Layer | HTTP/XCAP. Interaction with ASs: Management of information related to services and settings |

**Table 4.1** IMS SIP header extensions

| | | | |
|---|---|---|---|
| **P-Access-Network-Info** | 3GPP-UTRAN-TDD;utran-cell-id-3gpp=B31C4972E2 | RFC 3455 | Contains access-technology specific localization information |
| **P-Asserted-Identity** | 'John Doe'John@home.org;<tel:+1 212 123 4567> | RFC 3325 | Public identities verified by the P-CSCF |
| **P-Asserted-Service** | P-Asserted-Service: urn:urn-7:3gpp-service.ims.icsi.mmtel | draft | Draft-drage-sipping-service-identification-03. The SCSCF asserts the service that should be accessed by the user, after checking the HSS profile and the P-Preferred-Service desired by the user |
| **P-Associated-URI** | <sip :user-home@service.org>,<sip :+33 158713333@service.org;user=phone> | RFC 3455 | Lists the public identities that the UE is allowed to use, but which are not necessarily registered yet |
| **Path** | <sip :term@pcscf3.visited2.org; lr> | RFC 3327 | Before forwarding a REGISTER request, each proxy willing to be visited by any SIP request targeted to the UE which sent the REGISTER should add a Path header Before sending out requests to the UE, the S-CSCF will add the corresponding Route headers |
| **P-Called-Party-ID** | sip:user1-business@home.org | RFC 3455 | Contains the content Request-URI received by the final proxy before retargeting the destination |
| **P-Charging-Function-Addresses** | Ccf=192.1.1.1; ccf=192.1.1.2; ecf=192.1.1.3; ecf=192.1.1.4 | RFC 3455 | Preferred charging function addresses that proxies should attempt to use |
| **P-Charging-Vector** | Icid-value="zOrGlub34" | RFC 3455 | Globally unique charging vector for a given dialog (or an isolated transaction) |

| | | | |
|---|---|---|---|
| **P-Media-Authorization** | 003000100100200203040400444040e | RFC 3313 RFC 3520 | Authorization token obtained from the PDF |
| **P-Preferred-Identity** | 'John Doe'<John@home.org> | RFC 3325 | Public identity that the user agent is selecting as source of this request or dialog, among those previously registered |
| **P-Preferred-Service** | P-Preferred-Service: urn:urn-7:3gpp-service.ims.icsi.mmtel | draft | Draft-drage-sipping-service-identification-03. The client specifies the ICSI of the service it desires |
| **P-Private-Network-Indication** | P-Private-Network-Indication:company.com | draft | Draft-vanelburg-sipping-private-network-indication-03.txt Identifies private network traffic |
| **P-Profile-Key** | P-Profile-Key:<sip:chatroom-!.*!@example.com> | RFC 5002 | Provide SIP registrars and SIP proxy servers with the key of the profile corresponding to the destination SIP URI of a particular SIP request, e.g., when the destination is an AS handling a set of Public Service Identities which can be described as a regular expression. The AS profile is associated with the regular expression, specified in the P-Profile-Key as a Wildcarded Public Service Identity (3GPP TS 23.003) |
| **P-Served-User** | P-Served-User:<sip:user@example.com>; sescase=orig; regstate=reg | RFC 5502 | Contains the user id for which the AS must execute services. Fix introduced in 3GPP R8 for the cases (e.g., diversion) where the served user is not the user indicated in P-Asserted-Identity |

(*continued overleaf*)

Table 4.1 (continued)

| | | | |
|---|---|---|---|
| **P-Visited-Network-ID** | 'visited2 network' | RFC 3455 | String identifying the visited network |
| **Security-Client** | ipsec-3gpp; alg=hmac-sha-1-96; spi-c=12341344; spi-s=34563456;port-c=6789; port-s=4545 | RFC 3329 | Contains client side information that can be used to set up an IP-Sec association |
| **Security-Server** | ipsec-3gpp; q=0.1; alg=hmac-sha-1-96;spi-c=324534,spi-s=123434, port-c=1234; port-s=2345 | RFC 3329 | Contains client side information that can be used to set up an IP-Sec association |
| **Security-Verify** | ipsec-3gpp; q=0.1; alg=hmac-sha-1-96;spi-c=324534,spi-s=123434, port-c=1234; port-s=2345 | RFC 3329 | Information fed back over the IPSec channel to prove that the Security-Server header was not altered |
| **Service-Route** | <sip :orig@scscf.service.org; lr> | RFC 3608 | Present in the responses to REGISTER requests, this header lists a sequence of proxy URIs that any future request initiated by the UE will need to visit. Note the 'orig' keyword added by the S-CSCF, which allows the S-CSCF to know that it will need to evaluate the originating IFCs, not the terminating IFCs |

#### *4.4.2.16 Interconnection between networks*

TISPAN defines two types of Next Generation Networking Interconnection:

- **Service oriented Interconnection (SoIx):Service oriented Interconnection (SoIx)** Two network interfaces through SoIx implement end-to-end services (e.g., end-to-end QoS). The Ic (or Iw) service level interface and optionally the transport level Iz interface must be provided.
- **Connectivity oriented Interconnection (CoIx) Connectivity oriented Interconnection (CoIx):** An interconnection limited to the implementation of transport level interface Iz between the networks.

For instance, two networks may be connected directly through SoIx (limited to Ic) and indirectly through an intermediary network using CoIx (Iz).

### 4.4.3 Summary of SIP extensions required in an IMS network

3GPP TS 24.229 provides all the details of the 3GPP SIP profile for use with IMS networks. It introduces the following extension headers (Table 4.1):

The `Replaces` (RFC 3891) and `Referred-By` (RFC 3892) headers, which are not IMS specific, are also key in the mechanisms used by an IMS service platform.

## 4.4.2 Interconnection between networks

[illegible] two types of [illegible] Networking [illegible]:

- Service-oriented Interconnection (SoIX) [illegible] Interconnection [illegible] (QoS) [illegible] must be provided.
- Connectivity-oriented Interconnection (CoIX) Connectivity-oriented Interconnection [illegible] between the networks.

For instance [illegible] networks [illegible] directly [illegible]

## 4.4.3 Summary of SIP extensions required in an IMS network

3GPP TS 24.229 [illegible] SIP [illegible] for use with IMS networks [illegible] (Table [illegible]). [illegible] IMS [illegible]

[illegible] call control protocol and the media control protocol assume a stateful [illegible] endpoint [illegible] endpoint which implements its own call state machine.

# 5

# The Media Gateway to Media Controller Protocol (MGCP)

## 5.1 INTRODUCTION: WHY MGCP?

### 5.1.1 Stimulus protocols

SIP and H.323 are very similar session-based, stateful protocols. The similarity is hidden behind all the cosmetic differences due to different ways of serializing essentially the same information, but basically both protocols share the same characteristics:

- They are composed of a call control protocol (H.225.0, SIP) and a media control protocol (H.245, SDP offer–answer model), with the media control protocol encapsulated in the call control protocol.
- The call control protocol is a slightly simplified version of ISDN Q.931 (H.225.0), with a more basic way of closing connections (three messages in Q.931, only one message in H.225.0 and SIP—although this is likely to change since the single-message closing sequence causes some issues).
- Both the call control protocol and the media control protocol assume a stateful or 'intelligent' endpoint (i.e., an endpoint which implements its own call-state machine, and its own logic, such as for the handling of call waiting, providing a ring-back tone while off-hook, etc.)

From a marketing point of view, having an 'intelligent' client is always a good thing, since it seems so obvious that an intelligent client will be able to do 'more things' than a 'dumb' client.

*IP Telephony: Deploying VoIP Protocols and IMS Infrastructure, Second Edition* O. Hersent

The problem is that, when looking at the most sophisticated corporate phone installations today, *none* uses ISDN phones (the equivalent in traditional telephony of a smart phone)! In fact, most, if not all, of the corporate PBXs use another class of protocols called **stimulus protocols**, optimized for the control of dumb phones. It is easier to understand why with an analogy. If SIP or H.323 were programming languages, they would be very similar to the BASIC language. You can do a lot of things with BASIC as long as you do things for which BASIC has the proper instructions, but you can do many more things with the C language, or with an assembly language. If a stimulus protocol were a programming language, it would be a low-level assembly language: certain things take longer to code, but there is nothing you cannot do. For instance:

- When you pick up the handset of an H.323 or a SIP phone, you get ring-back. When you pick up the handset of a PBX phone, sometimes you get a message like 'you have voicemail'.
- On an H.323 or a SIP phone, you have feature buttons or lamps, hard-coded by the phone manufacturer, for hold, transfer, three-way calling, message-waiting indication, etc. On a PBX phone, you may want to assign any feature to any button, to control any lamp, exactly as you like.
- On an H.323 or a SIP phone (without proprietary extensions), you need to pick up the handset or press the loudspeaker button to get a call. On a stimulus phone, the loudspeaker can be remotely activated by the PBX.

A stimulus protocol carries lower level instructions than ISDN, H.323, and SIP. For an incoming call, all these protocols simply send a 'you have a new call' message, and the phone is expected to ring all by itself. It is also expected to send ring-back as soon as you pick up the handset. A stimulus protocol would send a 'ring with ring type X' command, then a 'notify me if someone picks up the handset' command (or it could send an 'activate loudspeaker' command directly). For an outgoing call, once notified that the handset is off-hook, the PBX would send a 'play dial-tone command', followed by a 'notify me of the digits that have been dialing' command (but it could also send a 'play this audio message command').

In general, stimulus-based protocols have the following attractive characteristics:

- They simplify the endpoint software design and therefore minimize the number of endpoint bugs that can affect a PBX application: for instance, a common bug in SIP or H.323 endpoints is the inability to alternate between normal ring-back and network-generated prompts, because the programmers did not think of this unusual, but nevertheless mandatory, transition. This is controlled by the PBX in a stimulus protocol, and therefore cannot be an endpoint bug.
- They facilitate management of large numbers of endpoints, by minimizing the problems caused by the diversity of software flavors deployed at endpoints.
- They facilitate the centralized deployment of new features or applications, even those interacting with the endpoint. Usually, such deployments do not require any change

in endpoint capabilities. Once the possibilities of the hardware endpoint have been properly mapped to stimulus commands, all services can be designed without requiring any addition to the device firmware.

- They make it easier to program applications or advanced services which require the co-ordination of multiple endpoints, by centralizing the state of all endpoints at the PBX. A typical example is the manager–secretary feature, where the manager screen needs to show that a call is coming (but does not ring), while the secretary phone rings. Such a service would require additions to the standard with H.323 or SIP endpoints.

The downside of stimulus protocols is that they absolutely require centralized resources: two stimulus protocol phones cannot communicate without a PBX. In addition, since the granularity of communications with the call controller is at a very low level, services require significantly more control messages than with more intelligent endpoints.

With only H.323 and SIP, VoIP would be lacking a stimulus-based protocol—MGCP fills the gap.

## 5.1.2 Decomposed gateways

In the early days of VoIP, most VoIP gateways were based on PCs, with some hardware boards handling media processing. Such gateways were already 'decomposed' in the sense that call control processing and media control resources were running on different modules, with some proprietary APIs between the telephony boards and the main PC-based gateway software.

The early fully embedded gateways usually retained this architecture, with a central processor handling call control, while dedicated **Digital Signal Processor (DSP)** boards handled media processing. But when the size of gateways grew to handle hundreds, or thousands, of channels, this architecture began to be problematic:

- Once the maximum number of DSP 'daughter boards' in a chassis was reached, another chassis needed to be installed with not only DSP boards, but also a new instance of the gateway call control software. This made it impossible to have centralized control of all the channels, and forced the duplication of call control resources.
- In the PSTN, carrier interconnections with thousands of channels typically use dozens of media-only trunks, and a single signaling-only channel (mostly SS7 ISUP) carrying the call control information for all the media trunks. If on the VoIP side the required capacity was big enough to require multiple gateway chassis, then with each gateway having its own local call control software, there would be a need for one SS7 call control-signaling link and ISUP stack per chassis (Figure 5.1), which is far too expensive.

One of the early proprietary protocols used to solve the problem was called Q.931+. One master call control device took SS7 ISUP signaling and distributed call control to each media gateway, in a Q.931-like form, over IP tunnels: this solution still required an instance of call control in each media gateway.

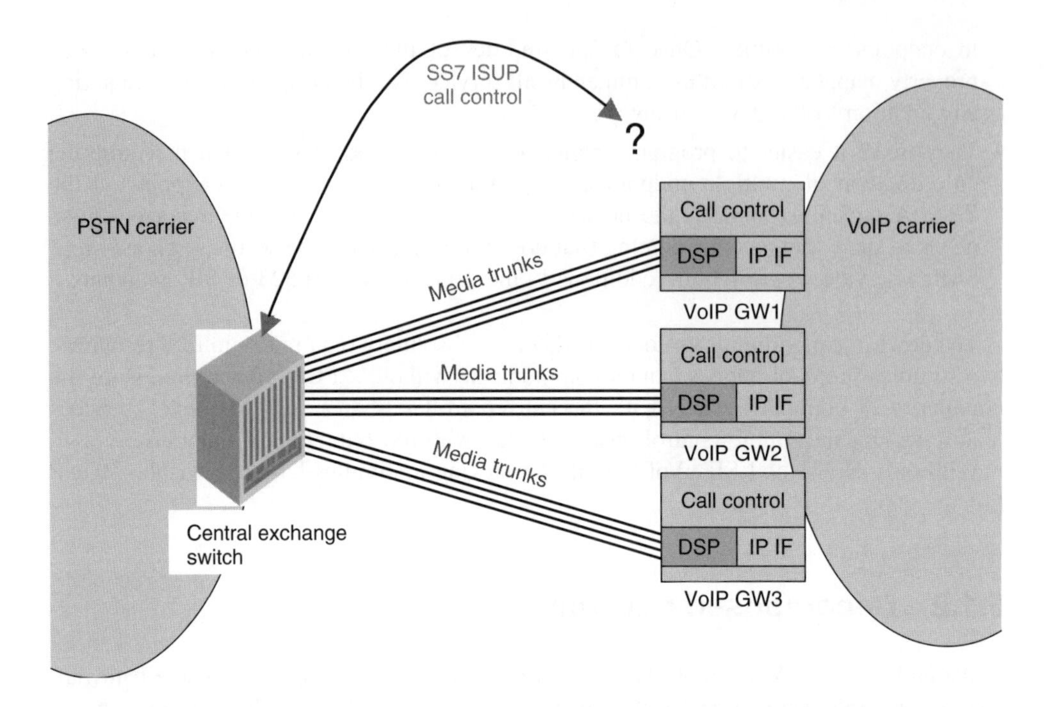


**Figure 5.1** Which box should receive SS7 signaling in an SS7 configuration?

Some vendors quickly found the best solution, which was to have one master call control module, and physically separate media-processing modules with only DSP resources, a TDM media-only interface, and an IP interface; this is a bit like the PC-based architecture with separate DSP boards and call control on the PC, except that since the modules are now physically separate, they do not communicate through an API as in the early PC days, but through a protocol (Figure 5.2). Since both the DSP modules and the call control module have an IP interface, logically the media resources remote control protocol had to be over IP as well.

In order to implement such a solution, there was a need for a standard protocol between a call control function and a media gateway with no call control. The de facto standard today is MGCP, logically named the 'Media Gateway Control Protocol'.

It seems surprising that the same protocol could be used to control stimulus phones and dense media gateways. In fact, a phone *is* a media gateway between a microphone + speaker and the VoIP media stream, plus some user interface components (a handset with a hook, a keypad, buttons, etc.). Therefore, an IP phone stimulus protocol should comprise a pure media gateway control portion, plus some user interface control optional commands. This is exactly what MGCP is: a core set of commands for pure media control (we will frequently use the term 'MGCP trunk', or MGCP/T, although this is not strictly correct), *plus* a set of optional commands (we will frequently refer to the MGCP protocol plus the extensions to control an IP phone as 'MGCP line', or MGCP/L).

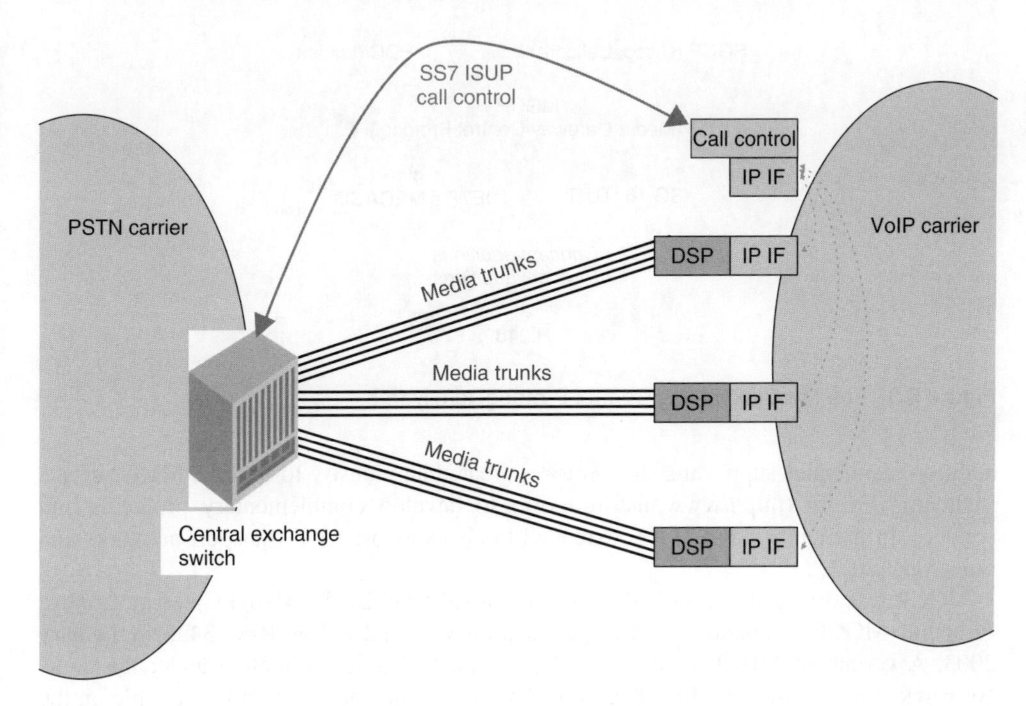


**Figure 5.2** Centralized ISUP control box, controlling remote media gateways via an IP protocol.

## 5.1.3 Some history

The first proposal came from Bellcore (now Telcordia) and Cisco to address the needs of cable operators that wanted to become competitive local exchange carriers (CLECs) by using VoIP on top of their HFC infrastructure. The **Simple Gateway Control Protocol** (**SGCP**) was introduced early May 1998 by Cisco during a PacketCable™ meeting (and in other standards bodies, IETF, ITU-T SG 16 and ETSI TIPHON) as a cost-effective alternative and better suited protocol to implement and deploy than the then-current H.323 implementations in the context of cable operators' market.

The second proposal, the **Internet Protocol Device Control** (**IPDC**) was presented to ITU-T SG 16, ETSI TIPHON and IETF a month later. IPDC addresses more or less the same requirements as SGCP but with a different transport approach. While SGCP relied solely on UDP, enhanced with application-level reliability features, IPDC proposed the use of **DIAMETER** (an extension and replacement of **RADIUS**) to carry **protocol data units** (**PDUs**) between respective entities.

It was not long before the forces behind these two protocols realized that by unifying their efforts they could get bigger consensus and foster the adoption of their position. Bellcore and Level3 played a key role in merging these two proposals into one (Figure 5.3): MGCP. MGCP was proposed to all main standards groups: the IETF's Media Gateway Controller (MEGACO) Working Group, ETSI TIPHON, and ITU-T SG 16. In

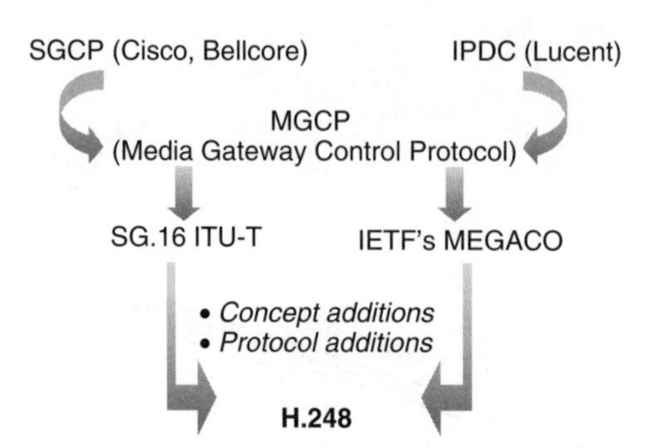


**Figure 5.3** The Media Gateway Control Protocol family tree.

addition, companies supporting this protocol created an industry forum, the Multi Service Switching Forum (http://www.msforum.org) to develop complementary protocols and services. In particular, MGCP was extended to also support ATM transport networks and voice on AAL2.

MGCP was originally published as informational RFC 2705, 'Media Gateway Control Protocol (MGCP)' version 1.0; the specification was updated as RFC 3435 in January 2003. A variant of MGCP is also used by the PacketCable™ initiative under the name **Network Based Call Signaling Protocol (NCS)**, and the specification is available on the PacketCable™ website (currently PKT-SP-EC-MGCP-I06-021 127).

Later the ITU began to work on a new generation of stimulus protocols, called H.248, which is functionally equivalent to MGCP.

Like most IETF media protocols, MGCP uses the SDP syntax to express the format, source, and destination of media streams.

Overall, the quality of the MGCP specification was way better than the quality of the initial specifications of H.323 and SIP. The protocol is simple, focused on a clear scope, well-structured, with a cleanly separate transport layer, a well-defined connection model, and very few bugs in the standard. Without much marketing buzz, MGCP has made its way into the world of VoIP protocols and it is today one of the most widely implemented protocols in all of its original target markets: residential gateways, IP phones, and large-scale trunk gateways.

## 5.2 MGCP 1.0

The Media Gateway Control Protocol was first specified in draft-huitema-MGCP-v0r1-00.txt and was finally published as MGCP 1.0 in RFC 2705. In January 2003 an updated version, which corrected some ambiguities and inconsistencies in RFC 2705, was published as RFC 3435. RFC 3435 is also known as MGCP 1.0bis; however, it is still MGCP version 1.0 and is fully compatible with RFC 2705 (except for error fixes).

MGCP is designed to interface a media gateway controller and media gateway and supports a centralized call control model. The protocol is text-based, offering a set of simple primitives. The media gateway controller is called the **call agent** in MGCP terminology and the media gateways can be of different types:

- VoIP gateways. **Residential gateways** are designed to be customer premises equipment, usually connected to a couple of analog phone lines. These gateways, besides their pure media-processing capabilities, are also able, when connected to analog phones or PBXs, to generate ring voltages, to send specific signals required to set message-waiting indication lamps or to send caller ID information to a phone. **Trunking gateways** are high-density gateways interconnecting TDM media trunks and a VoIP network, with media-processing capabilities only.
- **Network Access Servers** (**NASs**). The MGCP protocol includes some extensions which allow a call agent to control modem banks. The protocol is also capable of driving **universal port** gateways, which can behave as a voice gateway if the detected signal is voice, and can also locally terminate modem connections if they detect a modem signal. NAS extensions are no longer part of the base protocol in RFC 3435; instead, they are provided as a separate package.
- Voice over ATM gateways.

Unlike H.323 or SIP, MGCP is a master–slave protocol. As shown in Figure 5.4, the call agent is a central controller, and the media gateways are slave devices that can only report events requested by the call agent and execute the commands of the call agent.

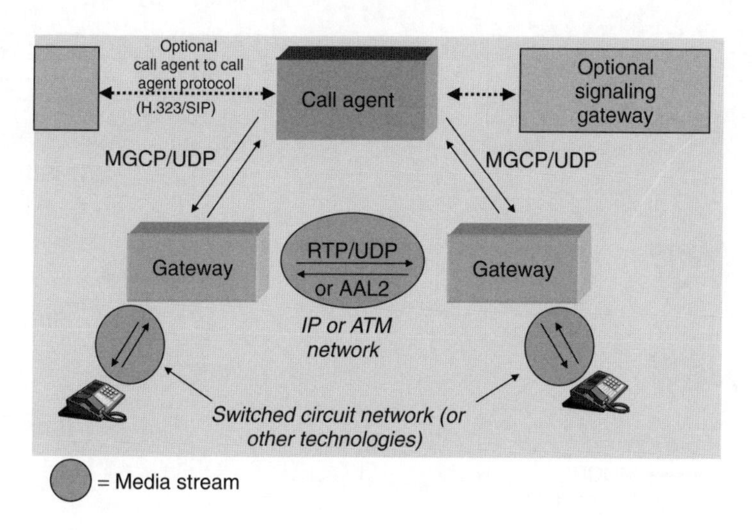


**Figure 5.4** The MGCP ecosystem.

There is often some confusion between H.323/SIP and MGCP. 'I don't want to use MGCP in my network' is a frequently heard sentence. With SIP or H.323, which are call control protocols, it is natural to select only one of these protocols in the core network. Although SIP/H.323 gateways exist, very few service providers have both protocols running in their core network. Many network engineers assume MGCP is the same kind of protocol, and if you begin to implement MGCP somewhere in the network, you must have MGCP everywhere in your network. But MGCP is not a peer-to-peer protocol: two MGCP call agents cannot communicate using MGCP. MGCP call agents can only be at the edge of a network and must communicate between one another using a call control protocol (e.g., H.323 or SIP). Therefore, MGCP should be seen:

- At the customer access edge of the network, as the stimulus-mode option to drive IP phones and IP residential gateways, SIP and H.323 being the call-stateful, ISDN-like alternative.
- At the PSTN interface edge of the network, as an internal protocol of large-scale decomposed trunk-side gateways.

MGCP is not an option for a core network call control protocol. It is an edge protocol as illustrated in Figure 5.5.

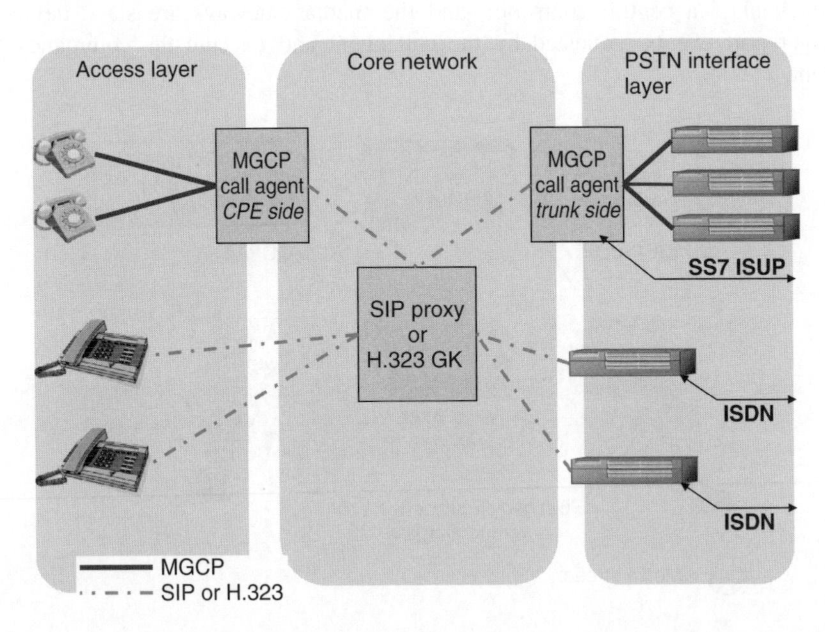


**Figure 5.5** MGCP is an edge protocol, while SIP or H.323 must be used in the core network.

## 5.2.1 The MGCP connection model

The core component of a traditional telephony switch is a TDM bus. Each interface channel is connected to a time slot on the bus. If a channel *A* on an interface sends a media signal on time slot 231 of the TDM bus (let's call it channel 231), then any channel *B* on any interface listening to time slot 231 will receive a copy of this media signal. A full duplex connection for a phone call between channels 231 and 308 is established if channel 231 listens to time slot 308 and channel 308 listens to time slot 231. The traditional switch can perform all the media-switching functions it needs only by transmitting 'send' and 'listen' commands to interface channel 'objects' identified only by a time slot number.

A packet-based switch is more complex because, instead of using a TDM bus as a switching matrix, it uses a packet network. Destinations on a packet network are identified by an address and some parameters, not by a simple integer. Also, the media type for a TDM switch is always 64 K G.711 encoded speech or clear data, whereas it can be anything for a packet-based switch.

The MGCP connection model is built on two objects:

- Endpoints, which can originate (media-source) and/or terminate (media-sink) a media flow. A circuit which can receive RTP packets, decode them, and send the resulting G.711 data to a time slot on a TDM trunk is an endpoint. An entity which can receive RTP packets and relay them to another destination is an endpoint. An IP-based media mixer (MP in H.323 terminology) which can receive RTP streams, mix them, and send back the resulting streams to other RTP sinks is an endpoint. Similarly, on ATM networks any entity which can receive or originate AAL2 media streams is an endpoint.
- Connections (Figure 5.6). Each endpoint can have one or multiple connections, each of which can be inactive, send media, receive media or both. Of course, a connection is useful only if it collects media from an endpoint and sends it to another endpoint (in MGCP, connections are each attached to a named endpoint, but send media to a SDP-defined address, so it is possible to build connections from an MGCP endpoint to an IP address that is not necessarily a declared MGCP endpoint). Connections can be point-to-point or point-to-multipoint. A point-to-multipoint connection will typically send to a multicast IP address.

This model is generally much more powerful and scalable than the TDM time slot model. However, MGCP (like all other VoIP protocols) has one weakness compared with the TDM model: in TDM, point-to-multipoint connections are native, very easy to establish (each destination 'listens' to the source time slot), and invisible from the source. In VoIP, they require endpoints to support multicast or need to be emulated through a special-purpose endpoint that receives media from the source, then duplicates and sends media to all the destinations. Because of this, features like silent monitoring in contact centers, or lawful interception in residential telephony, are *much* more difficult to build than in the TDM world.

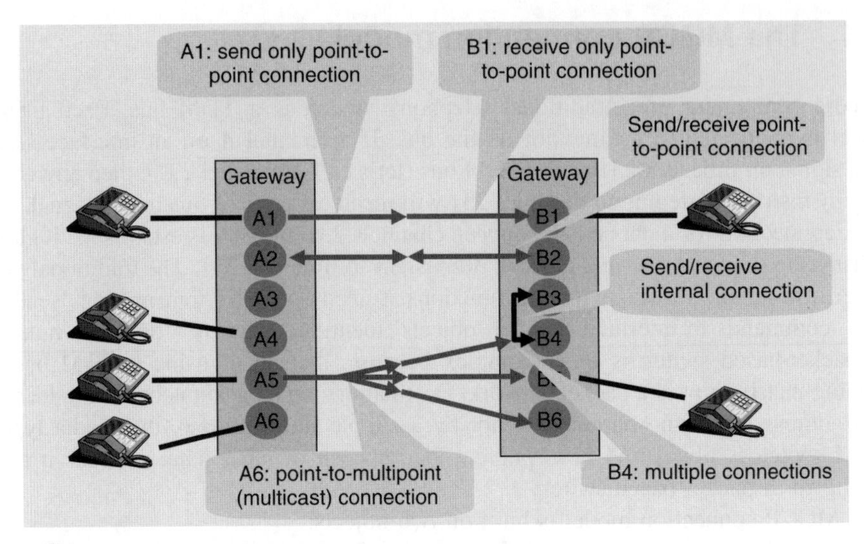


**Figure 5.6** MGCP endpoints and connections (the call agent controls connections on gateway endpoints).

MGCP uses a simple syntax for endpoint identifiers, with two components separated by the '@' character:

- A prefix which should be a unique identifier of the endpoint within the gateway. The prefix structure can be hierarchical (interface, channel on interface), with each component of the hierarchy separated by a slash ('/'). Local endpoint identifier components can be composed of any visible character except a space, '@', '/', '*', or '$'. '*' has the special meaning of '*all* defined values of this component', and '$' means '*any* of the values defined for this component'.
- The DNS domain name of the gateway. If the gateway is not registered in the DNS and the call agent expects non-DNS names, then the name should be any string composed of letters, numbers, and '.' or '-', as long as it is unique on the network. Some vendors also use the numeric IP address of the gateway, between square brackets (e.g., [10.10.10.10]).

A two-port analog gateway typically uses:

- aaln/1@analog-gateway.anydomain.org
- aaln/2@analog-gateway.anydomain.org

On a TDM trunk interface, where each interface handles multiple trunks and each trunk has multiple circuits, each with its circuit identifier (referenced by the ISUP CIC), vendor may use:

- IF3/2/1@large-trunk-gateway.anydomain.org for circuit 1 on trunk 2 of interface IF3.
- IF5/4/$@large-trunk-gateway.anydomain.org for *any* circuit on trunk 4 of interface IF5.

In practice, each vendor uses its own convention for local endpoint names; but, as long as sequences are identified by consecutive integers, it is very easy to configure the corresponding masks on a call agent.

## 5.2.2 The protocol

### 5.2.2.1 Overview

MGCP call agents and gateways exchange 'transactions', composed of one command, optional provisional responses, and a final response (Figure 5.7).

Commands and responses use a simple text format. MGCP 1.0 is composed of nine commands exchanged between the call agent and a media gateway or a NAS (generically called 'gateway' from now on). Each command is composed of a header, optionally followed by an empty line, and a session description. The header is composed of multiple lines (separated by a CR, an LF, or a CR + LF):

- A command line, composed of the command code, the transaction identifier, the target endpoint name (optionally with local endpoint identifier wildcards '*' or '$'), and the MGCP protocol version, separated by the ASCII space ($0 \times 20$) or tab ($0 \times 09$) character (e.g., the following is a valid command line 'RQNT 1207 endpoint/1@rgw-2567.anydomain.net MGCP 1.0'). Some MGCP gateways still use MGCP 0.1 (the MGCP version immediately after the merger of IPDC and SGCP), but the differences from MGCP 1.0 are minimal.
- A set of parameter lines, each using the 'name:value' format. Parameter names consist of one or two letters, followed by a colon (e.g., B:e:mu). The most common parameters are listed in Table 5.1.

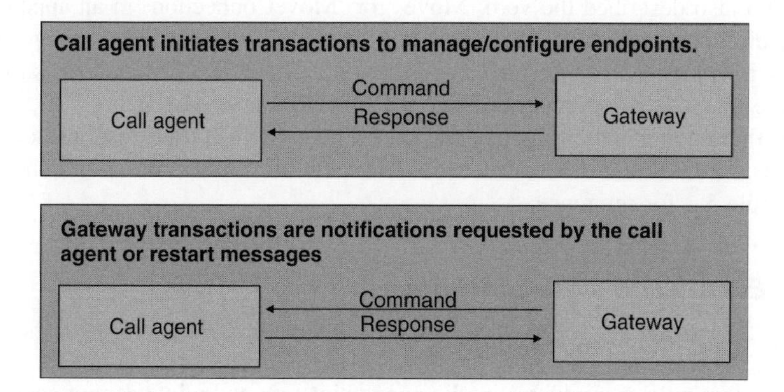


**Figure 5.7** MGCP commands and responses (each transaction refers to one or more gateway endpoints).

Table 5.1 Common MGCP parameters

| | |
|---|---|
| K | ResponseAck |
| B | BearerInformation |
| C | CallId |
| I | ConnectionId |
| N | NotifiedEntity |
| X | RequestIdentifier |
| L | LocalConnectionOptions |
| M | ConnectionMode |
| R | RequestedEvents |
| S | SignalRequests |
| D | DigitMap |
| O | ObservedEvents |
| P | ConnectionParameters |
| E | ReasonCode |
| Z | SpecificEndpointID |
| Z2 | SecondEndpointID |
| I2 | SecondConnectionID |
| F | RequestedInfo |
| Q | QuarantineHandling |
| T | DetectEvents |
| RM | RestartMethod |
| RD | RestartDelay |
| A | Capabilities |
| ES | EventStates |
| MD | MaxMGCPDatagram |

Command lines and parameter lines are case-insensitive. A sample command is shown in Figure 5.8. Each command targets one or more endpoints. Table 5.2 lists the verbs for each of the nine MGCP commands. Experimental verbs can also be added, whose names should start with an 'X'.

RFC 2705 also described the verb 'Move' for 'MoveConnection' in an appendix. This command disappeared in RFC 3435; it is now provided in a separate package (`draft-andreasen-mgcp-moveconnection-00.txt`). Also, RFC 3435 defines the 'Mesg' verb for the 'Message' command defined in RFC 3435, appendix B.

Each of these commands triggers a response, including a 3-digit response code. Commands and responses are associated by a TransactionID. The most common codes are listed in Table 5.3 for reference.

#### 5.2.2.2 Events and signal packages

##### 5.2.2.2.1 Definitions and syntax

A call agent that only manipulates media connections on endpoints cannot easily interact with media information (e.g., in order to send back a dial tone to an analog gateway, the call agent would need to connect the gateway endpoint with an IP-based dial tone

```
RQNT 1207 endpoint/1@rgw-2567.anydomain.net MGCP 0.1
X: 0123456789
R: hu
S: v
```

The call agent sends a NotificationRequest to the gateway:

- The NotificationRequest verb is RQNT.
- The TransactionId is 1207.
- The target endpointName is endpoint/1.
- The gateway domainName is @rgw-2567.anydomain.net.
- MGCP protocol version is 0.1.
- The requestIdentifier is 0123456789.
- The gateway will need to notify the call agent when it detects hang-up 'hu'.
- The gateway must immediately generate an alerting tone 'v'.

**Figure 5.8** A sample MGCP command.

**Table 5.2** MGCP command verbs

| Verb | Code | Direction: Call agent (→) Gateway (←) |
|---|---|---|
| EndpointConfiguration | EPCF | → |
| NotificationRequest | RQNT | → |
| Notify | NTFY | ← |
| CreateConnection | CRCX | → |
| ModifyConnection | MDCX | → |
| DeleteConnection | DLCX | → and ← |
| AuditEndpoint | AUEP | → |
| AuditConnection | AUCX | → |
| RestartInProgress | RSIP | ← |

generator). MGCP uses **signals** to provide a simpler way to allow a call agent to give instructions to endpoints that have some local signal generation capabilities. A signal is simply an identifier known to the gateway which corresponds to some action on the media bearer (e.g., playing the dial tone). Many applications also require the call agent to be aware of certain events that are present in-band (e.g., DTMF signals), or through some means accessible only to the gateway (e.g., an off-hook transition). The gateway can report this type of information to the call agent using MGCP **events**.

Various types of gateways can report different events and generate different signals. Some signals and events, however, are very common and most gateways support them. In order to give some flexibility and extensibility for MGCP to support new types of gateways that require specific events or signals, without interfering with already-defined names, MGCP introduced the notion of 'packages'. A package is simply a namespace,

**Table 5.3** Common MGCP response codes

| Code | Description |
|---|---|
| 100 | The transaction is currently being executed. An actual completion message will follow later (provisional response) |
| 200 | The requested transaction was executed normally |
| 250 | The connection was deleted |
| 400 | The transaction could not be executed, due to a transient error |
| 401 | The phone is already off-hook |
| 402 | The phone is already on-hook |
| 403 | The transaction could not be processed, because the endpoint does not have sufficient resources at this time |
| 404 | Insufficient bandwidth at this time |
| 500 | The transaction could not be executed, because the endpoint is unknown |
| 501 | The transaction could not be executed, because the endpoint is not ready |
| 502 | The transaction could not be executed, because the endpoint does not have sufficient resources |
| 510 | The transaction could not be executed, because a protocol error was detected |
| 511 | The transaction could not be executed, because the command contained an unrecognized extension |
| 512 | The transaction could not be executed, because the gateway is not equipped to detect one of the requested events |
| 513 | The transaction could not be executed, because the gateway is not equipped to generate one of the requested signals |
| 514 | The transaction could not be executed, because the gateway cannot send the specified announcement |
| 515 | The transaction refers to an incorrect connection-id |
| 516 | The transaction refers to an unknown call-id |
| 517 | Unsupported or invalid mode |
| 518 | Unsupported or unknown package |
| 519 | Endpoint does not have a digit map |
| 520 | The transaction could not be executed, because the endpoint is 'restarting' |
| 521 | Endpoint redirected to another call agent |
| 522 | No such event or signal |
| 523 | Unknown action or illegal combination of actions |
| 524 | Internal inconsistency in LocalConnectionOptions |
| 525 | Unknown extension in LocalConnectionOptions |
| 526 | Insufficient bandwidth |
| 527 | Missing RemoteConnectionDescriptor |
| 528 | Incompatible protocol version |
| 529 | Internal hardware failure |
| 530 | CAS signaling protocol error |
| 531 | Failure of a grouping of trunks (facility failure) |

identified by one or more letters. Events and signals can be defined within this namespace without risk of ambiguity with other namespaces. An event or signal name must be prefixed by its package name, separated by a slash (e.g., 'L/HU' refers to event 'HU' within package line 'L'). Package names and events are not case-sensitive (i.e., 'L/HU' is equivalent to 'L/hu' or 'l/hu').

There are some special cases in which the package prefix can be omitted:

- A call agent can omit the prefix for events and signals that are part of the default package, when sending commands to an endpoint that is known to support a default package (e.g., if the package line is the default package for an analog gateway, then dl and L/dl are equivalent signals).
- Digit events can be sent without a prefix, although it is recommended to send them with the appropriate prefix (e.g., L/8 for digit 8). Package names are not allowed to contain digits in order to prevent any ambiguity.

These exceptions are provided mainly to allow MGCP to remain tolerant of older implementations, but it is recommended to always include the package name prefix.

Events and signals are detected/applied by default on the bearer channels connected to the endpoints. It is also possible to request the event/signal to apply to a connection; in this case, the connection identifier is added after the event name, separated by an '@' sign (e.g., G/rt@234A2).

MGCP packages is an area where RFC 3435 expanded significantly on RFC 2705; many things (reason codes, actions, etc.) can be expanded in a package now, not just events and signals (see Section 6 in RFC 3435).

##### *5.2.2.2.2 Categories of signals*

There are three signal categories depending on the way they persist or not after being applied:

- **On–off (OO) signals** last until they are explicitly turned off by a NotificationRequest with an empty Signals line (or the endpoint restarts). These signals can be turned 'on' (resp. 'off') repeatedly, they simply remain 'on' (resp. 'off'). A message-waiting indication is a typical example.
- **Timeout (TO) signals** last until they are explicitly canceled or after a timer. An 'operation complete' event is generated when such a signal expires (e.g., the ring-back tone provided when a handset goes off-hook will typically expire after 3 min). Once applied, ring-back will stop (1) if canceled (new NotificationRequest without ring-back in the Signals line), (2) if an event requested by the NotificationRequest occurs (this is the default behavior, although it is possible to override it), or (3) if it times out. The timeout value must be defined by the package.

- **Brief (BR) signals**. These very short signals always complete once the endpoint has begun to execute them, regardless of subsequent events or Notification requests.

#### *5.2.2.2.3 Common packages*

Many packages have been defined. Here is a list of some of the reference documents:

- RFC 3064 (CAS)
- RFC 3149 (Business Phone)
- RFC 3441 (ATM)
- draft-foster-mgcp-basic-packages-10.txt
- draft-foster-mgcp-bulkaudits-08.txt
- draft-andreasen-mgcp-fax-01.txt
- draft-foster-mgcp-lockstep-00.txt
- draft-andreasen-mgcp-moveconnection-00.txt
- draft-aoun-mgcp-nat-package-02.txt
- draft-foster-mgcp-redirect-01.txt

There are still more on the way. The most common packages are listed in Table 5.4.

The following sections list the contents of some commonly used packages, defined in RFC 3660 (Basic MGCP packages). An X in the 'R' column denotes an event that can be requested by the call agent. The S column specifies the type of signal (on–off, Timeout, Brief).

##### *5.2.2.2.3.1 Generic media package*

As of RFC 3435, packages now have a version number, the original version is zero. The original version of the generic media package was version 0, the current version is version 1.

**Table 5.4** Main MGCP packages

| | |
|---|---|
| Generic media package | G |
| DTMF package | D |
| MF package | M |
| Trunk package | T |
| Line package | L |
| Handset package | H |
| RTP package | R |
| Network access server package | N |
| Announcement server package | A |
| Script package | S |

| Symbol | Definition | R | S |
|---|---|---|---|
| cf | Confirm tone or 'positive indication tone' of ITU E.182 | | BR |
| cg | Network congestion tone of ITU E.180/E.182 | | TO |
| ft | Fax tone detected (V.21 fax preamble and T.30 CNG tone) | X | |
| It | Intercept tone (ITU-T E.180 supplement 2) | | TO |
| ld | Long-duration connection (>1 hour) | X | |
| mt | Modem detected (V.25 ANSwer tone, V.8 modified answer tone) | X | |
| oc | Operation complete | X | |
| of | Report failure | X | |
| pat(&&&) | Pattern &&& detected (answering machine, tone, etc.) To be defined administratively on the gateway | X | OO |
| pt | Pre-emption tone (ITU-T E.180 supplement 2) | | TO |
| rbk(&&&) | Ring-back on connection | | TO (180 s) |
| rt | Ring-back tone or 'ringing tone' (ITU E.180 and E.182) | | TO (180 s) |

*5.2.2.2.3.2 DTMF package*

| Symbol | Definition | R | S |
|---|---|---|---|
| & | DTMF & | X | BR |
| * | DTMF * | X | BR |
| ... | ... | X | BR |
| 0 | DTMF 0 | X | BR |
| 9 | DTMF 9 | X | BR |
| A | DTMF A | X | BR |
| B | DTMF B | X | BR |
| C | DTMF C | X | BR |
| D | DTMF D | X | BR |
| DD | DTMF tone duration exceeded, or generate DTMF in TO mode | X | TO |
| DO | DTMF signal generated in OO mode | | OO |
| L | Long-duration indicator (over 2 s) | X | |
| oc | Operation complete | | |
| of | Report operation failure | X | |
| T | Interdigit timer: 4 s (T critical) if it causes a final match of the digitmap rule; 16 s (T partial) if there is still ambiguity and multiple rules still apply | | |
| X | Wildcard DTMF 0, 1, 2, 3, 4, 5, 6, 7, 8, or 9 | X | |

#### 5.2.2.2.3.3 *Trunk package*

| Symbol | Definition | R | S |
|---|---|---|---|
| as | Answer supervision | X | BR |
| bl | Blocking: bl(+) to block the circuit; bl(−) to unblock it | | BR |
| bz | Busy as defined in ITU E.180 | | TO |
| co1 | Continuity tone: 2,010 Hz (±30). When sending this tone during a continuity test, it is expected to receive this same frequency back | X | TO |
| co2 | Continuity test: 1,780 Hz (±30). When sending this tone during a continuity test, it is expected to receive the co1 frequency back | X | TO |
| lb | Loop-back | | OO |
| nm | New milliwatt tone (1,004 Hz) | X | TO |
| oc | Operation complete | X | |
| of | Report operation failure | X | |
| om | Old milliwatt tone (1,000 Hz) | X | TO |
| ro | Reorder tone (ITU E.182 congestion tone) | X | TO |
| tl | Test line 2,225 Hz (±25) | X | TO |
| zz | No circuit tri-tone | X | TO |

#### 5.2.2.2.3.4 *Line package*

The exact frequencies corresponding to some signals of the line package may vary in each country. Vendors usually provide localization of these signals through the gateway-provisioning interface.

| Symbol | Definition | R | S |
|---|---|---|---|
| adsi(string) | adsi display | | BR |
| aw | Answer tone | X | OO |
| bz | Busy tone (ITU E.180) | | TO (30 s) |
| ci(ti, nu, na) | Caller ID (time, calling number, calling name). If quoted, the string can be UTF-8 encoded | | BR |
| dl | Dial tone (ITU E.180) | | TO (16 s) |
| e | Error tone | X | BR |
| hd | Off-hook transition | X | |
| hf | Flash-hook | X | |
| hu | On-hook transition | X | |
| mwi | Message-waiting indication tone | | TO (16 s) |
| nbz | Network busy (fast cycle busy) | X | OO |

| Symbol | Definition | R | S |
|---|---|---|---|
| oc | Report on completion, may contain completed signal as a parameter | X | |
| of | Report failure | X | |
| osi | Network disconnect | | TO |
| ot | Off-hook warning tone (when phone has been left off-hook for too long, and there is no active call) | | OO ($\infty$) |
| p | Prompt tone | X | BR |
| r0 ... r7 | Distinctive ringing | | TO (30 s) |
| rg | Ringing | | TO (30 s) |
| ro | Reorder tone (congestion tone ITU E.182) | TO (30 s) | |
| rs | Ring-splash (reminder short ring for call-forwarded lines when a call is redirected) | | BR |
| s(&&&) | Distinctive tone pattern, to be defined on the gateway | X | BR |
| sit | Special information tone (ITU E.180) | | |
| sl | Stutter dial tone used to confirm an action and require additional input | | TO (16 s) |
| v | Alerting tone | | OO |
| vmwi | Visual message-waiting indicator | | OO |
| wt | Call-waiting tone (ITU E.180) | | TO (30 s) |
| wt1 ... wt4 | Alternative call-waiting tones (ITU E.180) | | TO (30 s) |
| y | Recorder warning tone (ITU E.180) | | TO |
| z | Calling card service tone | | BR |

###### *5.2.2.2.3.5 Handset emulation package*

This handset emulation package is the same as line package, but some handset-related events like 'off-hook' and 'on-hook' can be signaled as well as detected. This is useful to provide the automatic off-hook feature (activation of a speaker phone), for phones controlled via CTI (Computer Telephony Integration). This also allows providing features like paging or remote baby monitoring.

###### *5.2.2.2.3.6 RTP package*

These events can be used by a call agent to get a more dynamic view of gateway media processing: for instance, some gateways can automatically change their coders from a compressed low-bitrate coder to G.711 for fax; the UC event can be used to learn that this occurred.

| Symbol | Definition | R | S |
|---|---|---|---|
| UC | Used codec changed (e.g., UC(15) indicates the codec has changed to $\mu$-law). Codec numbers are according to RFC 1890 | X | |
| SR(&&&) | Sampling rate changed (e.g., SR(20) for 20 ms) | X | |
| JI(&&&) | Jitter buffer changed (e.g., JI(20) for 20 ms) | X | |
| PL(&&&) | Packet loss exceeded (e.g., PL(20) for 20 lost in 100,000) | X | |
| qa | Quality alert | X | |
| co1 | Continuity tone | X | TO |
| co2 | Continuity tone | X | TO |
| of | Report failure | X | |

#### 5.2.2.3 The MGCP transport layer over UDP

MGCP commands and responses are sent over UDP. The call agent MGCP default receive port is 2727 and the gateway default receive port is 2427.

UDP has many advantages, because it allows the call agent and gateways to control the retransimission of lost packets themselves, and therefore avoid the uncontrolled latency and head-of-line blocking issues associated with TCP. A call agent can be implemented using a single socket for multiple gateways and is not subject to operating system limitations on the number of sockets, which are unavoidable with TCP.

An MGCP stack must handle packet loss detection and retransmission, and must also detect the loss of a connection. The mechanism of MGCP is very sophisticated. It is based on retransmissions, but ensures that a given command cannot be executed twice ('at most once').

Each command may receive one or more provisional responses (1xx), and at most one final response. The command, provisional responses, and final response constitute a **transaction**. Each command contains a transaction identifier between 1 and 999,999,999 (both included), which is copied in all responses related to that command. Each MGCP entity maintains a list of recent commands in process of execution locally, and of recently sent responses. The processing of commands and responses is global for all the endpoints managed by the MGCP device (i.e., there is no separate transaction space or retransmission buffer for each endpoint: typically transactions are managed per MGCP gateway—identified by its gateway name).

New received commands, which are not in the 'command in process' list or which do not have any response in the pile of recent responses, are sent to the MGCP execution engine, which generates a response. The transaction identifier remains in the pile of commands in process until a final response has been generated. Once a final response has been sent, it remains in the 'recent responses' pile until it expires (see p 285 for details on the expiration). This command processing cycle is illustrated in Figure 5.9.

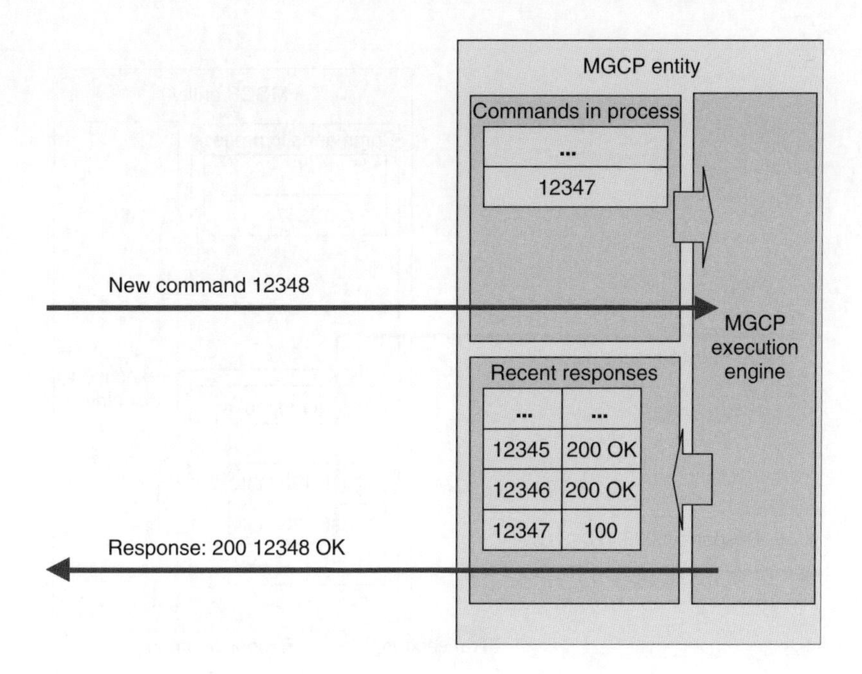


**Figure 5.9** Handling of a new command by MGCP.

If a command is repeated by a call agent after it has been fully executed, the corresponding response is already in the pile of recent responses. The command is not executed again, instead the stored response is sent as shown in Figure 5.10. If, on the other hand, the execution engine has not generated a response to the first command yet, the command is still in the commands in process pile, and a 100 PENDING provisional response is generated automatically (or 101 in the case of overload). As shown in Figure 5.11, the duplicate command is not sent to the execution engine.

A command in the commands in process pile expires as soon as the MGCP execution engine has generated a final response. An entry in the recent responses pile should not expire if there is still any chance that a duplicate command will be received. The expiration timer (T-HIST) should be greater than the maximum duration of a transaction which takes into account the maximum number of retransmissions, the delay between each retransmission, and the maximum propagation delay of a packet in the network. A typical value used is typically about 30 s; however, other values can be used as well, as long as the sender and receiver agree on the actual value.

MGCP also provides a way for the sender of a command to acknowledge previous responses: the response acknowledgment attribute contains a range of confirmed transactionIDs. In this case the corresponding response strings can be deleted immediately from the recent responses pile (there will never be a need to resend them), but the transactionID should still stay in the 'long-timer' pile in the unlikely case that some network element duplicates the UDP packet of the original command. This allows the MGCP entity to ignore the duplicate. No response at all is sent.

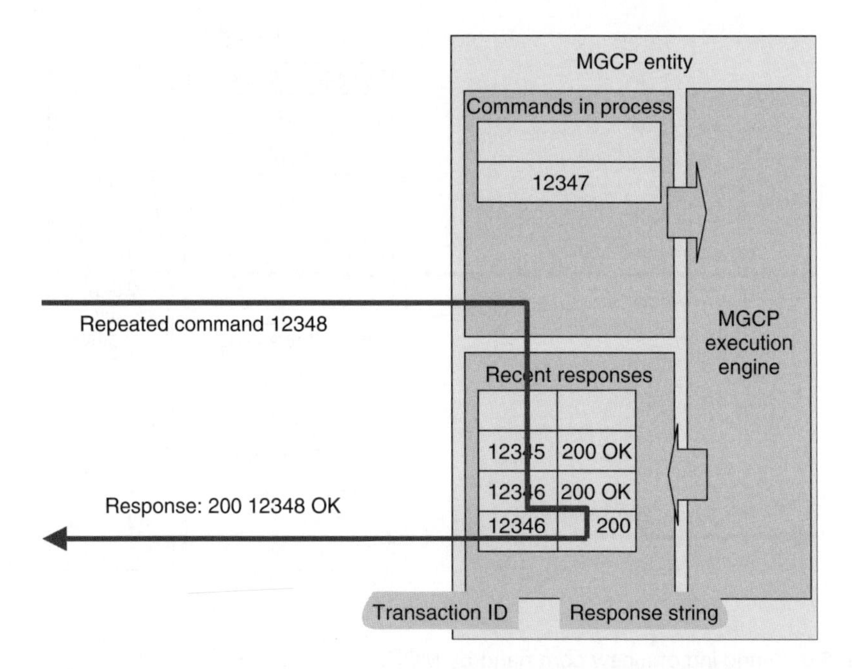


**Figure 5.10** Handling of a duplicate command already executed by MGCP.

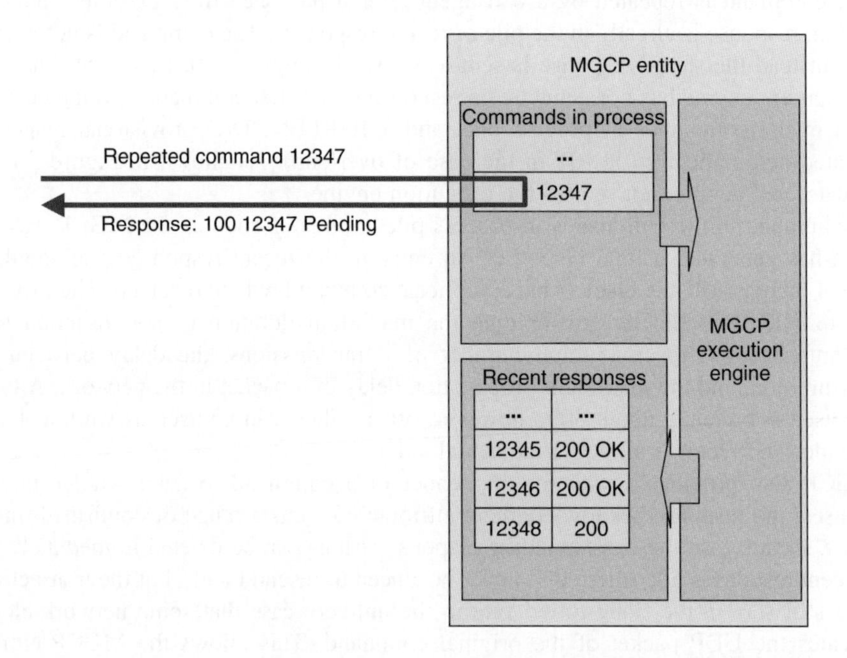


**Figure 5.11** Handling of a duplicate command already in execution.

MGCP entities are required to evaluate dynamically the network round trip time from the time elapsed between the sending of a command and reception of a response: for instance, they can evaluate the average acknowledgement delay (AAD) and the average delay deviation (ADEV). The first command retransmission timer can then be set to AAD + N * ADEV. Subsequent $k$th retransmission timers for the same command should be set to $AAD * 2^k + N * ADEV$ + random component (between 0 and ADEV), ensuring exponential back-off in case of network congestion, with an upper bound $B$ set to 4 s typically. Once the upper bound for the retransmission timer has been reached, the implementation should also limit the number $R$ of retransmissions. The recommended practice is to limit the total cumulated time during which the implementation attempts to resend. A complete retransmission scenario is shown in Figure 5.12.

In some special cases (e.g., transmission over satellite), the algorithm can also be modified to force a retransmission timer smaller than the round trip delay in order to ensure that the time to recover from packet loss is very small despite the link delay (but this uses more bandwidth). The sophisticated MGCP transport layer can therefore be tuned to show very network-friendly behavior, like TCP, but in a more application-controlled fashion, or to have a more aggressive behavior (very similar to the behavior of the SS7 transport layer MTP behavior over satellite links, called pre-emptive cyclic retransmission).

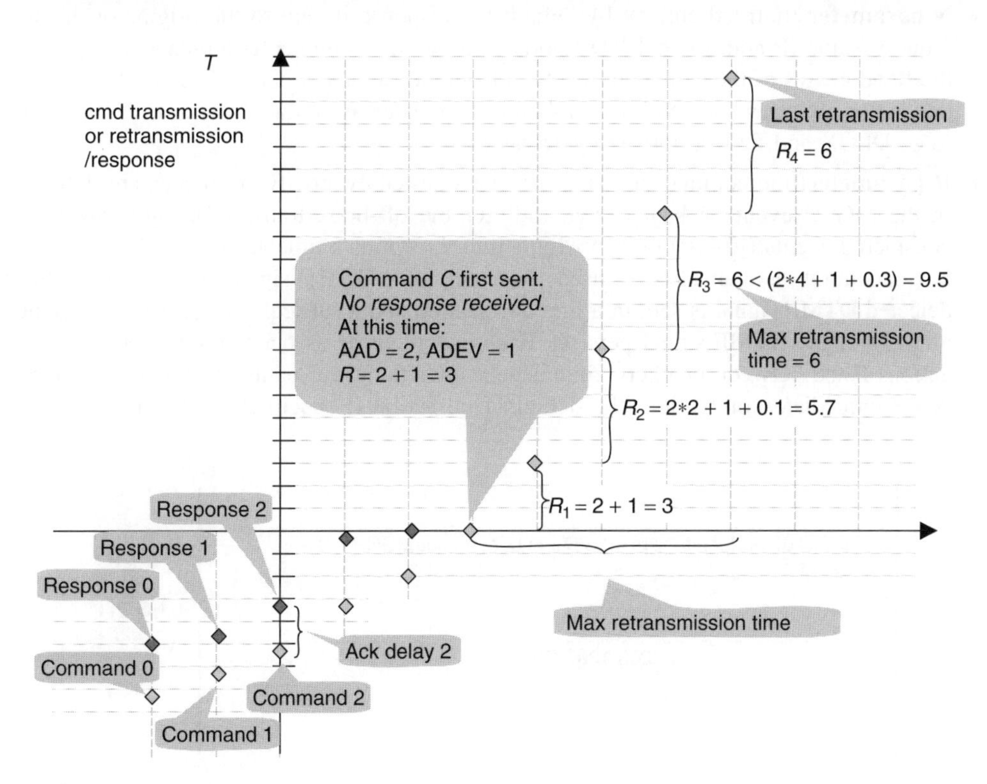


**Figure 5.12** MGCP command retransmission.

### 5.2.2.4 MGCP commands from the call agent to the gateway

#### 5.2.2.4.1 Endpoint configuration command (EPCF)

This command is sent by the call agent to the gateway to configure the type of bearer encoding ('B:') to expect and send on the line side (i.e., not on the VoIP side) of one endpoint or a range of endpoints. The two values defined so far are the G.711 A-law ('e:A'; e.g., Europe) or $\mu$-law ('e:mu'; e.g., in the US). The gateway simply responds with a return code (Figure 5.13).

#### 5.2.2.4.2 Notification request command (RQNT)

This very elaborate command requests the media gateway to watch for specific telephony events. These events can be detected in-band, such as fax tones, DTMF, continuity tones, or analog line status signals like off-hook, on-hook, flash-hook. An example of a RQNT command is given in Figure 5.14.

EndpointId and RequestIdentifier are mandatory parameters, RequestIdentifier is used to correlate the request and the notifications it triggers. In addition, the RQNT command will usually contain some of the following parameters:

- **N parameter** (notified entity): by default the response is sent to the originator of the request (same IP address and UDP port), and gateway-initiated commands are sent to the IP addresses of the call agent resolved from the call agent name. The NotifiedEntity parameter affects where notifications and other gateway-initiated commands are sent (e.g., DLCX and RSIP), until the gateway restarts.
- **R parameter:** the comma-separated list of requested events. Event names are defined in the MGCP event packages (e.g., 'hd' for the off-hook transition), digits are also considered events ([0–9#T] means digits 0 to 9, or # or a timeout of 4 s). The symbol L (long duration) can also be used to detect long DTMF signals. In this case the detected DTMF signal is sent in a first notification and a subsequent notification is sent after 2 s if the DTMF signal persists. Each event can be associated with one or more actions listed between brackets immediately after the event name. The actions can be N for immediate notification (the default if no action is specified), A for accumulate,

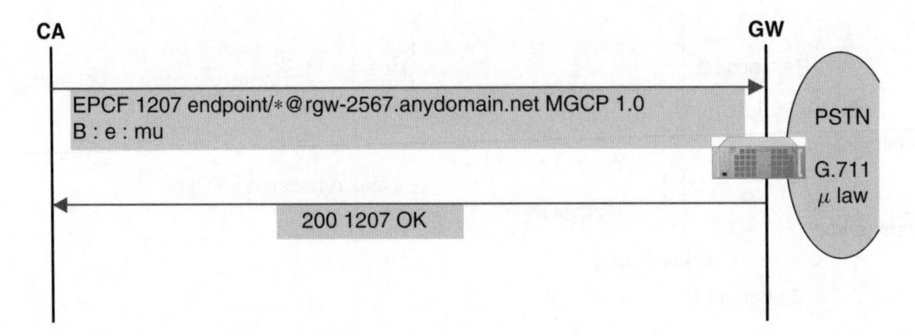


**Figure 5.13** EPCF command example.

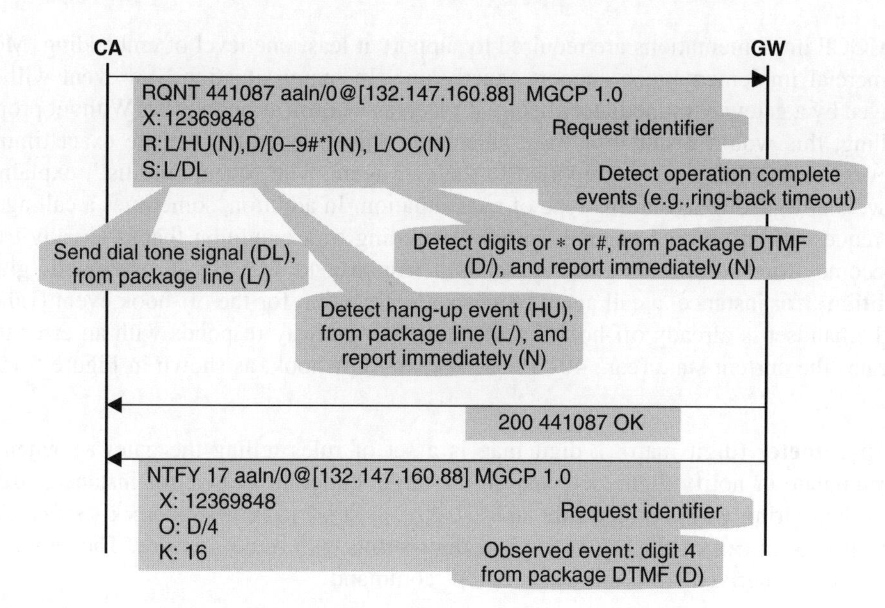


**Figure 5.14** RQNT/NOTIFY command example.

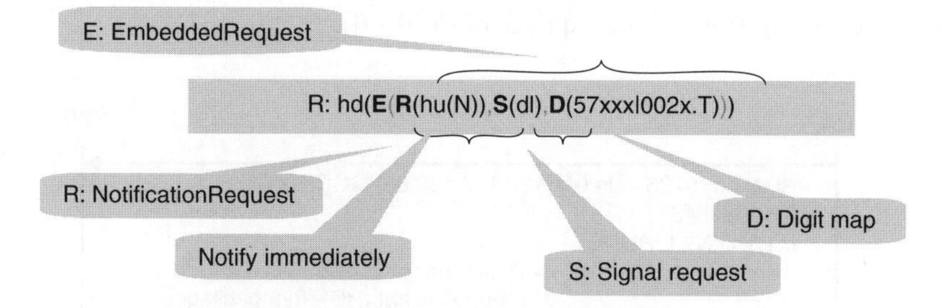


**Figure 5.15** Embedded signal request and digit map example.

D to accumulate according to the digit map (see Figure 5.15), S to swap the active media connection to the next one if there is any, I to ignore the event, K to keep the signal active (normally, the signals present in the S line of a NotificationRequest stop at the first detected requested event, see below), and E to enable (execute) an embedded notification request. An embedded notification request (Figure 5.15) applies to the same endpoint as the NotificationRequest and consists of:

- An optional embedded RequestedEvents parameter: R(embedded RequestedEvents line). For instance, `R([0-9#T] (D), hu (N))`.
- An optional embedded SignalRequests parameter: S(signal requests). For instance, `S(dl)`.
- An optional embedded DigitMap: D(digit map). For instance, `D([0-9]. [#T])`. If an embedded digit map is absent, the current value is used.

All MGCP implementations are required to support at least one level of embedding. Most commercial implementations support exactly one. In many situations, an event will be received by a gateway immediately before it receives a notification request. Without proper handling, this would result in a 'race situation' where, depending on the exact timing, the event may or may not be reported to the call agent. The 'quarantine list', explained below, is designed to avoid this type of race situation. In addition, sometimes a call agent will request to be notified of an event corresponding to a condition that is already true, or becomes true before the gateway sends a response to the RQNT command (**glare condition**): for instance, a call agent requests a notification for the off-hook event (L/hd), but the handset is already off-hook. In this case the gateway responds with an error that indicates the current state (e.g., 401 Phone Already Off-hook, as shown in Figure 5.16).

- **D parameter** (digit map): a digit map is a set of rules telling the gateway when to accumulate or notify digits detected on the target endpoint bearer: for instance, '`00T`' is a digit string with a single rule and '`(0T|00T|[1-7]xxx|8xxxxxxx|#xxxxxxx|*xx|91xxxxxxxxxx|9011x.T)`' is a digit string with multiple rules. The syntax of each rule is derived from the Unix 'egrep' command:
  - '`[1-4]`' matches any digit between 1 and 4, including 1 and 4.
  - '`[1-79BT]`' matches any digit between 1 and 7, or 9, or B, or a timeout.
  - 'x' matches any single digit (equivalent to [0–9]).

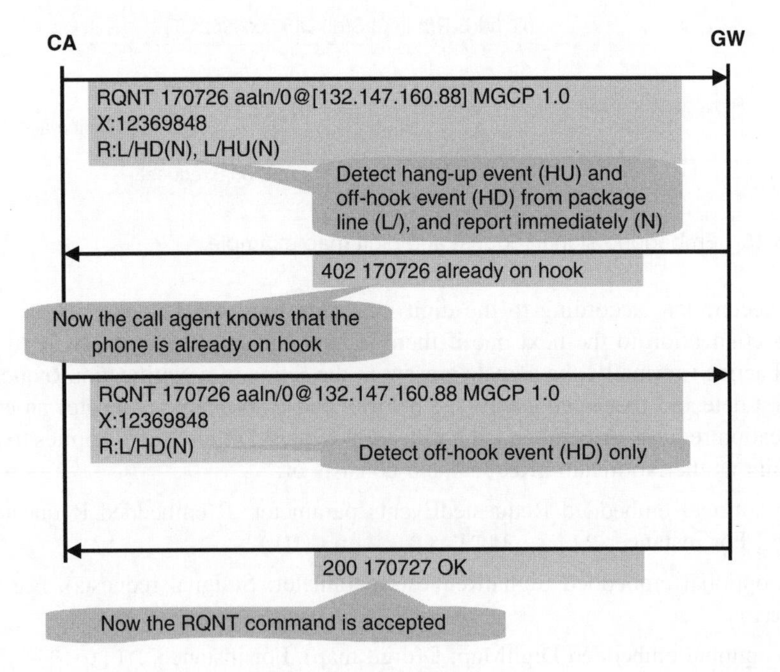


**Figure 5.16** Handling of a glare condition with MGCP.

- 'x.' (the dot operator) matches any positive number of occurrences of the previously specified symbol. Therefore, 'x.' or '[0-9].' means 'any positive number of digits'.

If the timer (T) symbol is the last event required to match a rule, the timeout event is triggered if the last detected tone occurred more than 4 s ago (**critical timer**). If there are more digits to match after the timeout event, or when at least one more digit is required to match any of the digit map rules, the timeout event is triggered after only 16 s (**partial timer**). A digit string is accumulated until it matches one of the rules, or if it can no longer match any of the rules (overqualification). The accumulation mechanism is illustrated in Figure 5.17.

- **S parameter:** a request to apply a signal to the endpoint (e.g., ringing or a ring-back tone). By default **timeout signals** stop as soon as one of the events in the list of requested events is detected (unless the event explicitly states through the K action that it should be kept active).
- **Q parameter** (quarantine handling): this describes how signals received just before the notification request (quarantine list, see Figure 5.18) should be handled. The default is to process them, but the call agent can specify that they should be ignored. The parameter also specifies whether only one notification should be sent or whether multiple notifications should be sent. As soon as a gateway has sent a NOTIFY, it transitions to the 'notification pending' state and begins to accumulate events in the quarantine list, a sort of buffer of events not yet processed. The gateway continues to accumulate events in the quarantine list until the response to the NOTIFY command is received (success or failure). If the call agent specified in the Q parameter that it wanted only *one* NOTIFY in response to a RQNT, then the gateway continues accumulating events until the next RQNT. On the other hand, if the gateway can send multiple successive NOTIFY commands, then it processes the list of quarantined events (Figure 5.19). The

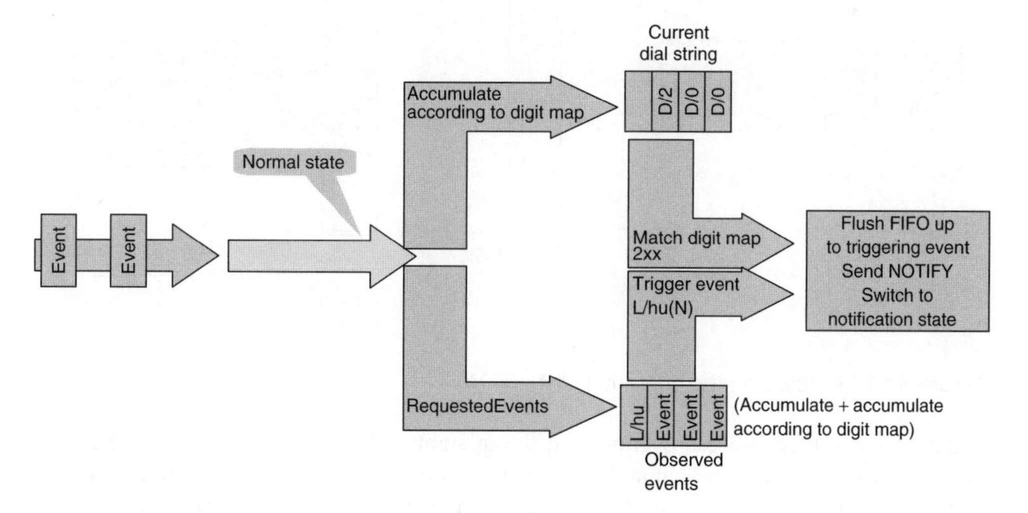


**Figure 5.17** MGCP event accumulation (normal state).

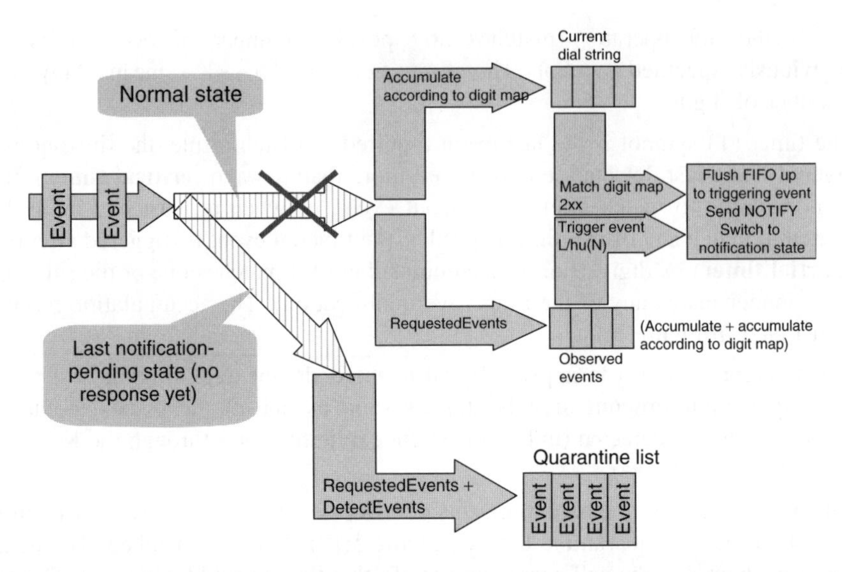


**Figure 5.18** Accumulation of events in the quarantine list. If CA expects no more than one notification, continue accumulating until the next RQNT. If multiple NOTIFY commands are allowed, process quarantined events and, if a triggering condition is met, send a NOTIFY and *either* leave unprocessed events in the quarantine list *or* empty the quarantine list. Then send NOTIFY with multiple Events and dial strings.

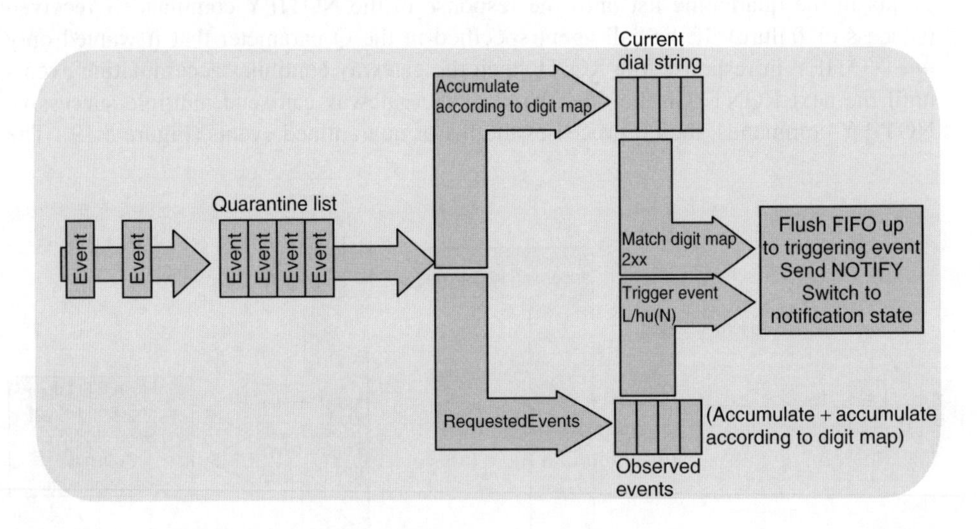


**Figure 5.19** Processing the quarantine list.

gateway can then process events from the quarantine list normally as if they had just been received, using the same list of requested events, and the same digit map (the corresponding FIFOs are empty). If a triggering condition is met, the gateway goes to notification state again. Optionally, the gateway can attempt to empty the quarantine list and transmit a single NOTIFY command with multiple events, up to the

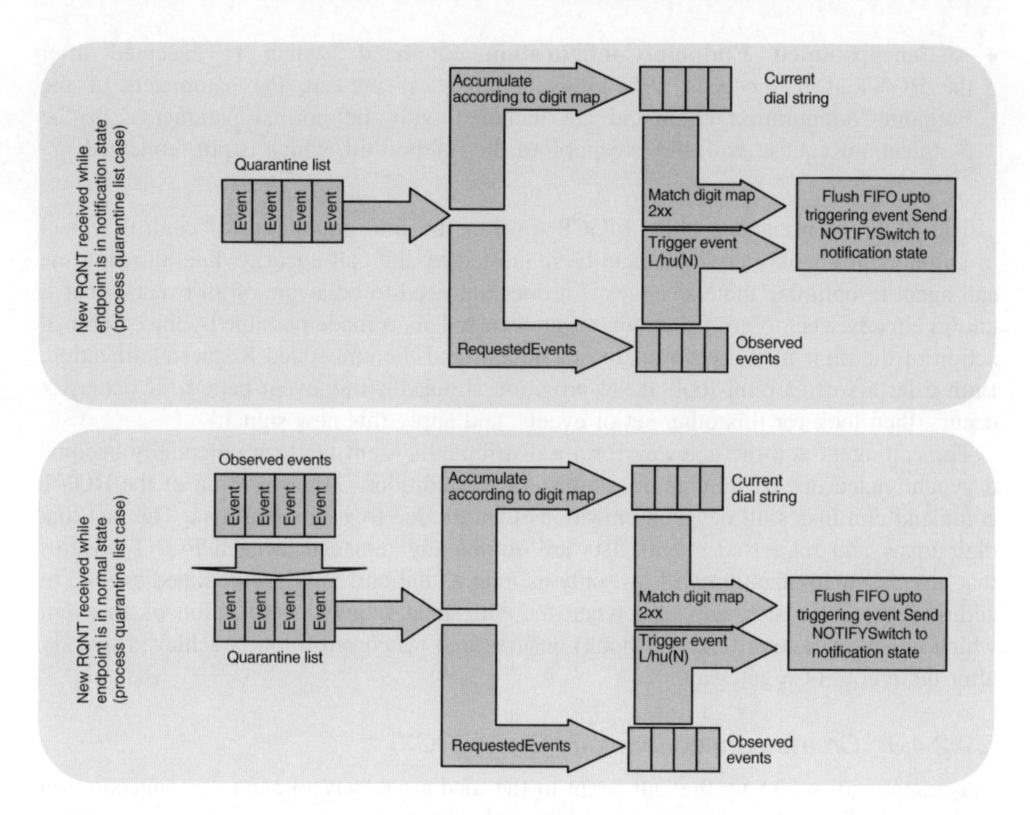


**Figure 5.20** RQNT processing according to current endpoint state.

last triggering event. The gateway goes back to normal state if the complete quarantine buffer is processed without encountering a triggering event. When a new RQNT is received following a notification, if the Q parameter states that the quarantine list should be ignored, then the quarantine list, dial string and observed event FIFOs are reset. Otherwise, the quarantine list must be processed (Figure 5.20):

- If the endpoint is in notification pending state, the dial string and observed event FIFOs are reset and the quarantine list is processed.
- If the endpoint is in normal state, the dial string is reset, the observed events list (remaining events accumulated that have not yet triggered a notification) is transferred to the quarantined FIFO, and the Quarantined FIFO is processed.

If a triggering condition is reached with the new RQNT, the gateway must ensure that any previous NOTIFY has been received by the call agent before sending a new NOTIFY. The easiest way to do so is simply to wait in notification state until an ACKNOWLEDGE of the previous NOTIFY has been received. Another way is to immediately resend the pending NOTIFY with the new NOTIFY piggybacked.

- **T parameter** (detect events): this adds some events that should be detected in the quarantine list extra to the events specified in the list of requested events.

- An **encapsulated EndpointConfiguration** command, which is executed after the RQNT if it succeeds. When this command is present, the parameters of the EndpointConfiguration command are included with the normal parameters of the NotificationRequest, with the exception of the EndpointId, which is not replicated.

The RQNT command enables MGCP to offer the fine granularity of control typical of stimulus protocols (any event can be requested by the call agent), while allowing the call agent to optimize the exchanges if it does not need to be aware of all events, or if it knows already what to do if a certain event occurs. This is made possible by the combined action of the digit map (accumulation of digits), and the embedded RequestNotification. Both offer a sort of mini look-ahead program: 'Look for this event pattern, if pattern *x* occurs, then look for this other set of events, and apply this new signal.'

The call agent and the gateway form a distributed system that can potentially become desynchronized due to failures or, simply, race conditions. The execution of the RQNT command eliminates all desynchronizations that are due to race conditions. The fact that digit maps and requested events lists are completely replaced at each RQNT ensures that any desynchronization will last only as long as the currently accumulated events. In addition, the error messages sent when the call agent requires notification of an event which is already active (e.g., off-hook) ensures that synchronization is achieved quickly after the reboot of a gateway.

### *5.2.2.4.3 Create connection command (CRCX)*

This command is sent by the call agent to the media gateway to create a connection on an endpoint. Several types of connections can be created.

#### *5.2.2.4.3.1 Connections to an external media source or sink described by SDP*

This is the most common case. The command (illustrated in Figure 5.21) contains the following parameters:

- A **CallId**: the call identifier is composed of up to 32 hexadecimal characters. It is unique to the call agent/gateway and identical on all connections that pertain to the same call. For the gateway, it is an opaque parameter that serves no operational purpose, but can be included in statistics and to facilitate troubleshooting.
- The target **EndpointId**. If the 'any of' wildcard is used by the call agent, the selected endpoint name will be included by the gateway in the SpecificEndpoint parameter of the response.
- Optionally, a new **Notified Entity** for the endpoint (where notifications and other gateway-initiated commands should be sent from now on).
- Optionally, **LocalConnectionOptions** which specify: the desired codecs (e.g., 'a: PCMU;PCMA;G726-32') in preference order; the MIME formats that are allowed (e.g., 'a:image/t38'); the desired packetization period in milliseconds (e.g., 'p:20–40'); the maximum bandwidth in kbps including IP/UDP/RTP overhead ('b:100–200');

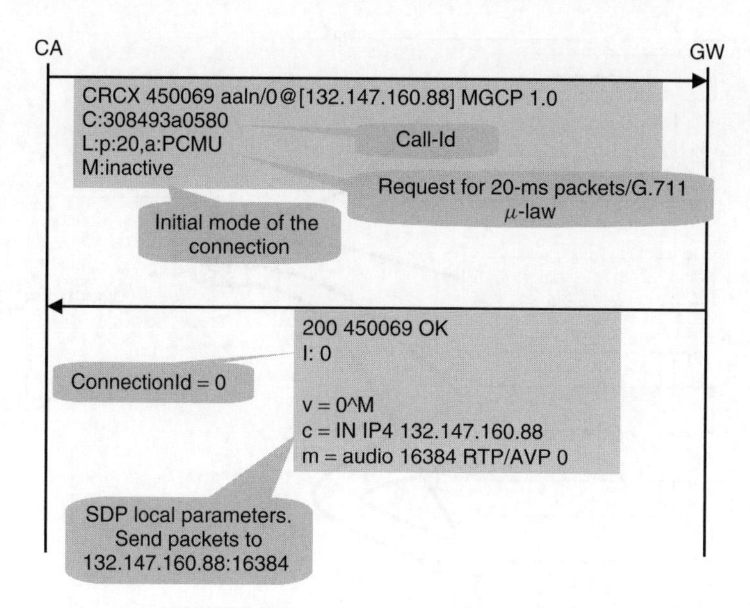


**Figure 5.21** CRCX command example.

the type of service ('t:a2', corresponding to the DiffServ[1] code point to use—the default is 0); the activation of the echo canceler (e.g., 'e:off'—the default is active); the activation silence suppression (e.g., 's:on'—the default is active); gain control (e.g., 'gc:auto'—the default is no gain control); the security key for RTP encryption (e.g., 'k:clear:mysecret', by default there is no encryption of RTP); the network type (e.g., 'nt:IN', most gateways support a single network type); resource reservation (e.g., 'r:g' for guaranteed service, 'r:cl' for controlled load[2]).

- The **mode** of the connection. The defined modes are 'sendonly', 'recvonly' (receive only), 'sendrecv' (send and receive), 'confrnce' (conference), 'inactive', 'loopback', 'conttest' (continuity test), 'netwloop' (network loop-back), and 'netwtest' (network continuity test). Figure 5.22 illustrates the relations between the media streams of each type of connection and the corresponding endpoint. The $\sum$ symbol means that signals are mixed before transmission. Signals received from 'conference' connections are sent to all other connections that are also in 'conference' mode, and are mixed before transmission. 'sendonly', 'sendreceive', and 'conference' connections require a RemoteConnectionDescriptor (or another endpoint identifier), otherwise the media cannot be sent. A loop-back connection returns any media that are received from the endpoint back to the same endpoint (standard ITU continuity test). A continuity test connection returns a 2,010-Hz signal if a 1,780-Hz signal is received from the endpoint (used in the US). A network loop-back connection returns media received from the

[1] See companion book, *Beyond VoIP Protocols*.

[2] See companion book, *Beyond VoIP Protocols*, for a detailed description of the RSVP resource reservation protocol.

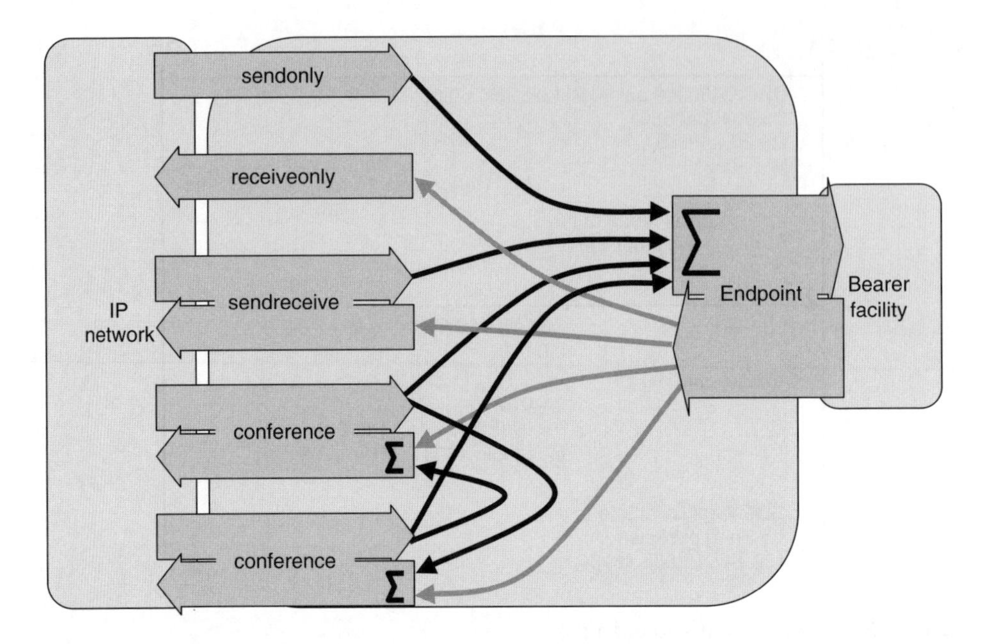


**Figure 5.22** MGCP connection types.

network back to the network. The network continuity test mode is obsolete. These connection modes are illustrated in Figure 5.23.

- Optionally, a **remote connection descriptor** that specifies using SDP where the media stream should be sent and the options for the codec to use.
- Optionally, an **encapsulated NotificationRequest command**. The parameters of the encapsulated command are simply added, apart from the EndpointID which is not duplicated.
- Optionally, an **encapsulated EndpointConfiguration command**. The parameters of the encapsulated command are simply added, apart from the endpointID which is not duplicated.

By sending an encapsulated NotificationRequest command the call agent has the ability to request the gateway to execute simultaneous actions. For example:

- Ask the residential gateway to prepare a connection, in order to be sure that the user can start speaking as soon as the phone goes off-hook.
- Ask the residential gateway to start ringing.
- Ask the residential gateway to notify the call agent when the phone goes off-hook.

This can be accomplished in a single CreateConnection command, by also transmitting the RequestedEvent parameters for the off-hook event and the SignalRequest parameter for the ringing signal. This combination dramatically reduces the number of round trips

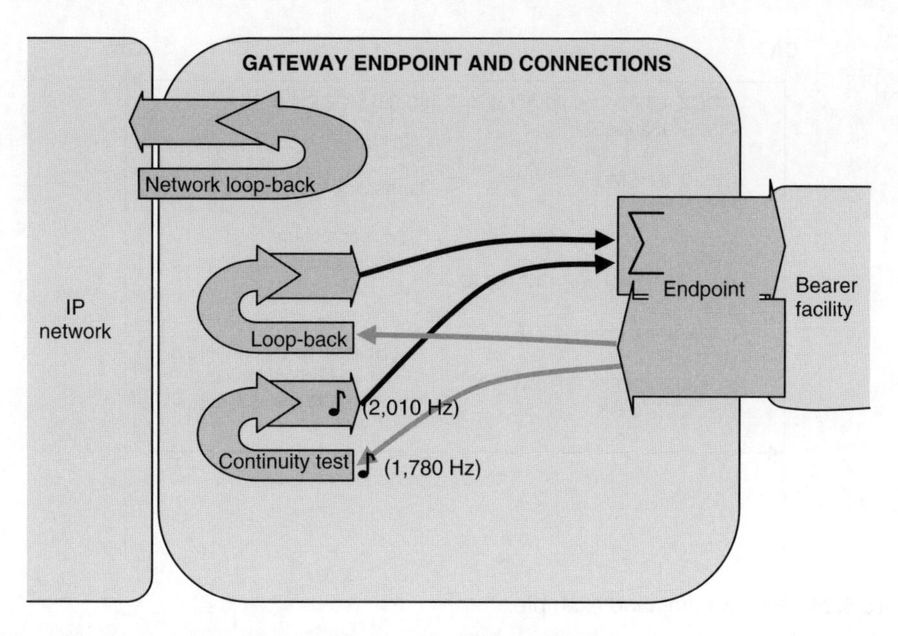


**Figure 5.23** Loop-back and continuity test connection modes.

necessary to establish a connection between two endpoints and provides more scalability options for the call agent.

After processing the CRCX command, the gateway returns:

- A ConnectionID.
- A LocalConnectionDescriptor specifying the local parameters of the connection using SDP.
- Optionally, a specific EndpointID if the endpoint was not specified in the command.

#### *5.2.2.4.3.2 Connections to another endpoint on the same gateway*

Connections to another endpoint on the same gateway occur frequently. Many manufacturers do not support any optimization for this type of connection and some events do not support it properly (this is one of the most common bugs in gateway implementations across all VoIP protocols), in which case the call agent uses the normal procedure, giving a local gateway IP address and port as the RemoteConnectionDescriptor.

But, some gateways allow endpoints to communicate locally without requiring packetization and switching through the IP network. This type of connection is similar to the previous type of connection, except that the call agent specifies a SecondEndpointID instead of a RemoteConnectionDescriptor. Such a command really creates two connections (one on each endpoint), the response provides the ConnectionIDs of both connections. The second connection is by default in sendrect mode.

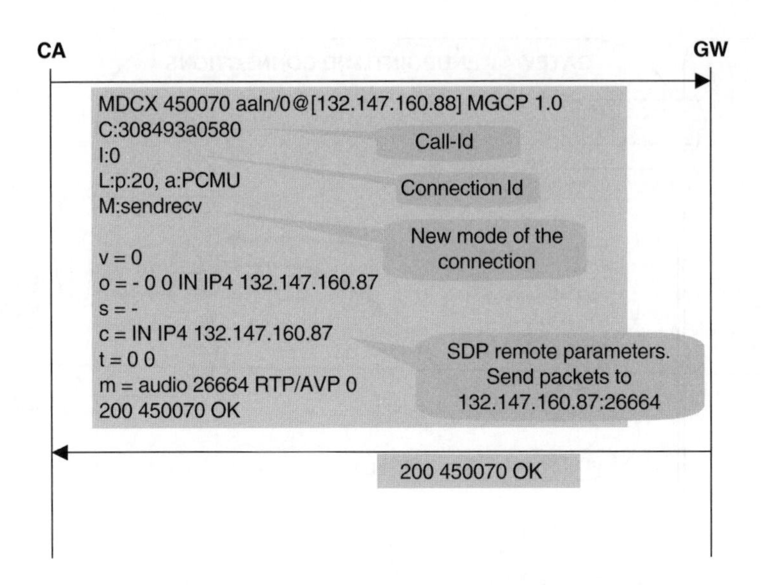
CA
GW
MDCX 450070 aaln/0@[132.147.160.88] MGCP 1.0
C:308493a0580
I:0
L:p:20, a:PCMU
M:sendrecv
v = 0
o = - 0 0 IN IP4 132.147.160.87
s = -
c = IN IP4 132.147.160.87
t = 0 0
m = audio 26664 RTP/AVP 0
200 450070 OK
Call-Id
Connection Id
New mode of the connection
SDP remote parameters. Send packets to 132.147.160.87:26664
200 450070 OK

**Figure 5.24** MDCX command example.

### *5.2.2.4.4 Modify connection command (MDCX)*

This command (illustrated in Figure 5.24) enables a call agent to modify a connection that has already been set up by the gateway. The parameters are the same as in the CreateConnection command, with the addition of the ConnectionID that serves to identify the target connection.

The ModifyConnection command can change all parameters of a connection: activation mode, codec, packetization period, etc.

### *5.2.2.4.5 Delete connection command (DLCX)*

This command enables the call agent to terminate a given connection. It should be noted that if there is more than one gateway involved in a call, the call agent sends the Delete-Connection command to each of the media gateways in order to fully tear down both ends of the call.

As shown in Figure 5.25, a nice functionality provided by MGCP is that the media gateway, on termination of a connection, has to send to the call agent the following information:

- Number of packets (RTP) sent.
- Number of octets sent.
- Number of packets (RTP) received.
- Number of octets received.
- Number of packets lost.

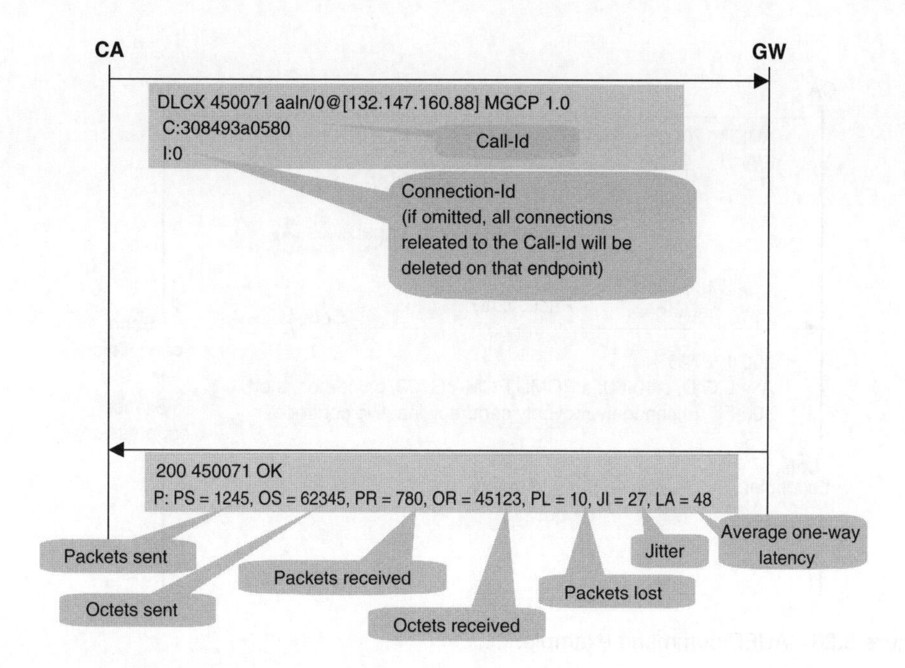


**Figure 5.25** DLCX command example.

- Interarrival jitter.
- Average transmission delay.

These parameters are calculated as for RTCP (see Chapter 1).

MGCP also allows a media gateway to clear a connection on its own (e.g., in the event of connection loss or a failure). In this context the media gateway sends a DeleteConnection command to the call agent including all the connection statistics.

It is also possible for a call agent to delete all the connections of a call at the same time by omitting the ConnectionID. Note that this command does not return any individual connection statistics or call parameters.

### *5.2.2.4.6 Audit endpoint command (AUEP)*

The call agent can use this command in order to check whether an endpoint is up and running, and to learn dynamically its capabilities (Figure 5.26). Using the 'all off' wildcard, the call agent can also learn the number of endpoints present on a given gateway.

### *5.2.2.4.7 Audit connection command (AUCX)*

This command enables the call agent to retrieve all the parameters attached to a connection identified by a ConnectionID on an endpoint identified by its EndpointID (Figure 5.27). This can be used by a call agent to check that a connection is still active: if no information is requested, the gateway simply responds with 200 OK if the connection exists.

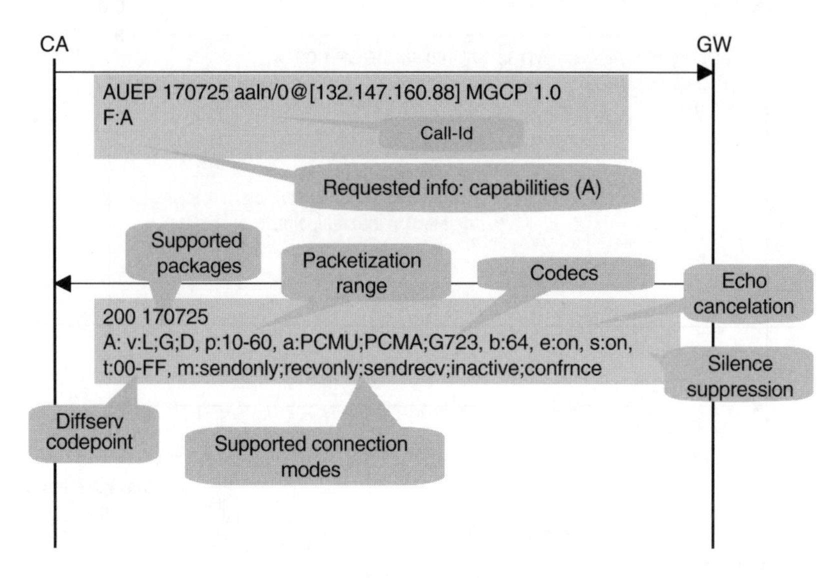


**Figure 5.26** AUEP command example.

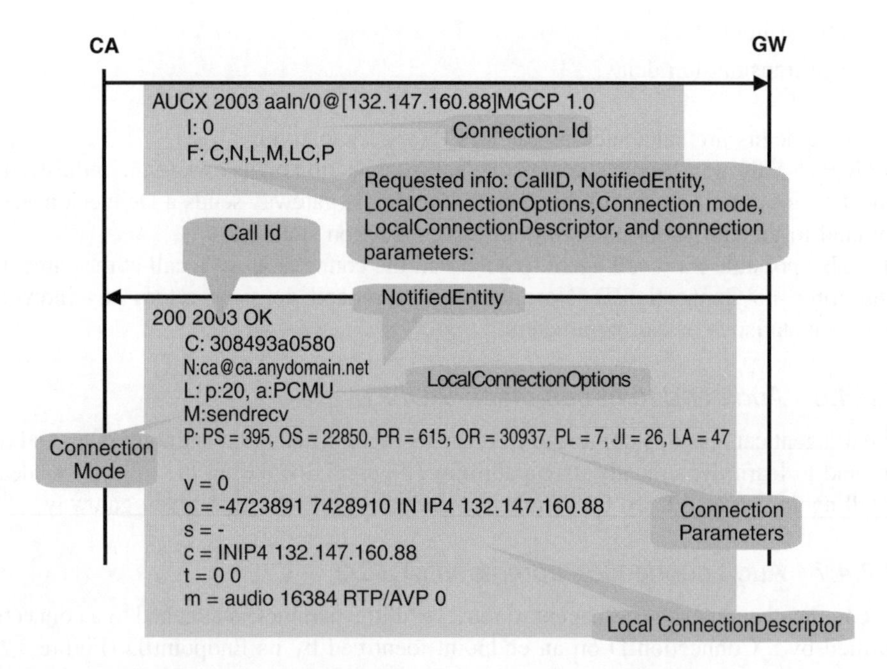


**Figure 5.27** AUCX command example.

### 5.2.2.5 MGCP commands from the gateway to the call agent

#### 5.2.2.5.1 Notify command (NTFY)

This command enables the media gateway to send back events that were requested by the media gateway controller. The media gateway can send one or several events in a NOTIFY command. Each notification reports events from a given endpoint (possibly a connection on an endpoint), listed in the endpoint part of the command header. The correlation between the request and the corresponding notification is provided by the RequestIdentifier (X parameter). The list of notified events is specified in the ObservedEvents (O) parameter, which is a comma-separated list of events. Events appear in the order in which they have been detected. The form of events can be:

- The event name, only if it is part of the default package (not recommended), such as hd.
- The package name and the event name: L/HD.
- The package name, event name, and ConnectionID for events detected on a connection: L/HD@134a23b.

When booting, some endpoints send without solicitation their current state (e.g., off-hook) with special request ID 0.

The NOTIFY command is acknowledged by a return code from the call agent.

#### 5.2.2.5.2 Restart in progress command (RSIP)

This command allows a gateway to make a call agent aware of an endpoint or a group of endpoints that are going to be taken out of service. In this case the restart method can be graceful (RM:graceful), can specify a delay (RD), or can be forced (connections are lost immediately).

The message is also sent by gateways when they boot, to make the call agent aware of their presence (Figure 5.28). In this case the restart method is 'restart', and a delay can be specified until the endpoints are operational (0 is the default value if nothing is specified). Restart method 'disconnected' can also be used to alert the call agent about potential state mismatch.

For gateways that acquire an address dynamically through DHCP, the call agent has three ways to learn the IP address of the gateway:

- By looking at the source IP address of the RSIP message. This is not always reliable if the RSIP message is relayed.
- If **Dynamic DNS** (**DDNS**) is used in conjunction with the DHCP server, the DNS name of the gateway as advertised in the RSIP message will resolve to the current IP address of the gateway. This is a robust method and also provides the ability to recontact the gateway immediately if the call agent reboots. On reboot, the call agent, if it knows about the gateway, queries the DNS and can send an AUEP to the current IP of the gateway.

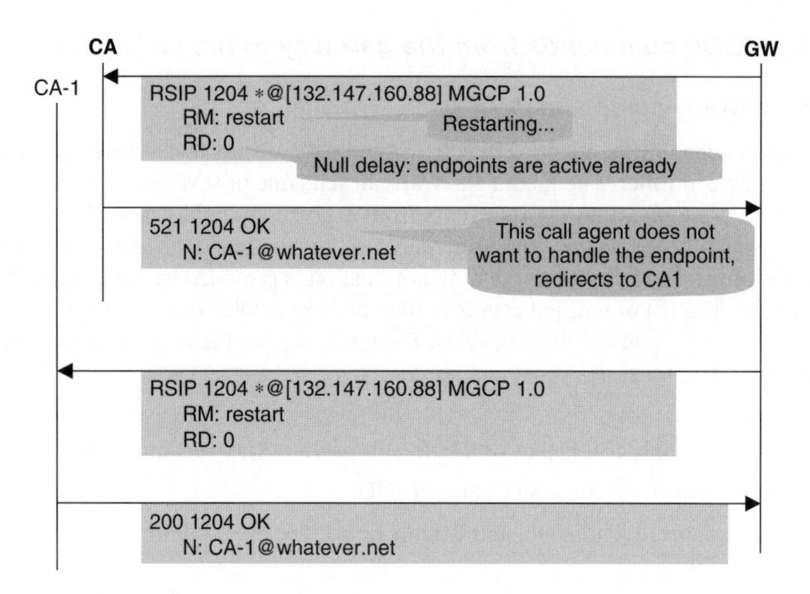


**Figure 5.28** RSIP and change of call agent.

- The gateway can include its current IP address as the gateway name. This works, but makes it difficult to keep track of the gateway since the name changes with the IP address. In addition, if the call agent reboots, it will be unable to reach the gateway unless it has saved the current IP address in persistent storage.

## 5.2.3 Handling of fax

A new package 'fxr' for fax is being defined (draft-andreasen-mgcp-fax-xx.html). The fxr package also uses extensions of SDP for negotiation (defined in RFC 3407, 'Session Description Protocol (SDP) Simple Capability Declaration', or **simcap**). The fxr package defines new local connection options:

- 'fxr/fx:t38' for strict handling of T.38. The gateway notifies the call agent that a T.30 fax preamble is detected ('fxr/t38(start)' event) and mutes the media channel. Before starting the T.38 procedure, the gateway will check that the remote party also supports the same variant of fax transport by checking its capabilities, expressed in an extension to SDP (see below for details). The call agent is responsible for switching the connection to T.38 fax mode by sending an MDCX with 'a:image/faxt38' in the LocalConnectionOptions (L:) and providing a RemoteConnectionDescriptor with the 'm = t38' media line. Should a failure occur during the fax call, it is indicated by the 'fxr/t38(failure)' event. The end of the fax is indicated by the 'fxr/t38(stop)' event.
- 'fxr/fx:t38-loose' for loose handling of T.38. The difference from strict handling is that no confirmation of common capabilities is required from the remote end. The fax transmission attempt starts as soon as a RemoteConnectionDescriptor with a media line indicating T.38 is received.

- 'fxr/fx:off' if there is no special handling of fax.
- 'fxr/fx:gw' if the handling of fax (not necessarily T.38) is left to the gateway. This is the default mode. The gateway will send a 'fxr/gwfax(start)' event if it begins a specific fax procedure, or 'fxr/nopfax(start)' if the gateway detected a fax but decided to take no action on it. In the case of 'fxr/gwfax(start)', the call agent should remain passive until it receives a 'fxr/gwfax(stop)' event.

In the following example (Figure 5.29) the call agent configures the gateway to use strict T.38 handling on line 'aaln/0'. The gateway returns local connection parameters, as well as 'a =' elements listing its capabilities. It was necessary to extend SDP to express capabilities (RFC 3407), because the normal way of expressing support for multiple codecs in SDP also implies that media can be received immediately on these coders. Here t.38 is supported, but still cannot be received. The capability set according to RFC 3407 is identified by a serial number, incremented each time the endpoint sends a new capability set (a = sqn:<serial number>). This attribute is immediately followed by capability lines (a = cdsc:<capability number><media type><transport><format list>).

Note that in Figure 5.29 the call agent still does not know the capabilities of the remote endpoint (not mentioned in CRCX). If a fax was received at this point, the t38 procedure would be delayed until a proper capability descriptor for the remote endpoint is received from the call agent.

Once the call agent knows the capabilities of the remote end, it sends these capabilities in the RemoteConnectionDescriptor of a MDCX command (Figure 5.30). Now that the

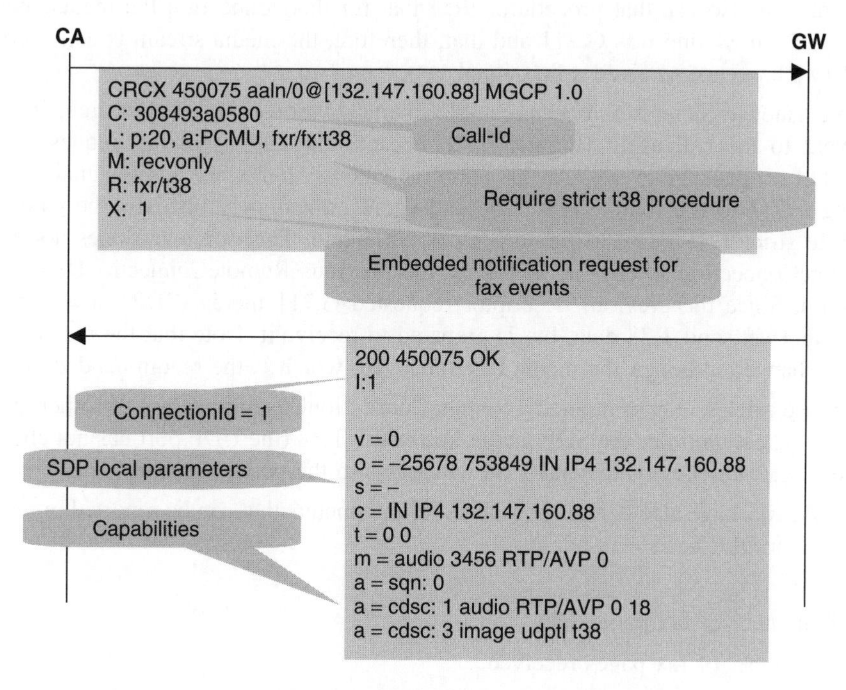


**Figure 5.29** CRCX command requiring usage of the strict T.38 procedure.

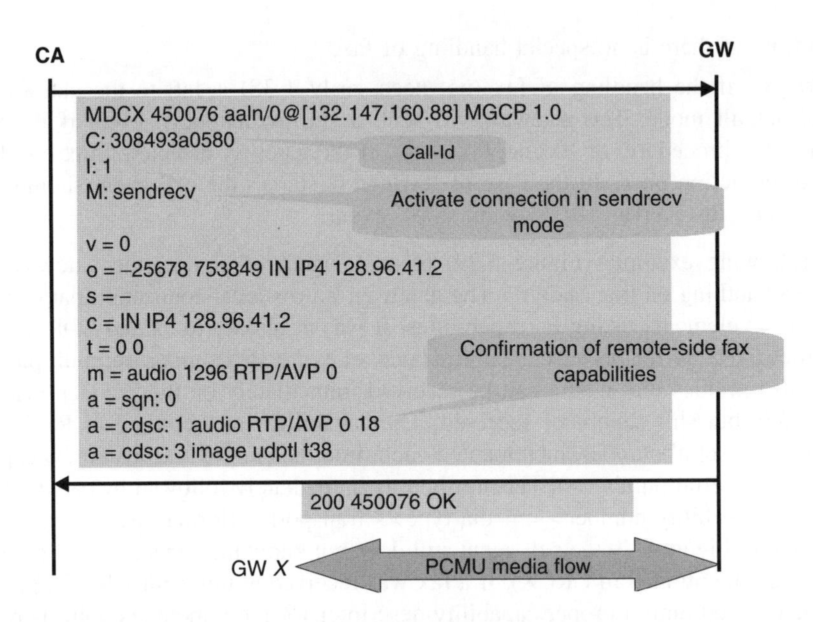
CA
GW
MDCX 450076 aaln/0@[132.147.160.88] MGCP 1.0
C: 308493a0580
Call-Id
I: 1
M: sendrecv
Activate connection in sendrecv mode
v = 0
o = -25678 753849 IN IP4 128.96.41.2
s = -
c = IN IP4 128.96.41.2
t = 0 0
Confirmation of remote-side fax capabilities
m = audio 1296 RTP/AVP 0
a = sqn: 0
a = cdsc: 1 audio RTP/AVP 0 18
a = cdsc: 3 image udptl t38
200 450076 OK
GW X
PCMU media flow

**Figure 5.30** Confirmation that the remote end supports T.38 fax.

local gateway knows that the remote endpoint also supports strict T.38 over UDP and, therefore, can also use that procedure. Note that for this connection the media specified in the SDP 'm =' line was G.711 and that, therefore, the media stream is activated with G.711 (we have not received a fax signal yet).

If the gateway detects a T.30 preamble characteristic of fax at any time, it reports the event to the call agent (Figure 5.31), because the call agent has requested to be notified of fxr package events. At this point the gateway mutes the audio signal and stops sending G.711 to the remote GW. The call agent immediately instructs the gateway to switch to strict T.38 mode using an MDCX command. This command does not contain a RemoteConnectionDescriptor; therefore, the previous RemoteConnectionDescriptor is still valid. Since the previous descriptor requested G.711 media (SDP 'm =' line), the GW cannot yet send T.38 data, but is prepared to receive it. Note that the reception port has not changed although the media have changed, which is the recommended behavior.

Once the call agent has obtained a RemoteConnectionDescriptor from the other gateway (Figure 5.32), it modifies the SDP media line to use T.38 (the UDP port has not changed). The local gateway begins to send T.38 datagrams to the port indicated.

The 'fxr' package also defines new statistics parameters that can be reported in response to AUCX or DLCX:

- PGS: number of fax pages sent.
- PGR: number of fax pages received.

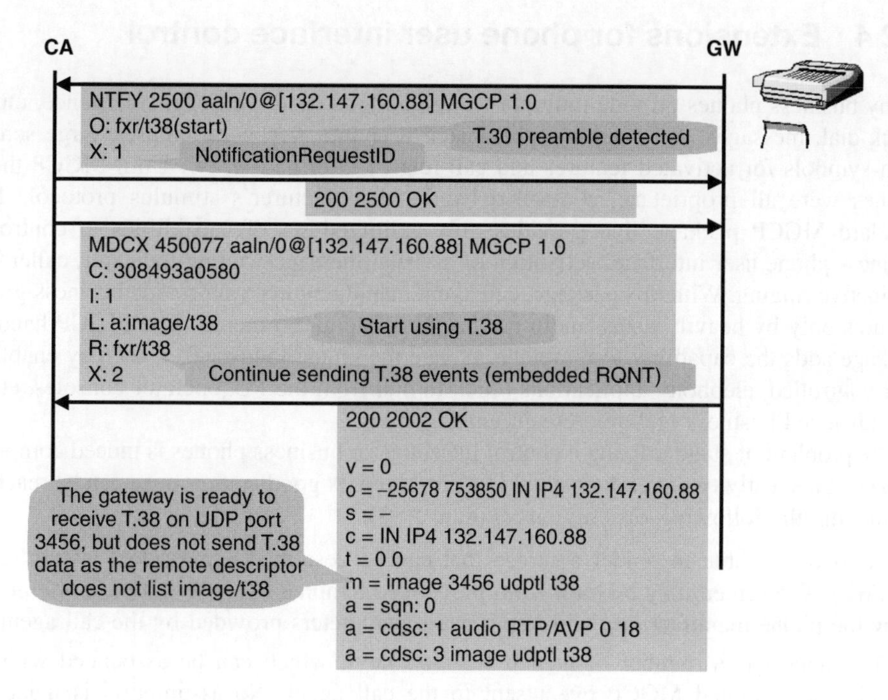


**Figure 5.31** Preparing the gateway to receive T.38 media data.

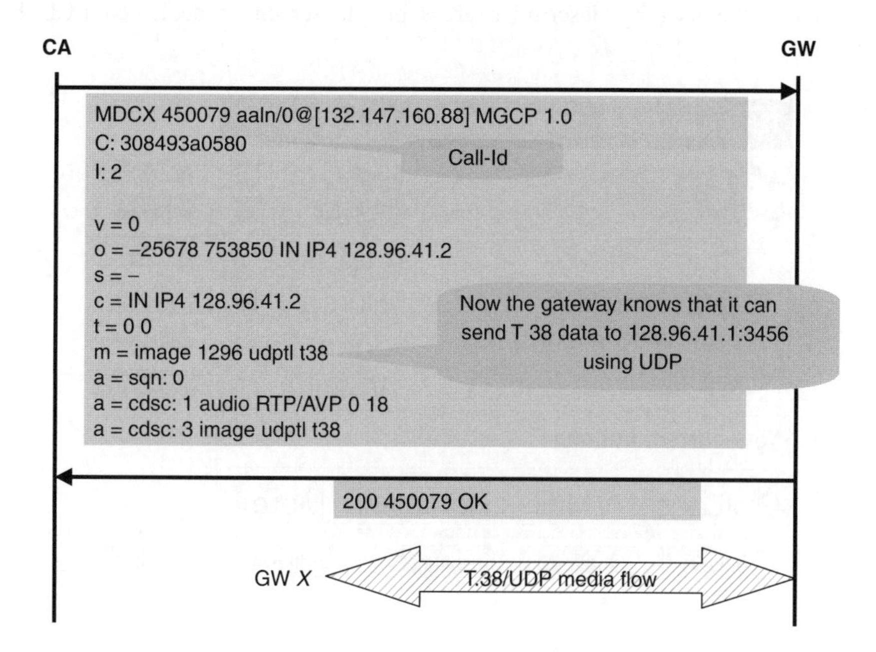


**Figure 5.32** The local gateway is instructed to send T.38 data.

## 5.2.4 Extensions for phone user interface control

Many business phones provide multiple feature buttons (hold, retrieve, conference, mute, quick-dial, messages, etc.) and a sophisticated user interface with lamps, a large screen with symbols for activated features and call-related information, etc. Until MGCP these phones were all proprietary, controlled by the manufacturer's stimulus protocol. The standard MGCP package line provides only a limited set of capabilities to control a business phone user interface: activation of a visual message-waiting indicator, caller ID, distinctive ringing. With this package, call agent manufacturers can provide business-grade features only by heavily using audio notifications and audio menus. The MGCP handset package adds the capability to remotely activate the phone loudspeaker, thereby enabling CTI-controlled telephony applications (click to dial from the PC, operator consoles, etc.), but advanced business features remain unaddressed.

The problem of standardizing a control interface for business phones is indeed complex, because the creativity of vendors should be preserved. A good compromise can be reached by making the following assumptions (Figure 5.33):

- The phone is able to render a screen that can be described using a text syntax (e.g., XML). The screen may be built from predefined templates (cards) stored in the phone by the phone manufacturer, with replaceable parameters provided by the call agent.
- The phone has a number of named function keys, which can be associated with an endpoint-generated MGCP event sent to the call agent. No assumption is made on the function of the key. Optionally, some keys can have their function dynamically described to the user by descriptive areas on the screen or dedicated LCD labels (softkeys).

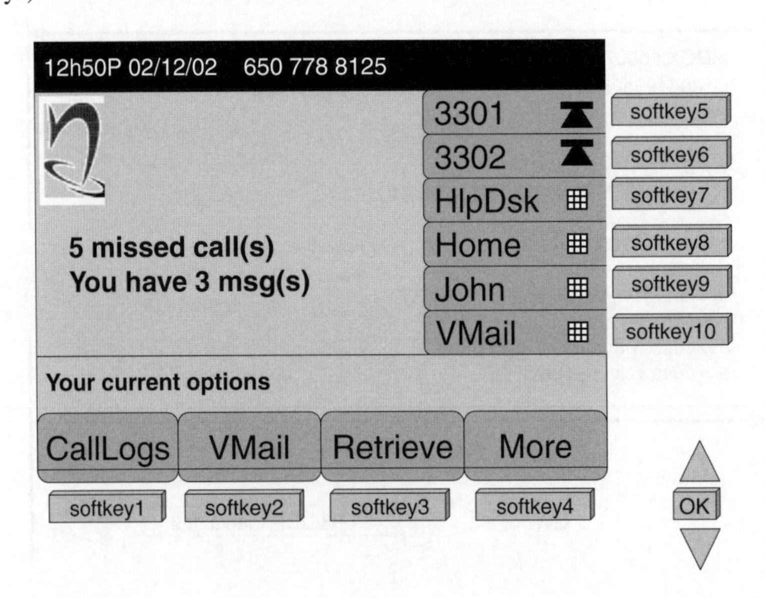


**Figure 5.33** Typical IP phone LCD screen with softkeys.

- Optionally, the phone can also provide a capability to navigate through menus and select an option, or can offer numeric or alphanumeric input fields.

To date, several vendors have implemented business phone control MGCP packages based on these assumptions:

- Cisco with the BTXML2 markup language.
- Polycom with the MGCP business phone packages documented in RFC 3149.
- Swiss voice with the MGCP business phone packages documented in RFC 3149.

RFC 3149 defines:

- A feature key package (KY) which describes signals to set the key label ('KY/ls(<KeyId,Label>)') and key activation state ('KY/ks(<keyId>,<KeyState>)'). The following states have been defined: en(enabled), db(disabled), id(idle), dt(dial tone), cn(connected), dc(disconnected), rg(ringing), rb(ring-back), ho(holding), he(held). 'S: KY/ks(5,en)' in a RQNT sets key fk5 to enabled state. MGCP events are used to report the key press events (KY/fk1 to KY/fk99). These events can be requested by sending a NotificationRequest with 'R:KY/fk'.
- A business phone package (BP) in which signals are used to force speaker phone activation ('BP/hd' for off-hook, 'BP/hu' for on-hook) or play a beep ('BP/beep').
- A display XML package (XML). An XML-format signal is used to render the screen. Events are used for user input or selection. Both are prefixed 'XML/xml'. The screen control feature of the display XML package uses a special endpoint name, derived from the phone endpoint name. If the phone is called ph1@anydomain.net, the screen endpoint name will be disp/ph1@anydomain.net. This separation avoids possible problems since events deactivate signals by default, which is not the desired behavior for screen navigation. In order to request events resulting from the selection of items on screen menus, the RQNT targeted at the display endpoint must contain 'R: XML/xml'.

The XML screen description syntax of RFC 3149 defines the following widgets: input box, enumerated list box, text box, and echo box. The XML screen template can contain replaceable parameters, or tags corresponding to dynamic content (e.g., time/date or call timer). The XML format also describes the event string to send back to the call agent for each possible selection. If the main phone keypad is used to select a choice on the screen menu, the event is reported to the call agent through the XML package on the screen endpoint: the screen has precedence and only passes unused key press events to the phone endpoint subsystem (the only exception is the echo widget, which displays events but does not consume them, and can be used to echo on the screen a dialing number).

The format of the XML/xml screen signal is as follows:

```
S: XML/xml
  (<url>?<card>?$<variable1>=<value$<variableN>=<value>)
```

The <url> can be http://screenserver.anydomain.net/deck1 if the set of screen templates (called a **deck**) must be fetched on an HTTP server, or any name if it is local to the phone (provisioned). The <card> component specifies the template to select within the deck. Usually, each phone state is associated with a specific card. The variables are replaceable parameters within the card template. For instance, if deck1 is:

```
<xml>
    <card id="one">
       <p>$line1</p>
       <timer value="2"/>
       <do type="ontimer">
       <go href="#two"/>
       </do>
    </card>

    <card id="two">
       <p>$line2</p>
    </card>

    <card id="home">
       <p mode="nowrap">$dn <time align="right"></time>
       <select type="item" name="Menu" iname="StrMenu">
       <option value="1" onpick="post?

       </select>
       </p>
    </card>

</xml>
```

If the applied signal is S: XML/xml(deck1?one?$line1=abc$line2=xyz), the phone would render:

```
<card id="one">
   <p>abc</p>
</card>
```

Then, after the timer:

```
<card id="two">
   <p>xyz</p>
</card>

<xml>
```

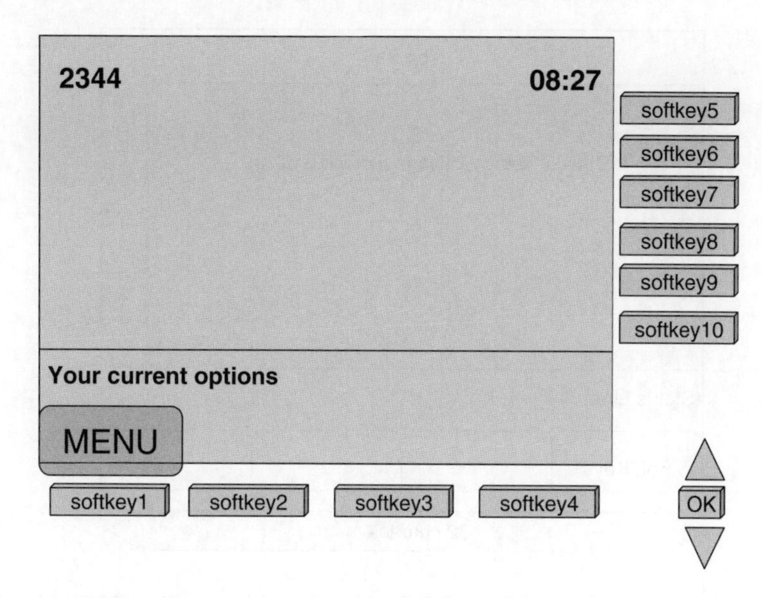


**Figure 5.34** IP phone screen appearance showing the 'home' card.

If the applied signal is `S: XML/xml(deck?home?$dn = 2344)`, the screen would be rendered by our sample phone as shown in Figure 5.34.

If softkey 1 is pressed, the following event would be reported in the NOTIFY:

```
O: XML/xml(post?basic?home?Menu=1)
```

In addition to the functions described above, some functions must be implemented locally on the phone, such as mute, volume control, contrast control, audio path control (handset/loudspeaker/headset). RFC 3149 assumes these functions come with their own screens defined by the phone manufacturer.

Cisco BTXML2 syntax was defined after RFC 3149 and is available in conjunction with the MGCP protocol on their IP phones. It is very similar to RFC 3149 (e.g., it likewise uses a separate endpoint for screen control prefixed by 'disp/'). The main difference is that the XML description also includes feature key event mappings (Cisco phones do not have separate LCDs for each button) and provides many more widgets than those defined in RFC 3149. The display is divided in zones (similar to HTML frames), which can be described separately (Figure 5.35).

Even though the industry has not yet agreed on a common XML description format, these open control interfaces are similar enough to make it relatively easy for call agent manufacturers to support business phones. In fact, the call agent does not need to be aware of the exact XML syntax used by the phone; it interacts with the phone only by calling predefined cards with replaceable parameters, and receives named events that it needs to map to call control actions. The customization of a call agent for a specific type of phone becomes straightforward. MGCP really invented the 'open business phone'!

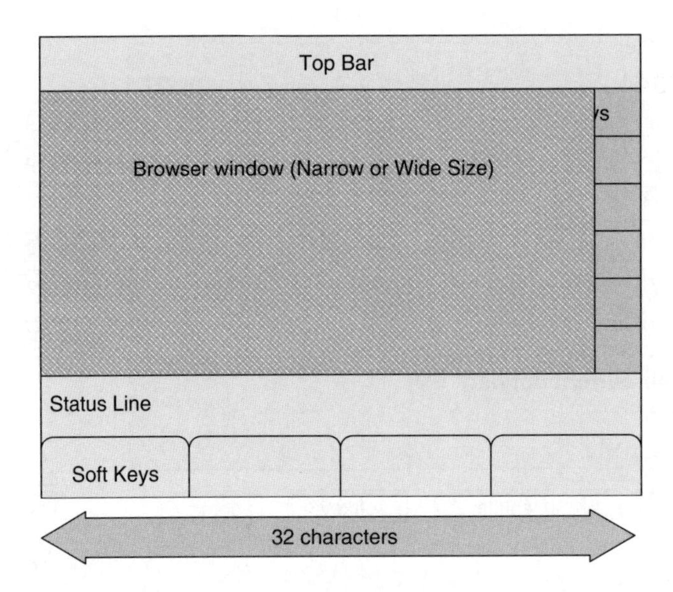


**Figure 5.35** Screen structure of Cisco BTXML2-capable phones (the 7960 is shown).

## 5.3 SAMPLE MGCP CALL FLOWS

### 5.3.1 Call set-up

In Section 5.1, we described the two main applications of MGCP (both at the edge of the network):

- Control of analog gateways or MGCP phones in customer premises.
- Control of trunking gateways in the service provider network.

The following call flow illustrates both cases. A call is received from an SS7 network signaling transfer point (STP) by an SS7 call agent (SS7_CA); the SS7 call agent sends the call to the core VoIP network using the H.323 protocol (it could also be SIP); the core network routes the call to a residential service call agent (R_CA), and the residential service call agent rings an analog phone on a residential gateway (R_GW).

SS7 ISUP messages are always associated with a **circuit identification code** (**CIC**), which relates call signaling to a specific media time slot on a given trunk (this relation is configured statically as part of the provisioning of TDM switches). The CIC enables the SS7 call agent to locate the proper media gateway and the endpoint on this gateway that terminates the specified trunk and time slot.

The first part of the call scenario is illustrated in Figure 5.36. The CRCX command instructs the endpoint to prepare for receiving G.711 $\mu$-law media from the IP network, with a 10-ms frame size. The gateway responds by giving an IP address and port where

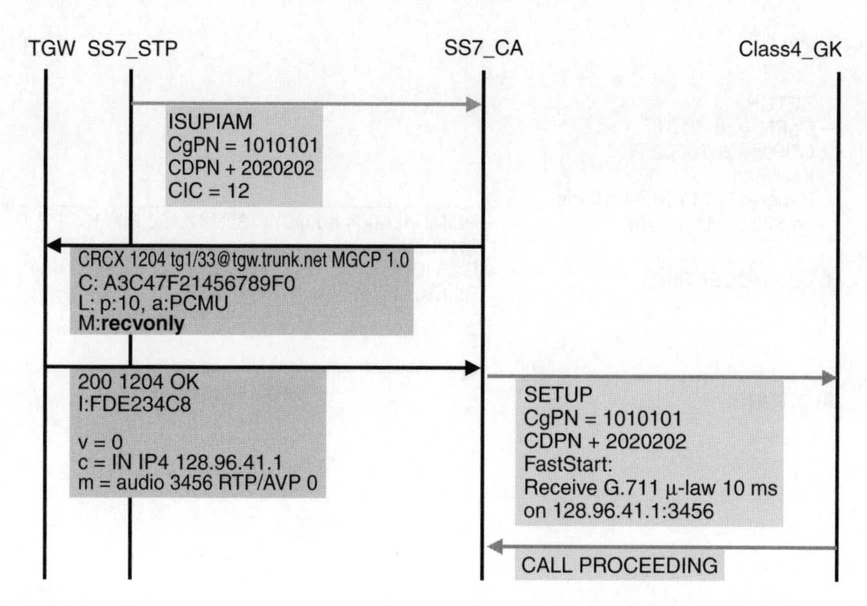


**Figure 5.36** New call received from SS7 network and transmitted to the VoIP core. IAM = initial address message.

the RTP stream should be sent. With this information, the SS7 call agent can now send an H.323 SETUP message to the VoIP network core-routing softswitch, in this case an H.323 gatekeeper. This core-routing softswitch is responsible for finding the proper egress route, or eventually for rerouting the call to back-up routes in case of congestion or any other problem. It will also translate the calling and called party numbers if necessary, as appropriate for the destination. The SETUP message contains the RTP IP address and port in the Fast Start element, in order to expedite the media connection.

The CALL PROCEEDING message indicates that the SETUP message has been received properly and that the dialing number is complete. If the number is not complete, the core softswitch would have sent a SETUP ACKNOWLEDGE message instead, to initiate a procedure known as overlap sending, in order to accumulate more digits.

The call flow is simple because there is only one voice coder in the VoIP network and codec negotiation is not required. Variants of this call flow are needed to use the negotiation capabilities of H.323: multiple 'inactive' connections could be set up on the media gateway for each voice coder, which would provide the list of proposed logical channels for the H.323 SETUP fast Start element. If the MGCP trunk gateway cannot open the multiple inactive connections of various coders, then AUEP codec information could be used to construct an H.245 CapabilitySet, and the codec would be negotiated through a normal H.245 exchange.

In our example (Figure 5.37) the called party is managed by a residential call agent. It immediately acknowledges the SETUP message with a CALL PROCEEDING, and also establishes the H.245 dialog either with the core softswitch, or directly with the source call agent (this is not shown in the diagram). The residential gateway (RGW) is instructed

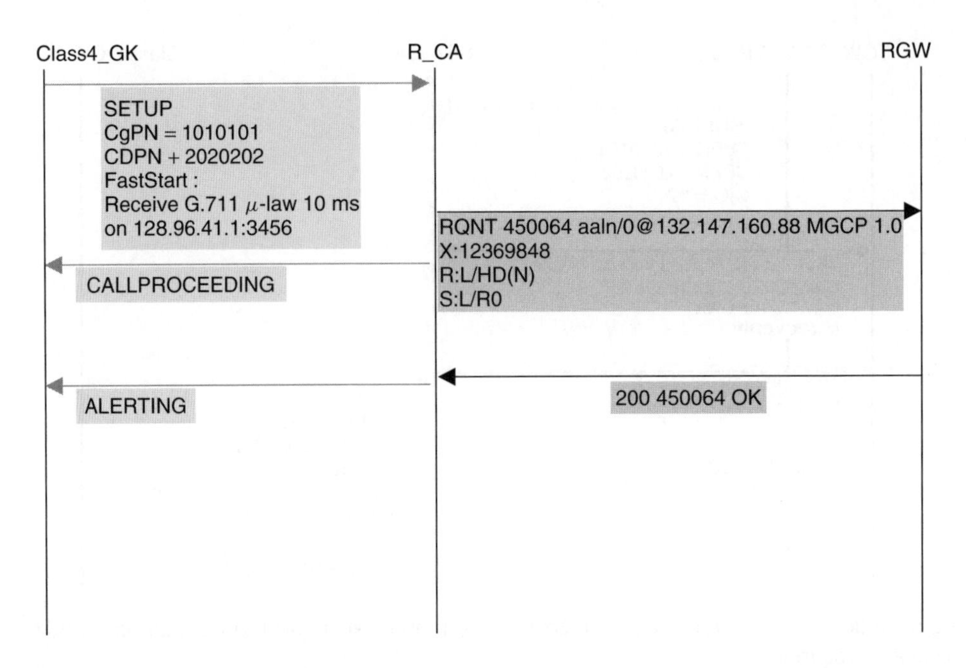


**Figure 5.37** The class 4 gatekeeper routes the call to the appropriate residential call agent.

to ring ('L/R0' signal) and to notify the off-hook event to the call agent. As soon as the residential gateway confirms that it is ringing, R_CA sends back the ALERTING message to the core softswitch. Note that within a national telephone network the actual ringing tone is never sent over the RTP media connection, it will be generated by the calling party switch when it receives the ALERTING message. However, it is still necessary to provide a receive-only media connection as soon as possible (if this can be done in the SETUP message), because in the case of a phone call to an international number the ring-back will be provided in-band through the RTP connection. In this case (not represented) the H.323 ALERTING message will contain a progress indicator (PI = 8), which instructs the originating switch not to play a local ring-back, but the remote ring-back instead.

The ALERTING message is relayed by the core softswitch to the originating SS7_CA, which sends back an address complete (ACM) ISUP message on the SS7 signaling link. In fact, the SS7_CA may also have sent the ACM immediately on receiving the CALL PROCEEDING message from the core softswitch: in SS7 ISUP the ACM means both that the number is complete and implicitly that the remote party phone is ringing. On SS7 networks the calling party may hear ring-back before the called party phone actually rings!

When calling an international network, the distant ring-back tone is sometimes provided (this also allows remote in-band error messages to be heard). In such a situation a PROGRESS or ALERTING message with a specific progress indicator (indicating that in-band audio information is being sent) would be received from the H.323 side, and

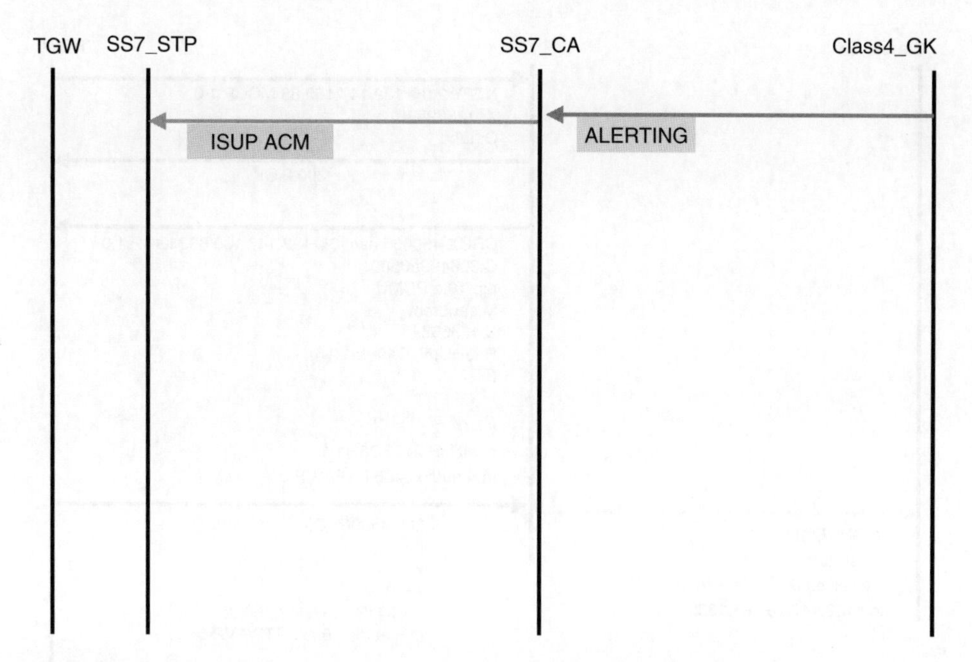


**Figure 5.38** ALERTING message converted to an ISUP ACM. ACM = address complete message.

the SS7 call agent should then set an equivalent indicator in the ACM or send a call progress (CPG) ISUP message. In modern telephony networks, this is normally the only situation where the ring-back tone is provided by the remote end. Figure 5.38 represents the normal case, with ringback provided by the calling party exchange. There is still no media exchanged on the VoIP network.

At this moment the called party answers the call. The residential gateway sends a NOTIFY back to the call agent (Figure 5.39). The call agent immediately creates a connection on the endpoint. Since the call agent already has received the voice coder settings, IP address, and port required to send media to the calling-side endpoint, this information is provided in the CRCX SDP. The residential gateway can immediately begin to send audio. In the answer to the CRCX, the residential gateway also provides an IP address and port where it will receive audio from the calling-side endpoint. This information is included in the FastStart element of the CONNECT message sent by the call agent to the class 4 gatekeeper.

The CRCX message also uses an embedded NotificationRequest instructing the residential gateway to immediately notify the call agent of any digit detected on the endpoint and, of course, if the user hangs up.

As soon as the SS7 call agent receives the CONNECT message, it relays the media information of the FastStart element to the trunk gateway, using a ModifyConnection message (Figure 5.40).

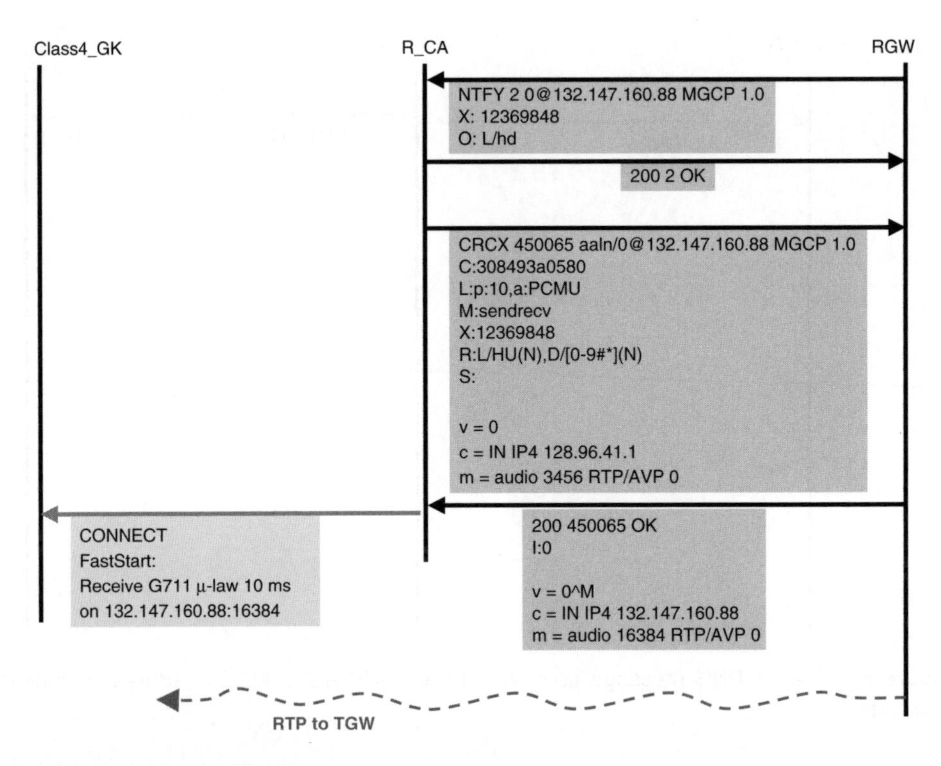


**Figure 5.39** The called user picks up the handset.

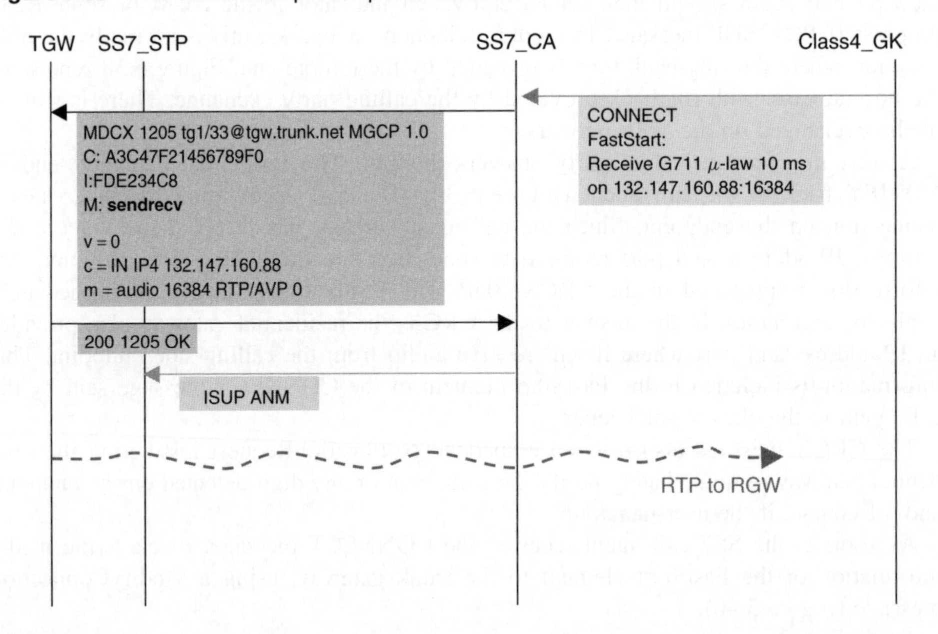


**Figure 5.40** The CONNECT message is converted to an ISUP ANM message. ANM = answer message.

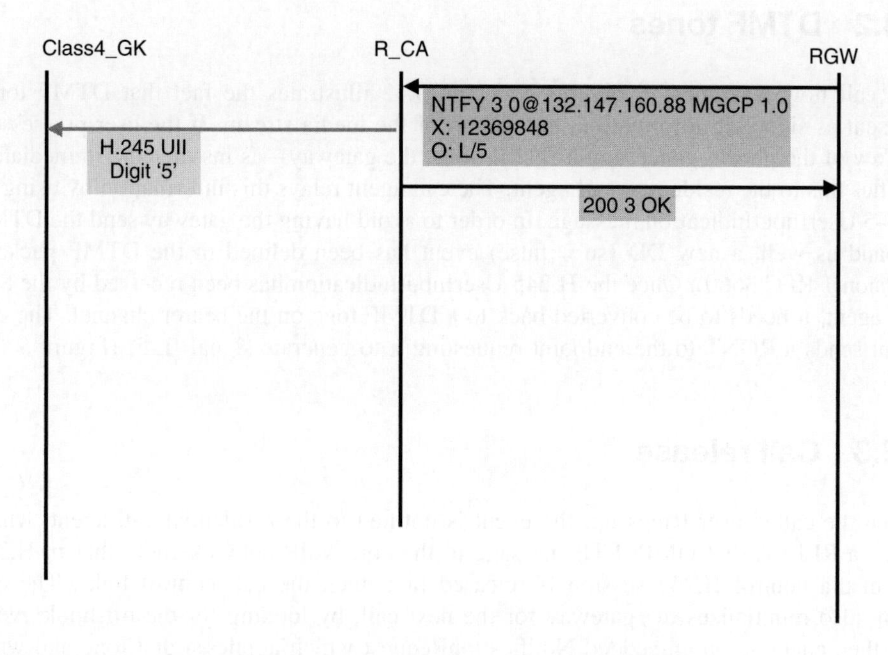


**Figure 5.41** Out-of-band DTMF handling. UII = user input indication.

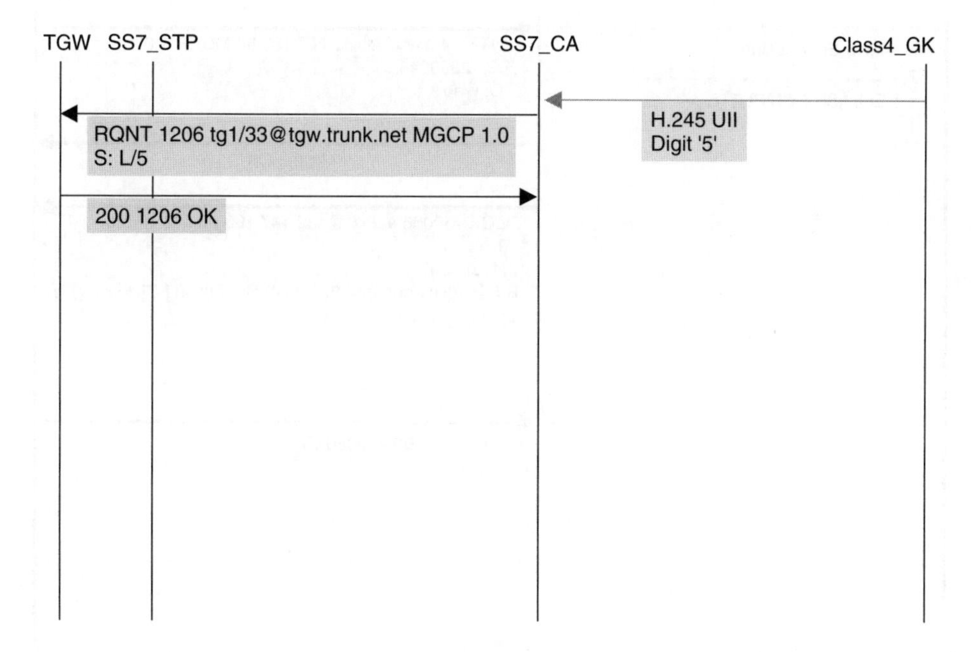


**Figure 5.42** DTMF regenerated by trunk gateway. UII = user input indication.

## 5.3.2 DTMF tones

The call flow described in Figures 5.41 and 5.42 illustrates the fact that DTMF tones are sent as signaling information, not as part of the media stream. If the user presses the '5' key of the phone, generating a DTMF tone, the gateway—as instructed—immediately notifies this to the residential call agent. The call agent relays this information by using an H.245 UserInputIndication message. In order to avoid having the gateway send the DTMF in-band as well, a new DD (su = false) event has been defined in the DTMF package (version 1 RFC 3660). Once the H.245 UserInputIndication has been received by the SS7 call agent, it needs to be converted back to a DTMF tone on the bearer channel. The call agent sends a RQNT to the endpoint requesting it to generate signal 'L/5' (Figure 5.42).

## 5.3.3 Call release

When the called user hangs up, the event is notified to the residential call agent, which sends a RELEASE COMPLETE message to the core VoIP network (note that in H.323 the media control H.245 session is released first, then the call control link). The call agent also reinitializes the gateway for the next call, by looking for the off-hook event, and then enabling an embedded NotificationRequest which applies a dial tone and waits

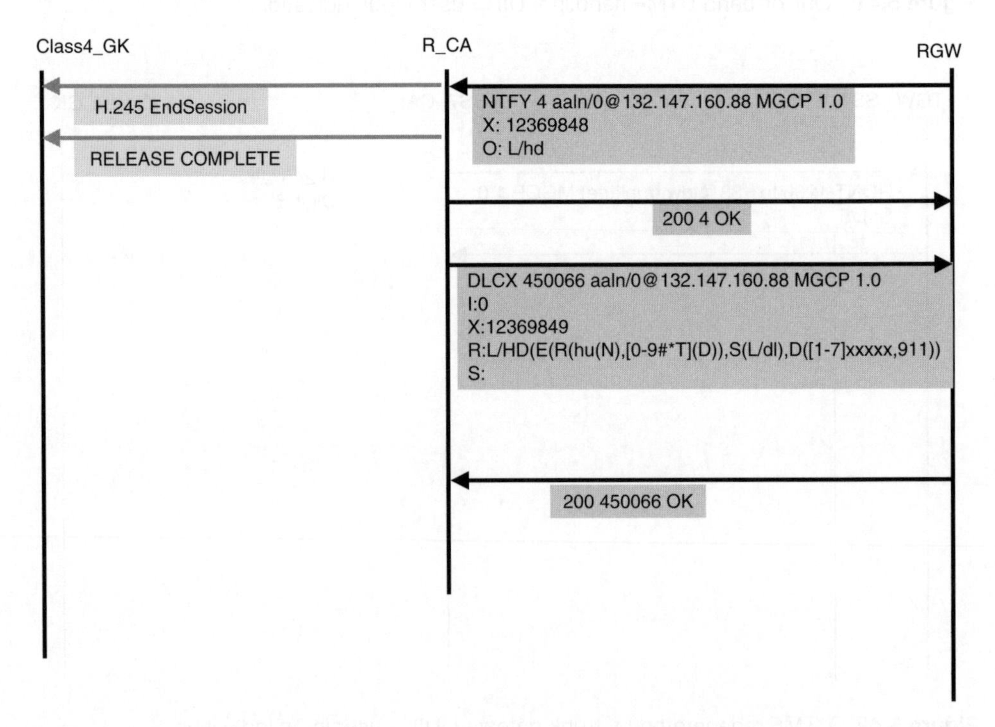


**Figure 5.43** An, end-user on the MGCP residential gateway hangs up the call.

for digits (Figure 5.43). The syntax of the requested events line means: 'accumulate all digits, *, #, and timer event according to the digit map.' The digit map is configured for a phone restricted to local service in San Jose, CA: it can only dial 6-digit numbers or an emergency number. Any other event not matching the digit map will trigger an immediate NOTIFY (e.g., '8' or '0'). A timeout of 16 s while still accumulating digits in a digit map will also trigger a NOTIFY (e.g., '123<16seconds>').

The ISUP disconnection sequence is more complex than the H.323 or SIP disconnection sequence, and typically requires at least two messages: REL (RELEASE) indicates that the called party hung up, while RLC (RELEASE COMPLETE) indicates that the calling party also hung up (Figure 5.44). This more complex sequence is used because some networks use in-band announcements at the end of certain calls (e.g., calling cards). The RLC message instructs the device sending the in-band information to stop sending it; this message also causes the release of all circuits.

## 5.4 THE FUTURE OF MGCP

The PacketCable NCS specification has been standardized as IPCableCom in ITU-T SG9. This specification is also an ANSI standard (via the Society of Cable Telecommunications Engineers). The IETF MEGACO or ITU are not actively working on MGCP since the IETF MEGACO Working Group decided to work jointly with the ITU on H.248.

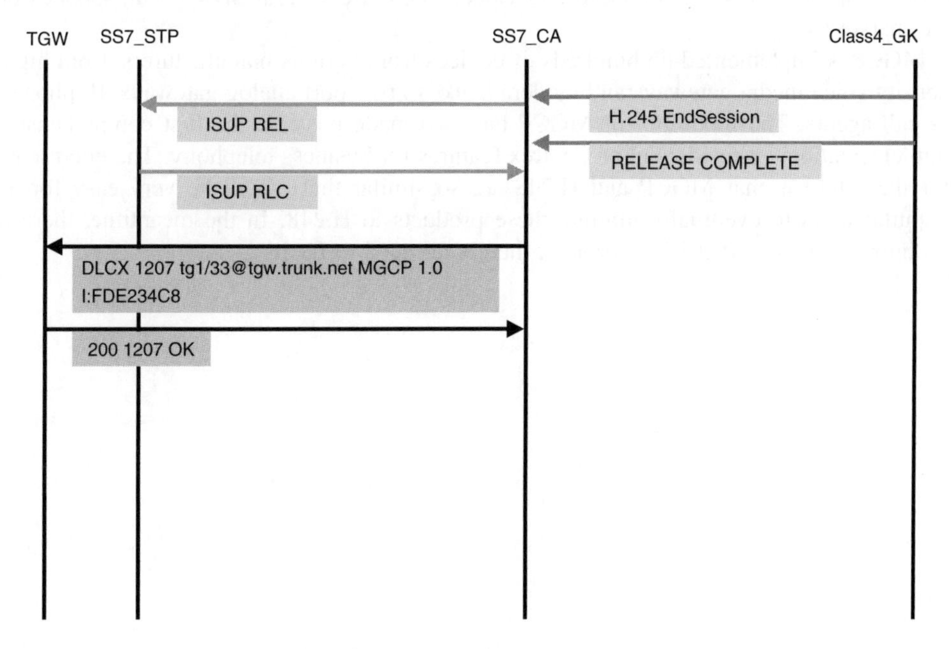


**Figure 5.44** RELEASE COMPLETE (RLC) message converted to an ISUP RELEASE (REL) message.

Despite this, MGCP is still alive and well. The reason is easy to understand: the authors of MGCP defined the scope of the protocol extremely well and, within this scope, managed to fulfill all the requirements. Manufacturers wanting to provide stimulus-controlled phones, or media gateways, cannot ask for more than MGCP provides.

In the midst of the heated debates about the best VoIP protocols with the accompanying tendency to greatly oversell or misrepresent the capabilities of each protocol, MGCP managed to stay aloof from this mixture of marketing and engineering characteristics that typified the telecom bubble era and jeopardized the quality of many specifications. Since the beginning, the development of MGCP has been driven by immediate customer requirements, while other standards were more driven by manufacturers. As a consequence, the quality of the MGCP specification is much better than that of the SIP specification or the early H.323 specifications. MGCP has been adopted by the cable industry and many manufacturers, and is currently deployed in many VoIP networks all over the world.

Although the quality of MGCP is a solid foundation for H.248, at the same time there is no great incentive for the industry to migrate to H.248, because there are very few features that the latter can provide beyond the MGCP capabilities available today. Even though RFC 3435 has introduced new extension capabilities, the few missing features can readily be added to MGCP as well. Initially, video was presented as the key feature added by H.248, with MGCP perceived as an audio-only protocol. This dates back to an early version of the RFC which stated that the protocol was not intended for video. In fact, since MGCP uses SDP, it can establish connections using any media that SDP can describe, including video. The latest versions of the MGCP RFC now acknowledge that MGCP can be used for both audio and video. There are several MGCP videophones on the market today.

MGCP is implemented on hundreds of devices from various manufacturers, from high-density voice media gateways and modem banks to two-port analog gateways, IP phones, or call agents. The flexibility of MGCP has also made possible the first comprehensive implementations of residential or Centrex features for business telephony. The good news for the future is that MGCP and H.248 are so similar that it will be very easy for all manufacturers to eventually migrate these products to H.248. In the meantime, there is no hurry to do so—if it is not broken, there's no need to fix it!

# 6

# Advanced Topics: Call Redirection

## 6.1 CALL REDIRECTION IN VoIP NETWORKS

### 6.1.1 Call transfer, call forward, call deflection

Call transfer, as opposed to call forward, is characterized by the timing of call redirection. Call transfer redirects the call after an initial connection with the called party. Call forward redirects the call before the call is connected to the initial called party. Two flavors of call transfers exist: in consultation call transfer the redirecting party talks to the redirected-to party before transferring the call; whereas in blind call transfer the transferring party transfers the call directly without verifying whether the redirected-to party is willing to/can accept the call.

Call deflection redirects the call after the call has been presented to the called party, but before the call connects: the initial called party never enters in a conversation with the caller.

Figure 6.1 illustrates the various types of call redirection.

The call forward service is relatively simple to provide over traditional or voice over IP (VoIP) networks. Call defection and call transfer services are much more complex to provide in any network technology, but they are even more difficult to provide in VoIP networks. They raise the following issues:

- Translation of the redirected to phone number from the redirecting party format into the format appropriate for the element performing call redirection.
- Dynamic redirection of the media streams from the initial called party to the redirected-to party.

*IP Telephony: Deploying VoIP Protocols and IMS Infrastructure, Second Edition* O. Hersent

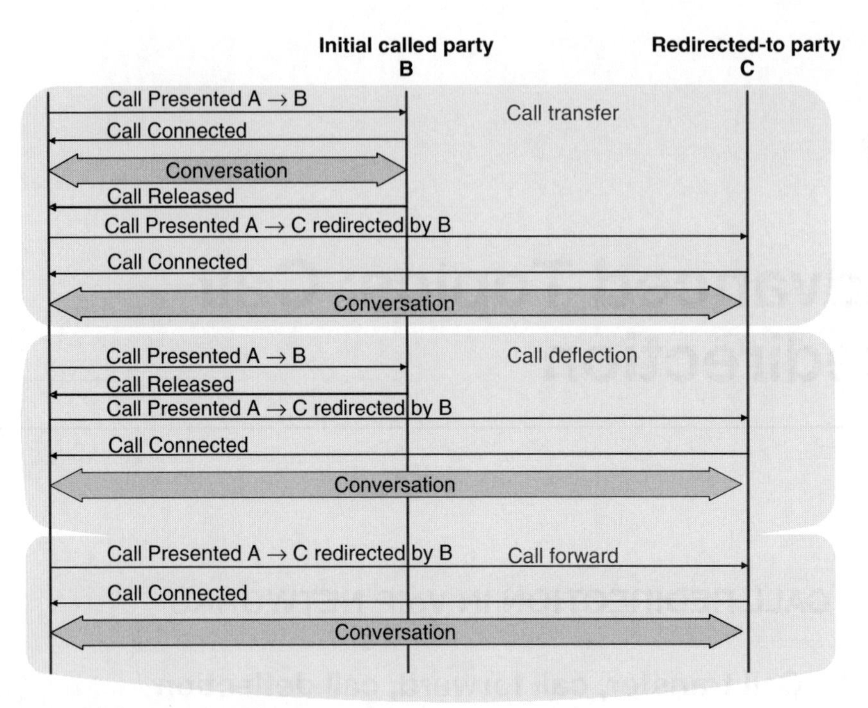


**Figure 6.1** Call redirection types.

- Correlation of the initial call and the redirected call in order to generate appropriate call detail records.

This section discusses in detail the various implementation choices for call redirection services, the difficulties that appear in the context of public networks, and how these challenges can be addressed.

## 6.1.2 Summary of major issues

### *6.1.2.1 Numbering formats*

Any phone call usually involves three distinct formats for calling and called party aliases:

- The **originating format**: this is the format of phone numbers that are familiar to the caller. For instance, the number of a phone in San Jose, CA will be known as 5217000 by a resident of San Jose, while the same phone will be known as 14085217000 by someone living in New York.
- The **pivot format**: this is the format used by the phone network billing system, and manipulated by the phone network administrators (e.g., when defining routes). A US

network may decide to use the full number including the area code, for instance, and may also include the country code.

- The **terminating format**: this is the format of phone numbers that are familiar to the called party.

Let's take an example to clarify: John, in San Jose, CA (+1 4085217000) calls Mark in Paris (+33158713333):

- The originating formats are: John 5217000, Mark 01133158713333.
- The pivot formats are (assuming the international format is used within the network): John 14085217000, Mark 33158713333.
- The terminating formats are: John 0014085217000, Mark 0158713333.

The issue as far as call redirection is concerned is the following: if John instructs the network to reroute the call to Mark, the network should properly understand the redirected-to phone number.

But, doesn't this seem trivial?

Look at the following example, using H.323, H.450, or SIP REFER. With these methods, a message (let's generically call it REDIRECT) is sent to the calling party, instructing him to make a new call to the supplied number.

Let's say Mark wants to redirect John to another phone number in Paris (+33158713300). For Mark, living in Paris, this phone number originating format is 0158713300, so Mark sends back to John's terminal a message REDIRECT 0158713300. John's terminal assumes this is an originating format for San Jose, CA and makes a new call to 0158713300, when it should have been dialing 01133158713300. The call transfer fails ... or, worse, succeeds, but to the wrong destination.

#### 6.1.2.2 Billing

Let's take the example of a call from A to B, with a duration of T1 seconds, redirected after the T1 seconds to C, for a duration of T2 seconds. In the case of call forward or call deflection obviously T1 = 0. How should we bill for this call?

One possibility is to bill the call as a call from A to B for T1 seconds, and a call from A to C for T2 seconds. This is the simplest choice for voice over IP where redirected calls frequently appear as two calls: A to B and A to C. Unfortunately, things are not so simple.

The reason is that such a way of billing gives B the opportunity of making a lot of money at other people's expense: he just needs to create a company that manages premium numbers C, and redirect all calls to B to this premium number C. Premium numbers are charged at a special rate, much more expensive than a traditional phone call, and the carrier shares revenues with the owner of the premium phone number. Now, anyone calling B, a 'normal' phone number, is redirected to C and actually pays for a much more expensive A to C communication! If you are not convinced of the magnitude of the potential fraud, just consider that many countries in the world get over half their state revenue from revenue sharing on international phone calls, and consider replacing C with an international phone number...

In the previous scenario, all phone companies charge the redirected call as follows:

- An A to B call lasting T1 + T2 seconds.
- A B to C call lasting T2 seconds.

In other words, B pays for the redirection, and A pays as if B had never redirected the call. In the case of multiple redirections, the same procedure applies. If the call is redirected to D for T3 seconds, it is charged as three separate calls:

- An A to B call lasting T1 + T2 + T3 seconds.
- A B to C call lasting T2 + T3 seconds.
- A C to D call lasting T3 seconds.

This requires the telephone switches to always be able to correlate all legs of a redirected call to compute the right sums. This is not always simple, especially with voice over IP protocols.

### *6.1.2.3 Call loops*

Call loops are a problem in any telephone network, because they can completely jam some trunks in a matter of seconds. Usually, the prevention of call loops uses machine-generated routes with automatic loop detection and a hop counter that is decremented by each telephone switch. SS7 ISUP messages have such a hop counter. In VoIP, some vendors (e.g., Cisco Systems), have added a proprietary hop counter to the H.323 LRQ message, making it possible to avoid loops between direct-mode gatekeepers. H.323v5 adds such a counter in SETUP messages. Counters are also included in SIP INVITE messages. Finally, a switch that may redirect calls must not authorize a redirection to a number, if it knows that the call has already been redirected by that number: in H.323, the number of the last redirecting party is stored in the Redirecting Number information field of the SETUP message.

Even with these improvements, call loops remain possible:

- Calls can be looped back into the VoIP network by the TDM network through SS7 gateways, in which case the counter can be lost (depending on SS7 gateway implementation).
- Edge devices connected through user interfaces (analog, ISDN) may loop calls back to the network, in which case the hop counter is always reset. This is one of the reasons the call forward service of external calls to external extensions is usually forbidden as part of the certification program of edge devices (call forward of internal calls to external extensions does not create a call loop problem).

Manual call transfer/divert does not present the same issues because it requires human intervention for each loop.

### 6.1.3 Reference network configurations in the PSTN

#### *6.1.3.1 Residential services: call forward only*

##### *6.1.3.1.1 Call forward*

PSTN residential services frequently provide the call forward service. The service is implemented in the last-hop class 5 switch. Most PSTN networks cannot optimize call forwards and, therefore, the last-hop switch reinitiates a new call for the redirected call and trombones the media stream.

With this configuration the issue of numbering formats is irrelevant since the redirected-to number is interpreted by the last-hop switch, using the same numbering conventions as the redirecting party.

Note that it is important to have the call forward service provided by the switch, not by the edge device, because of the call loop issue described above. If the call forward service is performed by the switch, the call hop information is not affected. It would be reset if the call forward was performed by the edge device, leading to potential call loops.[1]

##### *6.1.3.1.2 Call transfer/divert*

Residential services typically never provide the call transfer or call divert feature without strong restrictions. This is because residential users associate billing with 'my phone is off-hook'. But, as we have described above, in order to prevent fraud, when a call is transferred from a phone this phone is still charged for the transferred portion of the call, even if it is on-hook. And now the call is under control of the calling party, and can potentially last for hours or days (note that call forward is different, because it is assumed that you forward your line to yourself while away from your home, and therefore you remain in control of call duration). It is possible to imagine restricted versions of the call transfer service for residential users (e.g., to the mobile phone, fax number, or voicemail number of the served user only).

Call transfer/divert is the key feature that differentiates Centrex from residential services. Centrex users are not charged for internal calls and, therefore, are allowed to transfer calls to other internal extensions. This remains compatible with the perception, 'I don't pay if my phone is on-hook,' since the redirected portion of the call charged to the on-hook phone is a free call. Many countries required PBXs to block the transfer of calls to external numbers to get their type approval, but this is no longer strictly enforced.

Note that you can emulate a call transfer by creating a three-way A–B–C conference if the communication remains active when conference initiator B hangs up. For this reason

[1] Of course, the call forward service would also stop working when the phone is unplugged! Despite these problems, some badly engineered VoIP networks still use end point-based redirection today.

residential networks always require the conference to be dropped if the initiator of the conference drops the call.

### *6.1.3.2 Isolated PBX: forward/transfer by the PBX*

#### *6.1.3.2.1 Call forward*

Because of the loop problem described above, many countries restrict the ability for PBXs to forward external calls back toward the PSTN network. Call forward to internal extensions is usually performed by the PBX itself.

#### *6.1.3.2.2 Call transfer*

When a call transfer is performed by an isolated enterprise PBX, the call transfer is always performed locally by the PBX (Transfer by join).

The billing principles described above are still valid. For instance, if a call from random user A on the phone network is received by B in the enterprise (the communication lasts T1 seconds), and then redirected to random user C on the external phone network (this redirected communication lasts an additional T2 seconds), the phone network will see a call from A to B lasting T1 + T2 seconds (since the PBX does not notify the network in any way that a call redirection has occurred), and the network sees a call from B to C initiated by the PBX lasting T2 seconds. Therefore, the billing records generated by the public network are correct:

- Call A to B for T1 + T2 seconds.
- Call B to C for T2 seconds.

Note that if the PBX hides the real extension B, the network will use the PBX main number as the calling party number for the billing record of the redirected portion of the call. This does not create any potential for fraud, as the bill still goes to the same pocket.

### *6.1.3.3 Networked PBXs: network-optimized call transfer*

Many heterogeneous PBX networks use that call transfer by the PBX method that has been described in Section 6.1.3.2.2 (Figures 6.2). However, this method is not optimal as call media are 'tromboned' through the redirecting PBX, and this uses more bandwidth on the corporate intersite communications lines. If all PBXs are from the same brand, or if they support the QSIG extensions of ISDN (about one-third of PBXs are QSIG-capable), there is a possibility to optimize the call flow. The redirecting PBX will send a redirection message to the source PBX, and the source PBX is expected to re-establish the call directly to the redirected-to party (Figure 6.3). Now, the usage of transmission resources is optimized, but we must solve:

- The numbering format issue: all PBXs must operate under the same numbering format or be able to convert the redirection messages to the appropriate formats.

- The billing issue: if all internal calls are on a private network (leased lines), the service provider has no billing to do if redirected calls are restricted to internal extensions (call detail records or CDRs, for internal calls are generated directly by each PBX). If calls are redirected to the public network, the PBX will stop using QSIG and trombone the redirected call to the public network, returning us to the case of the previous paragraph (6.1.3.2, Transfer by join). If leased lines are not used, the public network may be unable to correlate the initial call and the redirected call, and the CDRs may be generated as A–B (T1 seconds), B–C (T2 seconds), instead of A–B (T1+T2 seconds), B–C (T2 seconds). This is not always a problem if the company is billed as a single entity and does not care whether A or B is billed for the redirected call. Note also that, if private accounting systems are used at each location, their records will be inaccurate, as the B to C leg of the call should be charged to site B. But, this leg of the call is now re-originated from site A and, therefore, site B has no information on it.

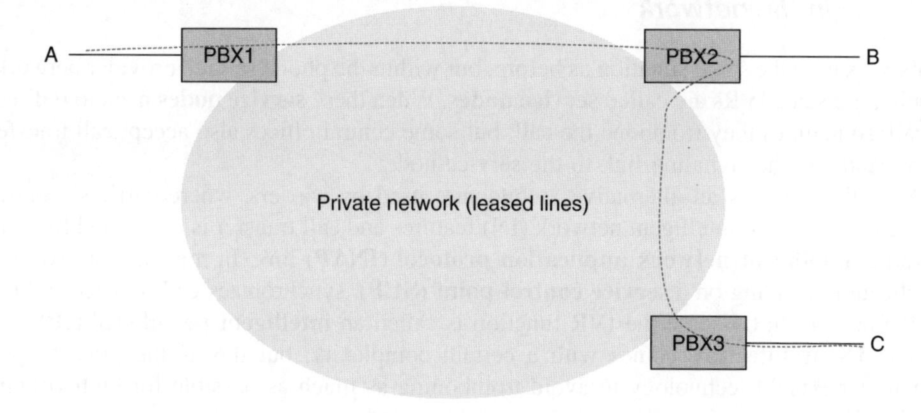


**Figure 6.2** Transfer performed by the transferring PBX (Transfer by join).

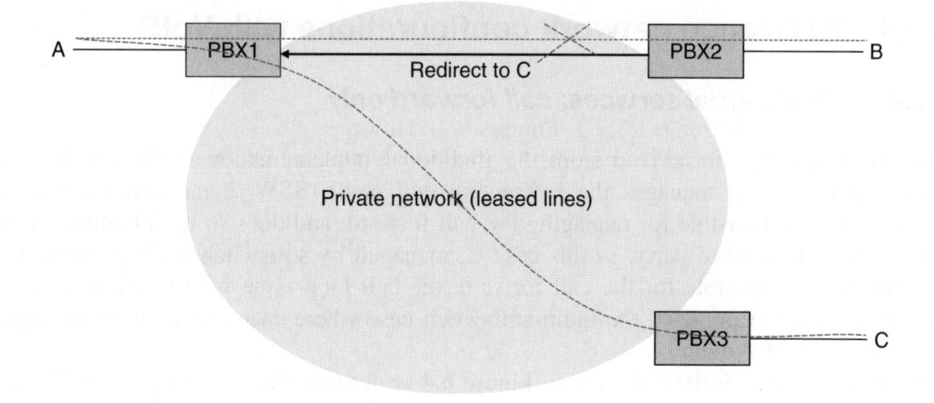


**Figure 6.3** Optimized transfer.

#### 6.1.3.4 *Voice resources connected to a QSIG-capable PBX: locally optimized call transfer at the edge*

In some cases (e.g., in call centers) a private installation routes calls first to a voice resource, such as an interactive voice response (IVR) server, and then the call must be redirected to another extension by the IVR. It would be a waste of resources to force the IVR to make a call to the extension and then bridge the incoming call with the new call to the extension. This would use two ports on the PBX.

Instead, most corporate IVRs are capable of sending the proper QSIG commands to the PBX, in order to ask the PBX to perform call transfer to the new extension. On analog connections, the IVR can alternatively use a DTMF command to perform call transfer. The call transfer is always performed locally by the PBX.

#### 6.1.3.5 *Network intelligent peripherals: locally optimized call transfer in the network*

This is exactly the same situation as before, but within the phone service provider network. Such large-scale IVRs are called **service nodes**. When these **service nodes** need to redirect a call, sometimes they trombone the call, but some central offices also accept call transfer commands on the signaling link to the service node.

Note that there is an alternative architecture used by carriers, where carriers' central offices make use of intelligent network (IN) features and call transfer is controlled from an external **intelligent network application protocol** (**INAP**) link. In this case an external application, residing on a **service control point** (**SCP**), synchronizes call transfer and the IVR function. In this case the IVR function is called an **intelligent peripheral** (**IP**).

The **IN architecture** comes with a certain complexity, but this is the price to pay when using TDM technology to avoid tromboning as much as possible for high-volume applications, such as hosted contact centers.

### 6.1.4 Reference network configurations with VoIP

#### 6.1.4.1 *Residential services: call forward only*

The implementation is derived from the traditional implementation in the PSTN. The edge softswitch that manages the redirecting end point (SSW_2 and end point B in Figure 6.4) is responsible for managing the call forward, and does so by initiating a new call to the redirected-to party, in this case C, managed by softswitch SSW_3. Since the softswitch is responsible for the call forward, the call loop issue is properly addressed. Figure 6.4 is an example of the multi-softswitch case where each end point is managed by a different softswitch.

Note that, although the call flow in Figure 6.4 is derived directly from the PSTN call flow, it is *much* more efficient: media flows are established directly between calling party A and redirected-to party C. Softswitch SSW_2 is only involved in the signaling path,

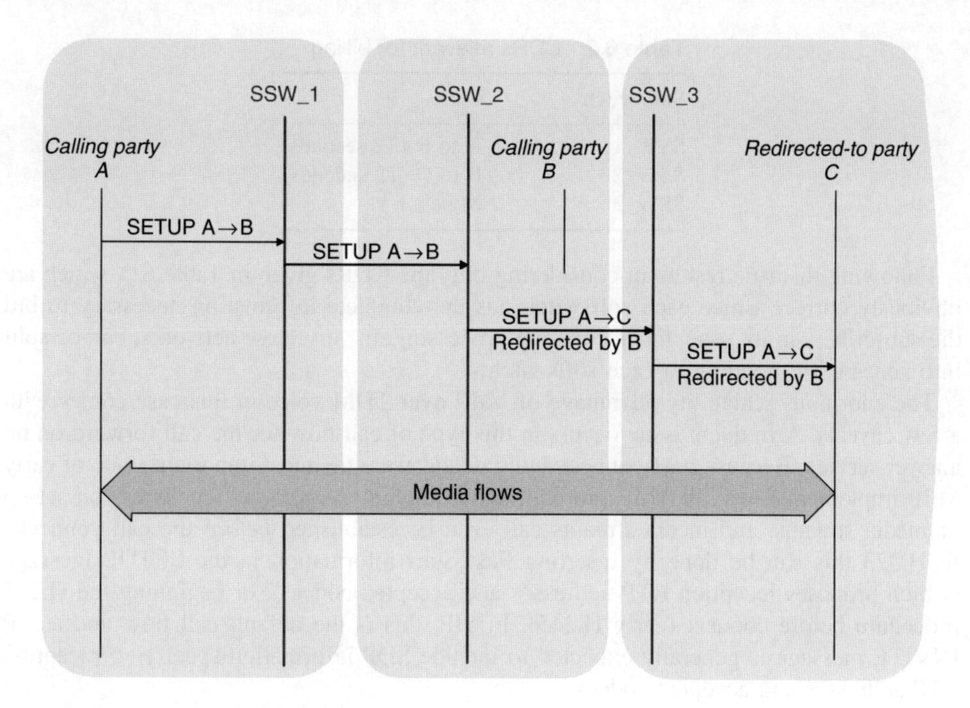


**Figure 6.4** VoIP call forward scenario.

it does not handle the media stream. If SSW_2 had been a PSTN switch, the redirected call would permanently use 128 kbit/s of transmission capacity to the switch point of presence, as opposed to exactly zero in the VoIP case.

At this point it is also interesting to examine how billing should be organized. Each softswitch will probably generate its own billing records (Table 6.1).

We see that we have duplicates, and also CDRs that, depending on the softswitch support of the redirecting number information (a field in H.323 indicating the identity of the redirecting party), may be correct or not. There is a simple way to extract the correct information for these CDRs, following this simple rule:

*For billing purposes, on softswitch SSW_i, keep only CDRs where the calling party belongs to the SSW_i zone.*

**Table 6.1** CDRs generated by a call forward in a multi-softswitch environment

| Softswitch | CDRs |
|---|---|
| SSW_1 | A to B (T2 seconds) |
| SSW_2 | A to B (T2 seconds) |
| | B to C (T2 seconds) |
| SSW_3 | A to C, redirected by B (T2 seconds) |

**Table 6.2** CDRs relevant for billing

| Softswitch | CDRs |
|---|---|
| SSW_1 | A to B (T2 seconds) |
| SSW_2 | B to C (T2 seconds) |
| SSW_3 | None |

Following this rule results in considering only the CDRs given in Table 6.2, which are obviously correct. Since each softswitch has the complete information necessary to bill the subscribers in its zone, the accounting processing, in very large networks, can be split into zones corresponding to each softswitch.

The enormous scalability advantage of VoIP over TDM voice in this case comes with a few caveats. A frequent issue occurs in this type of call flow for the call forward on no answer service. Because post-connect audio delays were the most important issue of early VoIP implementations, all VoIP protocols now implement ways to accelerate the set-up of media streams, and media streams can even be established *before* the call connects. In H.323 this can be done by inserting Fast Start information in the SETUP message (which proposes reception RTP addresses and accepted codecs), or beginning the H.245 procedure before connect (early H.245). In SIP, this is the default call flow and a SIP INVITE message is generally expected to include SDP information specifying reception RTP addresses and accepted codecs.

If this type of optimized call flow is used in the case of call forward on no answer, then A and B will establish media streams before B picks up the phone (B can be an IP phone or a PSTN number behind a gateway). But after the no answer timeout, the call will be redirected to C, and C will also start streaming audio toward A. This produces a variety of results, ranging from A crashing to A playing audio toward B instead of C. If the softswitch does not control the beginning of media streams, call forward on no answer will not work properly. Note that IP phones are usually smart enough not to start streaming audio until someone picks up the handset; so, the issue is really only with end points behind analog gateways and for audio streams from A to B instead of C.

What is the correct approach? The softswitch should delay all pre-connect media information from phones that have the switch-based call forward on no answer service activated until the phone actually connects. In H.323 this means the softswitch will capture the Fast Start information inserted by B in the CALL PROCEEDING or ALERTING message, and forward it to A only if B sends back a CONNECT message. Otherwise, if B does not answer, the softswitch will redirect the call to end point C and do the same with end point C. In SIP this means delaying SDP information until the 200 OK message. Some phones do this properly anyway, but it is a lot safer if the softswitch plans ahead and is prepared to delay any pre-connect information it receives; in particular, if B is behind a PSTN gateway. There is still the possibility of B starting to stream audio toward A before CONNECT (bad IP phone implementation, or a gateway which does not know that it is connecting a call to an end-user and has no support for the in-band audio indicators of the PROGRESS message in ISUP/ISDN). In H.323 the softswitch can explicitly request the gateways not to start streaming audio before CONNECT by positioning the MediaWaitUntilConnect parameter of the H.323 SETUP message.

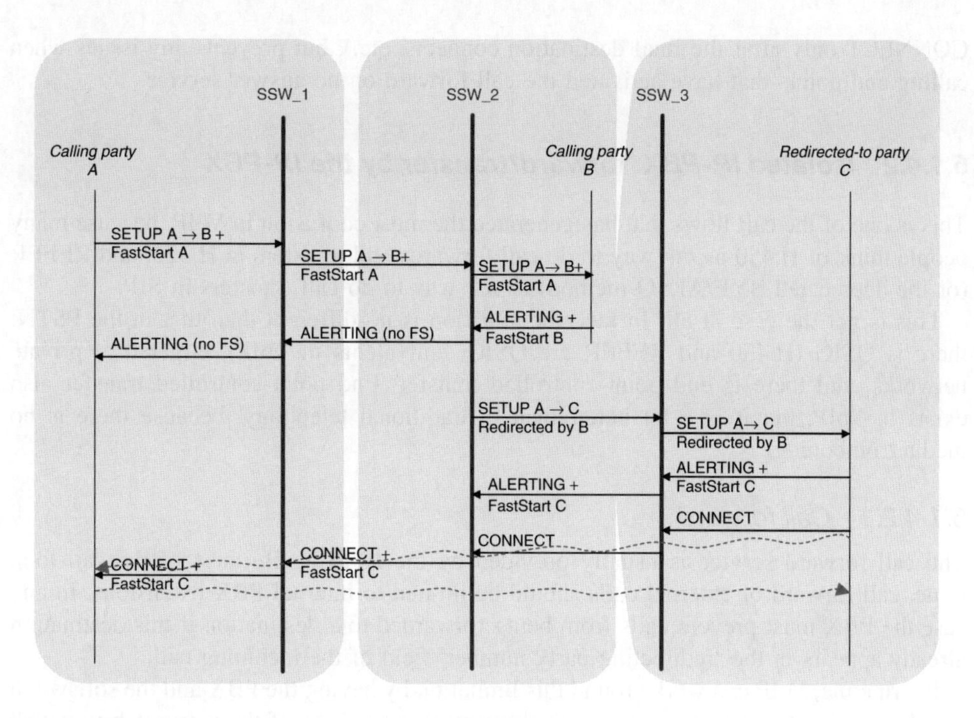


**Figure 6.5** VoIP call forward on no answer scenario.

This procedure now works correctly with the call forward on no answer service, and does not create any delay in post-connect audio (Figure 6.5):

- End point B has received the RTP transport addresses that can be used to stream audio toward A in the SETUP/INVITE message. So, as soon as B picks up the handset, 'Hello' can be transmitted toward A.
- End point A will stop playing ring-back tones as soon as it receives the CONNECT/200 OK message from B. In the same message it receives the ports to start talking with B, and therefore can answer B immediately.

The softswitch should therefore use the following rules to handle pre-connect media information:

- If it does not implement the call forward on no answer service for the destination, forward the media information 'as is'.
- If it implements the call forward on no answer service for the destination, delay the media information from the destination end point until CONNECT. In H.323 also position the MediaWaitUntilConnect parameter in the SETUP message.

These rules correctly enable the forwarding of pre-connect audio when calling PSTN destinations (congestion messages, some prepaid calling card services that send back a

CONNECT only after the final destination connects, etc.), but prevents any issues when calling end points that have activated the call forward on no answer service.

### 6.1.4.2 Isolated IP-PBX: forward/transfer by the IP-PBX

This is one of the call flows that has generated the most confusion in VoIP, because many people think of H.450 as *the* way to do call forwards and transfers in H.323, and REFER (or the deprecated BYE/ALSO method) as *the* way to do call transfers in SIP.

This is not the case at all. In fact, the situation is no different than that in the PSTN: there is QSIG (H.450 and REFER are QSIG equivalents in VoIP), targeted at private networks, and there is end point-controlled transfer. End point-controlled transfer also exists in VoIP, but it is a lot better than in traditional telephony, because there is no media tromboning.

#### 6.1.4.2.1 Call forward

The call forward service is usually provided by the IP-PBX. Because of the call loop issue, call forward of external calls should be limited to internal PBX extensions. In any case the PBX must prevent calls from being forwarded to a destination if this destination already appears in the 'redirecting party number' field of the incoming call.

In principle, VoIP can work around this limitation by having the PBX and the softswitch collaborate for the call forward service to external extensions. If the softswitch to which the PBX is connected supports the call forward service (e.g., if it supports the Call Processing Language), in principle the forward service of external calls can also be provided by the edge softswitch, as in the residential case, without creating a call loop issue. This softswitch-based call forward will not work for calls coming from internal extensions; so, the PBX should also forward internal calls to the external extensions, which does not create call loop issues.

#### 6.1.4.2.2 End point-controlled transfer with media optimization in H.323: NullCapabilitySet

In H.323 the call flow of Figure 6.6 can be used by any end point to control a call transfer and optimize the media path. In our example, A is the calling party, SSW_1 is the softswitch controlling end point A, B is the redirecting party (an IP-PBX in our case), and C is the redirected-to party. B is managed by SSW_2 and C is managed by SSW_3.

The call flow employs the third-party media control procedure, using a call flow known as **third-party-initiated pause and rerouting** or **NullTerminalCapabilitySet** (in short, **TCS = 0**), described in paragraph 8.4.6 of H.323v4. *All* H.323v2 (and above) end points are required to support receiving **TCS = 0** and correctly redirecting media streams as they are given a new CapabilitySet from the remote party. This is one of the most important tests to do on implementations that claim to be H.323-compliant. Not supporting this call flow is a major bug which does not allow the equipment to be connected on any H.323 network that implements transfer services (even if it is just other end points that need to do call transfers).

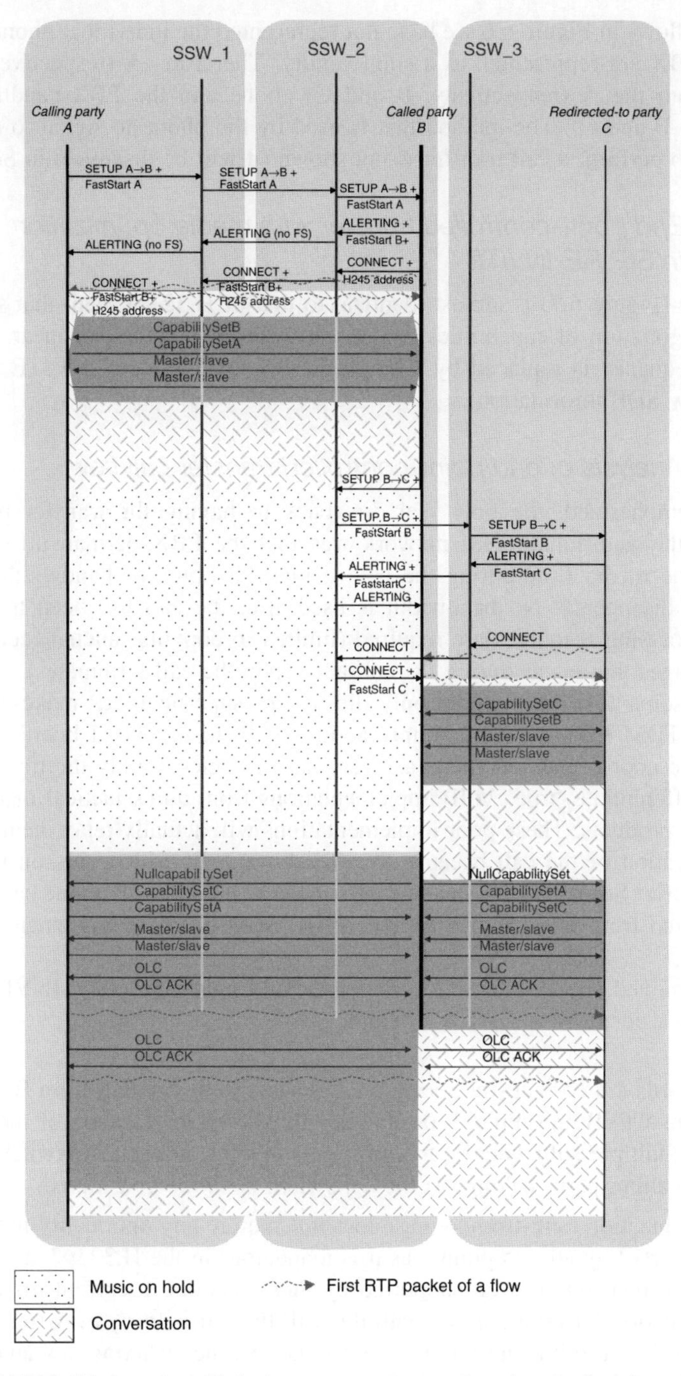


**Figure 6.6** H.323 third-party-initiated pause and rerouting (controlled transfer with consultation). Most ACK PDUs are not shown and all messages are in sequence for clarity.

In the call flows in Figure 6.6 we have not represented the individual phones; the phone and the IP-PBX are represented as a single entity. Therefore, A (respectively, B and C) represents both the A (respectively, B and C) phone and the PBX-handling phone A (respectively, B and C). The method that is used by the phone to signal to the PBX that it is willing to perform a call transfer is not shown (it will be discussed in Section 6.1.5).

###### *6.1.4.2.3 End point-controlled transfer with media optimization in SIP RE-INVITE*

The call flow (Figure 6.7) is almost identical to that of H.323, except that since SIP has no formal negotiation of capabilities, H.245 capability messages disappear. The NullCapabilitySet sequence is replaced by a single RE-INVITE (a new INVITE for the same call, with new SDP information).

###### *6.1.4.2.4 Analysis of end point-controlled transfer call flow*

As already emphasized, the only real drawback to having the transfer performed by the PBX in the traditional TDM network was that the PBX permanently had to relay (trombone) the media flow during the call, using 128 kbit/s of bandwidth. With VoIP, this problem disappears! No bandwidth is used after the call has been transferred and very few processing resources are required (limited to copying, unchanged, the few call control messages that occur during the call and when the call terminates).

Note that some IP-PBX vendors only did a very superficial 'IP make-up' on top of a traditional TDM PBX core, by simply adding VoIP to the TDM board on their PBX chassis. These poor implementations are very hard to detect—only the transfer call flow will really differentiate 'true' VoIP implementations from quick tactical marketing adaptations of old products. These poor implementations will actually relay media streams for the entire duration of the call because the switching itself still occurs on the old TDM switching matrix! Some IP-PBXs are optimized for IP phone to IP phone calls, but not for calls transferred from one PBX to another PBX. Such obsolete implementations should be avoided.

Having removed the tromboning issue, the NullCapabilitySet/RE-INVITE call flow becomes almost ideal for a service provider:

- Billing records are correct (the second call appears to be coming from B to C, the first call is from A to B and remains established for the entire duration of the call). In the context of multiple softswitches, the same rule applies: at each softswitch, only CDRs where the calling party belongs to the softswitch zone are considered.
- Media optimization (anti-tromboning) does *not* require any optionality in the standards and is supported by all end points (as it is mandatory in the H.323v2 and SIP baseline standard). In fact, this is well supported by most vendors in current implementations. Even if a vendor does not comply with this call flow, it is fairly easy to get the vendor to support it because it is nothing less than a bug that needs fixing, not an enhancement to the standard.

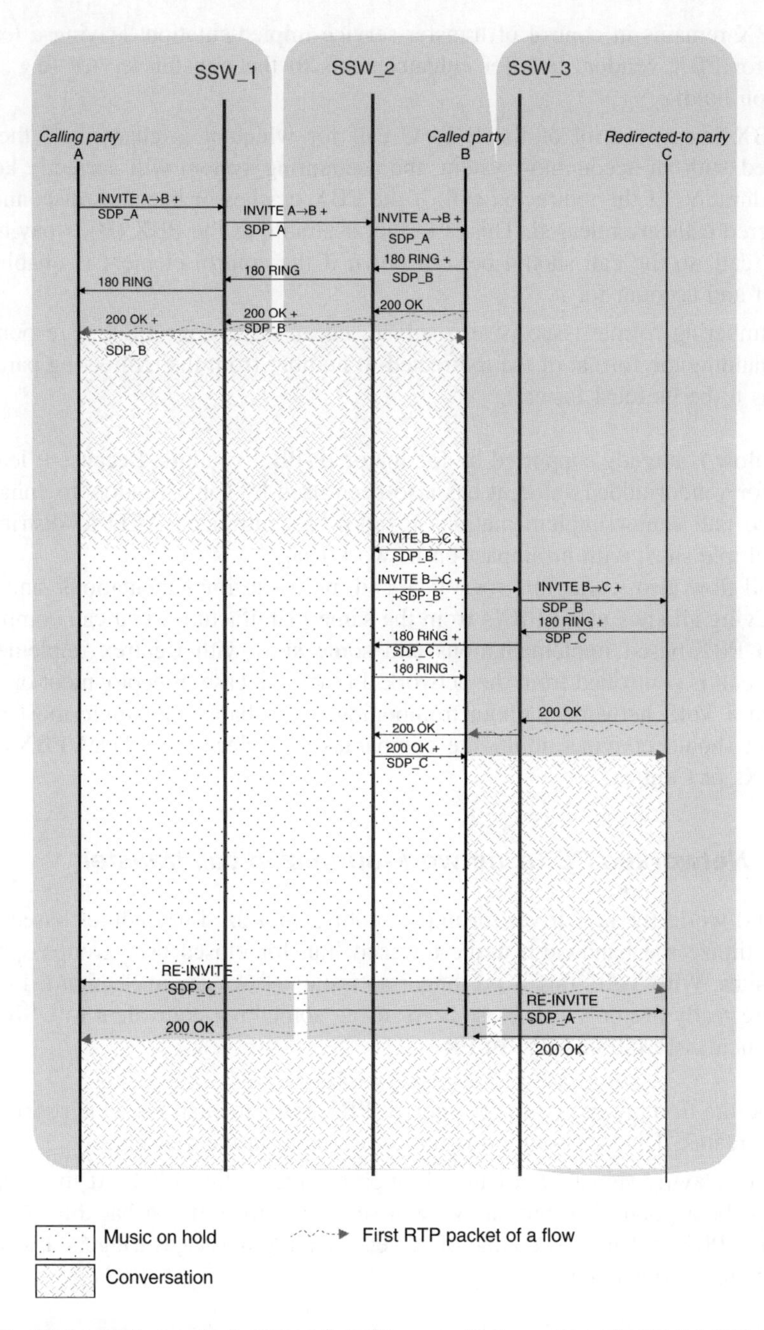


**Figure 6.7** Third-Party-controlled transfer with consultation using SIP RE-INVITE. *Most ACK PDUs are not shown and all messages are in sequence for clarity*.

- The PBX remains in control of transfer service implementation, leaving a lot of flexibility for PBX vendors to offer enhancements to the transfer service (e.g., personal music on hold).
- The PBX keeps control of the B to C call for which it is charged. If the PBX is equipped with an accounting system, the accounting system will correctly keep track of the duration of the redirected call. If the PBX crashes or becomes disconnected, all transferred calls are released. This is in fact desirable as the PBX (B) is paying for the B to C call, so the call should be torn down if the control element is unable to keep track of and account for it.
- The numbering format issue is also solved, because the edge PBX is responsible for understanding the format of the redirected-to number from the redirecting party (which happens to be its local format).

This call flow is already supported by the major IP-PBX vendors. Because it leaves a lot of room for vendor added value, it has also been used by some vendors to enhance their PBX-based call center implementations to allow a centralized ACD to distribute calls across multiple sites, with no impact on network usage!

This call flow also works fine for VPNs and multi-site implementations, and does not require having all sites using PBXs from the same manufacturer: you can combine some sites with PBX-based implementations and some sites with Centrex implementations, where the call is controlled from the network. In fact, the ease of deployment of advanced services in a VoIP network, without tromboning, is probably the single most important reason that should convince multi-site companies to replace their TDM PBXs with IP-based PBXs or Centrex.

#### *6.1.4.3 Networked PBXs: network-optimized call transfer*

In the TDM world the key driver for implementing optimized procedures based on QSIG was to optimize the bandwidth usage on each link for closed user groups spread over multiple sites. With VoIP, this is no longer an issue, as we saw in Section 6.1.4.2.

So, there really are only two reasons for using something equivalent to QSIG transfer in VoIP (such as H.450 or SIP REFER):

- To solve the tromboning issues of poor IP-PBX implementations which still trombone media streams. The best solution is not to use them in the first place!
- To avoid relaying signaling streams through the redirecting PBX. At first glance, this seems to be a good idea, but, as we will see, sometimes it can backfire: this attempt to get the PBX to handle a couple of sewer signaling messages triggers a whole lot of much more serious issues.

That being said, there is still a use for H.450/SIP REFER: this is to allow IP phones using the H.323 or SIP protocol to signal to the PBX that they are willing to perform a call transfer. This will be discussed in Section 6.1.5.

Let us describe the H.450/SIP REFER source-based call transfer call flows anyway.

#### 6.1.4.3.1 Call flow with H.450.2

The H.450.2 implementation of call transfer, because of the QSIG heritage, is well defined and there is a reasonable level of interoperability between vendors. Call flow is explained in detail in Section 6.1.4.3.3 and summarized in Figure 6.8. H.450.2 explicitly notifies transferred-to end point C that a call transfer service is being requested. In response, transferred-to end point C gives transferring end point A a reference for the call to be transferred, which is passed to transferred-to end point B. The call reference is explicitly included by transferred-to end point B in the new call generated to transferred-to end point C.

#### 6.1.4.3.2 Call flow with SIP REFER

The SIP implementation is almost identical to the H.450.2 implementation. The REFER message tells the calling end point to make a new call to C. The exact implementation is still, at the time of writing (Q2 2003), in a state of flux, and really only works after some vendor-to-vendor tuning. Details such as which phone releases the transferred call and when, how the transferred call notifies the transferred-to party that this is a replacement for the previous call, how the transferring phone is notified of the successful transfer, etc. are still not stable. A sample call flow between Cisco SIP phones is given in the Annex.

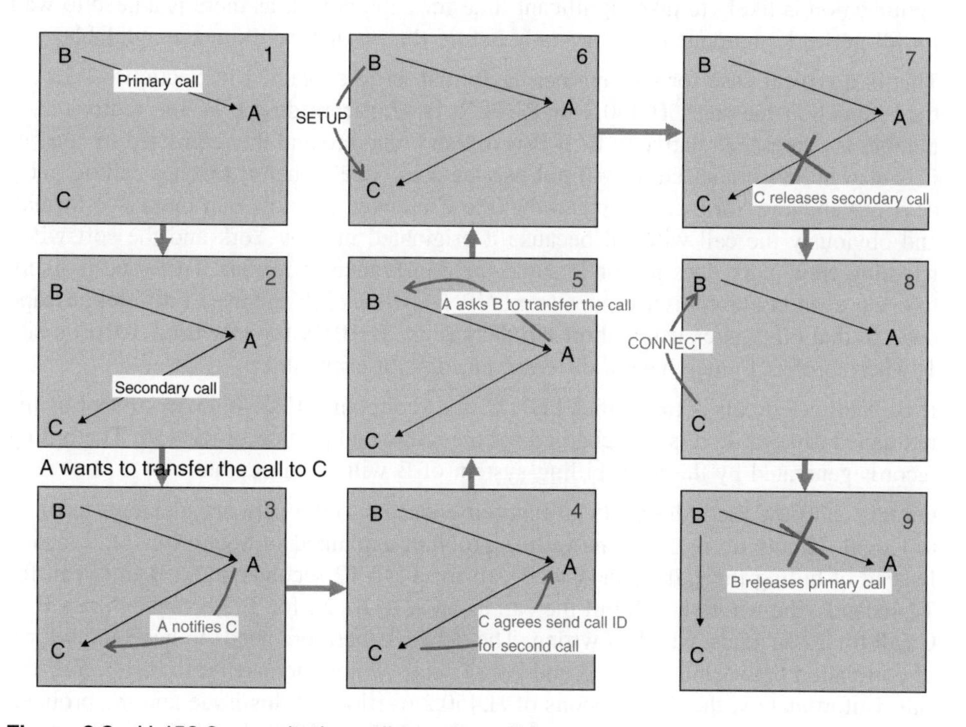


**Figure 6.8** H.450.2 consultation call transfer call flow.

#### *6.1.4.3.3 Analysis of the H.323/H.450/SIP REFER call flow in public networks*

As already discussed, there is really no fundamental advantage to using this procedure in the core network. It does not even save signaling messages as the H.450 activation itself requires a significant set of new messages and adds a lot of complexity to end point implementation (e.g., when implementing the second call presentation service, the end point should be careful not to accept any call if it is expecting a transferred call for which it has already given a call reference).

But there are a number of major issues raised by this call flow regarding public networks:

- It requires, obviously, all end points to support H.450.2/REFER and to have interoperable implementations. This may not be an issue in closed user groups (VPNs or networks of IP-PBXs), but this becomes a critical issue for a service provider: *all* VoIP devices in the network need to support the H.450.2/REFER feature, including SS7 call agents, IVRs, application softswitches, any customer equipment connected to the VoIP network, and even third-party VoIP networks interconnected to the network using H.450.2. This is obviously a challenge, as these features are options in the standards, likely not to be available by default in most products, and requiring an upgrade fee if they can be added. The consequence is that the introduction of the H.450.2/REFER optimization is likely to take significant time for deployment, as there is a need to wait for all network components to be ready before the service can be made available.
- This is a typical case for the numbering format issue explained in Section 6.1.2.1. If the redirection message, H.450.2 or REFER, is sent from device B, the redirected-to number is expressed relative to B. If B is in San Francisco and the redirected-to number C is also in San Francisco, it will not use the area code. When it reaches calling party device A in New York, A will try to dial the C number using its San Francisco format, and obviously the call will fail because it originated in New York and the softswitch handling New York does not understand the San Francisco format. The issue is likely to occur even in closed user groups (e.g., VPNs), when each site uses a different escape code to dial other sites using short numbers (e.g., 1–1010, for extension 1010 on site 1, where prefix '1' may have a different meaning at each site).
- If the redirected call is to another PBX C, the redirecting PBX B loses control of the redirected call, while it is still charged for the redirected portion of the call. The billing records generated by the local billing system of B will be inaccurate.
- On this call flow the redirected call naturally appears to the network elements as an A to C call. If nothing is done, the billing problem explained in Section 6.1.2.2 occurs. Instead of billing the call as an A to B call for T1 + T2 seconds and a B to C call for T2 seconds, the network will bill the call as an A to B call for T1 seconds, then a B to C call for T2 seconds, which is wrong. The network therefore needs to have some form of correlation between the A to B and A to C calls which indicates it to be a redirected call. Unfortunately, the first versions of H.450.2 overlooked this issue and the problem is now present in many implementations of H.450.2. In 2000 the H.323 implementer guides started introducing and documenting the usage of a new feature called 'call

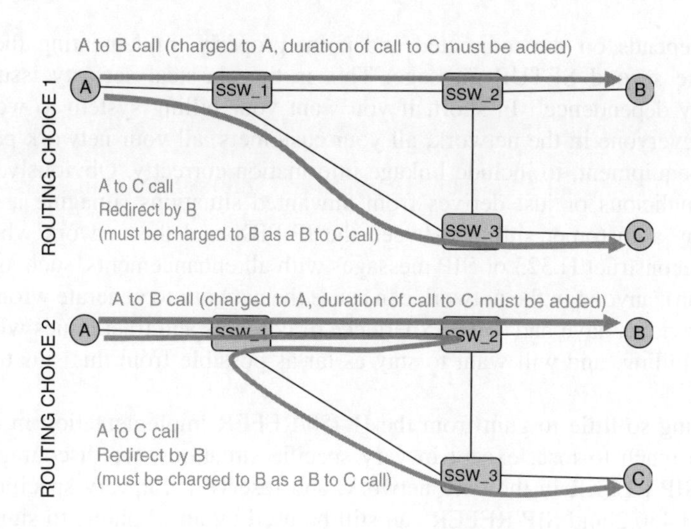


**Figure 6.9** Routing dilemma in a multi-softswitch network with source-based rerouting.

linkage' (using a new identifier called 'ThreadID'), which is used to associate one call with another and must be used in H.323v4. Unfortunately, this adds to the difficulty of introducing this feature in a multi-vendor network, as not only H.450, but the correct version of it, must be implemented everywhere. Most carriers' business plans require integration with as many PBX brands as possible, which makes this issue a critical one.

- If multiple softswitches control the network, the initial A to B call and the subsequent A to C call after redirection will be routed through a different set of softswitches. There are two possible routing choices, the simplest (choice 1 in Figure 6.9) is to route the A to C redirected portion of the call as if it was a normal call. In this case the problem is that the CDRs required to charge B, instead of being generated only by SSW_2 which is managing end point B, will be potentially generated by all softswitches in the network, requiring complex CDR reconciliation across all softswitches. This prevents any possibility to scale the billing system by isolating independent regions in the network corresponding to each softswitch. A further issue is that some services (e.g., legal call interception) are likely to be located in each regional softswitch for the subscribers in its zone; so, they would require complex inter-softswitch synchronization in order to be implemented in such a network. The second choice is to always reroute the redirected portion of the call along the same path as the initial A to B call, but then to send it back to the first softswitch for optimization of the routing. This call flow allows the generation of billing records to be segmented properly and facilitates legal interception. But, it also requires three times as many ports as other implementations, and really gets ugly if a call is transferred multiple times. Another issue is that B may be allowed to call (and therefore transfer calls to) destinations unreachable from A. In order to fix this, SSW_1 would need to be aware of all the restrictions applicable to B!
- A last issue, probably the most serious issue in the context of public networks, relates to the reliability of billing. For proper billing, it is important to be able to correlate the A to B and the redirected A to C calls. Resolution of this issue using end-to-end

H.450.2 depends on the redirected end point properly implementing the correlation field in the second SETUP message. This is the classical security issue known as 'third-party dependence'. In short, if you want your billing system to work properly, you need everyone in the network, all your customers, all your network peers, not just your own equipment, to include linkage information correctly. Obviously, whether the intent is malicious or just derives from unwanted situations (imagine a new firewall installed by a customer sitting between the IP-PBX and the network which does not properly reconstruct H.323 or SIP messages with all enhancements, such as call linkage information), anyone in the network can cause the network to generate wrong CDRs. All service providers have had some experience of conflicts and trust issues with customers related to billing, and will want to stay as far as possible from this situation.

Overall, having so little to gain from the H.450/REFER implementation in a public network and so much to lose, except in very specific situations, we discourage the use of H.450.2 or SIP REFER in the core network, and reserve it for very specific cases only. Obviously, H.450.2 and SIP REFER can still be used by an IP phone to signal the intention to make a call transfer to an IP-PBX; issues only occur with the use of H.450.2 or SIP REFER between customer devices and the network.

If the situation of having edge devices, using H.450/REFER redirection toward the network, cannot be avoided in an open environment, then the network softswitch should intercept the H.450/REFER message and execute the transfer locally using the NullCapabilitySet/RE-INVITE method. This works fine, but is a very intricate task for a core softswitch, actually requiring Centrex-type capabilities on the softswitch (Centrex is primarily defined by the availability of call transfer in the feature set of the switch).

#### 6.1.4.4 Voice resources connected to a H.450/SIP REFER-capable PBX: locally optimized call transfer at the edge

This is the same as the previously described case on TDM PBXs. It works fine and does not create any of the issues discussed in Section 6.1.4.3, because this is only a local call flow.

However, since this type of call flow was primarily intended to avoid media tromboning, and there is no media tromboning in VoIP if it is properly implemented, using this optimized call flow is a lot less useful than in the TDM world, unless the voice resource or the IP-PBX is not truly native VoIP implementation (e.g., those built on TDM cores using VoIP to TDM boards to connect to IP-PBXs). In most cases, the simple call flow, where the IVR bridges the two half-calls and uses TCS = 0 or RE-INVITE to optimize the media path, is just as good and will cause fewer interoperability issues.

#### 6.1.4.5 Network intelligent peripherals: locally optimized call transfer in the network

This is the same situation as above, but transposed into the public network. Again, the call flow works fine and does not create any of the numerous issues it creates when used between a user space device and a network device, but all this added complexity brings little benefit to VoIP networks.

It is expected that the service node model in VoIP networks (with the voice resource doing the call switching internally) will prevail over the intelligent network model (where the intelligent peripheral never switches the call, and only streams voice prompts). This would reverse the current situation on TDM networks, where the intelligent peripheral model dominates.

### 6.1.5 How to signal call transfer?

In the previous sections we have discussed in detail how to execute a call transfer. Of course, before executing a call transfer, some convention must be in place to signal to the switch executing the call transfer that the user is willing to transfer a call.

#### *6.1.5.1 H.323 or SIP phones and residential gateways*

Any device connected to a public network (IP phone, CPE gateway, etc.), if it is capable of transferring calls, must use the TCS = 0/RE-INVITE method and remain in control of the entire call during the transfer (this is the equivalent of multi-line high-end analog phones sold to small professionals). As explained in Section 6.1.3.1.1 this is because the call transfer function is never offered by the residential service provider (to avoid fraud). Therefore, transfer must be performed under the responsibility of the end point; obviously, the user interface to control this is entirely up to the end point. Note that in many countries it is even illegal to implement this service in a phone, and transfer is never offered by a public network to non-Centrex users.

When an H.323 or SIP phone is connected to an IP-PBX, the best solution is to use H.450/REFER messages to signal the transfer to the switch. Note that the switch itself can use the TCS = 0/RE-INVITE call flow toward the public network to EXECUTE the call transfer, or simply forward the H.450/REFER message to another PBX in a closed user group. The way the end point asks for a call transfer and the way the switch executes it are completely independent.

An alternative way of asking for a call transfer would be via DTMF tones (star key type of transfer code). In this case the switch would need to locally accumulate and analyse DTMF sequences. Although this is possible in theory, most SIP and H.323 IP phones have specific ‘transfer’ keys and use the H.450.2/REFER message to ask the switch to perform the transfer, in order to avoid having to make the appropriate transfer DTMF sequences for a given switch available in the phone. Note that choosing DTMF activation also necessitates having a specific escape sequence when transparent DTMF is required (e.g., when entering information into an interactive voice response server).

For H.323 or SIP residential gateways, the situation is a bit different and some vendors choose to let the IP-PBX perform the call transfer by analysing the DTMF digits, while other vendors analyse the DTMF digits locally and send the transfer requests to the IP-PBX using H.450/REFER.

#### 6.1.5.2 MGCP phones and residential gateways

In a public network the transfer service is typical of Centrex offerings. In many ways, MGCP phones and CPE gateways are better suited than SIP or H.323 at offering a Centrex type of service. MGCP makes fewer assumptions on the specific logic implemented in the phone and, therefore, is more manageable in a multi-vendor environment, making it straightforward to implement new services in a homogeneous way. In addition, MGCP offers more possibilities by signaling phone-related events even before the call is active (e.g., the off-hook event), enabling such features as the announcement of the number of pending voicemail messages as soon as the handset is picked up. Finally, MGCP is now a very mature protocol, with dozens of vendors and the endorsement of large organizations, such as PacketCable® (the organization that also drives the DOCSIS standard).

The most natural way of using MGCP (with an event package line) to signal call transfers is to use DTMF activation sequences, using MGCP digit maps. This works well across all vendors of IP phones and residential gateways.

Most IP phone vendors also offer shortcut keys for the most common call actions (three-way, hold, transfer, etc.). These phone vendors allow the call agent to associate an MGCP event with each phone key. This event can be parsed by the call agent and trigger a specific action (e.g., transfer). Because of this, seen from the user's perspective, the phone behavior is identical to an H.323 or SIP phone, except that it offers more services while off-hook and can be managed much more easily by the service provider.

Once the MGCP call agent has been notified that the phone is ready to perform a call transfer (if the transferred party is not controlled by the same call agent) the call agent will use the appropriate call flow on the network side (TCS = 0 or SIP REFER), depending on the inter-softswitch protocol it uses (H.323 or SIP). The call agent may use MGCP or H.323/SIP to establish the call to the transferred-to party, depending on whether the transferred-to party is on the same MGCP call agent or not.

The call flow illustrated in Figure 6.10 is an example showing end point B managed by an MGCP call agent, using H.323 as an inter-call agent protocol. The calling party, as well as the redirected-to party, are on separate switches in the example. In Figure 6.10 the music on hold is played by the call agent (it could also be streamed by an external announcement server controlled by the call agent).

### 6.1.6 VoIP call redirection and call routing

#### 6.1.6.1 Call redirection and routing in traditional voice

A TDM switch routes calls according to a route table. Most switches are able to route calls not only according to the call destination, but also according to the source of the call. In case there are multiple route tables, if a call from A to B is redirected to C, then the route table attached to call originator B will be used to route the redirected call to C.

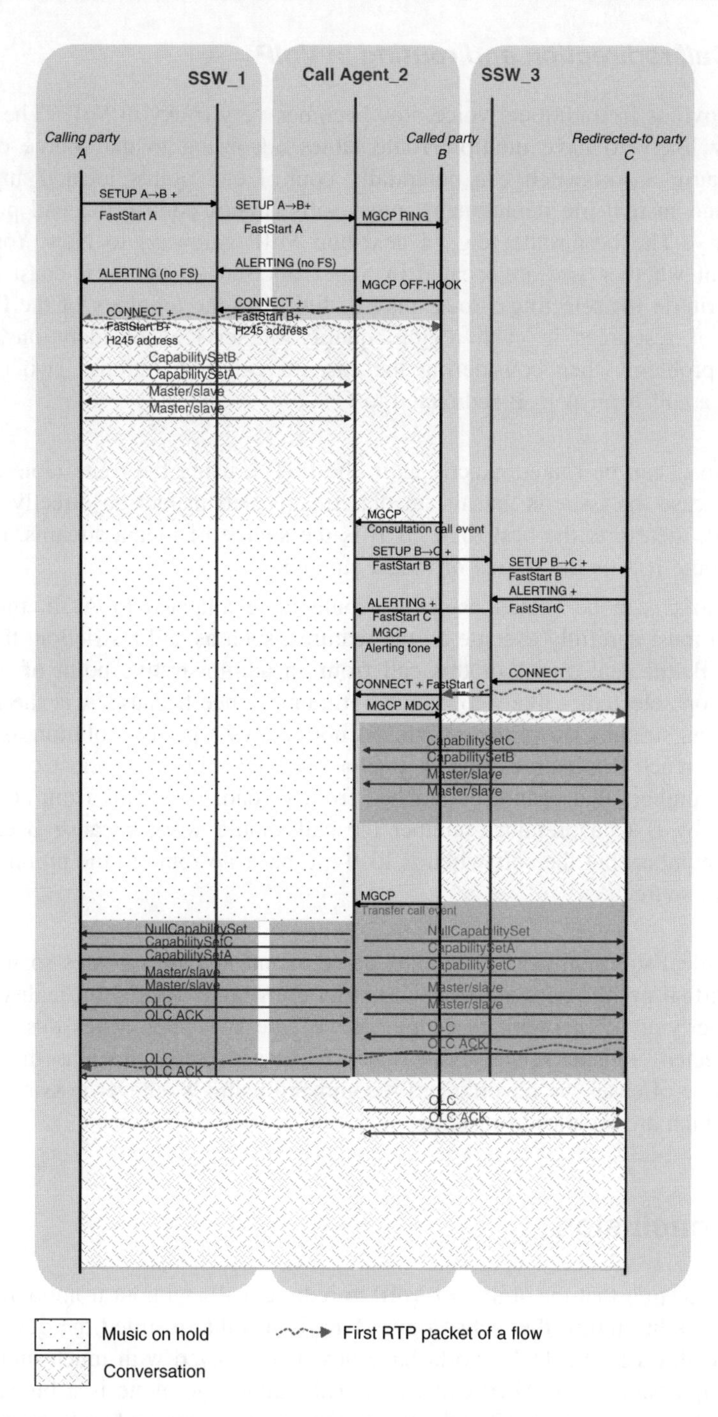


**Figure 6.10** Transfer triggered by an MGCP end point, executed using the H.323 NullCapabilitySet (MGCP PDUs not detailed). Most ACK PDUs are not shown and all messages are in sequence for clarity.

#### *6.1.6.2 Call redirection and routing in VoIP*

What was obvious in traditional voice now becomes very tricky in VoIP. The softswitch is now very likely to have multiple route tables according to the source of the call. This is because a softswitch can potentially control end points located anywhere on the planet and in real-life deployments most softswitches control the end points of an entire country. The best route (e.g., a next-hop VoIP gateway) to New York may be very different whether you are located in San Francisco on the west coast or in New York! The criteria for selecting a route may be linked to the topology of the IP network, availability of resources in each region, quality of service considerations, etc. This becomes a problem when considering the case of call redirections. Two choices are possible for a call from A to B redirected to C:

- The call to C can be routed exactly as in the PSTN, using the route table attached to B. In this case the issue is that the media streams will in fact go directly between A and C and, therefore, the best route, if B is the source of media streams, may not be the best route if A is the source of media streams.
- The call to C can be routed using the route table attached to A. In this case the softswitch must carefully execute all the features, such as call restriction, that must be linked to B, the real owner of this call from an administrative point of view. Now, the next hop selected is likely to be the optimum choice because A is the real source of the media streams. Nevertheless, this is not always a perfect solution as B may be allowed to reach destinations that A cannot reach. For instance, most countries have premium numbers that cannot be reached from abroad. If A calls from abroad and is redirected by B to a premium number the call should work because B can call the premium number, but the softswitch is likely to have no route to the premium number attached to source A.

There is no ideal solution. In any complex deployment, involving routes spanning across countries, virtual private networks, or with strict constraints related to quality of service and the underlying IP network topology, one of the two approaches described above must be selected, and the related issues need to be addressed on a case-by-case basis. In the majority of cases we nevertheless recommend using whenever possible traditional networks, which are simpler to manage.

### 6.1.7 Conclusion

There is no question that the ability of VoIP to redirect calls without tromboning media is the major breakthrough of the technology. Many issues that justified the development of complex call flows in the TDM world have now been solved with much more elegance by more simple and robust VoIP call flows. This advantage alone is a big incentive to migrate TDM networks to VoIP technology: the redirect service has been presented in a Centrex or corporate telephony context above, but in many countries VoIP will be an interesting solution to the issue of local number portability (LNP), which often uses the call forward service ('onward routing' technique).

As we have seen, various call flows and technologies can potentially be used both to request a call transfer and to execute a call transfer. The complex, source-based call transfer protocols, such as SIP REFER or H.450.2, are really useful only in the corporate space to allow IP phones to request a transfer to the switch, but in the network they create many more issues than they solve when also used to execute the transfer. With VoIP, the simplest transfer call flow (using TCS = 0, or SIP RE-INVITE) is also the best, because tromboning has been removed.

Most of the call flows discussed above also demonstrate the similarity of H.323 and SIP, both protocols are identical in design and performance when deployed as a core network protocol. H.323 and SIP also present the same issues when it comes to the deployment of Centrex-type services or the control of IP phones. As already demonstrated in the TDM world where all PBXs use a stimulus protocol to control IP phones, a protocol that does not need to make assumptions about the built-in logic of a phone or a CPE is much easier to deploy and operate. In VoIP the same advantages are obtained by using MGCP (a stimulus protocol) as the protocol for controlling Centrex end points at the edge of the network. Putting all these remarks together, we can summarize as follows:

- In core service provider networks, using SIP or H.323, support of the mandatory options of the standards that enable dynamic redirection of media streams (TCS = 0, RE-INVITE) is sufficient to support all the redirection services already deployed in the TDM world. The redirection services are much more efficient on VoIP networks.
- Regarding phones managed directly by a public infrastructure, H.323 or SIP are fine at the edge for IP phones or gateways providing residential telephony. But if Centrex features, among which is mid-call transfer, are also to be provided, the MGCP package line is a better choice at the edge to control IP phones and residential gateways. Of course, both solutions can be implemented at the same time on a network, and in all cases H.323/SIP remains the core protocol.
- If corporate PBXs are to be connected to the core public network, using the same call flows as the current TDM network (local transfer by the PBX) is the only practical solution because of the third-party dependence issue. H.450 or REFER can only be used in closed user groups, and even then do not bring any significant advantage when IP-PBXs are implemented properly and are capable of media anti-tromboning.

# 7

# Advanced Topics: NAT Traversal

## 7.1 INTRODUCTION TO NETWORK ADDRESS TRANSLATION

### 7.1.1 One-to-one NAT

Network Address Translation (NAT) was initially used to protect corporate networks from people attempting to access internal networks from the Internet. Many corporations decided to use private addresses internally (e.g., addresses in the 10.X.X.X range): such addresses cannot be routed on the public Internet and therefore it is impossible to send a packet to a private address through the Internet. Of course certain computers still need to be able to receive packets from the Internet (e.g., email servers). Such computers are given a public (also called routable) address on the Internet, and the site router or firewall has to translate this public address to the private address of the server on the fly.

For instance, in Figure 7.1 the computer with internal IP address 10.3.0.4 is a mail server and needs to receive traffic from the Internet. Its private IP address is mapped to the public address 162.167.3.14 on the Internet, and therefore any packet sent to this address on the Internet will reach the site access router, which will translate the destination address to 10.3.0.4.

Computers that need to send information to the Internet with a protocol that needs to receive an acknowledge message that the packets have been properly received (e.g., TCP) also need to have a mapping to a public address, otherwise the acknowledge message cannot reach the sender from the Internet. This is the case for computer 10.3.0.2. In Figure 7.1 it is sending a packet to a computer at 1.2.3.4. The source address of this

*IP Telephony: Deploying VoIP Protocols and IMS Infrastructure, Second Edition* O. Hersent

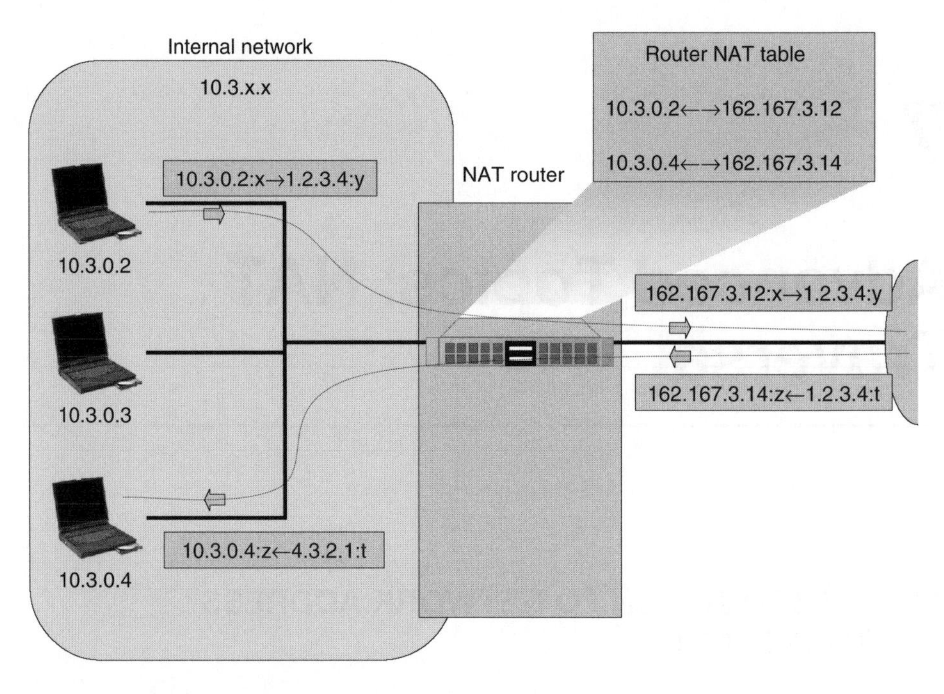


**Figure 7.1** One-to-one NAT.

packet is translated from 10.3.0.2 to 162.167.3.12 by the router, so that if the receiving computer needs to send an acknowledge message, it will be able to do so by sending it to IP address 162.167.3.12. The computer 10.3.0.3 has not been given any mapping to a public address, and therefore cannot be reached from the Internet. This also saves public IP addresses since only servers and computers reaching the Internet need them.

One-to-one NAT (in short, NAT; though NAT is commonly mistaken for NAPT) essentially gives to a set of computers a 'shadow' image on the Internet with a public address. NAT works with any protocol: in Figure 7.1 we inserted port numbers x, y, z, t in the IP packets, but NAT works just as well for protocols that do not have port numbers.

### 7.1.2 NAPT

With one-to-one NAT, each computer accessing the Internet needs to have one public address. With the advent of the WWW, this quickly became inconvenient as virtually anyone can browse the Internet, and this would require either an HTTP proxy that relays all queries to the Internet on behalf of the users, or a public IP address mapping for everyone.

The Network Address and Port Translation (NAPT) technique makes it possible for multiple users to access the Internet without an application-level proxy, for all applications that use transport protocols that have a port information to characterize a connection in addition to the IP address (e.g., UDP and TCP).

#### *7.1.2.1 Full-cone NAPT*

Full-cone NAPT is the most common implementation. In this implementation each outgoing stream of UDP or TCP data from a given IP address and port, *irrespective of its final destination*, is allocated a port on the router's public address. In Figure 7.2 the router public address in 162.167.3.1, and the router has allocated port 2002 for HTTP/TCP communication which sends packets from computer 10.3.0.2, port 3210. The router has allocated port 2004 for SMTP/TCP communication which sends packets from host 10.3.0.4, port 5678. The port (2002, 2004) is used as an index in the NAPT table by the router when it receives IP packets at IP address 162.167.3.1 and needs to forward them to the appropriate internal IP address and port. Each NAPT entry creates a **pinhole** through the router for incoming packets from the Internet and forwards these packets to the proper host.

In the case of TCP the NAPT entry is created in the router when the first TCP segment is sent from the internal host to the Internet, and deleted when the TCP communication is closed (FIN, FIN_ACK packets), or after a very long timeout (if one of the two computers crashes).

NAPT also works with UDP-based communications if the communication protocol responds to UDP packets received from a given IP address and port by sending response packets back to the *exact* same IP address and port. The NAPT entry in the router is then created when the first UDP packet is sent out by the internal host to the Internet, and remains for a relatively short period (about 30 s, typically). This period is extended each time a packet is sent or received corresponding to this NAPT entry.

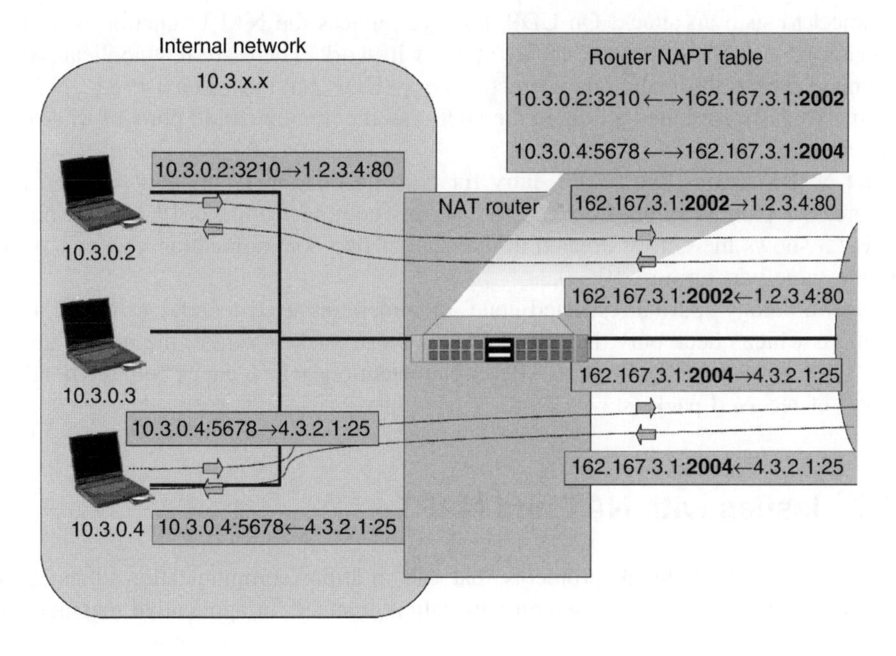


**Figure 7.2** Full-cone NAPT.

As already mentioned, a mapping entry in the NAPT table is created for each source IP and port, irrespective of the destination. This means that if host 10.3.0.2 reuses port 3210 to communicate with port 7000 on host 1.2.3.4, or even to another computer altogether, the same entry will be used. Therefore, the NAPT entry maps to a 'full cone' of connections, which all originate at the same internal IP address and port. This property is very important for some NAT workaround methods, such as **Simple Traversal of UDP Through Network Address Translators** (**STUN**).

#### *7.1.2.2 Strict NAPT*

Strict NAPT is used by some firewalls to prevent hosts on the internet from using the pinhole opened by NAPT entries, which could be used by a malicious user to send IP packets to internal computers. Let's assume that someone discovers a 'killer TCP segment or UDP packet' that crashes a computer or allows someone to take control of it. In the case of full-cone NAT, a malicious user knows that, since many people are browsing the Internet, many NAPT entries are active in the router. It is relatively easy to discover the public IP address of the router (e.g., looking at DNS entries): the attacker will then send the malicious packet to all ports at that address. At each port that corresponds to an active NAPT entry, the router will forward the packet to the internal host, leading to a successful attack. This works because full-cone NAPT does not check that the source IP address and port of the packet is indeed a server in active communication with an internal host.

Now, this is not as bad as it seems: there are random serial numbers in TCP that are very hard to guess, and a full-cone implementation, which would check these numbers, is not subject to such an attack. On UDP, however, unless the NAPT function is aware of the higher level protocol properties, the attack will work. There are few applications that succumb to potentially malicious instructions over UDP, but they do exist (e.g., in 2003 a virus successfully exploited a hole in the UDP-based communication ports of Microsoft's SQL server).

Strict NAPT creates one NAPT entry for each destination IP address (or even port); therefore, the pinhole is only opened for packets coming from this IP address on port. Figure 7.3 shows the entries created in the case of two communication channels opened from the same port on host 10.3.0.2.

The terminology 'partial/restricted cone' is sometimes used to refer to NAPT implementations which check only the source IP address of packets sent to the private domain, while 'symmetric cone' applies to NAPT implementations which check both the IP address and port of received packets.

### 7.1.3 Issues with NAT and NAPT

NAT and NAPT both break protocols that use multiple communication channels and transmit the IP addresses of these communication channels in applicative messages. For

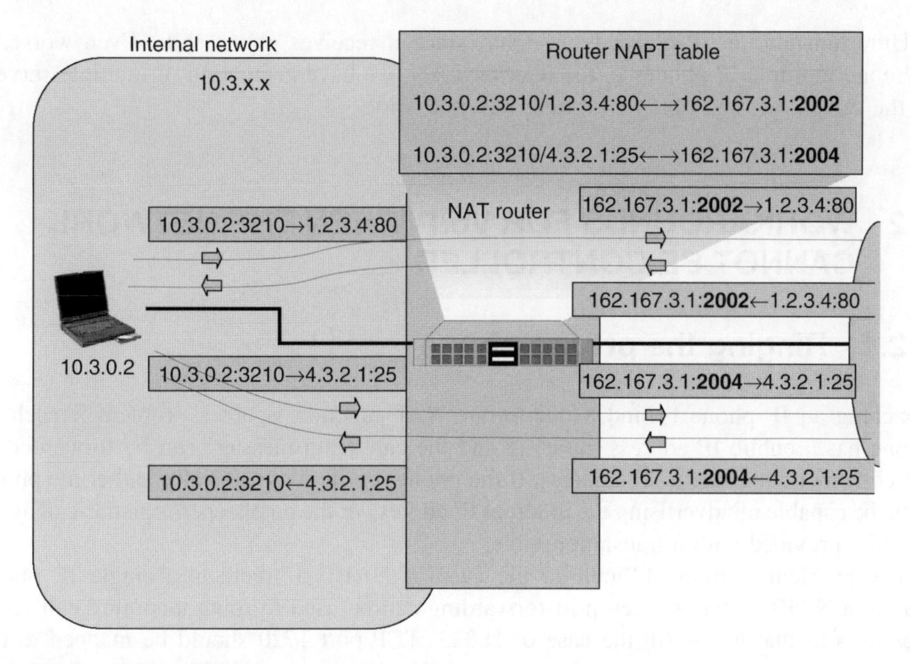


**Figure 7.3** Strict NAPT.

instance, SIP, H.323, and MGCP all use one communication channel for call control and several RTP communication channels for the media. The IP addresses of the RTP communication channels are transmitted on the call control communication channel.

In the example network of Figure 7.3, if computer 10.3.0.2 opens a VoIP communication with a computer on the Internet using a SIP INVITE or an H.323 SETUP with 'Fast-Connect', it will advertise one or more RTP reception ports on IP address 10.3.0.2. If the remote computer attempts to send RTP packets to this address, these packets will be dropped by the first router on the path, because they correspond to private addresses and cannot be routed on the Internet. The RTP packets in the other direction (from 10.3.0.2) will get through if the remote computer is on the Internet, and therefore we end up with a half-duplex conversation.

NAPT presents another problem: it cannot work with servers on a private network. A server is a computer that receives connections from clients' machines (e.g., a web server). Since NAPT creates forwarding entries only when packets are sent from the internal network to the Internet, there will be no active entry in the case of a new incoming connection, and it will fail. The router would need to know where to route these new incoming connections.

One workaround is called **port forwarding**: incoming connections that use the TCP or UDP protocol to the router IP address and a given port are forwarded by the router to a given internal host. Since most servers use a well-known port, this enables the use of exactly one instance of each type of server on the internal network.

Unfortunately, an IP phone is a server, since it receives phone calls. Even worse, if there are multiple IP phones in the internal LAN, we have a situation of multiple servers of the same type, and port forwarding will not work.

## 7.2 WORKAROUNDS FOR VoIP WHEN THE NETWORK CANNOT BE CONTROLLED

### 7.2.1 Ringing the proper phone

Reaching an IP phone behind a one-to-one NAT function is not a problem if each IP phone has a public IP address mapping and the call setup message can be forwarded to the corresponding internal IP address. If the phone registers dynamically, either the phone must be capable of advertising the external IP address or the gatekeeper/registrar/call agent must be provided with a translation table.

The problem is more difficult in the case of NAPT. If there is a single IP phone behind a NAPT function, then port forwarding can be used to route incoming call setup messages to that phone (in the case or H.323, TCP port 1720 should be mapped to the IP phone, whereas in the case of SIP over UDP, UDP port 5060 should be mapped to the IP phone). This solves the problem of being able to 'ring' the phone even when it is behind a NAPT function. This solution immediately extends to the case of an analog gateway with multiple telephone lines, since a single IP address is used for all the lines.

The case of multiple IP phones can be supported if the IP phones have configurable non-standard call-signaling ports (this is rarely possible), in which case each port can be mapped to a given phone.

The problem can be solved with another approach (using the pinhole maintained by the NAPT function) if the phone can maintain a permanent connection to the call control server. This can be done easily with MGCP or SIP, which are both UDP-based protocols, if the messages used for their registration function (REGISTER message in SIP, RSIP in MGCP) use the same source port as the one used to receive call control messages. This first registration message will reach the call control server, and by looking at the source IP address and port the server can learn the external IP address and port used for the NAPT mapping entry for each phone. In order to 'ring' this phone, the call server simply needs to send a call setup message (SIP INVITE, MGCP CRCX) to that self-same port (the pinhole for that phone), and the router will forward it to the correct phone. This behavior is standard in MGCP, but wasn't correctly specified in SIP RFC 3261; this was fixed in August 2003 in RFC 3581 by using a new 'rport' parameter of the Via header to force responses to the exact apparent receive port (see Section 3.2.2.1.1 and Figure 3.27 for more details). The only issue is to keep the pinhole open, which requires having some traffic on the signaling connection every couple of minutes. Some phones can be configured to do so (e.g., REGISTER refresh), otherwise the call server can periodically send a message to the phone that should cause the phone to respond (MGCP AUEP, SIP OPTIONS, or even malformed messages that should trigger an error response).

### 7.2.2 Using port forwarding to solve the wrong media address problem

These methods alone do not solve the problem of internal network IP phones advertising private IP addresses for the media stream. There are also several possible solutions to this problem:

- Some IP phones have a configurable field for the "external IP address" of the NAT function (the address that will be used by the NAT function when replacing the source IP address). These phones will open a reception RTP port on the local host (e.g., 10.3.0.2:2345), but will advertise it in the call control messages with the public external address instead (e.g., 162.167.3.1:2345, note that the port is the same). This works only if the NAPT router is configured with a default port forwarding to the IP phone (i.e., all packets to 162.167.3.1 that do not correspond to an existing active connection are routed to the IP phone). In order to avoid conflicts with existing mappings, these IP phones can usually also be configured with a restricted set of ports to use for inbound RTP connections. This also allows this solution to extend to the case of several IP phones if the NAPT router can map a specific range of ports to each IP phone.
- Some IP phones will automatically detect that the source of the RTP packets they receive from the internal host is not the same as the IP address advertised for the RTP reception port. In such a case they will assume that NAPT is in place and will ignore the IP address provided, sending their own RTP packets to the source IP address present in the packets received from the IP phone on the internal network. This works because most phones use the same sending and reception RTP ports, but, unfortunately, this is not a standards-compliant implementation (see Section 7.2.4). Obviously, both phones should not be behind distinct NAPT functions.

All these solutions were used in the first PC-to-phone implementations, with only one or a couple of IP software phones in a residential environment, and work well with 'techies' and early adopters who are not scared of reconfiguring their NAT router. Unfortunately, these methods cannot be used for the general public, or for such services as VoIP-based Centrex, which need to reach many phones behind a NAT function.

### 7.2.3 STUN

Many proposals have been made to facilitate or ideally automate the configuration of VoIP networks to successfully work even across NAT functions. The IETF Midcom Working Group is defining the specifications for a 'middlebox' access control device and a 'Midcom' control protocol. Most of these proposals work in an ideal world where all routers could be upgraded overnight, but are of little practical interest in current networks, given the size of the installed base of low-end cable or DSL NAT routers.

The STUN approach however stands out as it does not require any configuration or change of existing NAT/NAPT routers or existing call control servers. STUN is defined in

RFC 5389 (Simple Traversal of User Datagram Protocol Through Network Address Translators), and is a simple query response protocol encapsulated over UDP (some security primitives also use TCP connections).

STUN is available in more and more phones, and there are several public STUN servers on the Internet. STUN provides a way for the phone to dynamically learn the external IP address and port that will be used for each communication through the NAT function. STUN also allows a host to discover the type of NAT implementation (full-cone or strict).

Each time the IP phone knows it is about to advertise an address and port that cannot be reached through the NAT/NAPT function, it first sends a STUN query from that exact IP address and port to the STUN server on the public Internet (1.2.3.4:3478 in Figure 7.4, 3478 is the well-known port of STUN servers). The response packet of the STUN server simply indicates what is the apparent source address and port of the query (i.e., the external IP address and port allocated dynamically by the NAT function for this communication).

The IP phone is now ready to advertise the correct external address and port in the call control messages to the remote host (SIP INVITE or 200 OK response, MGCP 200 OK response to a CRCX or MDCX command, H.323 OLC or OLC ACK).

Note that this works only if the NAPT function is full-cone; otherwise, the external port allocated by the router for the STUN query will differ from the external port allocated for the RTP stream to the remote host, since they have different IP addresses. Most residential NAT functions are full-cone. In a bullet-proof implementation that works even with strict NAPT, a service provider can co-locate the STUN server with a call server that also

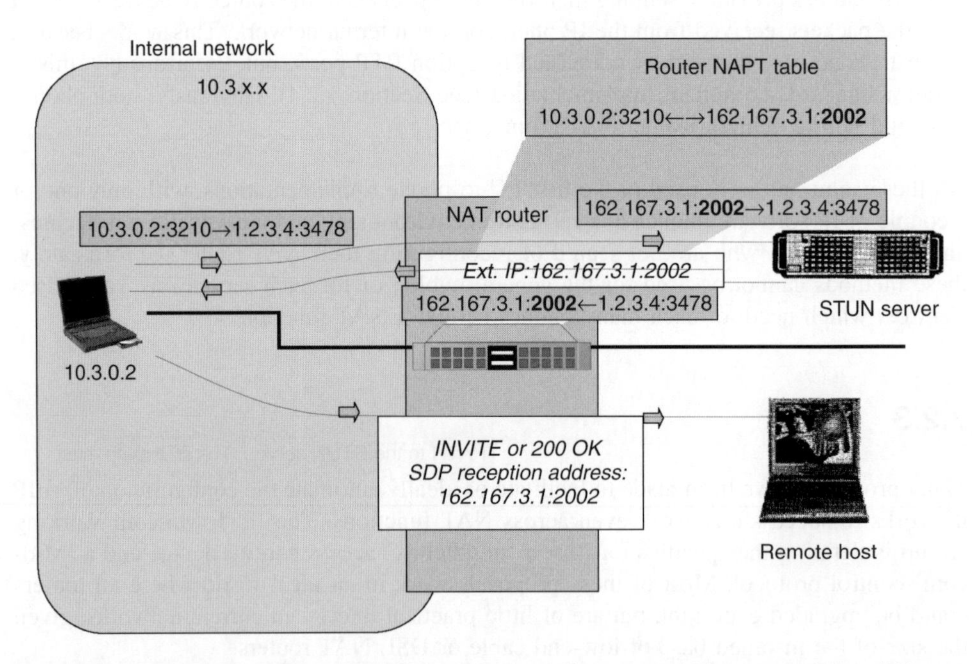


**Figure 7.4** Softphone using a STUN server to learn the external IP port used by a full-cone NAPT router for VoIP signaling or media flow.

serves as an RTP proxy. In this situation the RTP packets, and STUN queries go to the same destination and will be allocated the same external port by the router's strict NAT function. The only downside of this approach is that the RTP proxy will add some delay to the conversation and will impact the density of the call server (RTP processing is very CPU-intensive, a typical Linux kernel-mode implementation will route a maximum of about 500 media streams of 20-ms packets per GHz on a Pentium).

Another problem with STUN is that if there are multiple IP phones behind the NAPT function, these phones will advertise a public address for media streams even if one phone calls the other in the private network. This will force the router to relay media streams that could otherwise have been transported directly by the LAN. Usually, this will not create significant QoS problems as there are relatively few internal calls on small sites and the connections to the router are 10/100 Mbit/s Ethernet. However, this may create issues on large sites. This problem could easily be fixed by getting the IP phone to advertise two reception ports (one private, one public) and having the call server select the correct one on the fly (the call server knows which IP phones are behind the same NAT function); but, this would require small changes to the existing VoIP standards.

STUN will also introduce non-optimal media paths in some networks, as media are forced to flow through the routers closer to the STUN server, which are not necessarily in the shortest possible path for media streams. The same problem may happen for signaling (Figure 7.5), although this is much less critical as signaling does not have strong latency constraints.

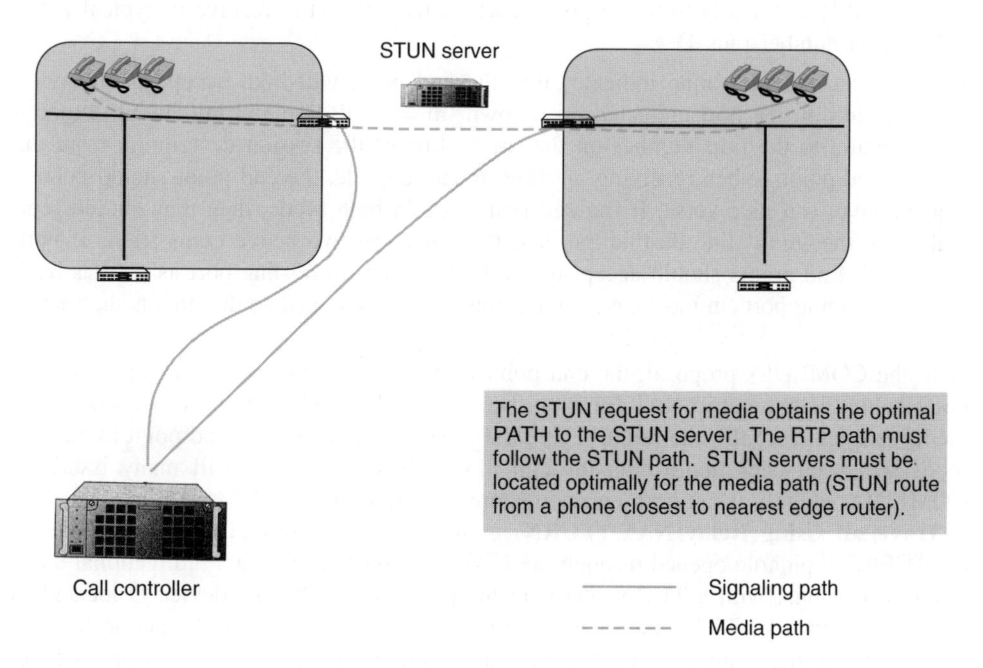
STUN server
The STUN request for media obtains the optimal PATH to the STUN server. The RTP path must follow the STUN path. STUN servers must be located optimally for the media path (STUN route from a phone closest to nearest edge router).
Call controller
Signaling path
Media path

**Figure 7.5** STUN may force utilization of non-optimal paths for media and/or signaling.

### 7.2.4 Other proposals: COMEDIA and TURN

**Connection-Oriented Media Transport in SDP (COMEDIA)** is described in RFC 4572 it enables traversal of symmetric NAT by allowing VoIP gateways to dynamically update their destination RTP port according to the source IP and RTP port detected in the received RTP packets, instead of continuing to use the port advertised in the remote SDP. COMEDIA uses a new SDP attribute 'a = direction:role', where role can take the following values:

- 'Active', which indicates that the end point will initiate a connection to the port number on the 'm =' line of the session description from the other end point. The port number on its own 'm =' line is irrelevant, and the opposite end point must not attempt to initiate a connection to the port number specified there; instead, it is prepared to receive media on the ports from which it sends. These end points should also immediately send some media to the ports indicated in the 'm =' line of the remote end point.
- 'Passive', which indicates that the end point will accept a connection to the port number indicated on its 'm =' line of the session description. The end point will not send any media (including control packets such as RTCP) from their passive ports until they receive a packet on these ports and record the source address and port of the sender. The passive end point then assumes that the first packet received corresponds to its active peer. From this point onward, passive end points must send UDP or RTP media from the same port as the port indicated in their 'm =' line (receive port). They must also send RTCP media from the port on which they expect to receive it (typically, the RTP port number plus 1).
- 'Both', the default value, indicates that the end point will both accept an incoming connection on the port indicated on its own 'm =' SDP line and initiate an outgoing connection to the port number on the 'm =' line of the session description from the other end point. When receiving an SDP in active mode, the end point should behave as passive, and vice versa. If the end points are in both mode, then they should send data on the 'm =' line destination, and there may be two active connections if both succeed. End points should accept media both on the 'm =' line port as well as back to the sending port (in most cases end points will be designed so that this is the same).

With the COMEDIA proposal, the end point in passive mode can send media to an end point behind a symmetric NAT function, because a UDP pinhole will be opened by the media sent out from the end point behind the NAT function, and the end point in passive mode will send back audio data through this pinhole. There are still many issues in COMEDIA, notably when both end points are behind separate NAT functions.

**Traversal Using Relay NAT (TURN)** is another NAT traversal approach that uses the TCP/UDP pinhole opened through the NAT function to establish a bidirectional communication tunnel with a TURN server in the public network. The device D located in the private network which requires to establish a communication with the public Internet first communicates with the TURN server using the TURN protocol. As a first step D requests an IP address and port for his own use on the turn server. The TURN server allocates an IP address and port (IP_t:port_t). Device D can then advertise this IP address and port to external IP devices that need to send packets to it. When the TURN server receives packets on IP_t:port_t, it simply forwards these packets to device D using the TURN protocol. The TURN protocol can traverse the NAT function because it is based on a permanent TCP connection between device D and the TURN server, or it uses symmetric UDP.

TURN can be seen as a way to obtain a 'remote network interface' outside the NAT domain. TURN is documented in RFC 5766.

#### *7.2.4.1 VoIP NAT traversal using an RTP relay*

When the network cannot be controlled, when end points do not implement any NAT traversal algorithm, and when NAT functions may be any combination of full-cone or strict NAT without any specific support for VoIP, the only possibility that remains to enable VoIP calls is to attempt to use the pinholes opened by all NAT functions when sending traffic to the public network.

As indicated in Section 7.1.2.2, even strict NAPT functions will accept, once a UDP packet from the internal network has been sent to a server, a response packet from that server to the exact port that was allocated by the NAPT function as a source port for the UDP packet initially forwarded.

This property can be used by a network-based entity E in order to get UDP-based VoIP protocols to work across the NAT function (Figure 7.6):

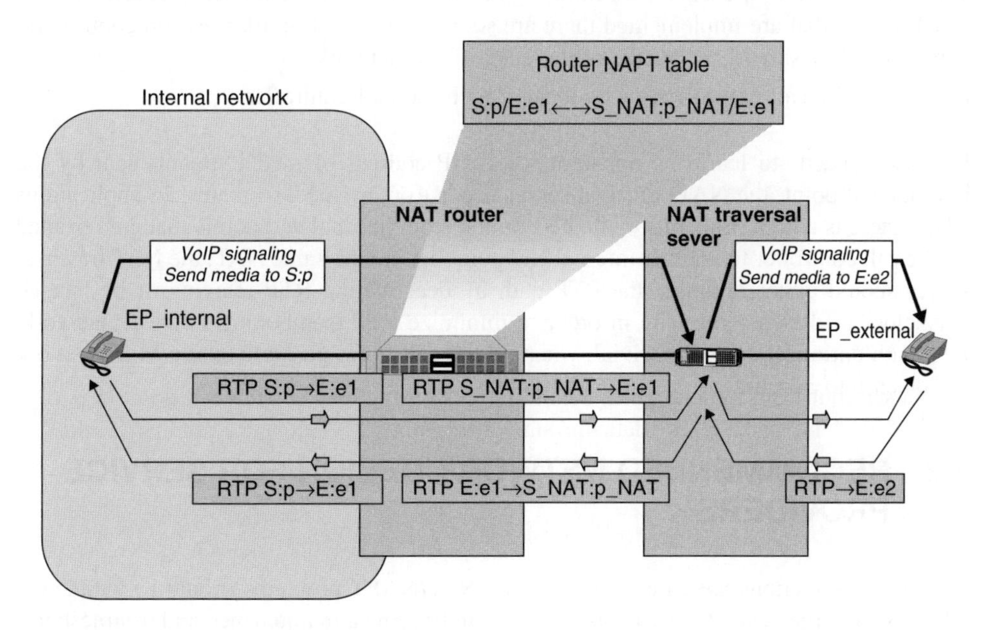


**Figure 7.6** NAT traversal using an RTP relay.

- Server E will receive all VoIP signaling from the end points (EP_internal) behind the firewall, and send all responses to the apparent source port of the UDP packets it receives. All protocol-level indications to send the responses to a different IP address are ignored. In addition, in order to keep the pinhole open, server E or the end point need to exchange a packet at least every 30 s. When the protocol is MGCP, this can be achieved by the server independently of the end point, using AUEP commands. When the protocol is SIP, this can be client-based (e.g., REGISTER messages) or server-based (e.g., OPTION messages).
- For media streams, server E needs to intercept all VoIP signaling commands where the end point advertises the RTP reception ports where it expects to receive media, and forward these commands to EP_external, indicating itself as the reception device. E will also put itself forward to receive all media streams sent by EP_internal, in order to analyze their apparent source address, as translated by the NAT function. As soon as the apparent source address and port S_NAT:p_NAT of a media stream sent by EP_internal is known, server E will be able to forward media streams sent by EP_external to the self-same address and port S_NAT:p_NAT. The NAT function will forward these packets to the original source address and port S:p used by EP_internal to send its RTP packets; this works because virtually all end points send and receive RTP streams for each media on the same port.

Such a NAT traversal server can be implemented on any network link where VoIP signaling can be intercepted. There are obviously two natural locations:

- At the IP point of presence concentrating the traffic from the customer. The NAT traversal servers that are implemented there are sometimes called 'border session controllers' because they should be located at the edge of the network.
- Close to the call controller, or co-located with the call controller.

Because it needs to learn the apparent source IP address of media streams sent by the internal end point, the NAT traversal server needs to relay media streams. In applications where there is a high probability of calls coming from internal end points that are rerouted by the call controller to another internal end point behind the same NAT, the NAT traversal server should either optimize the RTP path by deactivating RTP relaying or be located very close to the customer site in order to minimize RTP tromboning in the IP network. RTP path optimization is not trivial, when all the call flows possible over a VoIP network are taken into account.

## 7.3 RECOMMENDED NETWORK DESIGN FOR SERVICE PROVIDERS

The previous sections have made it clear that NAT/NAPT problems should be avoided at all costs for large-scale deployments, as they can become a maintenance and troubleshooting nightmare (this amounts to knowing in detail and sometimes debugging many different

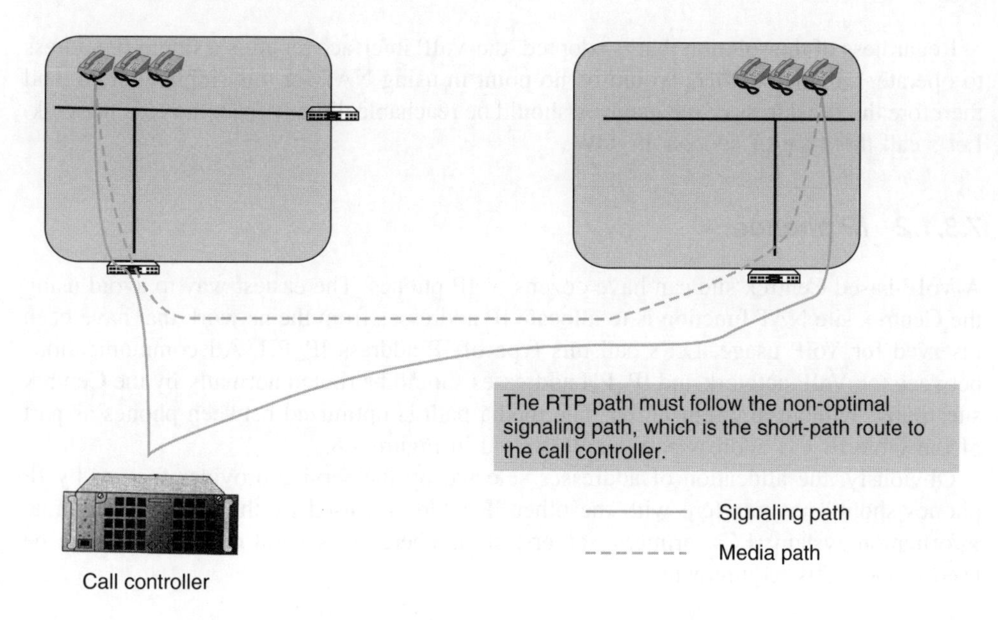
The RTP path must follow the non-optimal signaling path, which is the short-path route to the call controller.
Signaling path
Media path
Call controller

**Figure 7.7** VoIP NAT routers may lead to non-optimal media paths.

sorts of NAT and firewall implementations). It is likely that enterprise routers, and obviously residential routers, will be unable to provide adequate support for VoIP protocols before 2005. Even when such methods work, it is likely that they will not be optimal in some networks, because most NAT routers will need to route both signaling and media, as illustrated in Figure 7.7.

All the traversal methods described above will work, but they are really workarounds and hardly capable of sustaining robust network deployment. However, it would obviously be very costly to allocate a routable, public IP address for every IP phone.

The strategies we describe below are what we believe to be the best engineering options for a service provider wanting to implement business-grade VoIP on a large-scale network.

## 7.3.1 Avoid NAT in the customer premises for VoIP

### *7.3.1.1 Business trunking and connection to PBXs and IP-PBXs*

Most large corporate sites have an existing PBX and do not wish to switch to a pure internal VoIP network before the PBX has been fully depreciated. Nevertheless, such sites can benefit from a VoIP network to carry communications between corporate sites (voice VPN), or to the PSTN (arbitrage, least cost routing). In most cases the easiest way is to connect the PBX to a CAS or 5ESS (in the US) or PRI (most of Europe) VoIP gateway. More and more PBXs also have optional VoIP trunk boards that can be purchased and will packetize voice without a need for an external VoIP gateway.

Regardless of the solution that is adopted, the VoIP interface requires a single IP address to operate. Obviously, there would be no point in using NAT for this single address, and therefore the IP address of the gateway should be reachable directly from the core network. Let's call this type of address IP_GW.

### *7.3.1.2 IP phones*

A VoIP-based Centrex site can have dozens of IP phones. The easiest way to avoid using the Centrex site NAT function is to allocate IP addresses from the network that have been reserved for VoIP usage. Let's call this type of IP address IP_PH. All communications between the VoIP network and IP_PH addresses should be routed normally by the Centrex site router, without any translation. The media path is optimized between phones as part of the same IP_PH address pool, as illustrated in Figure 7.8.

Obviously, the allocation of addresses selected by the service provider for use by IP phones should not interfere with the other IP addresses used by the existing corporate information system (PCs, printers, servers, etc.). There are several methods that can be used to meet this requirement:

- The most trivial way is to allocate IP_PH addresses in a private address pool (such as 10.X.X.X, 192.168.X.X, or 172.X.X.X). However, the customer may also be using private addresses, and therefore the service provider should select IP addresses in the private address pool that do not conflict with the network currently in place. Depending on the service provider, this may be easy or very difficult. Some service providers have a fully packaged SME IP connectivity offer where the corporate NAT router is always

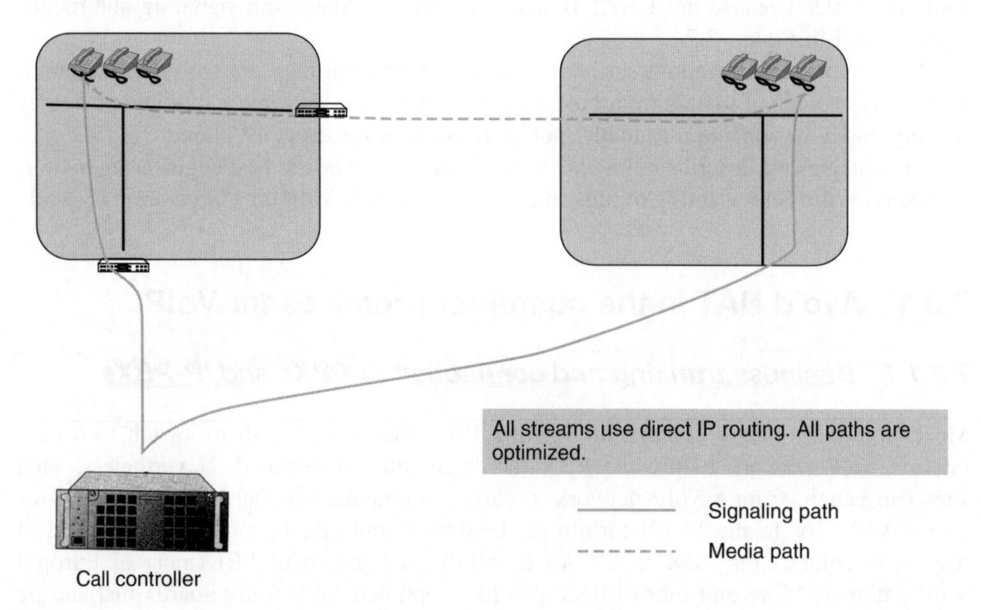
All streams use direct IP routing. All paths are optimized.
Signaling path
Media path
Call controller

**Figure 7.8** Optimized media path between IP phones as part of the same address pool.

configured to allocate private IP addresses to internal PCs in the same address pool (say, 10.1.X.X). On such networks, all other blocks of IP addresses, (i.e., 10.2.X.X, 10.3.X.X, ..., 192.168.X.X, and 172.X.X.X) can be used safely without creating conflicts. Unfortunately, many service providers didn't plan ahead for this type of problem and allow their customers to use any set of private IP addresses they like. It can then become cumbersome to ask each customer to select IP addresses that do not conflict with the network in place. For these service providers the following two alternative approaches will work.

- Instead of allocating the IP_PH addresses in a private pool, one sure way to avoid any conflict is to allocate these IP addresses in a pool of public IP addresses. For example let's select 162.168.X.X. This block is large enough for about 65,000 phones. Apparently, this defeats the object of not using one public address per IP phone. Fortunately, this is not the case as we will see below, as this pool of IP addresses can be reused in the network as many times as necessary, providing unlimited scalability with just one class B address block.
- Another option is to put all IP phones on a VLAN. The router is then instructed to route all incoming packets from the VoIP backbone (e.g., arriving in an IP tunnel), to that VLAN, resolving de facto any addressing ambiguity. Similarly, any packet sent from the VLAN of IP phones should be sent to the VoIP backbone without translation. All other data flows from other VLANs should be routed to the router NAT function and undergo normal processing. Unfortunately, this is only possible on relatively sophisticated routers.

A summary of the optimal-routing configuration for a site is given in Figure 7.9.

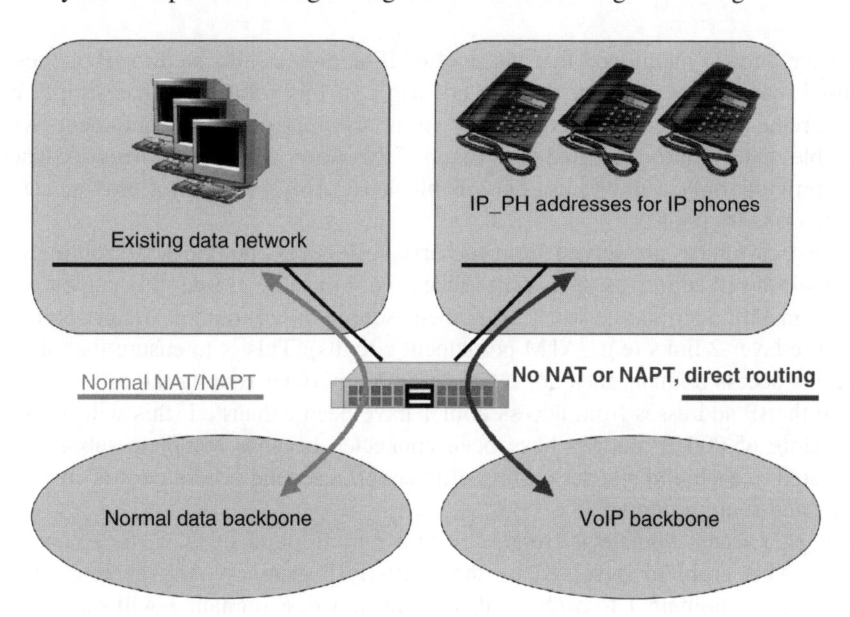


**Figure 7.9** Separation of VoIP and other data flows.

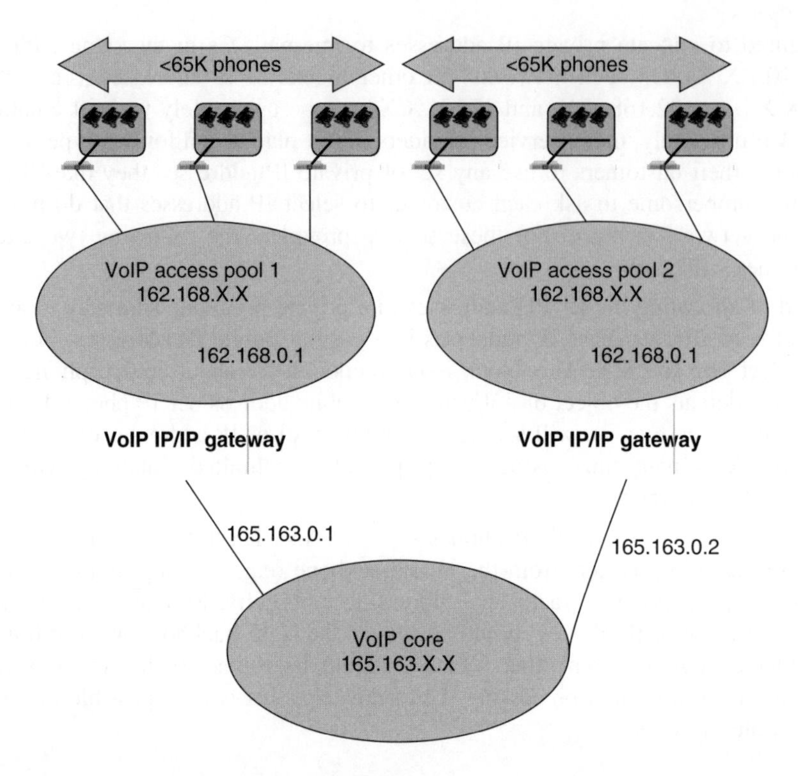


**Figure 7.10** Reusing access pool IP addresses.

We have already mentioned that the pool of IP addresses allocated to IP phones could be reused many times. In fact, there is no magic in this solution, it has simply pushed the requirement for a NAT function away from customer premises equipment (which is of variable quality) into the backbone. Figure 7.10 shows an example network where the service provider only wants to use class B block 162.168.X.X and yet provide service to more than 65,000 phones.

The first customers are served from the first VoIP access pool. The IP routing domain between all the IP addresses of this first access pool must be closed; this can be achieved by using an MPLS virtual network, IP tunnels, source-based routing, router subinterfaces on specific layer 2 links (e.g., ATM permanent circuits). This is to ensure that all routing within this access domain does not interfere with any other routing table.

When the IP addresses from access pool 1 have been exhausted (this will probably be much before 65,000 IP phones have been connected because complete subnets need to be allocated to each end site set routing efficiency), a second access pool is created in its own isolated routing domain.

Since each access domain is isolated from a routing point of view, they cannot communicate. This problem is solved by the VoIP IP/IP gateway. Any phone call from a phone in access domain 1 to a phone that is not in access domain 1 will reach the VoIP IP/IP gateway, which will terminate it locally using an IP address in the access pool (e.g., 162.168.0.1), then re-originate the call using a single public IP address. In essence, the VoIP IP/IP gateway *summarizes* the complete access pool in a single public IP address. If the call needs to reach a phone in the second access zone, it will be routed by the VoIP core to the VoIP IP/IP gateway of access pool 2, which will re-originate the call within the access 2 domain using IP address 162.168.0.1.

This hierarchical access network scales indefinitely to arbitrarily large VoIP networks, by just adding more access pools. The VoIP IP/IP gateway is not a trivial function (we will discuss it in Section 7.3.2).

The method also has the advantage of cleanly separating the VoIP network and the regular data network, which helps delineate the respective responsibilities of the enterprise and the service provider in terms of security (this will be described in Section 7.3.3).

#### 7.3.1.3 Software IP phones and PC-based videoconferencing devices

Unfortunately, the case of PC-based VoIP equipment cannot be solved by the previous method. If an IP address of type IP_PH is allocated to the PC, then any data application also running on the PC may stop working because the VoIP network isn't designed for it and will probably block non-VoIP communications. Even if it worked, this would still allow the VoIP network to reach corporate PCs, which would make the service provider responsible for possible breaches of security on corporate MIS equipment.

Therefore, PCs should be allocated IP addresses normally by the corporation, without any specific restriction for VoIP. The NAPT problem cannot be avoided. The solutions described in Section 7.2 must be used. If the corporation uses full-cone NAPT, then STUN is the best solution. It is supported by a number of VoIP software manufacturers (e.g., EyePmedia at www.eyePmedia.com). If the corporation uses strict NAT, either the company must implement a VoIP IP/IP gateway in the premises (more and more firewalls provide this feature) or the service provider must implement an RTP relay.

All solutions will end up in the same final situation; the audio or video calls originated from PCs will be mapped to VoIP calls that appear to originate from routable, public IP addresses. Similarly, it will be possible to reach all PCs by placing a VoIP call to a routable IP address (either the call control server with an RTP relay, or a public address and port on the corporate site router selected by the router NAT function—the STUN case).

Communications between PCs and IP phones can occur by routing the call to the appropriate VoIP IP/IP gateway (the gatekeeper or SIP proxy required to do this is not shown), which relays the call to the proper IP phone (as shown in Figure 7.11).

Note that the customer access router in Figure 7.11, is shown as having one link to the VoIP access pool domain and one link to the normal Internet backbone of the service provider; this was done for clarity. It can be done with a single link either using IP tunnels or more simply by carefully configuring service provider concentration routers to route all packets with source IP addresses of type IP_PH to the VoIP access pool domain (e.g., an MPLS domain), and all other addresses according to normal Internet routing tables.

As the destination of voice IP packets is to a private address (in fact, the change to existing concentration router configurations is quite minimal), a path to 162.186.x.x is added which routes by default all these packets to the closed routing domain (e.g., an

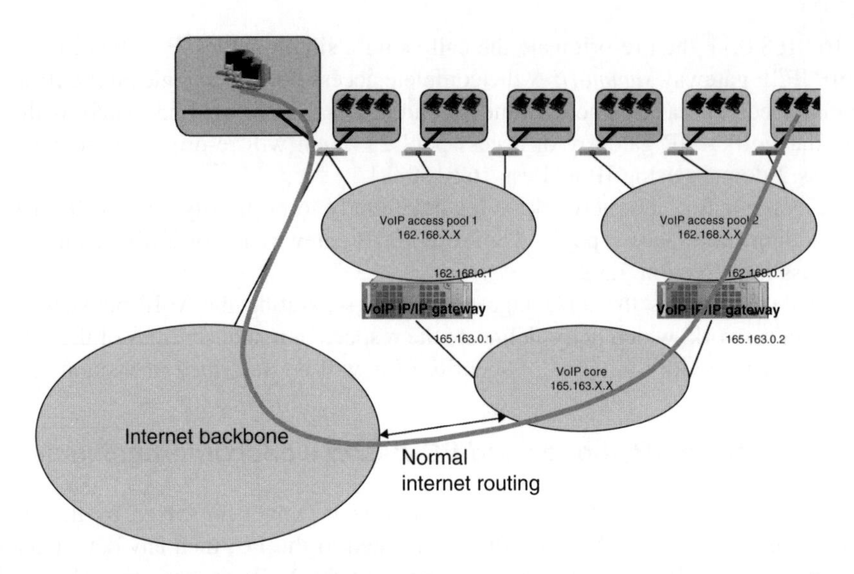


**Figure 7.11** Mixed call scenario: STUN softphone to IP phone in a dedicated address pool.

MPLS domain), instead of the default Internet route. For each VoIP customer connected to this concentration router and using, for instance, 162.168.Z.x, a static route must be added to the subnet 162.168.Z.8 with the customer access router as the next hop.

### 7.3.1.4 The case of data VPNs

Large multi-site corporations using a data VPN pose a specific problem, because in most cases all corporate sites communicate on an isolated routing domain (e.g., an MPLS domain), which usually has a couple of controlled access points to the Internet, though a firewall.

From a VoIP perspective, the entire data VPN can be seen as a large site, with the access routers as the data VPN-controlled access points to the Internet.

Deploying a VoIP service in such networks (e.g., to provide a VoIP short-numbering service between sites, or least cost routing) requires some configuration:

- The service provider softswitch must be located in a specific 'resource domain', which can be reached from all VPN sites and can reach all VPN sites. Most service providers already have such a domain for their DNS and email servers. Note that the resource domain cannot be used to communicate across data VPNs, only communications to and from the shared resource domain and each data VPN is allowed. An MPLS domain can be used for this purpose.
- All VoIP gateways connected to sites' PBXs must be able to communicate with the service provider softswitch. This can be achieved simply by allocating each VoIP gateway an IP address from the resource domain. There are relatively few gateways, but this does not pose an IP address depletion problem.

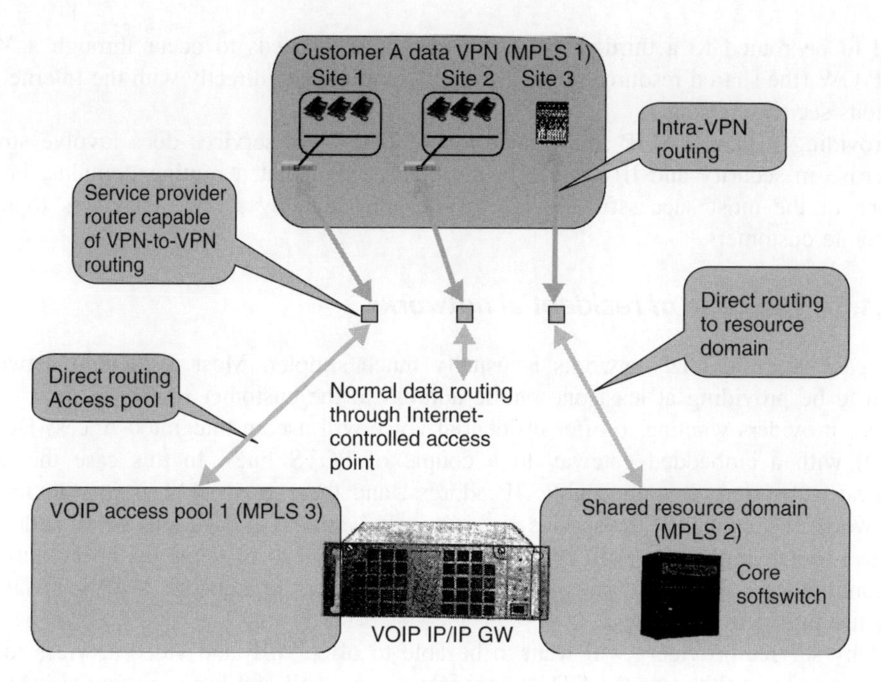


**Figure 7.12** Usage of shared resource domain and VoIP access IP address pool in a VPN service context.

- All IP phones must be provided with IP addresses in the IP_PH pool, allocated as described above. The IP_PH addresses allocated to the data VPN must be routed within this data VPN. All controlled Internet access points of this data VPN must route packets from these IP_PH addresses to the appropriate IP VoIP access pool domain (e.g., this can be a separate MPLS domain).

Figure 7.12 shows the routes that need to be enabled between the various MPLS domains typically found in a VPN environment:

- One MPLS domain per VPN customer.
- One shared resource MPLS domain.
- One MPLS domain for each VoIP access domain (shared across multiple customers).

VPN-to-VPN phone calls over IP can also be enabled using the VoIP IP/IP gateway: since IP communications cannot occur between each customer's data VPN, the VoIP call is first terminated to the VoIP IP/IP GW (communications to the access/shared resource domains are allowed), then the VoIP IP/IP gateway re-originates the call and routes it to the destination data VPN (communications from the access/shared resource domain to any data VPN are allowed).

If the service provider wants to provide PSTN connectivity through shared VoIP gateways, these VoIP gateways must also be part of the shared resource domain. If the calls need to be routed to a third-party VoIP network, this needs to occur through a VoIP IP/IP GW (the shared resource domain cannot communicate directly with the Internet for obvious security reasons).

Providing managed VoIP services on top of data VPN services does involve strong expertise in security and IP routing in the context of isolated routing domains; but, it is one of the most successful service bundles provided by service providers to large corporate customers.

#### 7.3.1.5 The case of residential networks

The case of residential networks is usually much simpler. Most residential networks seem to be providing at least one public address to the customer router. In most cases service providers wanting to offer residential voice will use an Integrated Access Device (IAD) with a embedded gateway to a couple of POTS lines. In this case the VoIP gateway subsystem uses the public IP address, and there is no NAT issue. Obviously, this works just as well if the service provider only wants to allocate private IP addresses to each router; but, some VoIP IP/IP gateways may need to be deployed in the network to communicate with third-party VoIP service providers or network VoIP gateways if they use public IP addresses.

Many service providers will want to be able to offer VoIP and video services to PC users as well. In this case the STUN approach works well and has very few drawbacks, as it is highly unlikely that two PC phones behind the same residential NAT router will want to communicate. Some IAD devices also include a VoIP-friendly NAT function. In this case using STUN is obviously no longer necessary.

#### 7.3.1.6 Smooth deployment scenario

At first it may seem scary to implement the full approach of VoIP access domains for an initial trial. This approach was introduced only to enable the reuse of one pool of IP addresses indefinitely, in order to scale the network. For small trials, however, there is no need to reuse IP addresses, and the VoIP access domain can be merged with the IP core domain. In other words, VoIP core domain components (such as a central gatekeeper or SIP proxy) can be located in the first VoIP access domain, no VoIP IP/IP gateway is initially required.

The expansion to a larger network will require the formal creation of a VoIP core domain (the core softswitch will need to be relocated) and the introduction of additional VoIP access domains. All of this can be done without having to change the IP addresses already allocated to existing IP phones, by substituting the core softswitch by an IP/IP gateway. This method enables smooth expansion of the network.

### 7.3.2 Media proxies

Today, many service providers still use two VoIP gateways connected back to back to provide the media proxy function. Indeed, every VoIP call will terminate at the VoIP address of the first gateway and be re-originated from the second VoIP gateway. However, this crude design is not a viable solution:

- VoIP gateways always decode the media stream to the G.711 format. If voice was originally encoded, the decoding and re-encoding of voice (known as 'tandeming') will significantly reduce the quality of voice (typically, 0.5 MOS points).
- VoIP gateways have a jitter buffer in order to prevent any gap in the TDM media stream. This jitter buffer adds a significant delay (typically, 50 ms) to the media path and adds to jitter buffer delay at the destination.
- In some call scenarios, a call may be routed from one IP domain to the other, then routed back to the original IP domain (call forward, local number portability, call transfer, etc.). The back-to-back gateway in such circumstances will continue to 'route' the media stream, although this is clearly not the optimal path through the IP network! Again this adds unnecessary delays and jitter to the VoIP network.
- VoIP gateways only support voice (actually, H.320 video gateways do exist, but they are expensive and would degrade the video quality that can be achieved on IP networks).
- Last but not least, every call through the back-to-back gateway uses two gateway ports, which is expensive.

There are many providers of dedicated media proxies (VoIP IP/IP gateways, RTP proxies, etc.). Ideally they should support the following features:

- Support the relaying of media streams without requiring decoding and re-encoding in order to minimize delays.
- They should not have jitter buffer.
- They should add a minimal delay overhead for media processing. This does not necessarily require dedicated hardware implementations (actually, kernel implementations on operating systems, such as Linux, and high-performance network interfaces now have a performance comparable with most router).
- They should support many types of media streams (audio, T.38 fax, T.120, video).
- They should support all the call flows found in the network, not just basic calls. This includes dynamic media redirection (H.323 NullCapabilitySet, SIP RE-INVITE).
- They should automatically detect calls that are looped back to the originating IP domain and optimize the media path to stop using the proxy. This is not trivial and requires relatively sophisticated algorithms.
- They should support multiple network interfaces in order to facilitate connectivity to multiple separate networks. These interfaces can be physical or virtual (VLANs, IP tunnels).

Other nice-to-have features include some denial-of-service protection (call rate limiters, checking of media streams' token bucket profiles, mapping of DiffServ codepoints, etc.).

The media proxy function does not need to be a stand-alone product; in fact, this function works even better when combined with a call controller, because the call controller can know many more properties of the end points, such as which end points are on the same site and need to have an optimized RTP path, or which end points need to have the RTP stream fully relayed through the proxy.

Figure 7.13 shows a sample network with two enterprises A and B. Enterprise A has two sites A1 and A2. An MGCP call agent provides the IP Centrex service and includes an RTP proxy function. The call agent controls IP phones using MGCP and, when calls need to be transferred to the backbone, uses the H.323 or SIP protocols.

Since all IP phones are located on the same IP address plane (access pool 1), in theory any call from an IP phone in this domain to another IP phone in this domain can use direct RTP routing. This can be seen when a phone on site A1 calls a phone on site A2: the MGCP call controller provides each IP phone with the address of the other IP phone, so the media can be routed directly by the IP network.

However, for security reasons, enterprise B does not want to receive media streams directly from any other enterprise, it only wants to receive media streams from the service provider (only packets from IP address 162.168.0.1 are allowed). In this case the MGCP

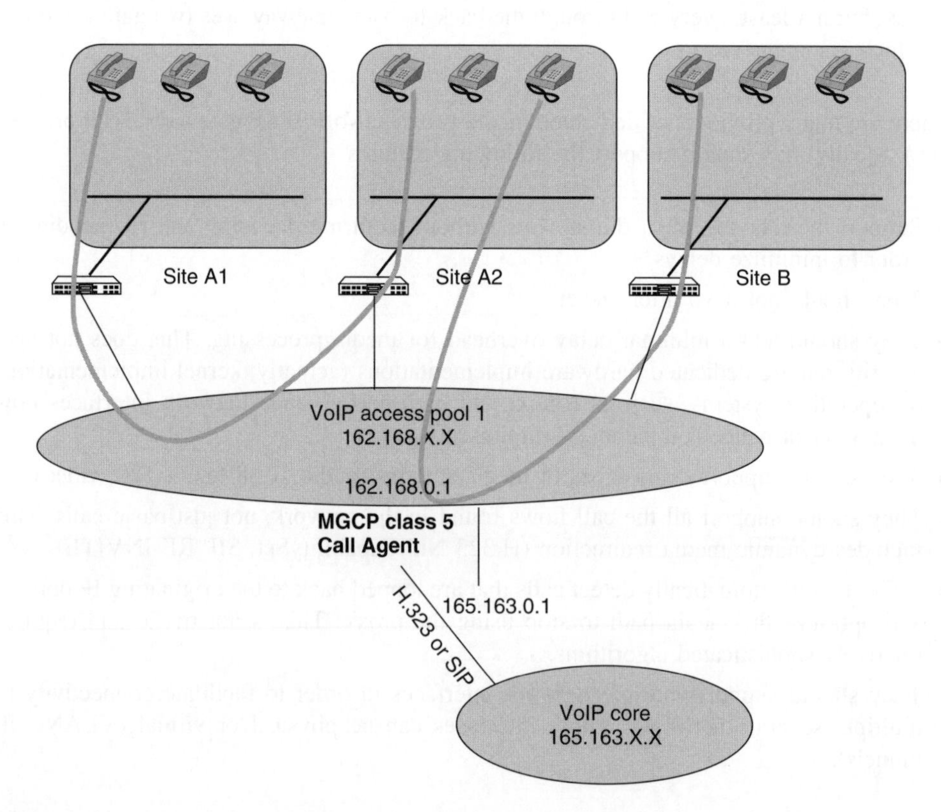


**Figure 7.13** Optimization of RTP proxy usage when the softswitch has access to administrative data.

call agent will also act as an RTP proxy and terminate, then re-originate the media stream. This is shown in the case of a call between a phone on site A2 and a phone on site B.

The example shows that the association of a call controller and a media proxy can be powerful. A separate media proxy does not have enough application-level information and would have optimized the media streams in all cases, routing RTP packets directly between enterprise B and site A1.

### 7.3.3 Security considerations

The allocation of controlled IP addresses to IP phones not only helps to solve the NAT issue, it also clarifies the potential security responsibility issues that can arise in a managed telephony deployment inside the enterprise. By definition, a managed VoIP service must be able to initiate IP communications to servers (IP phones) *inside* the corporate private network.

Our experience with initial deployments of IP Centrex show that some corporate MIS managers may be tempted, if they find that their corporate network has been compromised, to blame the managed telephony service provider. Therefore, it is very important for a service provider to be able to propose some mutually acceptable, simple to understand, security rules that completely isolate IP phones from the PCs on the internal network.

This is very difficult if IP phones are allocated IP addresses at random inside the corporation, but becomes much simpler if the IP addresses of all IP phones can be identified easily, as proposed.

The corporate firewall/access router should be configured to:

- Accept VoIP signaling only from the IP address of the service provider call controller.
- Accept media traffic only from IP addresses of the IP_PH pool *and* only to the IP addresses allocated to the IP phones (with an RTP filter if available in the router/firewall, otherwise a UDP filter on ports higher than 1024 should be used). This rule can be made stricter if the call controller is capable of enabling an RTP proxy for media calls coming from third-party sites, in which case the source IP address of the media streams can be restricted to just the IP address of the media proxy (or proxies).
- If the IP phones are located on a VLAN or only accessible through a specific router interface, all inbound media traffic should only be routed to this interface.

IP phones should not be allowed to communicate with any IP address that is not part of the IP_PH pool. This can be achieved by placing the phones on a separate LAN or VLAN. This prevents the unlikely but potential threat of having an IP phone compromised and serving as a relay to other machines on the LAN.

These rules are relatively simple and make it impossible for any potential attacker on the VoIP backbone to reach the computers of the enterprise by using the VoIP NAT bypass route through the corporate router. Any attack must come through the regular router NAT function (or other security policy set for calls coming from the public IP network), and therefore is not under the responsibility of the service provider. Our recommendation to service providers is to include a detailed description of the security policy and have all IP Centrex customers signing acceptance of this security policy, thereby clearing the service provider from potential future accusations.

## 7.4 CONCLUSION

It is frequently heard that, until IPv6 is adopted, VoIP cannot be deployed due to the lack of IP addresses. This is clearly wrong. The sophistication of the current tools allowed by IPv4 routing and the use of application-level IP/IP gateways make it possible to use only a restricted set of addresses and yet provide a service to a virtually unlimited number of users. Due to the smaller size of IPv4 packets, latency on such networks, especially on access links, will be better than on an equivalent IPv6 network and the IP overhead is also much better. Given the fact that all the tools required to provide quality of service (DiffServ, RSVP) perform just as well on IPv4 or IPv6 networks, there is really no reason today for a service provider to wait for the deployment of large-scale VoIP networks on currently deployed IP networks.

# Annex

Here is the call flow between Cisco SIP phones, showing which implementation choices have been made between Cisco phones. Cisco SIP phones reply with a 'trying' message after each request, for clarity these have been removed in the traces listed below.

**Phone A (5559000) calls phone B (5555000)**

INVITE sip:5555000@172.18.192.230 SIP/2.0
Via: SIP/2.0/UDP 172.18.192.218:5060
From: "A Phone" <sip:5559000@172.18.192.230>;tag=00070e8b5777000339a3170e-74ee1c83
To: <sip:5555000@172.18.192.230>
Call-ID: 00070e8b-577708ee-5fbe5e66-7f70fecd@172.18.192.218
Date: Thu, 13 Jun 2002 16:04:46 GMT
CSeq: 101 INVITE
User-Agent: Cisco-SIP-IP-Phone/3
Contact: sip:5559000@172.18.192.218:5060
Expires: 180
Content-Type: application/sdp
Content-Length: 271
Accept: application/sdp

v=0
o=CiscoSystemsSIP-IPPhone-UserAgent 18338 11953 IN IP4 172.18.192.218
s=SIP Call c=IN IP4 172.18.192.218
t=0 0
m=audio 29304 RTP/AVP 0 8 18 97
a=rtpmap:0 PCMU/8000
a=rtpmap:8 PCMA/8000

*IP Telephony: Deploying VoIP Protocols and IMS Infrastructure, Second Edition* O. Hersent

a=rtpmap:18 G729a/8000
a=rtpmap:97 telephone-event/8000
a=fmtp:97 0–15

SIP/2.0 180 Ringing
Via: SIP/2.0/UDP 172.18.192.218:5060
From: “A Phone”<sip:5559000@172.18.192.230>;tag=00070e8b5777000339a3170e-74ee1c83
To: <sip:5555000@172.18.192.230>;tag=003094c2e691000357031309-41c9de44
Call-ID: 00070e8b-577708ee-5fbe5e66-7f70fecd@172.18.192.218
Date: Thu, 13 Jun 2002 16:04:36 GMT
CSeq: 101 INVITE
Server: Cisco-SIP-IP-Phone/3
Contact: sip:5555000@172.18.192.221:5060
Record-Route: <sip:5555000@172.18.192.230:5060;maddr=172.18.192.230>
Content-Length: 0

SIP/2.0 200 OK
Via: SIP/2.0/UDP 172.18.192.218:5060
From: “A Phone”<sip:5559000@172.18.192.230>;tag=00070e8b5777000339a3170e-74ee1c83
To: <sip:5555000@172.18.192.230>;tag=003094c2e691000357031309-41c9de44
Call-ID: 00070e8b-577708ee-5fbe5e66-7f70fecd@172.18.192.218
Date: Thu, 13 Jun 2002 16:04:39 GMT
CSeq: 101 INVITE
Server: Cisco-SIP-IP-Phone/3
Contact: sip:5555000@172.18.192.221:5060
Record-Route: <sip:5555000@172.18.192.230:5060;maddr=172.18.192.230>
Content-Type: application/sdp
Content-Length: 220

v=0
o=CiscoSystemsSIP-IPPhone-UserAgent 11411 26110 IN IP4 172.18.192.221
s=SIP Call c=IN IP4 172.18.192.221
t=0 0
m=audio 24396 RTP/AVP 0 97
a=rtpmap:0 PCMU/8000
a=rtpmap:97 telephone-event/8000
a=fmtp:97 0–15

ACK sip:5555000@172.18.192.230:5060 SIP/2.0
Via: SIP/2.0/UDP 172.18.192.218:5060
From: “Kazoo-9 Phone”<sip:5559000@172.18.192.230>;tag=00070e8b5777000339a3170e-74ee1c83
To: <sip:5555000@172.18.192.230>;tag=003094c2e691000357031309-41c9de44
Call-ID: 00070e8b-577708ee-5fbe5e66-7f70fecd@172.18.192.218
Date: Thu, 13 Jun 2002 16:04:50 GMT
CSeq: 101 ACK
User-Agent: Cisco-SIP-IP-Phone/3
Route: <sip:5555000@172.18.192.221:5060>
Content-Length: 0

**A puts B on hold:**

```
INVITE sip:5555000@172.18.192.230:5060 SIP/2.0
Via: SIP/2.0/UDP 172.18.192.218:5060
From: "A Phone"<sip:5559000@172.18.192.230>;tag=00070e8b5777000339a3170e-74ee1c83
To: <sip:5555000@172.18.192.230>;tag=003094c2e691000357031309-41c9de44
Call-ID: 00070e8b-577708ee-5fbe5e66-7f70fecd@172.18.192.218
Date: Thu, 13 Jun 2002 16:04:53 GMT
CSeq: 102 INVITE
User-Agent: Cisco-SIP-IP-Phone/3
Contact: sip:5559000@172.18.192.218:5060
Route: <sip:5555000@172.18.192.221:5060>
Content-Type: application/sdp
Content-Length: 263

v=0
o=CiscoSystemsSIP-IPPhone-UserAgent 15014 5663 IN IP4 172.18.192.218
s=SIP Call c=IN IP4 0.0.0.0
t=0 0
m=audio 29304 RTP/AVP 0 8 18 97
a=rtpmap:0 PCMU/8000
a=rtpmap:8 PCMA/8000
a=rtpmap:18 G729a/8000
a=rtpmap:97 telephone-event/8000
a=fmtp:97 0-15

SIP/2.0 200 OK
Via: SIP/2.0/UDP 172.18.192.218:5060
From: "A Phone"<sip:5559000@172.18.192.230>;tag=00070e8b5777000339a3170e-74ee1c83
To: <sip:5555000@172.18.192.230>;tag=003094c2e691000357031309-41c9de44
Call-ID: 00070e8b-577708ee-5fbe5e66-7f70fecd@172.18.192.218
Date: Thu, 13 Jun 2002 16:04:44 GMT
CSeq: 102 INVITE
Server: Cisco-SIP-IP-Phone/3
Contact: sip:5555000@172.18.192.221:5060
Record-Route: <sip:5555000@172.18.192.230:5060;maddr=172.18.192.230>
Content-Type: application/sdp
Content-Length: 213

v=0
o=CiscoSystemsSIP-IPPhone-UserAgent 11411 26110 IN IP4 172.18.192.221
s=SIP Call c=IN IP4 0.0.0.0
t=0 0
m=audio 24396 RTP/AVP 0 97
a=rtpmap:0 PCMU/8000
a=rtpmap:97 telephone-event/8000
a=fmtp:97 0-15
```

ACK sip:5555000@172.18.192.230:5060 SIP/2.0
Via: SIP/2.0/UDP 172.18.192.218:5060
From: “A Phone”<sip:5559000@172.18.192.230>;tag=00070e8b5777000339a3170e-74ee1c83
To: <sip:5555000@172.18.192.230>;tag=003094c2e691000357031309-41c9de44
Call-ID: 00070e8b-577708ee-5fbe5e66-7f70fecd@172.18.192.218
Date: Thu, 13 Jun 2002 16:04:54 GMT
CSeq: 102 ACK
User-Agent: Cisco-SIP-IP-Phone/3
Route: <sip:5555000@172.18.192.221:5060>
Content-Length: 0

**A calls C:**

INVITE sip:5551000@172.18.192.230 SIP/2.0
Via: SIP/2.0/UDP 172.18.192.218:5060
From: “A Phone”<sip:5559000@172.18.192.230>;tag=00070e8b577700045e92261b-204bc3c6
To: <sip:5551000@172.18.192.230>
Call-ID: 00070e8b-577708ef-746a6bfe-6c96b214@172.18.192.218
Date: Thu, 13 Jun 2002 16:04:59 GMT
CSeq: 101 INVITE
User-Agent: Cisco-SIP-IP-Phone/3
Contact: sip:5559000@172.18.192.218:5060
Expires: 180
Content-Type: application/sdp
Content-Length: 271
Accept: application/sdp

v=0
o=CiscoSystemsSIP-IPPhone-UserAgent 27275 16432 IN IP4 172.18.192.218
s=SIP Call c=IN IP4 172.18.192.218
t=0 0
m=audio 29306 RTP/AVP 0 8 18 97
a=rtpmap:0 PCMU/8000
a=rtpmap:8 PCMA/8000
a=rtpmap:18 G729a/8000
a=rtpmap:97 telephone-event/8000
a=fmtp:97 0–15

SIP/2.0 180 Ringing
Via: SIP/2.0/UDP 172.18.192.218:5060
From: “A Phone”<sip:5559000@172.18.192.230>;tag=00070e8b577700045e92261b-204bc3c6
To: <sip:5551000@172.18.192.230>;tag=003094c25d94001039653a76-0c6588ad
Call-ID: 00070e8b-577708ef-746a6bfe-6c96b214@172.18.192.218
Date: Thu, 13 Jun 2002 16:04:57 GMT
CSeq: 101 INVITE
Server: Cisco-SIP-IP-Phone/3
Contact: sip:5551000@172.18.192.220:6062
Record-Route: <sip:5551000@172.18.192.230:5060;maddr=172.18.192.230>
Content-Length: 0

```
SIP/2.0 200 OK
Via: SIP/2.0/UDP 172.18.192.218:5060
From: "A Phone"<sip:5559000@172.18.192.230>;tag=00070e8b577700045e92261b-204bc3c6
To: <sip:5551000@172.18.192.230>;tag=003094c25d94001039653a76-0c6588ad
Call-ID: 00070e8b-577708ef-746a6bfe-6c96b214@172.18.192.218
Date: Thu, 13 Jun 2002 16:05:00 GMT
CSeq: 101 INVITE
Server: Cisco-SIP-IP-Phone/3
Contact: sip:5551000@172.18.192.220:6062
Record-Route: <sip:5551000@172.18.192.230:5060;maddr=172.18.192.230>
Content-Type: application/sdp
Content-Length: 221

v=0
o=CiscoSystemsSIP-IPPhone-UserAgent 5685 4962 IN IP4 172.18.192.220
s=SIP Call c=IN IP4 172.18.192.220
t=0 0
m=audio 16394 RTP/AVP 18 97
a=rtpmap:18 G729a/8000
a=rtpmap:97 telephone-event/8000
a=fmtp:97 0–15
```

```
ACK sip:5551000@172.18.192.230:5060 SIP/2.0
Via: SIP/2.0/UDP 172.18.192.218:5060
From: "A Phone"<sip:5559000@172.18.192.230>;tag=00070e8b577700045e92261b-204bc3c6
To: <sip:5551000@172.18.192.230>;tag=003094c25d94001039653a76-0c6588ad
Call-ID: 00070e8b-577708ef-746a6bfe-6c96b214@172.18.192.218
Date: Thu, 13 Jun 2002 16:05:02 GMT
CSeq: 101 ACK
User-Agent: Cisco-SIP-IP-Phone/3
Route: <sip:5551000@172.18.192.220:6062>
Content-Length: 0
```

**A puts C on hold**

```
INVITE sip:5551000@172.18.192.230:5060 SIP/2.0
Via: SIP/2.0/UDP 172.18.192.218:5060
From: "A Phone"<sip:5559000@172.18.192.230>;tag=00070e8b577700045e92261b-204bc3c6
To: <sip:5551000@172.18.192.230>;tag=003094c25d94001039653a76-0c6588ad
Call-ID: 00070e8b-577708ef-746a6bfe-6c96b214@172.18.192.218
Date: Thu, 13 Jun 2002 16:05:06 GMT
CSeq: 102 INVITE
User-Agent: Cisco-SIP-IP-Phone/3
Contact: sip:5559000@172.18.192.218:5060
Route: <sip:5551000@172.18.192.220:6062>
Content-Type: application/sdp
Content-Length: 263

v=0
o=CiscoSystemsSIP-IPPhone-UserAgent 22866 2538 IN IP4 172.18.192.218
```

s=SIP Call c=IN IP4 0.0.0.0
t=0 0
m=audio 29306 RTP/AVP 0 8 18 97
a=rtpmap:0 PCMU/8000
a=rtpmap:8 PCMA/8000
a=rtpmap:18 G729a/8000
a=rtpmap:97 telephone-event/8000
a=fmtp:97 0–15

SIP/2.0 200 OK
Via: SIP/2.0/UDP 172.18.192.218:5060
From: "A Phone"<sip:5559000@172.18.192.230>;tag=00070e8b577700045e92261b-204bc3c6
To: <sip:5551000@172.18.192.230>;tag=003094c25d94001039653a76-0c6588ad
Call-ID: 00070e8b-577708ef-746a6bfe-6c96b214@172.18.192.218
Date: Thu, 13 Jun 2002 16:05:04 GMT
CSeq: 102 INVITE
Server: Cisco-SIP-IP-Phone/3
Contact: sip:5551000@172.18.192.220:6062
Record-Route: <sip:5551000@172.18.192.230:5060;maddr=172.18.192.230>
Content-Type: application/sdp
Content-Length: 214

v=0
o=CiscoSystemsSIP-IPPhone-UserAgent 5685 4962 IN IP4 172.18.192.220
s=SIP Call c=IN IP4 0.0.0.0
t=0 0
m=audio 16394 RTP/AVP 18 97
a=rtpmap:18 G729a/8000
a=rtpmap:97 telephone-event/8000
a=fmtp:97 0–15

ACK sip:5551000@172.18.192.230:5060 SIP/2.0
Via: SIP/2.0/UDP 172.18.192.218:5060
From: "A Phone"<sip:5559000@172.18.192.230>;tag=00070e8b577700045e92261b-204bc3c6
To: <sip:5551000@172.18.192.230>;tag=003094c25d94001039653a76-0c6588ad
Call-ID: 00070e8b-577708ef-746a6bfe-6c96b214@172.18.192.218
Date: Thu, 13 Jun 2002 16:05:06 GMT
CSeq: 102 ACK
User-Agent: Cisco-SIP-IP-Phone/3
Route: <sip:5551000@172.18.192.220:6062>
Content-Length: 0

**A transfers B to C:**

REFER sip:5555000@172.18.192.230:5060 SIP/2.0
Via: SIP/2.0/UDP 172.18.192.218:5060
From: "A Phone"<sip:5559000@172.18.192.230>;tag=00070e8b5777000339a3170e-74ee1c83
To: <sip:5555000@172.18.192.230>;tag=003094c2e691000357031309-41c9de44
Call-ID: 00070e8b-577708ee-5fbe5e66-7f70fecd@172.18.192.218
Date: Thu, 13 Jun 2002 16:05:06 GMT

CSeq: 103 REFER
User-Agent: Cisco-SIP-IP-Phone/3
Contact: sip:5559000@172.18.192.218:5060
Route: <sip:5555000@172.18.192.221:5060>
Content-Length: 0
Refer-To:
sip:5551000@172.18.192.230?Replaces=00070e8b-577708ef-746a6bfe-6c96b214%40172.18.192.218
%3Bto-tag%3D003094c25d94001039653a76-0c6588ad%3Bfrom-tag%3D00070e8b577700045e92261b-204bc3c6
Referred-By: "A Phone" <sip:5559000@172.18.192.230>

SIP/2.0 202 Accepted
Via: SIP/2.0/UDP 172.18.192.218:5060
From: "A Phone"<sip:5559000@172.18.192.230>;tag=00070e8b5777000339a3170e-74ee1c83
To: <sip:5555000@172.18.192.230>;tag=003094c2e691000357031309-41c9de44
Call-ID: 00070e8b-577708ee-5fbe5e66-7f70fecd@172.18.192.218
Date: Thu, 13 Jun 2002 16:04:56 GMT
CSeq: 103 REFER
Server: Cisco-SIP-IP-Phone/3
Contact: sip:5555000@172.18.192.221:5060
Record-Route: <sip:5555000@172.18.192.230:5060;maddr=172.18.192.230>
Content-Length: 0

**Phone B calls C**

INVITE sip:5551000@172.18.192.230:5060 SIP/2.0
Via: SIP/2.0/UDP 172.18.192.221:5060
From: "5555000"<sip:5555000@172.18.192.230>;tag=003094c2e69100040d6ba94e-5f32d799
To: <sip:5551000@172.18.192.230:5060>
Call-ID: 003094c2-e69100af-66c9e3b4-04b3f0e8@172.18.192.221
Date: Thu, 13 Jun 2002 16:04:56 GMT
CSeq: 101 INVITE
User-Agent: Cisco-SIP-IP-Phone/3
Contact: sip:5555000@172.18.192.221:5060
Referred-By: "Kazoo-9 Phone" <sip:5559000@172.18.192.230>
Replaces:
00070e8b-577708ef-746a6bfe-6c96b214@172.18.192.218;to-tag=003094c25d94001039653a76-0c6588ad;from-tag=00070e8b577700045e92261b-204bc3c6
Expires: 180
Content-Type: application/sdp
Content-Length: 270
Accept: application/sdp

v=0
o=CiscoSystemsSIP-IPPhone-UserAgent 19502 5249 IN IP4 172.18.192.221
s=SIP Call c=IN IP4 172.18.192.221
t=0 0
m=audio 24398 RTP/AVP 0 8 18 96
a=rtpmap:0 PCMU/8000
a=rtpmap:8 PCMA/8000

a=rtpmap:18 G729a/8000
a=rtpmap:96 telephone-event/8000
a=fmtp:96 0–15

SIP/2.0 200 OK
Via: SIP/2.0/UDP 172.18.192.221:5060
From: "5555000"<sip:5555000@172.18.192.230>;tag=003094c2e69100040d6ba94e-5f32d799
To: <sip:5551000@172.18.192.230:5060>;tag=003094c25d9400111fc7caba-03f8330a
Call-ID: 003094c2-e69100af-66c9e3b4-04b3f0e8@172.18.192.221
Date: Thu, 13 Jun 2002 16:05:05 GMT
CSeq: 101 INVITE
Server: Cisco-SIP-IP-Phone/3
Contact: sip:5551000@172.18.192.220:6062
Record-Route: <sip:5551000@172.18.192.230:5060;maddr=172.18.192.230>
Content-Type: application/sdp
Content-Length: 223

v=0
o=CiscoSystemsSIP-IPPhone-UserAgent 18795 10818 IN IP4 172.18.192.220
s=SIP Call c=IN IP4 172.18.192.220
t=0 0
m=audio 16396 RTP/AVP 18 96
a=rtpmap:18 G729a/8000
a=rtpmap:96 telephone-event/8000
a=fmtp:96 0–15

ACK sip:5551000@172.18.192.230:5060 SIP/2.0
Via: SIP/2.0/UDP 172.18.192.221:5060
From: "5555000"<sip:5555000@172.18.192.230>;tag=003094c2e69100040d6ba94e-5f32d799
To: <sip:5551000@172.18.192.230:5060>;tag=003094c25d9400111fc7caba-03f8330a
Call-ID: 003094c2-e69100af-66c9e3b4-04b3f0e8@172.18.192.221
Date: Thu, 13 Jun 2002 16:04:57 GMT
CSeq: 101 ACK
User-Agent: Cisco-SIP-IP-Phone/3
Route: <sip:5551000@172.18.192.220:6062>
Content-Length: 0

**B notifies A that communication with C is active:**

NOTIFY sip:5559000@172.18.192.218:5060 SIP/2.0
Record-Route: <sip:5555000@172.18.192.230:5060;maddr=172.18.192.230>
Via: SIP/2.0/UDP
172.18.192.230:5060;branch=855cf819-524d09ad-dbaea7b3-dc1ce9c9-1
Via: SIP/2.0/UDP 172.18.192.221:5060
From: <sip:5555000@172.18.192.230>;tag=003094c2e691000357031309-41c9de44
To: "A Phone" <sip:5559000@172.18.192.230>;tag=00070e8b5777000339a3170e-74ee1c83
Call-ID: 00070e8b-577708ee-5fbe5e66-7f70fecd@172.18.192.218
Date: Thu, 13 Jun 2002 16:04:57 GMT
CSeq: 101 NOTIFY
User-Agent: Cisco-SIP-IP-Phone/3

Event: refer
Content-Type: message/sipfrag
Content-Length: 14

SIP/2.0 200 OK

SIP/2.0 200 OK
Via: SIP/2.0/UDP
172.18.192.230:5060;branch=855cf819-524d09ad-dbaea7b3-dc1ce9c9-1,SIP/2.0/UDP
172.18.192.221:5060
From: <sip:5555000@172.18.192.230>;tag=003094c2e69100035703l309-41c9de44
To: “A Phone” <sip:5559000@172.18.192.230>;tag=00070e8b5777000339a3170e-74ee1c83
Call-ID: 00070e8b-577708ee-5fbe5e66-7f70fecd@172.18.192.218
Date: Thu, 13 Jun 2002 16:05:07 GMT
CSeq: 101 NOTIFY
Content-Length: 0

**C releases the call from A:**

BYE sip:5559000@172.18.192.218:5060 SIP/2.0
Record-Route: <sip:5551000@172.18.192.230:5060;maddr=172.18.192.230>
Via: SIP/2.0/UDP
172.18.192.230:5060;branch=fd11b5e1-aee38059-c54b46b5-f5f62a06-1
Via: SIP/2.0/UDP 172.18.192.220:6062
From: <sip:5551000@172.18.192.230>;tag=003094c25d94001039653a76-0c6588ad
To: “A Phone” <sip:5559000@172.18.192.230>;tag=00070e8b577700045e92261b-204bc3c6
Call-ID: 00070e8b-577708ef-746a6bfe-6c96b214@172.18.192.218
Date: Thu, 13 Jun 2002 16:05:05 GMT
CSeq: 101 BYE
User-Agent: Cisco-SIP-IP-Phone/3
Content-Length: 0

SIP/2.0 200 OK
Via: SIP/2.0/UDP
172.18.192.230:5060;branch=fd11b5e1-aee38059-c54b46b5-f5f62a06-1,SIP/2.0/UDP
172.18.192.220:6062
From: <sip:5551000@172.18.192.230>;tag=003094c25d94001039653a76-0c6588ad
To: “A Phone” <sip:5559000@172.18.192.230>;tag=00070e8b577700045e92261b-204bc3c6
Call-ID: 00070e8b-577708ef-746a6bfe-6c96b214@172.18.192.218
Date: Thu, 13 Jun 2002 16:05:07 GMT
CSeq: 101 BYE
Server: Cisco-SIP-IP-Phone/3
Content-Length: 0

**A releases the call to B:**

BYE sip:5555000@172.18.192.230:5060 SIP/2.0
Via: SIP/2.0/UDP 172.18.192.218:5060
From: “A Phone”<sip:5559000@172.18.192.230>;tag=00070e8b5777000339a3170e-74ee1c83

To: <sip:5555000@172.18.192.230>;tag=003094c2e691000357031309-41c9de44
Call-ID: 00070e8b-577708ee-5fbe5e66-7f70fecd@172.18.192.218
Date: Thu, 13 Jun 2002 16:05:07 GMT
CSeq: 104 BYE
User-Agent: Cisco-SIP-IP-Phone/3
Content-Length: 0
Route: <sip:5555000@172.18.192.221:5060>
SIP/2.0 200 OK
Via: SIP/2.0/UDP 172.18.192.218:5060
From: “A Phone”<sip:5559000@172.18.192.230>;tag=00070e8b5777000339a3170e-74ee1c83
To: <sip:5555000@172.18.192.230>;tag=003094c2e691000357031309-41c9de44
Call-ID: 00070e8b-577708ee-5fbe5e66-7f70fecd@172.18.192.218
Date: Thu, 13 Jun 2002 16:04:58 GMT
CSeq: 104 BYE
Server: Cisco-SIP-IP-Phone/3
Content-Length: 0

# Index



*IP Telephony: Deploying VoIP Protocols and IMS Infrastructure, Second Edition* O. Hersent